# SHELF SANDS AND SANDSTONE RESERVOIRS

organized, compiled and edited by

R.W. Tillman

D.J.P. Swift

R.G. Walker

SEPM SHORT COURSE NOTES NO. 13

Printed in U.S.A.

© Copyright, SEPM, 1985. All rights reserved.

# PREFACE

Shelf sandstone reservoirs are becoming a more and more common exploration target. What they are, how they may be characterized, and how they differ from shoreline and deep-water deposits is the subject of this publication.

Shelf sands and sandstone reservoirs are among the more poorly understood types of sandstones. Continental, shoreline and deep water sandstones have all been studied in much more depth than have shelf sands and sandstones. However, during the last fifteen years significant progress has been made in understanding shelf sands and sandstones.

Studies of modern sediments have allowed us to understand many of the depositional processes active on the shelf. Sedimentological studies have documented a variety of sedimentary features which aid in recognizing these processes in shelf sandstone reservoirs. Storm dominated shelf sandstones can commonly be contrasted with subtidal tidal-current deposited reservoirs, and the overprint of other shelf processes can also be observed.

This book, along with SEPM Special Publication No. 34, Siliclastic Shelf Sediments, is intended to be an up-to-date summary of shelf processes and products. The papers in this book are intended for those new to shelf sands and sandstones as well as the shelf specialist. The topics included in this book were carefully selected to give the broadest possible overview of shelf sands and sandstones.

The introductory paper, a Spectrum of Shelf Sands and Sandstones, gives an overview of the sedimentary processes active on various parts of the shelf. A major in-depth review of modern shelf processes, including fair weather processes such as tides, waves and wind, and storm processes is included. The products of these shelf processes are also discussed.

Ancient examples of the products of each of the major processes are given. Specific producing shelf sandstones such as the Cardium, Shannon, Tocito and Woodbine Sandstones are discussed in detail.

Core studies form the basis for recognition of shelf environment of deposition in many subsurface reservoirs. Cores from the Shannon Sandstone of Wyoming, the Cardium Sandstone of Alberta, the Woodbine Sandstone of the Gulf Coast and the Tocito Sandstone Lentil of New Mexico are illustrated and discussed.

A brief field guide to the Shannon Sandstone outcrops in the Powder River Basin of Wyoming provides the types of descriptions necessary for field examinations on one type of shelf-sand ridge.

CONTENTS

PREFACE .................................................................. i

## MODERN SHELF SANDS

A SPECTRUM OF SHELF SANDS AND SANDSTONES ....................R.W. Tillman          1

FLUID AND SEDIMENT DYNAMICS ON CONTINENTAL
SHELVES......................................D.J.P. Swift and A.W. Niedoroda       47

RESPONSE OF THE SHELF FLOOR TO FLOW............................D.J.P. Swift      135

COMPARISON OF SAND RIDGES ON THE NEW JERSEY CONTINENTAL SHELF,
U.S.A. (abs.)....................R.W. Tillman, J.M. Rine, W.L. Subblefield       242

## ANCIENT SHELF SANDSTONES

GEOLOGICAL EVIDENCE FOR STORM TRANSPORTATION
AND DEPOSITION ON ANCIENT SHELVES................................R.G. Walker     243

ANCIENT EXAMPLES OF TIDAL SAND BODIES FORMED IN OPEN, SHALLOW
SEAS.............................................................R.G. Walker     303

THE SHANNON SHELF RIDGE SANDSTONE COMPLEX, SALT CREEK ANTICLINE AREA,
POWDER RIVER BASIN, WYOMING (abs.)..........R.W. Tillman and R.S. Martinsen      343

HARTZOG DRAW, A GIANT OIL FIELD (abs.)......R.M. Martinsen and R.W. Tillman      349

FACIES AND RESERVOIR CHARACTERISTICS OF A SHELF SANDSTONE: HARTZOG DRAW
FIELD, POWDER RIVER BASIN, WYOMING (abs.)...R.S. Martinsen and R.W. Tillman      351

CARDIUM FORMATION 4. REVIEW OF FACIES AND DEPOSITIONAL PROCESSES IN THE
SOUTHERN FOOTHILLS AND PLAINS, ALBERTA, CANADA....................R.G. Walker    353

THE TOCITO AND GALLUP SANDSTONES, NEW MEXICO, A COMPARISON.....R.W. Tillman      403

COMPARISON OF SHELF ENVIRONMENTS AND DEEP BASIN TURBIDITE
SYSTEMS..........................................................R.G. Walker     465

SHELF SANDSTONES IN THE WOODBINE-EAGLE FORD INTERVAL, EAST TEXAS: A
REVIEW OF DEPOSITIONAL MODELS.......S. Phillips, D.J.P. Swift, C.T. Siemers      503

CONTENTS (cont.)

## SHELF SAND-RIDGE CORE STUDIES

| | | |
|---|---|---|
| TOCITO SANDSTONE CORE, HORSESHOE FIELD, SAN JUAN COUNTY, NEW MEXICO | R.W. Tillman | 559 |
| SHANNON SANDSTONE, HARTZOG DRAW FIELD CORE STUDY | R.W. Tillman and R. S. Martinsen | 577 |
| CARDIUM AND VIKING SANDSTONE CORES | R.G. Walker | 645 |
| WOODBINE CORE (HINTON DORANCE 7A) (abs.) | C.T. Siemers | 677 |
| SHANNON SHELF RIDGE SANDSTONES FIELD TRIP | R.W. Tillman, R.M. Martinsen, N. Gaynor, and D.J.P. Swift | 683 |

# A SPECTRUM OF SHELF SANDS AND SANDSTONES

R. W. Tillman[1]

Cities Service Oil and Gas Corporation, Tulsa, Oklahoma

## ABSTRACT

A wide variety of processes have operated on the seabed of the shelves of the world in the past including perhaps storms, permanent currents, wind induced alongshore currents, wave modified currents, subtidal tidal currents and turbidity currents. These processes generated sand bodies with different geometries which commonly contain different sedimentary structures or different sequences of sedimentary structures. On ancient shelves the most common sedimentary structures observed in vertical sections of sandstones are planar-tangential to planar-tabular cross beds, horizontal to subhorizontal laminations, current ripples, wave rippleswave modified current ripples and burrowed and bioturbated (>75% burrowed) sandstones. This sequence is in approximate order of decreasing energy (fluid power). Where consistent vertical sequences of sedimentary structures are observed, one of the most common reflects upward increase in depositional energy. However, a sequence reflecting upward increasing energy and consequent increase in grain size is not unique to shelf sandstones; a similar coarsening upward pattern is reflected in subsurface log patterns in both river- and wave-dominated deltas and in beach/barrier dominated shorelines.

Ancient sandstone examples used to characterize a variety of these processes, geometries, and shelf locations include the "Gallup" (Tocito), Shannon, Fales and Frontier Sandstones from the Cretaceous of the Western Interior. In addition modern Atlantic shelf and North Sea systems are discussed.

Shelf sandstones may be classified on the basis of their position on the shelf (shoreface-attached, inner shelf, middle shelf, outer shelf) and on the basis of whether they are deposited during a transgression, regression, or a stillstand. Both vertical and lateral sequences of lithologies vary with position on the shelf, processes of deposition, and position within a transgression-regression spectra. On the middle and outer shelf, shelf sandstones are almost always surrounded by shale.

On the inner shelf, and where attached to the shoreface, shelf sandstones overlie a variety of lithologies (sandstone, siltstone and shale) dependant in part on whether they were deposited during a transgression, regression or stillstand. Lithologies deposited lateral to shelf sandstones also vary with the position of the sand body within the spectra of transgression-regression. Vertical and lateral sequences of lithologies are probably the most variable on the inner shelf.

1. Present address: Consulting Sedimentologist, 4555 South Harvard, Tulsa, Oklahoma 74135

Local topography also may affect the distribution of shelf sandstones. Winnowing of the seabed in areas which are topographically high may concentrate sand into sand ridges. Depressions in the shelf sea floor may result from erosion of lithologies which have different susceptabilities to erosion (i.e., strike valley sands), during sea level drops, by shoreface retreat or as a result of submarine erosion. Depressions on the sea floor may fill with fine- to coarse-grained sand.

## INTRODUCTION

Shelf sandstones were for many years neglected by exploration geologists as reservoir targets. In contrast, shoreline and deepwater sandstones have both been explored for actively. Until the mid 1970's the area between the shoreline and the slope was considered to be primarily shale and carbonate with very few sandstones.

The purpose of this paper is to outline the physical processes which act on various parts of the shelf and distribute sand into potential hydrocarbon reservoirs. Variations in process intensities within the inner, middle, and outer shelves and during periods of transgression, regression, and stillstand will be examined.

The shelf in modern oceans was originally defined to extend from the low water line to approximately 600 feet (100 fathoms) depth (Wiseman and Ovey, 1953). However, in modern oceans the shelf-slope break may be as shallow as 120 feet or as deep as 1800 feet (Emery and Uchupi, 1984). According to Shepard (1973) the shelf-slope break in modern oceans averages 432 feet (132 m).

Many geologists and oceanographers divide the shelf into three segments: inner, middle, and outer shelf (Fig. 1). The inner shelf lies seaward of the shoreface and for purposes of discussion is arbitrarily assigned to extend to a depth of about 100 feet. The middle shelf ranges from 100 to 300 feet in depth, while the outer shelf extends from about 300 feet to the shelf-slope break which is rarely as deep as 600 feet. When the shelf-slope break is shallower than 600 feet, as it is for almost all the shelves around the modern Atlantic Ocean, all the other depths for subdivision of the shelf would also be diminished in proportion.

Areas of the shelf where the following processes are active will be investigated: tidal currents, wind-generated alongshore currents, wave modified currents, storm currents, permanent or semi-permanent currents, and turbidity currents. These processes vary in terms of location on the shelf and in terms of intensity (Fig. 2). Tidal currents effecting the sea bottom below low tide level occur most commonly in shoreface-attached and inner shelf areas. On the other end of the spectrum are "shelf-turbidites"; they are most likely to occur on the middle or outer shelf. Other processes occur in varying water depths and with varying intensities.

# SHELF SUBDIVISIONS

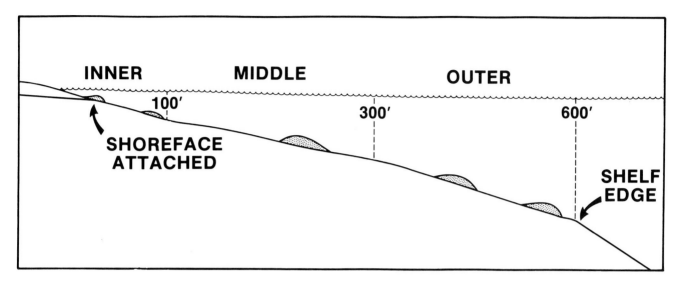

Figure 1. Subdivisions of the shelf where sand may accumulate.

## SHELF PROCESSES

|  | Tidal Currents | Wind Induced Alongshore Currents | Wave Modified (Combined flow) Currents | Storm Currents | Permanent or Semi-Permanent Currents | Turbidity Currents |
|---|---|---|---|---|---|---|
| Shoreface Attached | X | X | X | X | X | — |
| Inner Shelf | X | X | X | X | X | ? |
| Middle Shelf | ? | ? | X | X | X | X |
| Outer Shelf | — | — | — | X | X | X |

Figure 2. Shelf processes and their distribution on the shelf.

Figure 3. Producing sandstones known to be at least in part shelf sandstones.

---

Shelf Reservoirs

A number of reservoirs produce from sandstones deposited on the shelf. Some of the better known reservoirs include the Cardium and Viking Sandstones of Canada, the Shannon, Sussex and Tocito ("Gallup") Sandstones of the Rocky Mountains and the Morrow, Woodbine and Cotton Valley of the mid-continent (Fig. 3). The Shannon Sandstone, which produces in Hartzog Draw Field in the Powder River Basin of Wyoming, contains 350 million barrels of oil. Some of these shelf sandstone reservoirs were inadvertantly found; others were found by purposely exploring for shelf sandstones.

In Canada, the Cardium and Viking sandstones have been explored for some time and many explorationists interpret the units to be primarily shelf deposits. The Hygiene Sandstone in the Denver Basin is a shelf sandstone deposit. Shelf facies of the Morrow Sandstone produce in the Anadarko Basin in the western part of Oklahoma and in Colorado. Many of the Morrow sands are not shelf sandstones, however, the Morrow sandstones that are shelf sandstones yield significant production. The "Gallup" or Tocito Sandstone Lentil in New Mexico is currently an exploration target. The Woodbine Sandstone is in part a shelf sandstone and in part a slope to deep water turbidite type deposit. The Cotton Valley is not well documented but, it also appears to produce, in part, from shelf sandstones.

## Characteristics of Shelf Sands and Sandstones

The characteristics most commonly found in shelf sandstones are summarized in Figure 4. Since there is a wide spectrum of shelf sandstone types, this list of characteristics should not be considered typical of <u>all</u> shelf sandstones. However, when the suite of characteristics listed in Figure 4 is found together shelf sandstones should be considered as a possible depositional environment.

Many shelf sandstones display a coarsening-up sequence, and generally are coarser grained than their time equivalent shoreline deposits. Shelf sandstones commonly are linear in map view and vary in thickness from 0.5 to 80 feet. Shelf sandstones are commonly surrounded by shale; this lithologic association may be observed on electric logs, however, on seismic sections sand bodies less than 100 feet thick may be difficult to recognize.

## COARSENING-UP SEQUENCES

A coarsening-up sequence is one which many geologists who have studied shelf sands have emphasized very strongly. Figure 5 illustrates a coarseningup sequence consisting, in vertical succession, of shale, interbedded sandstone and shale, cross-bedded sandstone, and at the top of the sequence a highly cross-bedded, coarse-grained sandstone; this whole sequence is overlain by shale. Although this type of sequence is not unique to shelf sandstones it does characterize shelf sandstones such as those described below.

A typical coarsening-upwards shelf-ridge complex that produces is the Shannon Sandstone in Hartzog Draw field in Wyoming. The facies recognized in the Shannon and several other sandstones on logs and cores are shown in Figures 6A and 6B. These facies are described in detail in Tillman and Martinsen (1979 and 1984). The basal portion is a marine shelf shale which is somewhat silty and contains certain characteristics which are recognized as being more typical of shelf shales than nearshore shales. The <u>Inter-Ridge Facies</u> is composed primarily of rippled fine-grained interbedded sandstone and shale. It is not a reservoir unit, and it commonly occurs between or at the base of shelf ridges. Overlying the <u>Inter-Ridge Facies</u> is commonly what is termed a <u>Ridge-Margin Facies</u>. <u>Ridge Margin Facies</u> may be a <u>High-Energy</u> or a <u>Low-Energy Ridge-Margin Facies</u> depending on the location on the shelf ridge. <u>High-Energy Ridge-Margin Facies</u> tend to be primarily cross bedded sandstone and are located on the side of the ridge on which depositional currents most commonly impinge. The <u>Low Energy Bar Margin Facies</u> tend to be on the downstream (lee) side of the sand ridges and are formed by interbedded cross beds and ripples. The <u>Central Ridge Facies</u> is the major producing facies in Hartzog Draw field and a number of the other fields which produce from Shannon or similar type sandstones. This facies is usually cross-bedded, fairly glauconitic, well sorted and has good porosities and permeabilities (Martinsen and Tillman, 1978 and in preparation). This coarsening-up sequence is typical of one or more types of shelf sandstones. Notice that the sandstone at the top of the sequence has a very sharp upper contact with the overlying marine shale.

# CHARACTERISTICS OF SHELF SANDSTONES

1) Coarsening upward sequence

2) Coarser grained than shoreline deposits

3) "Surrounded" by shale

4) Generally linear

5) Vary in thickness from 0.5 to 80 feet

Figure 4. General characteristics of most shelf sandstones.

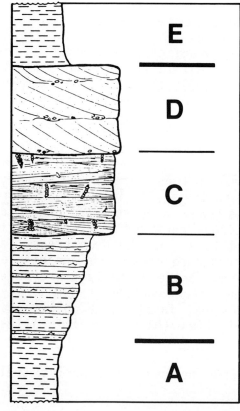

**COARSENING UP SEQUENCE**

Figure 5. Coarsening up sequence typical of shelf as well as other sandstone sequences.

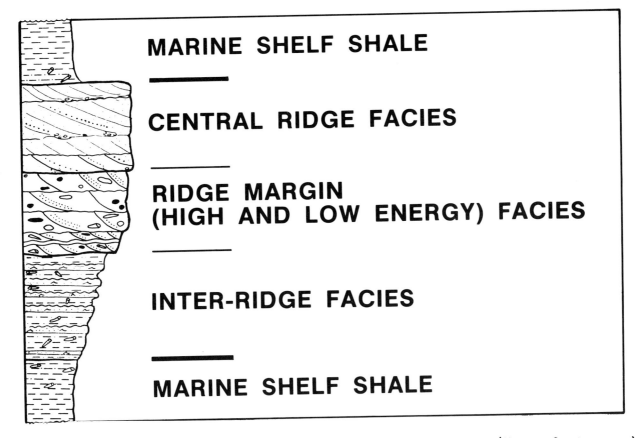

Figure 6A. Facies types recognized in Shannon Sandstone (Upper Cretaceous) shelf-ridges in the Powder River Basin of Wyoming. Note general coarsening-up sequence.

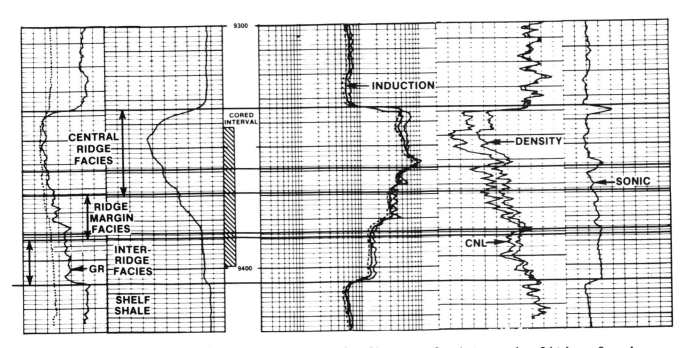

Figure 6B. Coarsening up sequence in Shannon Sandstone in Cities Service Federal AK-1 (5292), Hartzog Draw Field, Wyoming.

Another kind of coarsening up sequence commonly seen on logs and in cores, but which is not encased by marine shale and is not a shelf sandstone is a delta distributary channel complex (Fig. 7). A typical deltaic sequence, from bottom to top includes a pro-delta shale at the base. A distributary mouth bar or shoreface sandstone commonly occurs above the prodelta shales, and above that the distributary channel may be as much as 40 to 50 feet in thickness. Overlying the channel may be fresh to brackish water shales. This deltaic sequence is a thickening and coarsening-up sequence which might be misinterpreted as a coarsening-up of shelf-sandstone sequence.

A third type of coarsening up sequence is produced by regressive shoreline sandstones (Fig. 8). Marine shales occur at the base and are overlain by lower shoreface burrowed to bioturbated silty shale and sandstones. The lower shoreface coarsens upward into an upper shoreface which may be sandier and exhibits more physcial structures and less burrowing. At the top of the sequence is commonly a foreshore sandstone deposit (subaerial beach). The foreshore sandstones have low angle laminations and are generally well sorted. The foreshore may be overlain by either a marine shale or lagoonal deposits depending on what follows deposition of the foreshore.

## INNER SHELF

A variety of processes operate on the inner shelf during transgression, regression, and stillstand (Fig. 9). During transgression, the shoreline usually moves landward and the sea deepens. During a regression (or progradation) the shoreline moves seaward. During a stillstand, sea level remains relatively stationary. These relative sea level changes may be a result of tectonic effects, progradation of a delta or shoreline, or actual sea level changes. It is not always possible to isolate which of these effects is causing the transgression, regression or stillstand. But if we can determine whether relative sea level is rising, falling, or remaining static then we may be able to predict the effects of the various processes in a variety of general depositional settings such as (1) river dominated deltas, (2) wave dominated deltas, (3) beaches and barrier islands and (4) offshore sand-ridges. Shelf sandstones associated with these depositional settings vary according to variations in types and intensities of processes.

Where on the shelf would we expect to see coarsening-upward sequences, fineing-upward sequences, or an absence of sequences? Figure 9 illustrates the types of sequences which may be expected under a variety of conditions on the inner shelf. Several sequences are expected to form during transgression. In a river dominated delta, coarsening and/or fining-upward sequences may form during a transgression. In contrast, sand-ridge coarsening upward sequences are commonly in sand ridges deposited during both transgression and during a stillstand.

Special conditions may be required to form a coarsening up shelf sand sequence during a regression. Where beaches or barrier islands form during a transgression, fining-upward sequences may be deposited.

# DELTA DISTRIBUTARY CHANNEL COMPLEX

**SHALLOW MARINE SHALE**

---

**DISTRIBUTARY CHANNEL**

---

**DISTRIBUTARY MOUTH BAR OR SHOREFACE**

**PRO-DELTA SHALES**

---

**SHALLOW MARINE SHALE**

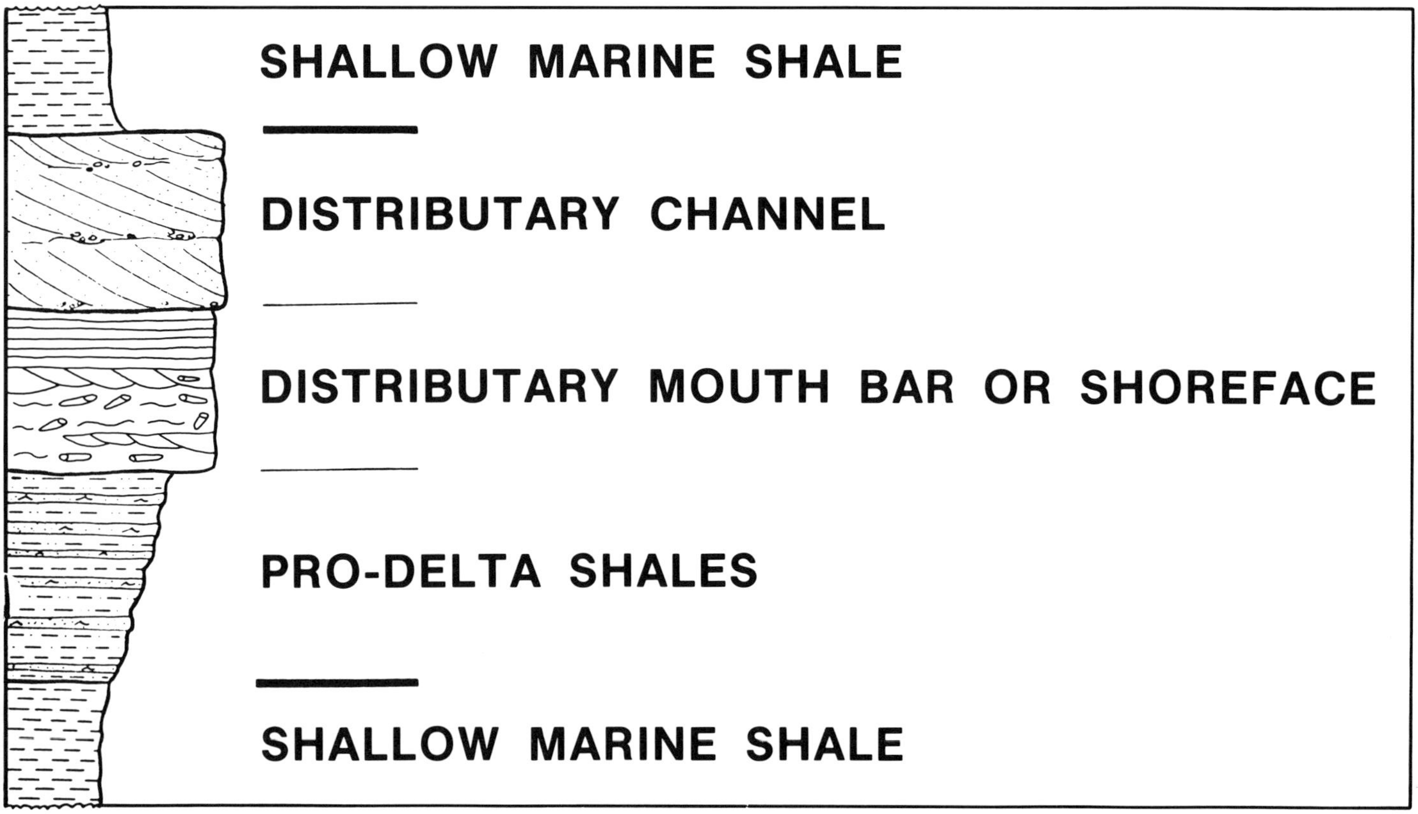

Figure 7.  Coarsening up sequence of distributary channel complex.

---

Sub-tidal tidal current sand deposition

The processes associated with sub-tidal tidal-current flow are important in understanding the origin of inner shelf sand bodies.  Sub-tidal tidal currents affect sand distribution most commonly in areas of shoreface-attached sandbodies and on the inner shelf (Fig. 10).  Tidal currents move sediment on the middle shelf in areas such as the North Sea.  Areas of movement or deposition of sand on the middle shelf probably would depend in part on local topography.

An important point to note is that this discussion will not be concerned with intertidal deposits as much as subtidal tidal-current deposits. Sedimentologists who work with carbonates are more commonly concerned with intertidal deposits.  Geologists who study shoreline deposits also more commonly observe intertidal deposits (Fig. 11).  The effectiveness of tidal currents in forming preservable sand deposits in the subtidal areas is only poorly understood.  Subtidal-tidal currents are known to flow at rates in excess of 20 cm/sec (Walker, 1985) and they commonly move fine to coarse sand.

# REGRESSIVE SHORELINE SANDSTONE

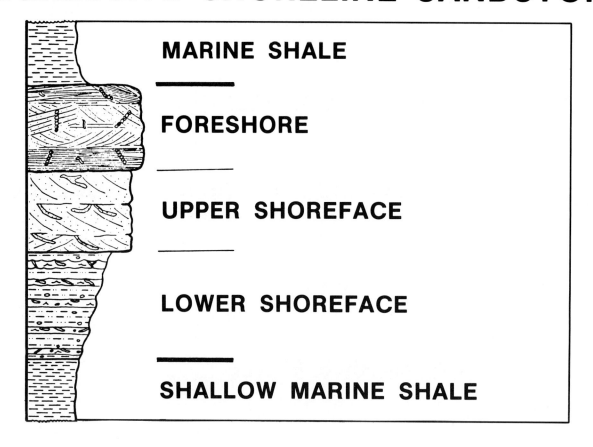

Figure 8.  Coarsening up sequence in regressive shoreline sandstone.

---

One of the areas where subtidal-tidal currents have been studied in detail is Holland.  Holland, of necessity, is a country of many dikes, a long series of embayed shorelines, and substantial areas of artificially made land. The Osterschelde estuary in Holland (Fig. 12) has been studied for several years by Visser (1980), Nio, and others associated with Utrecht University. They documented in their work a suite of characteristics which are unique to subtidal deposits.  During spring tides, tidal ranges and tidal current velocities are much stronger than during neap tide periods (Fig. 13).  Consequently, during spring-tide periods more sand is moved than during neaptide periods.  Also the amount of clay-size material is more widely distributed during spring tide periods of faster flow; as a result clay-size material is diminished to absent within sand laminasets deposited during spring-tide periods. In contrast Neap tide deposits contain a higher proportion of clay and thinner deposits.

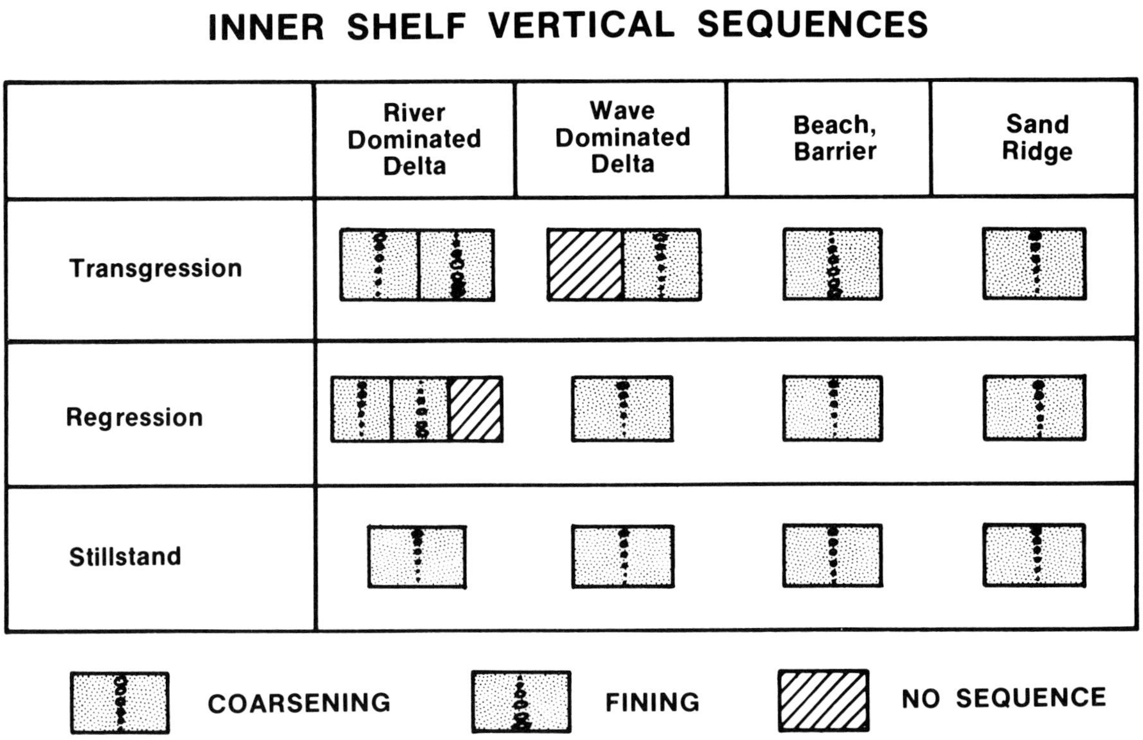

## INNER SHELF VERTICAL SEQUENCES

|  | River Dominated Delta | Wave Dominated Delta | Beach, Barrier | Sand Ridge |
|---|---|---|---|---|
| Transgression | coarsening / fining | no sequence / fining | fining | fining |
| Regression | coarsening / fining / no sequence | coarsening | coarsening | coarsening |
| Stillstand | coarsening | fining | fining | fining |

Legend: COARSENING | FINING | NO SEQUENCE

Figure 9. Distribution of Inner Shelf vertical sequences during transgression, regression, and stillstand.

---

Late Holocene subtidal tidal current deposited sand deposits have been studied in large excavations made in the process of building a sort of "sieve"-type dam which will prevent flooding of the Osterschelde (Fig. 12) during stormy high-water periods (Visser, 1980). Sub-tidal deposits in the Osterschelde commonly form multiple stacked sequences exhibiting similar flow directions (Fig. 14A). Each of these sequences begins with a series of relatively thin laminasets containing more clay than subsequently deposited thicker laminasets. These laminsets have been termed "bundles" by Nio (personal communication) and Visser (1980). The origin of the variations in the sequence of bundles is an important question. How are the thinner bundles arranged compared to the thicker ones? A bundle represents the material deposited on the slipface of a bedform during the dominant tidal current stage (Boersma, 1969). The bundle may be bounded below by slack water deposits which were deposited after flow of the subordinate current. The bundle is bounded above by thin, clayey laminae deposited during slack water following deposition by the dominant current (Visser, 1980). A diagram constructed by Visser (Fig. 14B) shows the variation in thicknesses of bundles. Shown in this figure are a number of cycles that reflect the sequence of neap to spring tides. The thicker, more sandy bundles are deposited during spring tide. The bundles deposited during the period from spring to neap tide thin progressively to the right. This sequence of thin to thick to thin bundles when observed in ancient rocks may help differentiate tidal current deposits from other types of deposits.

## SHELF PROCESSES

|  | Tidal Currents | Wind Induced Alongshore Currents | Wave Modified (Combined flow) Currents | Storm Currents | Permanent or Semi-Permanent Currents | Turbidity Currents |
|---|---|---|---|---|---|---|
| Shoreface Attached | X | X | X | X | X | — |
| Inner Shelf | X | X | X | X | X | ? |
| Middle Shelf | ? | ? | X | X | X | X |
| Outer Shelf | — | — | — | X | X | X |

Figure 10. Distribution of tidal currents on the shelf.

# INTERTIDAL vs. SUBTIDAL

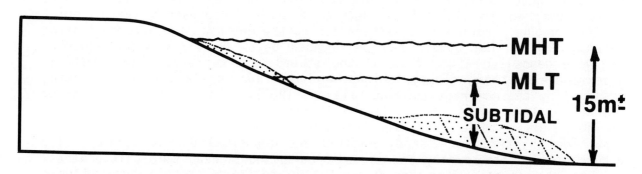

Figure 11. Contrast between intertidal and sub-tidal sand deposits.

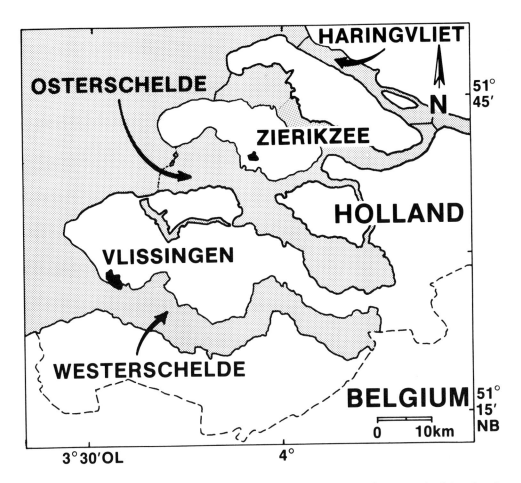

Figure 12. Estuaries of southwest Netherlands. Osterschelde is location of study by Visser (1980) of subtidal sand deposits.

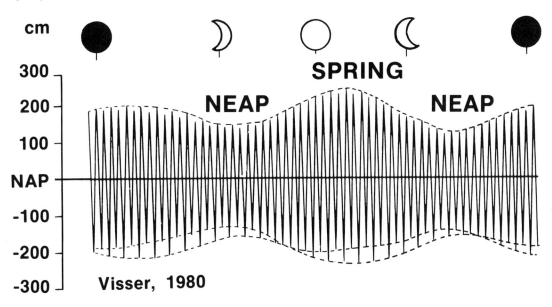

Figure 13. Tidal pattern at Vlissingen, Westerschelde, Netherlands. Tidal range (vertical scale) and current speeds increase in periods of spring tides and are less during periods of neap tide.

Figure 14A. Bundles of tidal deposits in pit in Osterschelde estuary, Netherlands. Thin bundles (dark area to right of pole) correspond to neap tide; thick bundles (top four units of pole) correspond to spring tide.

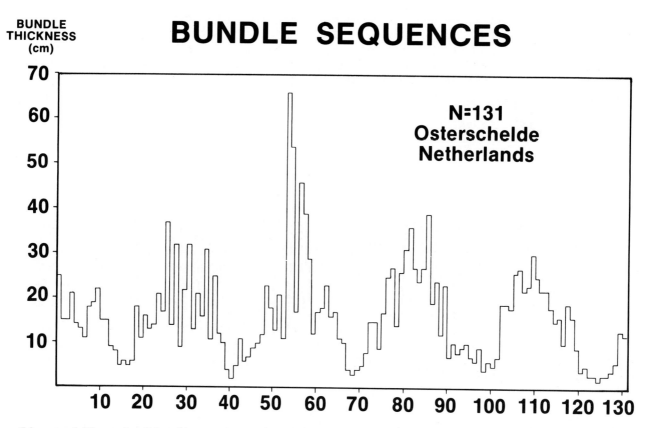

Figure 14B. Cyclical bundle sequences for tides at Vissingen, Netherlands (Visser, 1980). The thickness of individual bundles is indicated on the vertical scale. Individual bundles are numbered from left to right.

Differentiation of subtidal from intertidal sand deposition is important in reconstructing facies distribution and in identifying inner shelf deposits. The problem is difficult because the same types of currents are responsible for depositing both. One possible partial solution to differentiating the two are the mud layers which are deposited during periods of slack tidal current flow and have been described by Visser (1980). During periods of high (dominant) flow, sand waves will prograde over the bottom and most of the mud in the system remains in suspension. However, following formation of the sand waves during periods of slack tide, mud will be deposited as drapes over the sand waves (Fig. 15). During this period of mud-drape deposition, the tide would be in the process of changing from the "dominant" flow direction, to a significantly different "subordinate" flow direction. This reversed or "subordinate" current may then deposit a thin sand on top of the mud layer on the lower part of slipface of the sand wave. The mud layer located above the dominant-current sand deposit is commonly preserved and extends as a layer from subtidal depths shoreward into the intertidal zone. However, more commonly, portions of the mud layer which extend into the intertidal zone are eroded during flow of the subsequent current which forms the thin ripples and discontinuous flasers of partially eroded clay drape. Following the deposition of the subordinate current sand deposits a second slack water period occurs, and a second mud is deposited forming a mud couplet (two clay drapes separated by a thin [< 3 cm] sand) is formed. Only in the subtidal environment do we expect to commonly preserve mud couplets; in the intertidal area only one mud layer is commonly preserved.

Sigmoid cross bedding is another criteria which is considered to be of significance in interpreting tidal deposits. Among the first to discuss the significance of sigmoid bedding was Mutti (personal communication). Portions of the Jurrasic Curtis Formation in Utah (Fig. 16) have been interpreted by Kreisa, (personal communication) partially on the basis of extensive sigmoid bedding, as subtidal sand deposits. He observed very large sigmoid ("S") shaped crossbed sets (Fig. 17A). He also observed thin shale drapes interbedded with sandstone in cyclical bundled sequences very similar to those observed in the Osterschelde area in Holland. In the Curtis Sandstone Kreisa measured the thickness of individual bundles (Fig. 17B) and recognized that the bundles varied from thicker to thinner to thicker in a cyclical manner similar to that observed by Visser (1980) in Holocene deposits in Holland. These characteristics, in addition to an understanding of the regional geology allowed them to identify these sandstones as ancient subtidal tidal deposits.

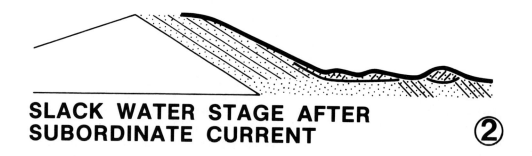
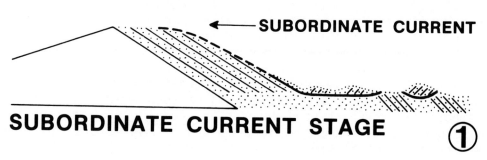

Figure 15. Subordinate and slack water tidal current stages. (1) Slack water mud drapes form over foresets formed during dominent tidal stage and are pre-preserved only in subtidal area. Minor ripples of sand are deposited during subbordinate current following slack Water Stage. (2) Second slack water mud drape preserved on top of subordinate current deposits; typical of subtidal (not intertidal) deposits (Visser, 1980).

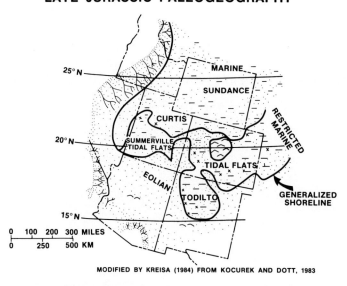

Figure 16. Study area of Curtis tidal flat deposits and Late Jurassic paleogeography, modified from Kocurek and Dott, 1983 (courtesy of R. Kreisa).

Figure 17A. Sigmoid shaped coset typical of Curtis tidal deposits in Summerville area, Utah (courtesy of R. Kreisa).

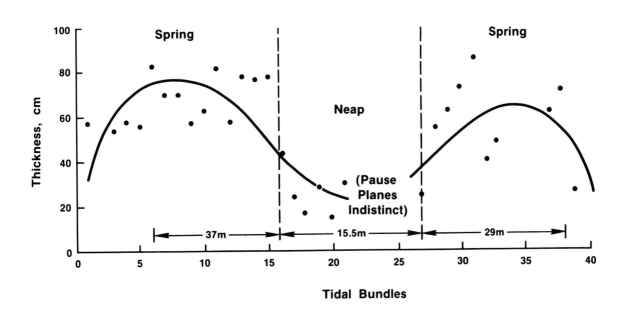

Figure 17B. Curtis neap/spring tide cycles. Thicker cosets are attributed to spring tide periods and thinner cosets are attributed to neap tide periods (R. Kreisa and R. J. Moiola, in press, Geological Society of America Bulletin).

The Tocito Sandstone Lentil in New Mexico is recognized as being primarily an inner shelf deposit (McCubbin, 1969; Molenaar, 1973). The Tocito Sandstone Lentil (or "Gallup" as it was formerly designated), in my estimation, was not primarily deposited by subtidal tidal currents; however, within the Tocito there are units which were deposited by subtidal tidal currents. Thick sets of cross-bedded sandstones (Figures 18A and 18B) showing bundling sequences are locally present in the Tocito. The bundles are moderately coarse grained and contain interlaminated shales. The bundle sequences, instead of having a 28day monthly cycle as observed in the Holocene by Visser, comprise a 12 to 15 bundle cycle. This shortened cycle may occur for a variety of reasons most of which still are compatable with subtidal deposition (Walker, 1985).

### Alongshore Wind Generated Current Deposits

Longshore (shore-parallel to shore-oblique) wave generated currents have a strong influence on nearshore sand deposition (Fig. 19). In the area where shoreface-attached ridges are deposited on the inner shelf, wind generated longshore currents are effective in moving sediment; however, it is not certain how effective these currents are or even if they occur on the middle shelf during fair-weather periods.

The New Jersey shelf area on the west side of the Atlantic Ocean is an excellent area to study nearshore sand ridge formation. Since 1979, a consortium of scientists have examined four different sand ridge areas on the New Jersey shelf (Fig. 20A). Ongoing work is focusing on three of these areas-one of which is Peahala Ridge, a shoreface attached ridge (Fig. 20B). Halsey (1978) and Figueiredo (1984) discussed possible origins for ridges of this type. Apparently tidal currents, longshore currents as well as storm currents influenced the formation of ridges such as Peahala Ridge (Duane et al, 1972 and Swift et al, 1972). In this area, net longshore current flow is probably from north to south as it is in the area to the north which was studied by Vincent et al. (1983). Peahala Ridge is an elongate topographic feature formed at an oblique angle to the shoreline (Fig. 20B). The sands deposited on this ridge are post-Pleistocene in age. Seismic sections perpendicular to the ridge indicate a fairly uniform interior (Fig. 21); however, some markers within the sand body are useful in making sedimentologic interpretations. Several vibracores taken in the line of the seismic section will allow determination of the sedimentary structures and sequences necessary to interpret the origin of the ridge.

Figure 18A, B. Tidal bundles in upper bed of Tocito (Gallup) Sandstone, San Juan Basin, New Mexico (courtesy of D. Nummedal).

## SHELF PROCESSES

|  | Tidal Currents | Wind Induced Alongshore Currents | Wave Modified (Combined flow) Currents | Storm Currents | Permanent or Semi-Permanent Currents | Turbidity Currents |
|---|---|---|---|---|---|---|
| Shoreface Attached | X | X | X | X | X | — |
| Inner Shelf | X | X | X | X | X | ? |
| Middle Shelf | ? | ? | X | X | X | X |
| Outer Shelf | — | — | — | X | X | X |

Figure 19. Distribution of wind induced longshore current deposits on the shelf.

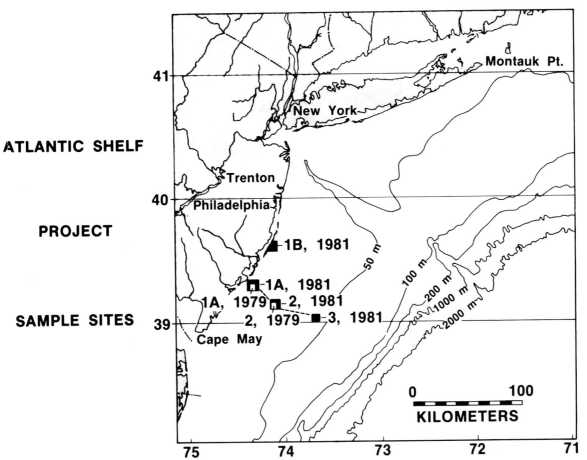

Figure 20A. Locations of New Jersey shelf ridge areas (1A, 1B, 2, 3) studied by consortium of companies 1979-1986.

## BOTTOM TOPOGRAPHY / PEAHALA RIDGE, NEW JERSEY

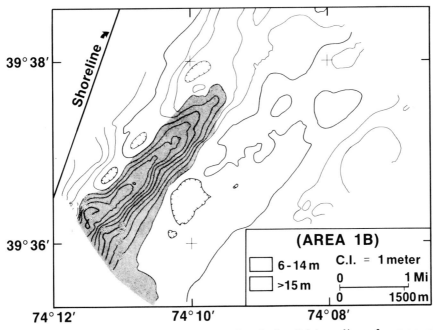

Figure 20B. Shoreface-attached ridge, Peahala Ridge New Jersey shelf. Shoreline intersects ridge near bottom of figure.

Another type of deposit which is interpreted to be the result of longshore currents are the deposits of the type which occur "downcurrent" of the Nile Delta in the Mediterranean (Fig. 22; Coleman et al, 1981 and Sharaf Al Din, 1977). This particular type of deposit has been designated by Boyles and Scott (1983) as a "plume" deposit. In this setting sediment is fed into the marine environment (in this case, the Mediterranean Sea) by the distributary channels of the delta. Longshore currents redistribute the sand from the distributary-mouth bar to form a ribbon or "plume" of sand extending along the shore in a downstream direction. The sand is transported from the delta-mouth bars to form an elongate sand body on the inner shelf. The mechanism of alongshore transport due to strong fresh-salt water density contrasts off the Nile is discussed by Phillips, et al (this volume).

Longshore currents, such as those believed to occur off the Nile, may have been active during the past in depositing "plume" type shelf reservoirs. In the Wind River Basin of Wyoming (Fig. 23) the Fales Member of the Mesaverde Sandstone is a relatively thin sandstone located between two shales (Fig. 24). Where the Fales Sandstone produces at West Poison Spider field it has been interpreted as a "plume" deposit (Barrett, 1982 and Boyles and Scott, 1983); (Fig. 25). A "plume" deposit can be recognized in the geologic record by evidence of strong currents flowing on the inner shelf and parallel to the shoreline. Current deposits may be indicated by cross bedding in outcrops or cores. Outcrops of the Fales (Fig. 26) suggest that a large north flowing deltaic complex existed west-southwest of West Poison Spider field (Barrett, 1982). In cores from the subsurface at the West Poison Spider field evidence of strong current flows and possible unidirectional flow may be observed (Fig. 27). The location of this field relative to the delta, which appears to have supplied it with sand, also supports the "plume" model of deposition. In the outcrop we see evidence for flow of distributary channels toward the north; we also see evidence for probable longshore flow toward the east. In the subsurface this plume of sand and silt extending eastward has been mapped by Barrett (1982). Union Oil Company is now producing from and continuing to drill wells for production from the Fales Sandstone at 13,000 feet in this area. The origin of this type of reservoir is attributed to deposition on the inner shelf by lonhshore currents which flowed parallel to the contours initially and then formed a "plume" of sediment in a more offshore position farther down the coast. Work on the Fales was done primarily as a masters thesis at the University of Texas by Barratt; the study was later discussed by Boyles and Scott (1983).

Water depth on the inner shelf and in the shoreface attached ridges can be recognized in cores and outcrop by using certain criteria (Fig. 28A). Heavy mineral concentrations often occur at very shallow depths on the foreshore or beach. Ophiomorpha burrows occur most abundantly in the upper shoreface. Asterosoma burrows are most common in slightly deeper water (middle shoreface). The base of tidal inlets or channel cuts can commonly be recognized on shallow seismic. These channel cuts may extend down below the level of abundant Ophiomorpha. In the central part of the inner shelf hummocky cross stratification (HCS) is a common sedimentary feature (Fig. 28B).

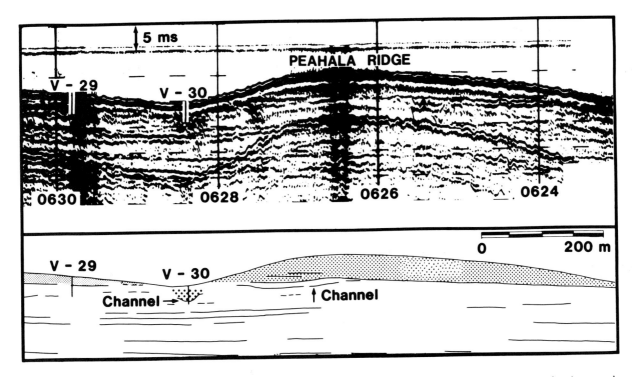

Figure 21. Sparker profile through Atlantic shelf shoreface-attached sand ridge. Active ridge is shaded area in diagram. Channels below sand ridge are earlier deposits. (Figuieredo, 1984)

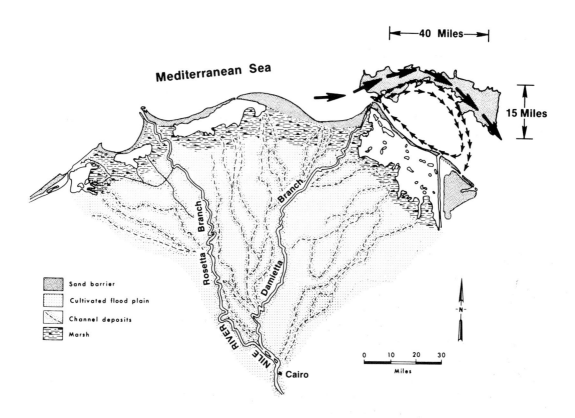

Figure 22. Nile delta and shelf sand bodies formed into a "plume" by longshore currents (Coleman, 1981).

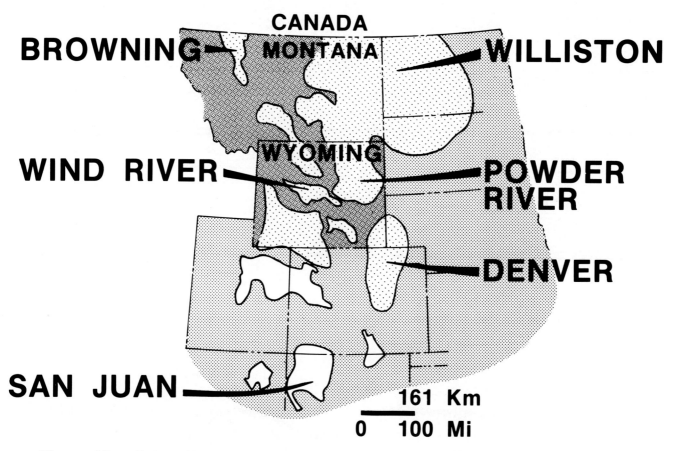

Figure 23. Major Rocky Mountain basins. Fales Sandstone is located in the Wind River Basin of Wyoming.

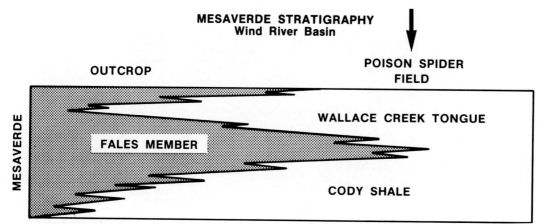

Figure 24. Mesaverde stratigraphic section Wind River Basin, Wyoming (courtesy of M. Boyles). Section is oriented west (left) to east. Fales Sandstone is separated from other Mesaverde sandstones by shale tongues. Location of Poison Spider Field is indicated.

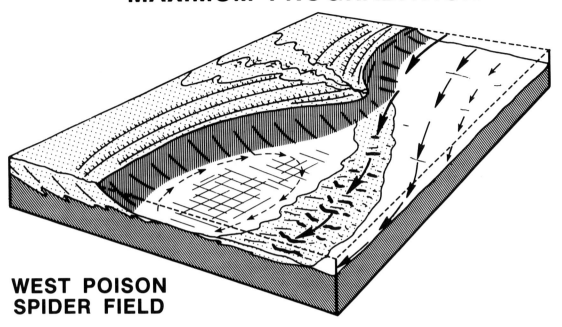

Figure 25. "Plume" of Fales sandstone formed by longshore current distribution of sand that was fed into a shallow sea from a delta which outcrops (Barratt, 1982). West Poison Spider Field produces from the Fales Sandstone plume.

Figure 26. Troughs in outcrop of lower unit of Upper Cretaceous Fales Sandstone in Wind River Basin, Wyoming.

Figure 27. Trough and planar-tabular bedding, Fales Sandstone, Union Oil No. 25 10,702-03' (from Barratt, 1982).

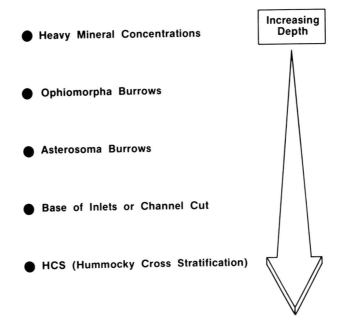

Figure 28A. Features which may be used on the inner shelf as depth markers.

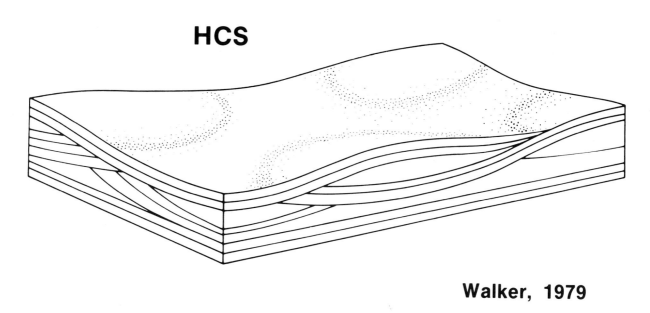

Figure 28B. Hummocky Cross Stratification (HCS).

HCS is believed by many workers to result from settling out of sediment following the passage of storms; others believe that bottom currents are responsible for HCS. One of the most conspicuous aspects of HCS are laminations in the form of half domes which are convex upward. Additional characteristics of HCS are described by Dott and Bourgeois (1982). HCS is commonly deposited and preserved between fair-weather and storm-wave base.

Until recently most sedimentologists had difficulty in recognizing hummocky cross stratification in cores; however, excellent examples are observed in the Frontier Formation from western Wyoming and the Viking and Cardium Sandstones in Canada (Figs. 29A and 29B). As is shown in these figures, when appropriately wide (4" diameter) slabbed cores are available very sharp lower contacts of HCS and the concave-downward or convex-upward geometries of HCS laminations can be recognized.

## MIDDLE-SHELF

Coarsening-upward sequences form the major sand accumulations on the middle shelf (100-300 feet water depth) during transgression, regression, and stillstand (Fig. 30). On the middle portion of a wide shelf during transgression several types of coarsening-upward sequences may be deposited. Under certain conditions deltas may prograde seaward into water depths as great as those observed on the middle shelf. During a stillstand, if rapid progradation were taking place, coarsening-upward deltas may have their most seaward deposition (prodelta) in water as deep as the middle shelf. During periods of regression and/or progradation coarsening-upward sequences resulting from deposition of deltas may be found on the middle shelf. Deltaic coarsening-up sequences on the middle shelf are more probable in river than wave dominated deltas. The number of types of coarsening-upward sequences is diminished on the middle shelf when compared to the inner shelf. Upward-coarsening shoreline deposits such as barrier island sequences occur only shoreward of or on the inner shelf.

A sedimentary feature that may be common on the middle and (inner?) shelf is wave formed horizontal laminations (Fig. 31). These laminations form under certain conditions when wave orbital flow encounters the bottom during storms. These horizontal laminations may appear to be very similar to hummocky cross stratification; however, they are horizontal on a much larger scale. In cores (Fig. 32), it may be very difficult to separate these two sedimentary structures. Wave generated horizontal laminations may be eroded if they are deposited above storm wave base; they may also be destroyed by burrowing. Wave formed laminations are usually parallel to horizontal bedding contacts (Fig. 33). They commonly occur in thin beds up to two feet in thickness. Laminasets composed of horizontal laminations formed by fast flowing currents, such as those observed in the North Sea (Tillman and Reineck, 1975) are commonly thinner than wave-laminated beds associated with sand ridges (Tillman and Martinsen, 1985). The horizontal laminations may be interrupted by burrowed to bioturbated intervals, and thin layers of shell hash may be scattered throughout or concentrated near the bottom of the units.

Figure 29A. Hummocky cross stratification (HCS) in core of Frontier Formation inner shelf deposit, Amoco Whiskey Buttes No. 6, 11,159', Sweetwater County, Wyoming.

Figure 29B. Hummocky cross stratification (HCS) in core of Gulf Oil Canada, Ltd., Gulf Stettler (LSD 07-22-38-20-W4M) at 1157.5 meters.

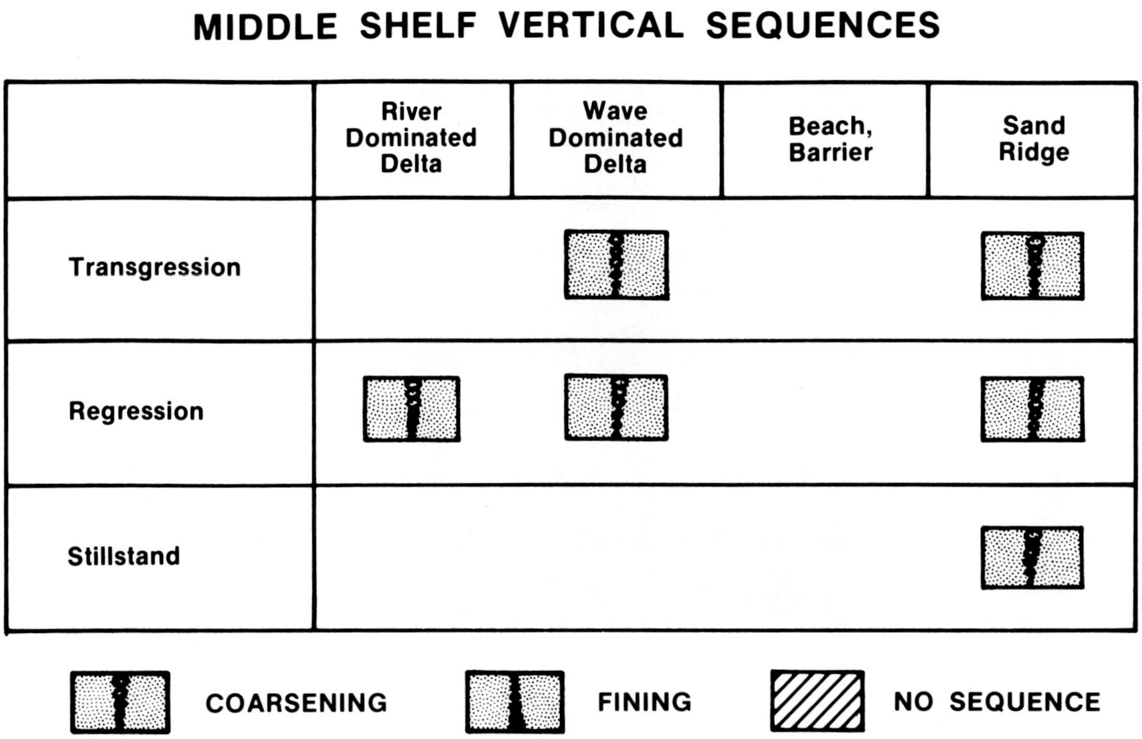

## MIDDLE SHELF VERTICAL SEQUENCES

|  | River Dominated Delta | Wave Dominated Delta | Beach, Barrier | Sand Ridge |
|---|---|---|---|---|
| Transgression |  | fining |  | fining |
| Regression | coarsening | coarsening |  | coarsening |
| Stillstand |  |  |  | coarsening |

Legend: COARSENING | FINING | NO SEQUENCE

Figure 30. Distribution of middle-shelf vertical sequences during periods when shoreline is regressing, transgressing or is at stillstand.

---

Storm currents are probably the most important processes for distributing sand on the middle shelf, although storm currrents can have an effect anywhere on the shelf, and the resultant products from storm processes may be similar on all parts of the shelf. Spearhead Ranch field, which produces from the Frontier Formation in Wyoming (Fig. 34A), is an example of storm sands deposited on the middle shelf. The field, which was discovered in the mid-1970's and produces from a series of thin marine shelf ridges (bars) occuring above a submarine (?) unconformity. Producing wells in the field typically have a 4- to 10-foot-thick producing section (Fig. 34B) initial production rates of 2,000 to 3,000 barrels per day. The reservoir sandstones are relatively coarse grained, glauconitic and poorly sorted. The amount of diagenesis in the post-unconformity sand ridges contrasts strongly with the underlying deposits and greatly influences the productive capabilities of the unit. In the cross-bedded marine-shelf sandstones below the sandstones, production is poor to nil in Spearhead Ranch field. The upper sandstone is porous in part because of the growth of chlorite rims on the quartz grains. The chlorite prevented significant quartz overgrowths and porosity is maintained even after the porosity of the adjoining subjacent shelf facies is completely occluded (Tillman and Almon, 1979). The lower percentage of chlorite in the subjacent facies results in part from the finer-grained nature of these cross-bedded sediments.

## SHELF PROCESSES

|  | Tidal Currents | Wind Induced Alongshore Currents | Wave Modified (Combined flow) Currents | Storm Currents | Permanent or Semi-Permanent Currents | Turbidity Currents |
|---|---|---|---|---|---|---|
| Shoreface Attached | X | X | X | X | X | — |
| Inner Shelf | X | X | X | X | X | ? |
| Middle Shelf | ? | ? | X | X | X | X |
| Outer Shelf | — | — | — | X | X | X |

Figure 31. Probable distribution of wave-modified current (combined flow) laminations on segments of the shelf.

Figure 32A. Horizontal laminations which may have been formed by wave modified (combined-flow) currents. Frontier Formation, Spearhead Ranch Field, Wyoming. Mountain Fuel Spearhead Ranch No. 4, SE SW Sec. 19 T39N R74W, 12,544.0'.

Figure 32B. Horizontal laminations in core of Gallup Sandstone, San Juan Basin, New Mexico. These laminations may have been deposited by combined flow, wave modified, currents. Tenneco #1, Dry Creek, NWSE Sec. 5 T24N, R10W, 6098.1'. (Core stored in U.S.G.S. Core repository in Denver, Colorado).

**Characteristics of Wave Formed Laminations**

- Horizontal to subhorizontal lamina

- Horizontal to subhorizontal bedding

- Thin beds (< 2')

- Interrupted by burrowed to bioturbated intervals

- Thin layers of shell hash accumulations

Figure 33. Characteristics of wave formed laminations.

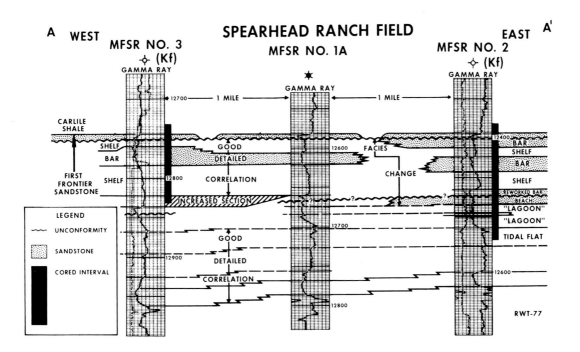

Figure 34A. Frontier Formation cored intervals (black columns) and correlations at Spearhead Ranch field. Production in the field is primarily from the shelf ridge sandstone above an unconformity at the top of the First Frontier Sandstone (from Tillman, 1979).

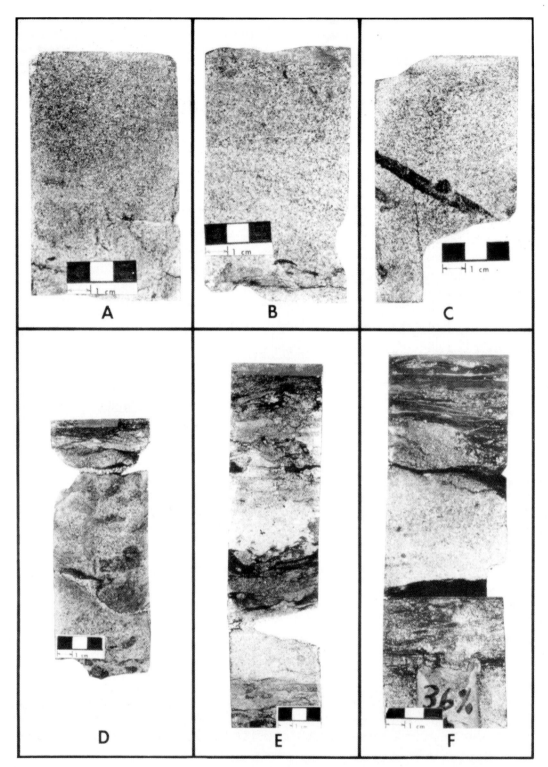

Figure 34B. Frontier Sandstone shelf-ridge (bar) facies. A. "Massive appearing" medium-grained shelf sandstone. Salt and pepper appearance in cherty facies; reservoir sandstone. Uppermost sandstone of First Frontier Sandstone. MFSR No. 4, 12,532.1 feet. B. Medium-grained sandstone with abundant black chert grains; "massive apearance", MFSR No. 2, 12,406.2 feet. C. <u>Asterosoma</u> burrow, concentric clay laminated at a typical angle of 45° from horizontal. Top of First Frontier Sandstone. MFSR No. 3, 12,771.7 feet. D. Bioturbaded medium-grained sandstone at top of First Frontier Sandstone. Contact with subhorizontally laminated Cody (Carlile) Shale at top of core slab. Shelf sand-ridge facies, MFSR, No. 3, 12,769.7 feet. E. sandstone at the top of the sand ridge. Contact with shelf siltstone near top of core slab. Shelf sand-ridge. MFSR No. 2, 12,474.8 feet. F. Cody (Carlile) Shale (upper 4 cm) lying on top of medium grained shelf sand-ridge deposit, MFSR No. 2, 12,404.8 feet.

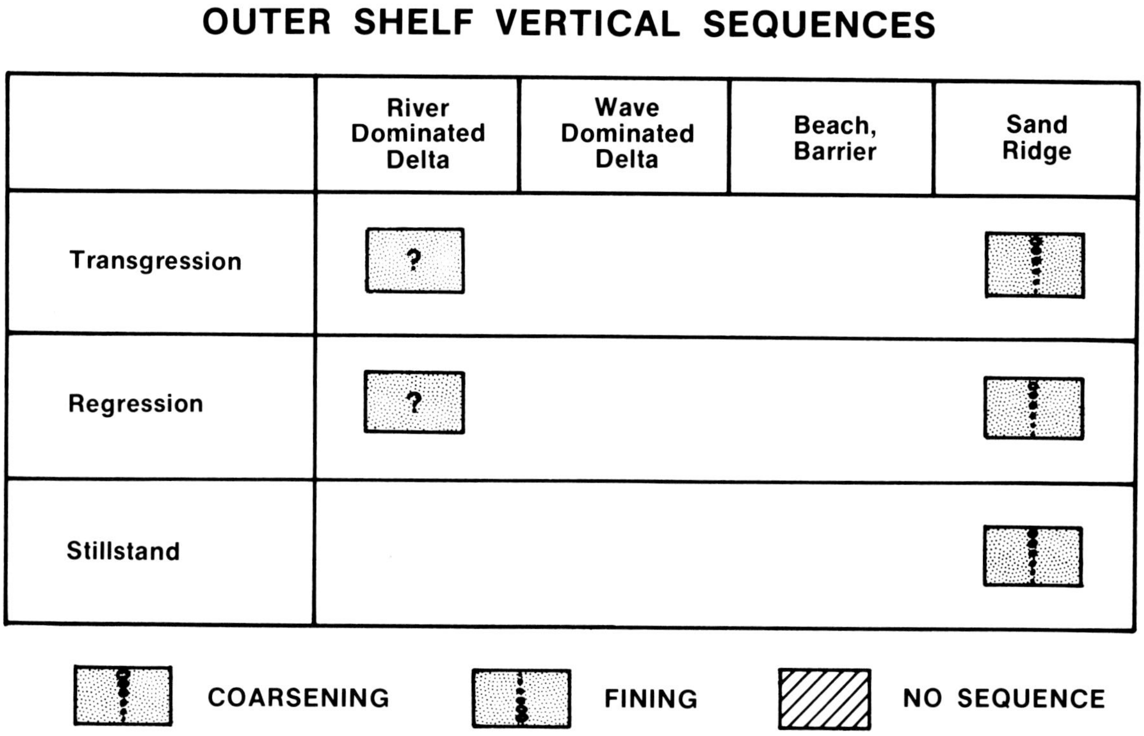

## OUTER SHELF VERTICAL SEQUENCES

|  | River Dominated Delta | Wave Dominated Delta | Beach, Barrier | Sand Ridge |
|---|---|---|---|---|
| Transgression | ? |  |  | fining |
| Regression | ? |  |  | fining |
| Stillstand |  |  |  | fining |

COARSENING    FINING    NO SEQUENCE

Figure 35. Distribution of outer shelf vertical sequences during periods when shoreline is regressing, transgressing, or is at stillstand.

---

## OUTER SHELF

On the outer shelf, sand sequences of all types are less common than on the inner or middle shelf. River dominated deltas may prograde out and directly overlie deposits of the outer shelf during periods of very rapid progradation during a rapid sea level drop. There also may be a remote possibility that a river dominated delta which is well established could survive a transgression and deposit sediments at outer-shelf depths. Otherwise, on the outer shelf there will be few sandy deposits associated with the first three environments in Figure 35.

### Shelf Sand Ridges

The most common sand deposits on the middle and outer shelves are sand ridges. Many shelf sand ridges are coarsening-up sequences. If a coarsening-up sequence can be identified as being on the outer shelf on the basis of paleontology or regional geology, it will almost always be a sand ridge.

On the outer shelf, in addition to storm currents, we may expect to see deposits of (wind-forced?) semi-permanent currents (Fig. 36). Those contour parallel currents could have contributed significantly to deposition of the Shannon Sandstone shelf ridges at Hartzog Draw Field in Wyoming (Fig. 37).

## SHELF PROCESSES

| | Tidal Currents | Wind Induced Alongshore Currents | Wave Modified (Combined flow) Currents | Storm Currents | Permanent or Semi-Permanent Currents | Turbidity Currents |
|---|---|---|---|---|---|---|
| **Shoreface Attached** | X | X | X | X | X | — |
| **Inner Shelf** | X | X | X | X | X | ? |
| **Middle Shelf** | ? | ? | X | X | X | X |
| **Outer Shelf** | — | — | — | X | X | X |

Figure 36. Distribution of storm and offshore permanent and semi-permanent current deposits on the shelf.

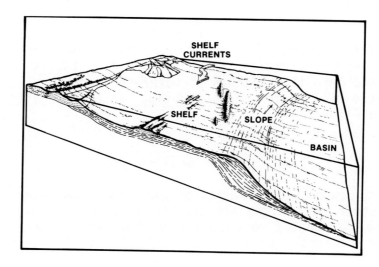

Figure 37. Wide shelf model for Shannon Sandstone, Powder River Basin, Wyoming. Hartzog Draw Field produces from sand ridge oblique to the shelf-slope break.

The source of the sediments for Hartzog Draw and other Shannon sandstones is probably from areas far to the north in Montana (Tillman and Martinsen, 1984). In South Central Montana deltas have prograded out and distributed sand onto the shelf and, with time, permanent or semipermanent currents have moved sand south with the help of storm induced currents. These combined currents probably were responsible for deposition of the series of oblique Shannon Sandstone deposits. Hartzog Draw (Fig. 38) is the largest of the oil fields which produce from the Shannon Sandstone (Tillman and Martinsen, 1979).

## Transport Directions

One of the most important pieces of information required to interpret shelf sandstones are the flow directions of the currents which were instrumental in depositing sand ridges or other types of shelf sands. Flow directions at False Cape (Swift et al, 1972) and several other ridges along the Atlantic coast indicate that the currents which are assumed to form the ridges flowed obliquely across the ridges. Flow directions over the shoreface-attached Peahala Ridge off New Jersey (Fig. 20B) have not yet been determined.

In subsurface deposits such as Hartzog Draw field (Tillman and Martinsen, 1979), a limited amount of oriented core data allows determination of the current flow directions over the central portion of the field. In the Cities Service AS #1 Federal, using a conventionally oriented core, 132 measurements of trough cross bed flow directions were measured (Fig. 39). The trough cross bedding in this well has a southerly mean flow direction and the flow directions are concentrated within a 40° arc. One standard deviation (66% of the data) falls within a range of 20° on either side of the mean. In diagrams such as Figure 39 individual flow measurements are indicated by spots; the middle of the diagram represents 90° dip and the dip angle decreases to 0° at the outside. Most of the flow directions are toward the southeast and have a dip of about 30° or less. It is important to notice that the sand body trends northwest/southeast (Fig. 38) and the flow direction is oblique to the shelf ridge and flows nearly straight south. Other oriented core data (Hearn, 1984) supports the results indicated in the Cities Service AS No. 1; however, most of the orientation data is in the central portion of the field. It is not known what the current flow directions were in the south and western quadrants of Hartzog Draw field.

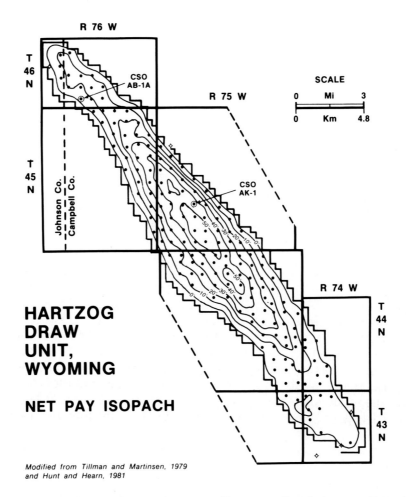

Figure 38. Net sandstone isopach map, Shannon Sandstone, Wyoming.

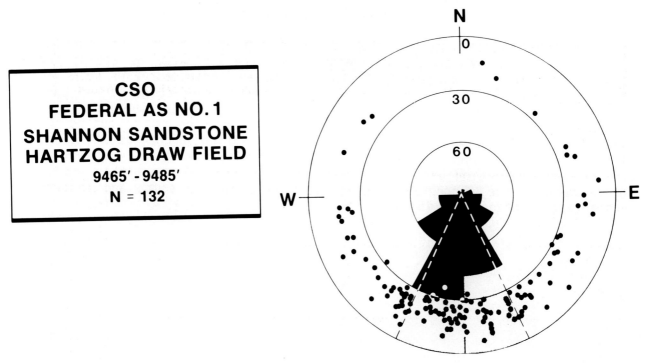

Figure 39. Transport direction of trough cross beds from Cities Service Federal AS-1, Hartzog Draw Field, Wyoming. Two-thirds of data points (one standard deviation) are concentrated between dashed lines. Mean transport direction is to the south.

Turbidites

Turbidity currents are an important process on portions of the shelf. I believe that the occurance of turbidites on the middle and outer shelf (Fig. 40) can be documented. Caution should be exercised in calling all graded beds on the shelf turbidites (Fig. 41). Graded beds can form from non-turbidite shelf currents which carry sediment in suspension. However, if complete or nearly complete, Bouma sequences (Fig. 42), can be identified, shelf turbidites probably should be interpreted to be the depositional mechanism. In areas where deltas prograde into middle (or outer) shelf environments there is a potential for significant relief. It is also possible that storms or other catastrophic events may stir up sand and form true turbid suspensions which because of the local relief move rapidly out onto the middle to outer shelf. The existence and/or distribution of shelf turbidites is not a settled issue, but I think that in time we will have more evidence to indicate that turbidites do occur in certain shelf settings. Turbidites are normally considered to form fans in deep water (Fig. 43); however, on the shelf or the slope, adequate topographic relief between the proximal and distal portions of a deposit may not always exist. Shelf turbidites probably form sheet-like deposits. The Upper Cretaceous Panther Sandstone in Utah contains complete Bouma sequences (Fig. 44) which are interpreted by a number of workers including Balsley (1982), W. Abbott and M. Link (personal communications, 1985) to be turbidites.

## SHELF PROCESSES

|  | Tidal Currents | Wind Induced Alongshore Currents | Wave Modified (Combined flow) Currents | Storm Currents | Permanent or Semi-Permanent Currents | Turbidity Currents |
|---|---|---|---|---|---|---|
| Shoreface Attached | X | X | X | X | X | — |
| Inner Shelf | X | X | X | X | X | ? |
| Middle Shelf | ? | ? | X | X | X | X |
| Outer Shelf | — | — | — | X | X | X |

Figure 40. Distribution of turbidity current deposits on the shelf.

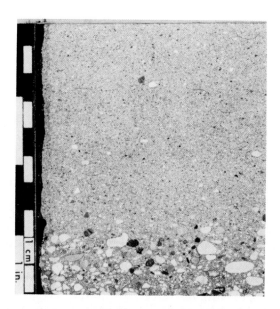

Figure 41. Graded bed in Cardium Sandstone, Alberta, Canada. Gulf et al, Ricinus, LSD 13-26-36-9W5, 9015.4 feet (courtesy of R. Walker).

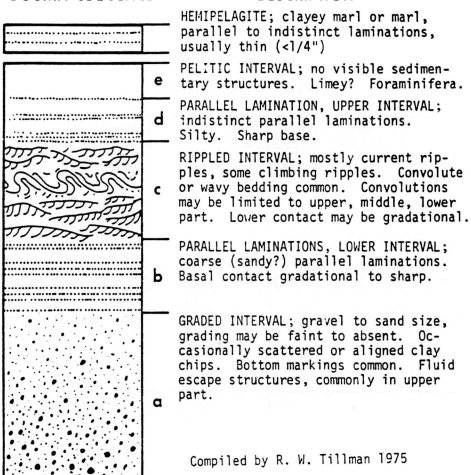

| BOUMA SEQUENCE | | DESCRIPTION |
|---|---|---|
| | | HEMIPELAGITE; clayey marl or marl, parallel to indistinct laminations, usually thin (<1/4") |
| | e | PELITIC INTERVAL; no visible sedimentary structures. Limey? Foraminifera. |
| | d | PARALLEL LAMINATION, UPPER INTERVAL; indistinct parallel laminations. Silty. Sharp base. |
| | c | RIPPLED INTERVAL; mostly current ripples, some climbing ripples. Convolute or wavy bedding common. Convolutions may be limited to upper, middle, lower part. Lower contact may be gradational. |
| | b | PARALLEL LAMINATIONS, LOWER INTERVAL; coarse (sandy?) parallel laminations. Basal contact gradational to sharp. |
| | a | GRADED INTERVAL; gravel to sand size, grading may be faint to absent. Occasionally scattered or aligned clay chips. Bottom markings common. Fluid escape structures, commonly in upper part. |

Compiled by R. W. Tillman 1975

Figure 42. Bouma sequence.

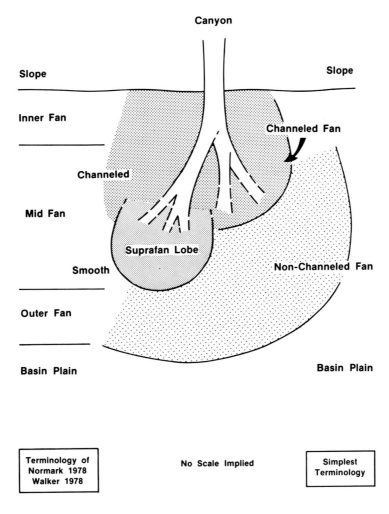

Figure 43. Deepwater turbidite fan terminology. Simplest terminology for a fan is given on right side of diagram. Slightly more complex fan terminology given on left (courtesy of R. Walker).

Figure 44. Complete Bouma sequences in the Panther Sandstone. These sequences are interpreted to be delta front turbidites. (courtesy of M. Link).

## DISCUSSION

What can be concluded about shelf sand deposits? (1) The shelf is a viable exploration area, (2) the shelf is an area which requires more study, (3) the shelf is an area where seismic techniques may have only limited success in finding reservoirs, but an area where sequences may be recognized and the relationship of the shelf sands then may be predictable, (4) it is an area in which careful paleogeographic reconstruction will allow the locations of reservoirs to be predicted (Fig. 45), (5) it is an area where one can expect initial production from rates ranging from 100 to 3,000 barrels of oil per day, and (6) it is an area which may yield additional production within basins which have already been explored.

Production from shelf sandstones may be developed in areas where drilling has been sparse or where wells have drilled through shaly sections and the sandier units within the shales have been ignored. By understanding the processes and the various effects which can modify the intensity of these processes that form sandstone deposits on the inner, middle, and outer shelf (Fig. 46), better shelf exploration models can be developed.

## ACKNOWLEDGEMENTS

Discussions with and reviews by Don Swift, Roger Walker, Djin Nio and others have helped to formulate the major points of this report. Any errors, however, are the responsibility of the author. The manuscript was also reviewed by Robert Wright and Marty Link. Illustrations were supplied by Ron Kreisa, Alberto Figuieredo and Dag Nummedal. Drafting of figures was done by Cities Service; typing was done primarily by Janice Brewer.

# SHELF SUBDIVISIONS

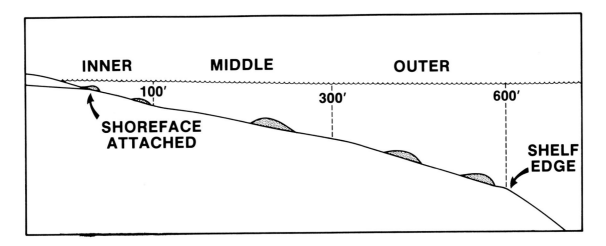

Figure 45. Subdivisions of shelf where sand may accumulate.

## SHELF PROCESSES

|  | Tidal Currents | Wind Induced Alongshore Currents | Wave Modified (Combined flow) Currents | Storm Currents | Permanent or Semi-Permanent Currents | Turbidity Currents |
|---|---|---|---|---|---|---|
| Shoreface Attached | X | X | X | X | X | — |
| Inner Shelf | X | X | X | X | X | ? |
| Middle Shelf | ? | ? | X | X | X | X |
| Outer Shelf | — | — | — | X | X | X |

Figure 46. Shelf processes and their distribution on the shelf.

# REFERENCES CITED

Balsley, J. K., 1982, Cretaceous Wave-dominated Delta Systems: Booh Cliffs, East Central Utah: American Association of Petroleum Geologists Guidebook, 219 p.

Barratt, J. C., 1982, Fales Member (Upper Cretaceous) Deltaic and Shelf-Bar Complex, Central Wyoming: M.S. Thesis, Univ. of Texas, 120 p.

Boersma, J. R., 1969, Internal structures of some tidal meagaripples on a shoal in the Westerschelde estuary, the Netherlands: Geologie en Mijnbouw, v. 48, p. 409-414.

Boyles, J. M. and A. J. Scott, 1983, Shelf-bar variations and a model for their prediction in the subsurface (abs.): Program and abstracts, The Mesozoic of middle North America, Calgary, Alberta, p. 26.

Coleman, J. M., H. H. Roberts, S. P. Murray, and M. Salama, 1981, Morphology and dynamic sedimentology of the eastern Nile delta shelf: v. 42, p. 301-26.

Dott, R. H., Jr. and J. Bourgeois, 1982, Hummocky stratification: significance of its variable bedding sequences: Bulletin of Geological Society of America, v. 93, p. 663-680.

Duane, D. B., M. E. Field, E. P. Meisburger, D. J. P. Swift and S. J. Williams, 1972, Linear Shoals on the Inner Continental Shelf, Florida to Long Island, in D. J. P. Swift, D. B. Duane and O. H. Pilkey (eds): Shelf Sediment Transport: Process and Pattern, Stroudsburg, PA, Dowden, Hutchinson and Ross, p. 447-499.

Emery, K. O. and E. Uchupi, 1984, The Geology of the Atlantic Ocean, Spinger-Verlag, 1050 p.

Figueiredo, A. G., Jr., 1984, Submarine Sand Ridges: Geology and Development, New Jersey, U. S. A.: PhD Dissertation, University of Miami, 408 p.

Halsey, S. D., 1978, Late Quarternary Geologic History and Morphologic development of the barrier island system along the Delmarva Peninsula of the Mid-Atlantic Bight: PhD Dissertation, Universtiy of Delaware, 592 p.

Hearn, C. L., W. J. Ebanks Jr., R. S. Tye and V. Ranganathan, 1984, Geological Factors Influencing Reservoir Performance of the Hartzog Draw Field, Wyoming: Journ. of Petrol. Technology, August 1984, p. 1335-43.

Hunt, R. D. and C. L. Hearn, 1982, Reservoir Management of the Hartzog Draw Field: Jour. Petroleum Tech., p. 1575-1582.

Kocurek, G. and R. H. Dott, Jr., 1983, Jurassic paleogeography and paleoclimate of the central and southern Rocky Mountain Region, in Reynolds, M. W. and E. D. Dolly (eds), Mesozoic paleogeography of west-central United States: Rocky Mountain Section, Society of Economic Paleontologists and Mineralogists, Denver, Colorado, p. 101-116.

McCubbin, D. G., 1969, Cretaceous strike valley sandstone reservoir, northwestern New Mexico: American Association of Petroleum Geologists Bulletin, v. 46, p. 546-561.

Molenaar, C. M., 1973, Sedimentary facies and correlation of the Gallup Sandstone and associated formations, northwestern, New Mexico, in J. E. Fasset (ed), Cretaceous rocks of the southern Colorado Plateau: Four Corners Geological Society Memoir, p. 85-110.

Normark, W. R., 1978, Fan valleys, channels and depositional lobes on modern submarine fans; characters for recognition of sandy turbidite environments: American Assoc. Petroleum Geologists Bull., v. 62, p. 912-931.

Sharaf Al Din, S. H., 1977, Effect of the Aswan Dam on the Nile flood and on the estuarine and coastal circulation pattern along the Mediterranean Egyptian coast: Limnology and Oceanography, p. 194-207.

Shepard, F. P., 1973, Submarine Geology, 3rd Edition, New York, Harper & Row, 517 p.

Swift, D. J. W., D. W. Stanley and J. R. Curry, 1971, Relict sediments on continental shelves: a reconsideration: Journal of Geology, v. 79, p. 322-46.

Swift, D. J. P., B. W. Holliday, N. F. Avignone and G. Scheidler, 1972, Anatomy of a shoreface ridge system, False Cape, Virginia: Marine Geology, v. 12, p. 59-84.

Tillman, R. W. and H. Reineck, 1974, Discrimination of North Sea Sand Environments with population-derived grain size parameters, Theme 6 IX$^{me}$ Congress International de Sedimentologie, Nice p. 217-20.

Tillman, R. W. and W. Almon, 1979, Diagenesis of Frontier Formation offshore Bar Sandstones, Spearhead Ranch Field, Wyoming, in P. A. Scholle and P. R. Schluger, Aspects of Diagenesis: Society of Economic Paleontologists and Mineralologists Special Publication No. 26, p. 337-378.

Tillman, R. W. and R. S. Martinsen, 1979, Hartzog Draw Field, Powder River Basin, Wyoming: in R. W. Flory (ed.), Rocky Mountain High, Wyoming Geol. Assn. 28th Annual Meeting, Core Seminar Core Book, p. 1-38.

Tillman, R. W. and R. S. Martinsen, 1985, Upper Cretaceous Shannon and Haystack Mountains Formation Field Trip No. 11, in Guidebook for Society of Economic Paleontologists and Mineralogists Mid-Year Meeting, Golden, Colorado, p. 97.

Visser, M. J., 1980, Neap-spring cycles reflected in Holocene subtidal large scale bedform deposits: A preliminary note: Geology, v. 8, p. 543-546.

Vincent, C. E., R. A. Young, and D. J. P. Swift, 1983, Sediment transport on the Long Island shoreface, North American Atlantic Shelf: role of waves and currents in shoreface maintenance: Continental Shelf Research, v. 2, p. 163-181.

Walker, R. G., 1978, Deep water sandstone facies and ancient submarine fans; models for exploration for stratigraphic traps: American Assoc. Petroleum Geologists Bull., v. 62 p. 932-966.

Walker, R. G., 1985, Shelf and shallow marine sands, in Walker, R. G. (ed.), Facies Models Second Edition: Geoscience Canada Reprint Series 1, Geological Association of Canada, p. 141-170.

Wiseman, J. D. H., C. D. Ovey, 1953, Definitions of features on the deep-sea floor: Deep Sea Research, v. 1, p. 11-16.

# FLUID AND SEDIMENT DYNAMICS ON CONTINENTAL SHELVES

Donald J. P. Swift, ARCO Exploration and Technology Company, 2300 W. Plano Parkway, Plano, Texas, 75075, and Alan W. Niedoroda, R. J. Brown Associates, Suite 200, 2010 N. Loop, West, Houston, Texas 77018

## INTRODUCTION

What do we need to know about shelf currents?

As an introductory step, this paper defines continental shelves and briefly discusses their origin and evolution. Most of the paper is concerned with the large-scale tidal and storm-driven fluid circulation patterns of the continental shelves and the manner in which these flows entrain and move sediment. It is essential to understand these circulation patterns in order to understand the distribution of facies on continental shelves. However, oceanic currents on a rotating planet are complex and their pattern is not intuitively obvious. Therefore, a considerable portion of the chapter is devoted to an analysis of the mechanisms of shelf flow, and the importance of these mechanisms in determining shelf sediment transport. Storm-driven and tidal currents are considered in turn. The shoreface and inner shelf together constitute a gateway through which all shelf sediments must pass, and the complex flows of the shoreface and inner shelf are described in detail. Finally, fluid and sediment dynamics at the shelf edge are reviewed.

## Modern Versus Ancient Shelves

Our awareness of shelf sedimentary sequences in the rock record is a relatively recent thing. We have long been familiar with shoreface and deltaic deposits on one hand, and deepwater depositional systems on the other, but the transition zone between them has been largely ignored. An important recent petroleum discovery which has served to focus interest on shelf facies has been the Hartzog Draw field, a giant field in the Upper Cretaceous of the Powder River Basin (Tillman and Martinsen, this volume).This sandstone body is encased in shale and lies 160 km east of the time-equivalent coastal facies, yet it was probably deposited in a little less than 100 m of water. Since then, the characteristics of a distinctive shelf depositional system have begun to emerge, and are described in detail in these notes.

Part of the confusion surrounding shelf sedimentary sequences stems from a perceived difference in modern versus ancient depositional environments (Shaw, 1964). Ancient shallow marine deposits are seen as the product of epicontinental seas, while modern shallow marine deposits are seen as deposited mainly on continental margins. This view has been rendered obsolete by the surge in petroleum exploration of continental margins; a large portion of the known rock record is now located in such areas. However, as will emerge in the pages of these notes, the textures, primary structures, and facies patterns of epicontinental sea and continental shelf deposits are fundamentally similar despite the differing dimensions and boundaries, because the processes which create them are similar.

## Definitions

In order to continue the discussion of shelf sequences, some definitions are required. The physiographic term continental margin includes at least two contrasting classes. Continental terraces consist of flat upper surfaces (the continental shelf) and more steeply inclined lower surfaces (the continental slope and continental rise.) A second category of Continental margin, the continental borderland consists of a series of basins and plateaux or ridges that descend from shelf to abyssal depths, as is the case with the transcurrent-faulted California Borderland. The average shelf break on modern shelves occurs at 132 m (Shepard, 1963). But the ice-loaded shelves of Antarctica extend to depths greater than 300 m, while the insular shelves of the Bahamas terminate at depths as shoal as 20 m.

Epeiric or epicontinental seas are seas contained partially or entirely within a continental land mass such as the modern Baltic, North, and Yellow Seas. Epeiric seas are generally at shelf depth, and will be included in this discussion together with continental shelf seas. Many ancient epeiric seas however, had basins sufficiently deep to have been flanked by their own epicontinental shelves, as in the case of the Cretaceous Western Interior Basin (Asquith, 1970), the Appalachian Basin of North America (Woodrow and Isley, 1983) and the Vienna Basin of Europe (Homewood and Allen, 1981). These epicontinental shelves within the continents are to be contrasted with the pericontinental shelves that rim the continents.

Until recently, the most comprehensive classifications of continental shelves have been the physiographic classification of Shepard (1963) and the structural scheme of K. O. Emery (1968). Emery saw the shelf sediment pile as free standing, or else dammed by fault blocks, diapirs, reefs, or salt domes. Inman and Nordstrom (1971) have recently presented a classification for continental margins keyed to plate tectonics; a margin is seen as convergent or passive.

More recent application of plate tectonic theory to basin evolution suggest that we should classify shelves primarily on the basis of the nature of subsidence. Passive margins that have rifted and are thermally subsiding continental margins such as

### A. PASSIVE MARGIN

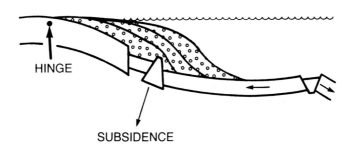

### B. CONVERGENT MARGIN

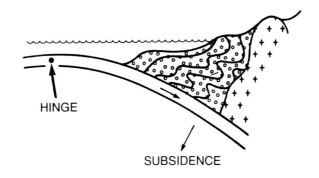

### C. FORELAND BASIN

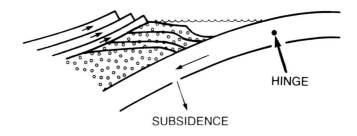

Figure 1. Structural settings of continental shelves; A: Passive Margin; B: Convergent Margin; C: Foreland.

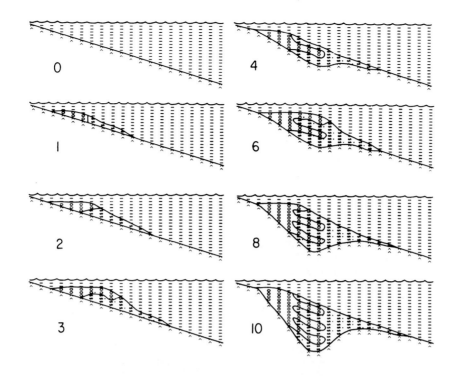

Figure 2. Numerical model for the development of a continental shelf. Sediment is added at shoreline during each time step. It moves seaward until it reaches a station below the depth assigned to wave base. Each of two sediment classes (coarse and fine) has a wave base value. Subsidence is lagged 3 time steps behind sedimentation. From Harbaugh and Bonham-Carter, 1977.

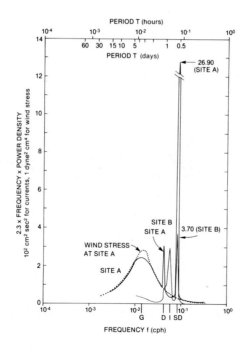

Figure 3. Spectral analyses of fluid power at two stations on the Inner New Jersey Shelf based on data from a long term current meter mooring (modified from Beardsley et al., 1976). The diagram indicates the relative amounts of energy expended in fluid motion of different frequencies. G: geostrophic frequency, D: diurnal tidal frequency, I: inertial frequency, SD: semidurnal tidal frequency.

those of the Atlantic continental margin are hinged at their landward side margin (Pitman, 1978; sediment may be delivered to such passive margin shelves by, large, integrated drainage systems (Fig. 1a). On convergent margins, subsidence is the direct result of subduction and the hinge also occurs at the landward edge. On convergent margins, subsidence is the result of subduction, and sediment loading. Convergent margin shelves tend to be narrow, wave-cut terraces, but if the sediment supply is sufficient, a seaward-thickening shelf prism may build seaward over the subduction zone (Makran coast of Iran, Harms et al., 1982; Fig. 1b, this paper). In foreland basins, subsidence is induced by loading of the continental craton by successive thrust slices (Jordan, 1982; see Fig. 1c). In this case the hinge is on the seaward side of the sediment pile and the greatest subsidence is beneath its landward margin. In both convergent margins and foreland basins, the deepwater basin may receive lateral fill from a large antecedent river system. However, much of the area on these tectonic shelves is fed by a series of small drainage basins providing relatively coarse sediments from along the tectonic front.

The shelf sedimentary piles of the structurally diverse environments described above are very similar in their gross geometry of shelf, slope, and rise, because subsidence has occurred in a marine setting in which the water column has a characteristic long-term velocity field. Tidal and wind-driven currents are topographically steered; they tend to parallel the shoreline, the contours of the shelf surface, and the shelf edge. They sweep sediment along the shelf and in combination with the wave-orbital velocity build a flat shelf surface. In the wave-orbital velocity field, the frequency and intensity of wave agitation is high at the water surface, but decreases rapidly with depth. The depositional surface can build up into the wave field to a level determined by the sedimentation rate; the surface is at equilibrium when it has reached the level where the wave build is intense enough to bypass downshelf and seaward as much sediment as is supplied to it. (see review of the "equilibrium profile concept in Swift, 1976a, and of the graded shelf concept in Swift, 1976b). Simple numerical experiments by Harbaugh and Bonham Carter (1977, Fig. 2) have shown that in such an aggrading sedimentary pile (continental shelf pile), a secondary surface develops (slope) in which gravity powered processes prevail. Thus shelf topography tends to converge towards a common pattern from a variety of initial states.

## Shelf Fluid Regimes

The character of the fluid dynamical regime is a major variable controlling the nature of shelf deposits. For geological purposes, there are three important fluid regimes on continental shelves. Shelves may be storm-dominated as is the Atlantic Shelf of North America (Swift et al., 1981). This category includes 80% of modern shelves. Shelves may be tide-dominated, as is the North Sea (McCave, 1971); 15% of the world's shelves fall into

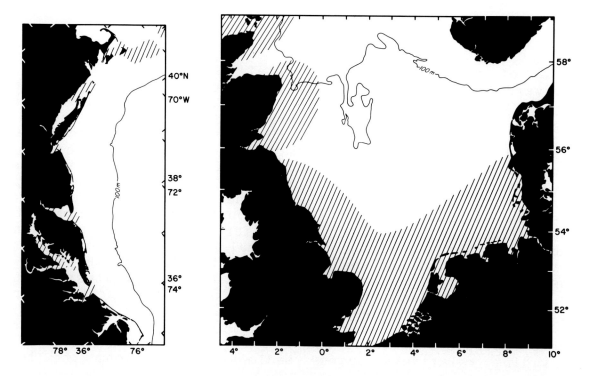

Figure 4. Maps of the Atlantic Continental Shelf, a storm-dominated shelf; and the North Sea Continental Shelf, a tide-dominated shelf. Shaded areas indicate areas in which mid-tide surface velocities exceed 25 cm sec$^{-1}$. From Swift et al., 1981.

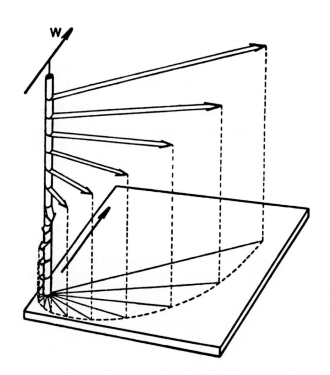

Figure 5. The Ekman spiral of current velocities in a wind-induced current. Wind direction is indicated by "W." Water velocities are indicated by open arrows. After Ekman, 1905.

this category. A few shelves (5%) are dominated by intruding anguilhas oceanic currents as is the Southwest African shelf which is in contact with the Anguilhas Current (Flemming, 1980). Transitional cases occur, where tide-dominated or storm-dominated shelves are bordered by strong oceanic currents. Meanders in the currents become closed loops, and drift over the adjacent shelf as high velocity eddies.

Physical oceanographers separate these classes of fluid motion by means of spectral analysis, in which energy density in $cm^2\ sec^{-1}$ is plotted against frequency. Figure 3 describes three kinds of low-frequency fluid motions as determined by current meter observations at an inshore station (A) on the Atlantic Continental Shelf of North America, and at an offshore station (B). The narrow band, high amplitude peaks at .081 and .042 cycles per hour (12.4 hour and 24 hour periods) are due to the diurnal and semidiurnal tidal current components, and are labeled D and S, respectively.

The broad peak at about 0.02 cycles per hour is labeled G (for geostrophic current, a class of wind-driven currents) and is coincident with the peak for wind stress. The currents in this frequency band are wind driven currents associated with low pressure disturbances, which generally form over the southeastern United States and propagate up along the eastern seaboard with characteristic periods of two to four days. In this diagram, the area beneath each peak is proportional to the contribution of that frequency range to the total variance of the current record. While the peak for wind-generated flow at the inshore station is lower than the peak for semi-diurnal tidal flow, it is much broader and the area under it is greater. Thus, the Central Atlantic Shelf is a storm-dominated shelf. Surface tidal currents whose surface mid-tide velocities exceed 20 $cm\ sec^{-1}$ occur mainly at estuary and inlet mouths, or over offshore banks. In the North Sea, by way of contrast, tidal currents occur over much wider areas (Fig. 4); the North Sea is a tide-dominated shelf.

The peak at 0.05 cycles per hour at station B is due to the inertial oscillations associated with storm flows on the outer shelf (see later discussion), and is labeled I.

This paper describes shelf fluid regimes in terms relevant to sediment transport. Storm and tidal currents are shown to be the two most significant transport agents. Storm-driven sedimentation is illustrated by data from the Atlantic Continental Shelf, a modern storm-dominated shelf whose fluid and sediment dynamics have been studied in great detail. Tide-driven sedimentation is illustrated by data from the North Sea, a well-studied example of a modern tide-dominated shelf.

# STORM CURRENTS ON CONTINENTAL SHELVES

## The Storm Fluid Regime

Circulation on storm-dominated shelves is now reasonably well understood, largely as a consequence of large-scale environmental impact surveys over the last 10 years, funded by the Bureau of Land Management in preparation for offshore leasing for petroleum exploration on U.S. Continental shelves (Summary in Allen et al., 1982) or by the National Oceanic and Atmospheric Administration in order to monitor marine pollution (Swift et al., in press). Storm circulation patterns are complex and their dynamics are not intuitively obvious; because they occur on a rotating planet, flows at the scale of the continental shelf generally go at right angles to the direction that the initiating force would seem to require. Since a knowledge of storm current patterns is essential to the understanding of shelf facies patterns, it is worth taking a close look at their mechanisms.

Wind-driven storm currents on continental shelves do not necessarily flow parallel to the wind, but are usually parallel, or nearly parallel to the shoreline, the shelf edge, and the contours of the shelf floor. A fundamental reason for this is the fluid dynamical concept of continuity; an application of the law of conservation of energy and mass. Shelf water can't flow away from the beach in large quantities because more water couldn't easily get in to replace it. Similarly, it can't flow towards the beach in large quantities because it would pile up. Partly for these reasons, winds will drive shelf water primarily alongshore, in whichever of the two coast-parallel directions for which they have a velocity component.

A second, more important reason for alongshore flow on continental shelves during storms is a consequence of the way that the wind-stressed water mass reacts to the shoreline, and to the earth's rotation. This pattern is basically the three-layer coastal flow system first predicted by Walfred Ekman in 1905, and is a major mechanism for the transport of sediment by storms on shelves. Storm winds blowing from the northeast along a shelf such as the North American Atlantic Shelf entrain surface water, which moves landward, to the right of the wind. This is because in the northern hemisphere, water moving toward the south is moving onto a part of the earth that is spinning more rapidly (the velocity of the equator is several thousand kilometers per hour while the velocity at the pole is zero). To an observer looking down the water path, the water seems to veer to the right, as the planet spins out from under it. If water moves in a northerly direction, it also moves to the right since it carries with it momentum from the region of faster rotation. The rule of thumb is that large-scale moving objects veer to the right in the northern hemisphere and to the left in the southern hemisphere. In

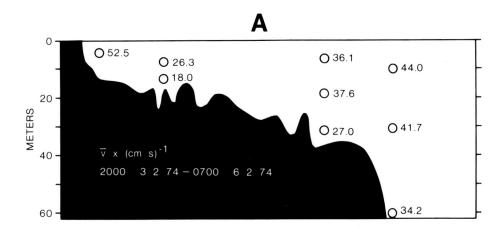

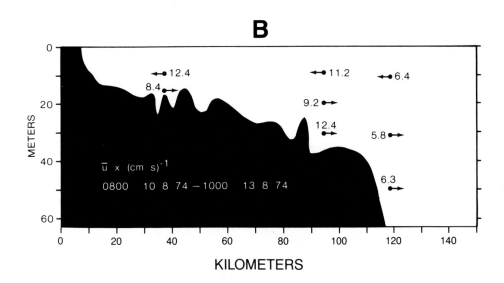

Figure 6. Storm velocity field on the Virginia shelf. A: Alongshelf (Southward) flow component during the storm of March 3, 1974. The high coastal velocity (52.5 cm sec$^{-1}$) is the consequence of a coastal jet. B: Unusually well-developed cross-shelf component of flow illustrating both upper (onshore) and lower (offshore) flow levels, during the storm of August 13, 1974. From Boicourt and Hacker, 1976.

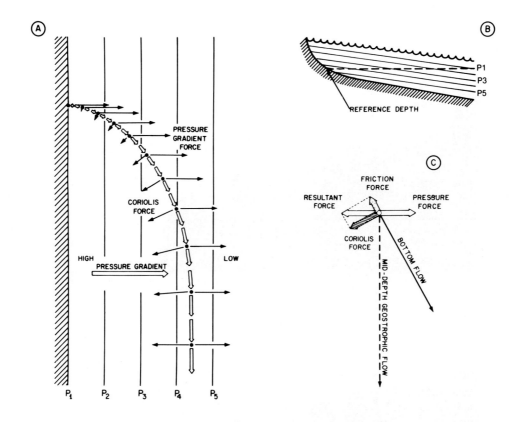
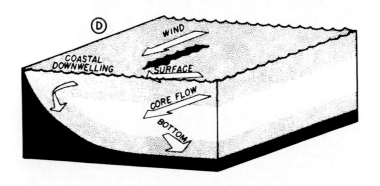

Figure 7. Structure of a geostrophic shelf current. A: Trajectory of a water particle in response to pressure and coriolis forces. Initial, states of offshore motion exist at time space scales of minutes and meters, respectively, and have not been observed. Measurable geographic flows are equilibrium alongshore currents. B: Cross-section through the coastal pressure field, showing reference level. C: Balancce of forces near the bottom. D.: Block diagram indicating current pattern. Modified from Strahler, 1963.

Newtonian terms, the water is said to be responding to an apparent force, the Coriolis force. The Coriolis force is a very weak force; our bodies experience it, but it is far too weak for us to detect among the many other accelerations to which we are subject at any moment. Fluids, however, are by definition substances that will deform (flow) in response to any force of finite magnitude. In large scale-fluid motions, all of the causative forces are weak, and the Coriolis force component becomes important.

In water, Coriolis veering increases with depth (Fig. 5). A thin layer of surface water entrained by wind stress moves off along a trajectory that is deviated 45° to the right of the wind by Cariolis force. It drags along a layer beneath it, which is deviated to the right of the layer almost, and so on down through the water column until the deepest, most slowly moving layer, tens of meters down, has been rotated 180° from the surface direction. If this spiral flow is averaged over depth, transport (Ekman transport) is oriented at 90° to the right of the wind that induces it.

During the Atlantic Shelf storm of August 13, 1974 (Fig. 6) transport of the surface water to the right of the wind caused sealevel to rise by as much as 50 cm along the coast. The response of the shelf water column to this sort of "coastal set-up" is complex, and needs to be considered in detail. The sloping sea surface creates an onshore-offshore pressure gradient, so that every water particle experiences an force driving it seaward (Fig. 7a). However, the movement of each water particle in response to this force results in a Coriolis acceleration to the right of the trajectory, that increases in strength as the particle accelerates. The offshore-flowing water therefore veers to the right and the veering angle increases with time. Eventually equilibrium is attained between a seaward-directed pressure gradient force and a landward-directed Coriolis force, with depth-averaged flow trending down-coast, parallel to the shore. The shelf current thus trends in the direction of the wind, and is wind-caused, but the causation is complex. It should be noted that the initial offshore motions diagrammed in Fig. 7 exist at time and space scales of minutes and meters and have never been measured. Observed wind-driven flows are equilibrium flows and trend alongshore. Shelf flows for which the pressure term are balanced in the equation of motion by the Coriolis term in this fashion are called geostrophic flows. On the Atlantic Shelf, north-trending geostrophic flows may be similarly driven by winds from the south, that push water away from the coast; in this case the coastal pressure field is the consequence of coastal setdown, and is directed onshore.

Coastal setup induced by storm winds has an interesting self-regulating mechanism built into it. As the setup increases, the slope of the sea surface increases and the pressure gradient intensifies, which in turn drives a faster alongcoast geostrophic flow. Near-bottom flow, however, experiences bottom friction; a friction term enters the equation of motion, and in the resulting three-way balance of forces, bottom water veers offshore from the alongshore, purely geostrophic flow of the fluid interior (Fig. 7c). The stronger the alongshore flow in the fluid interior, the greater the veering of the bottom flow. Eventually, the onshore transport of surface water by the wind (Ekman transport) is balanced by coastal downwelling, offshore bottom veering, and offshore bottom transport. The setup can no longer increase, and the alongcoast velocity becomes constant. Thus, setup is a dynamic rather than a static configuration, and there is a constant throughput of water.

### Where do storm flows go? Steady state flow directions

The continuity principle has been cited in this paper as a reason for along-shelf flow; water follows its coastal boundary. The Ekman model for geostrophic flow on continental shelves provides an additional mechanism for constraining currents to flow in an alongshore direction. Strong winds will entrain surface water and in the northern hemisphere, move it to the right. However, the water must pile up against the shoreline in order to develop the internal pressure field that is required for top-to-bottom geostrophic flow. Coastal setup (or setdown) can only create an offshore-onshore aligned sea surface slope, and therefore only along-coast flows. Wind direction may vary through 360°, and in a storm, which is a large rotary wind system, it usually does. However the geostrophic mechanism rectifies the oceanic response, so that shelf flows trend primarily along the shelf.

There is a third, more complex mechanism which causes shelf flows to parallel the shore, and more specifically, to follow contours of constant depth. This is the principle of conservation of vorticity. Should an alongshelf flow, following a contour, lose contact with it because the contour has swung inshore to wrap around a shelf valley, then the flow, finding itself in deeper water, will expand and decelerate. A cube of water will, in this process, reduce its horizontal dimensions and increase its vertical dimension, becoming an oblate rectangular solid. Like a spinning skater who pulls his arms in, the cube will experience an increase in angular momentum per unit volume, to the point where its angular momentum is higher than that of the surrounding fluid, and its path will curve back to the contour of origin. Similarly, if the fluid should stray inshore from the contour, then it will have less angular momentum than the surrounding water, and will not curve inshore as fast; thus, it will run into the contour of origin which is also curving inshore (See discussion of conservation of vorticity in Csanady, 1982).

Because of these mechanisms, the shelf constrains flows, directing them alongshore (continuity principle, geostrophic mechanism) and along the contours of the shelf floor (conservation

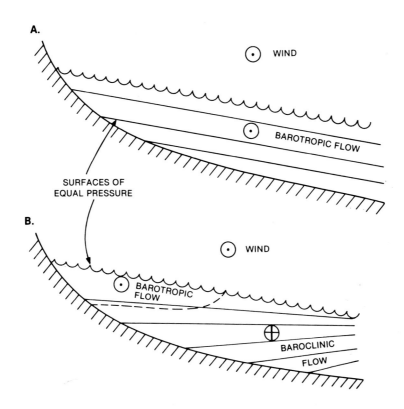

Figure 8. Role of stratification in shelf currents. A) wind blows along the shelf when the water column is stratified. Setup uniformly tilts pressure surfaces and the resulting geostrophic flow is barotropic in nature. B). water column is stratified; internal pressure surfaces are not uniformly tilted and the current has a baroclinic component. See Text for further explanation.

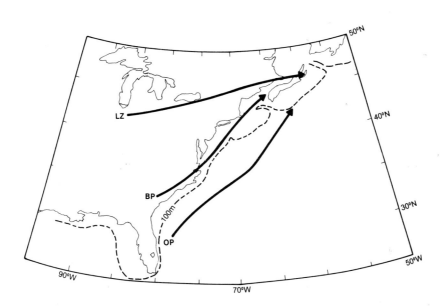

Figure 9. Classification of storm tracks on the North American Atlantic coast. See text for explanation of storm track labels. From Mooers et al., 1976.

of vorticity). Under steady state conditions, deviations from this principle are mainly due to frictional veering in the bottom boundary layer. Such deviations are generally low in angular value, short in duration, or both. However, brief intense flows nearly normal to shore do occur on the inner shelf and shoreface during storms (Niedoroda et al., 1984) and are associated with significant sediment transport.

## Density stratification and its effect on flow

In order to understand other kinds of shelf currents which move sediment, it is necessary to look at the role of density stratification in continental shelf currents. In the summer, the surface is warmed by the sun and warm winds, and is diluted by run-off from the land. As a result, the water column becomes density stratified, with warm, less salty, buoyant, surface water overlying cool, saline, dense bottom water. The gradient is not uniform; wind and wave stirring of the upper two meters or tens of meters renders an upper zone homogeneous. This upper mixed layer is separated from the underlying, uniformly stratified water by a zone of rapid density change known as the pycnocline (thermocline, if temperature is the dominant control of density or halocline, if salinity is dominant). As winter approaches, solar insolation and air temperatures decrease, and less thermal energy is available to maintain the density gradient. More importantly, winds intensify and the surface mixed layer slowly deepens. Turbulence generated by storm currents shearing over the sea floor creates a bottom mixed layer, and the two mixed layers thicken and merge, resulting in a homogeneous, neutrally buoyant water column.

During periods when the shelf water column is density stratified, there is an additional mechanism available for creating currents and transporting sediment. When winds blow along the shelf under such conditions and drive surface water landward (Ekman transport), this results in not only setup of the sea surface, but also a thickening of the surface mixed layer and depression of the thermocline as more buoyant surface water is blown onshore. As a result, horizons of equal pressure (isobaric surfaces) within the water column not only begin to tilt down toward the land, in response to the inclination of the sea surface, but fan apart as well, as a consequence of the thickening of the upper layers (Fig. 8). In addition to a barotropic current component (due to the sea surface tilt) there is now a baroclinic current component (due to the tilt of internal pressure surfaces). Since these two current types are usually due to pressure surfaces tilting in opposite directions, these current components generally flow in opposite directions. At any given point the stronger one cancels out the effect of the weaker.

Stratification is also important in that it tends to decouple surface water movements from bottom water movements. When the

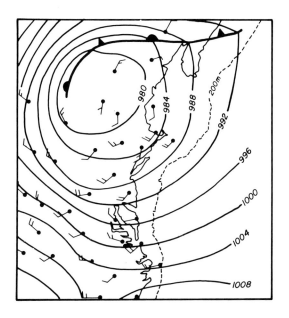 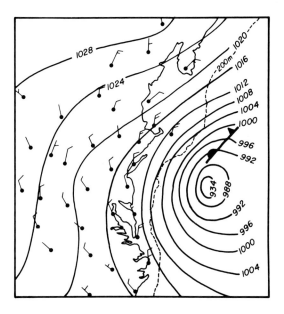

Figure 10. The storms of March 18 (left) and March 22 (right) 1973, and their relation to the Atlantic coastline of the United States. Modified from Beardsley and Butman, 1974.

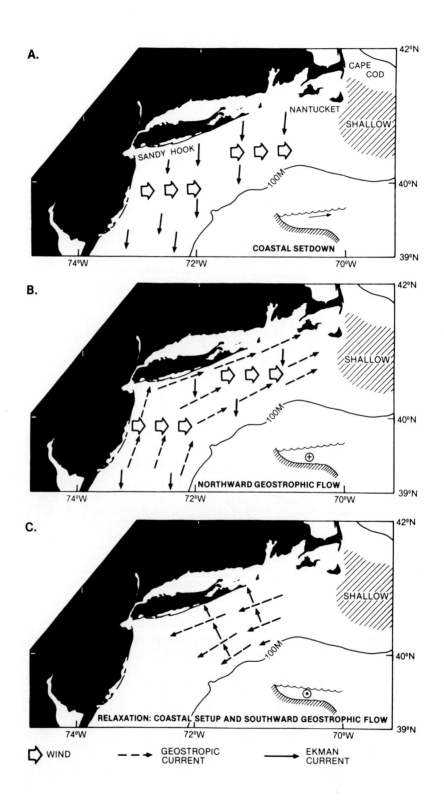

Figure 11. Diagram indicating sequence of time-averaged, depth-averaged fluid motions during the storm of March 15, 1973. From Beardsley and Butman, 1974, and Han, Pers. Comm., 1983.

water column is stratified, waves may occur within the water column as well as on its surface. Such "internal" waves can be observed from a moored ship which has lowered a vertical array of sensors. They appear as a periodic rise and fall of temperature and salinity surfaces. The orbital velocity and speed of propagation of internal waves is a function of the vertical density gradient and is usually less than that of surface waves of comparable wave length. However, internal waves are often very long waves, and consequently have strong orbital currents associated with them.

Internal waves have been observed on the Gulf Continental shelf and Atlantic continental shelf with velocities more than sufficient to transport bottom sediment (McGrail and Rezak, 1977; Butman et al., 1979).

### Time-dependent behavior; a storm-generated shelf wave

Storm currents on the Atlantic shelf are not simple square-wave phenomena, in which a current is switched on, flows in accord with the Ekman three-layer model, then is switched off. Storms move across the Atlantic shelf with differing results in terms of fluid and sediment transport. Mooers et al. (1976) have developed a three-letter classification for shelf storms, in which storms are labeled upper L, B, or O depending on whether the storm tracks pas primarily along the land, along the middle Atlantic Bight, or offshore; S or W depending on whether they are strong or weak, and P ;or Z, dependent on whether they parallel the coast or are zonal (parallel to lines of latitude; see Fig. 9).

The complex behavior response of the Atlantic shelf water mass to BSP storms was first described by Beardley and Butman (1974). In the LSP storm of March 15, 1973 (Fig. 10) strong (10-20 $sec^{-1}$) winds from the west drove Atlantic shelf surface water to the south (Ekman transport), down the north-south New Jersey coast. This surface flow paralleled the shore. Off the east-west Long Island coast however, flow was directed offshore, with the result that the Long Island coast was set down (Fig. 11a). As coastal sealevel dropped and a land-facing sea surface slope began to propagate seaward, an eastward geostrophic flow began in the fluid interior (Fig. 11b). However, the Middle Atlantic shelf essentially terminates against the Cape Cod Coast, and within 5 hours water began to pile up against this barrier. Sea level at Nantucket Island on the northern edge of the Middle Atlantic Bight rose 60 cm above that at Sandy Hook, near the center of the area, so that a west-facing sea surface slope began to form in addition to the northward (landward) facing slope.

The pile-up of water against the eastern end of the middle Atlantic Shelf can be viewed in wave dynamical terms as "forced shelf wave"; (Mooers, 1976; Bennett and Magnell, 1979; Mayer et al., 1981). It is a forced wave in that its rate of propagation

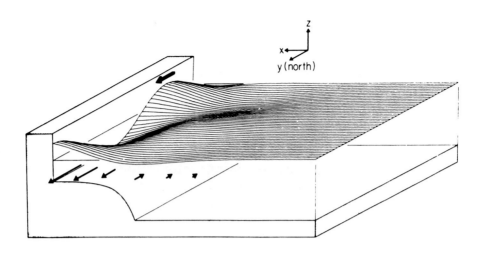

Figure 12. Diagram illustrating the form of a barotropic continental shelf wave (Rossby wave). The step beneath the wave represents the continental shelf. From Cutchin and Smith, 1973. Arrows indicate currents associated with the trough and crest of the loave.

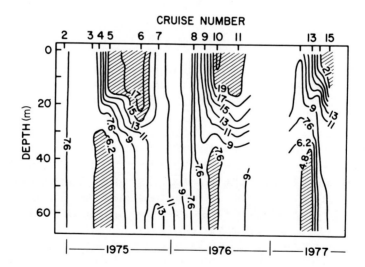

Figure 13. Cycle of stratification and mixing of the shelf water column at a representative station on the New York Shelf. Horizontal isotherms (°C) occur during the summer period of gentle winds and solar heating; vertical isotherms are the result of winter cooling and mixing of the water column by strong winds. From Han and Niedrouer, 1981.

is directly controlled by the wind, and a "shelf" or "long" wave in that its spatial scale is large. It is only a few decimeters high, but its wavelength is many tens of kilometers. When the winds shut off after having built such a forced shelf wave, the water bulge began to roll back southwestward along the shelf as a "fee wave"; free in that its rate of propagation was now that of a true wave, controlled by the acceleration of gravity and the Coriolis parameters (the restoring forces).

As the water began to flow back down the down-to-the-west sea surface slope, the Coriolis term in the equation of motion dropped to zero, then developed a large negative value, as the flow veered to the right. The across-shelf slope reversed and setup against the Long Island coast in response to a new geostrophic balance (Fig. 11c).

Such storm-initiated southwestward-traveling free waves are common events on the middle Atlantic shelf, and create some of the most intense flows that this area experiences. They are also called topographically trapped waves, because they keep veering against the coast in response to the Coriolis force, and therefore must propagate along shelf, rather than out to sea (Fig. 12). Families of topographically trapped shelf wave types exists, the members of which are classified on the basis of whether they are deformations of the sea surface (barotropic shelf waves) or of a density-stratified water column (baroclinic or interval shelf waves), and whether gravity is the main restoring force (Kelvin waves) or Coriolis force is the main restoring force (Rossby waves).

Storm-induced shelf waves are characteristic of the world's shelves. Shelves serve as wave guides for such long waves. Because of the role of Coriolis force in their propagation, they must travel counterclockwise around the margins of the basin in the northern hemisphere (southward on the U. S. Atlantic shelf; northward on the Pacific shelf; Mooers, 1976). Consequently, it is possible to predict dominant transport directions on modern and ancient shelves if the geography or paleogeography is known.

<u>Time dependent behavior; A scale-matching storm</u>

In the BSP storm of March 22, 1973; the Atlantic shelf water column followed a very different sequence of movements (Fig. 12B). The flows that resulted were somewhat less intense but resulted in far greater fluid transport, and in greater sediment transport as well (Vincent, in press). Winds in this second storm were more intense (15-30 m sec$^{-1}$), but the greater transports were due more to the matching of the storm's geometry with that of the Middle Atlantic Bight, than to wind speeds. The storm crossed slowly over the center of the bight so that for several days the isobars of atmospheric pressure matched the curve of the shoreline, and the winds (nearly parallel to the isobars) blew down the arc of the shelf. Ekman transport of the surface water by the wind was everywhere to the right of the wind and towards the shore. Sea level rose up to 60 cm along the entire coast, so that the sea surface sloped up toward the beach along the length of the middle Atlantic shelf.

Physical oceanographers, studying such large-scale storm-induced geostrophic flows in the Middle Atlantic Bight, have been impressed by their coherence (a mathematical measure of consistency of velocity values over distance). Because set-up occurs simultaneously along the entire Middle Atlantic Bight in "scale matching" storms like the March 22, 1973 storm (Fig. 10b), the response is a massive, "slab-like" geostrophic flow along the entire Middle Atlantic Shelf; mean velocities (1 hour averages) of 40 cm sec$^{-1}$ are reported and are commonly sustained for several days (Beardsley and Butman, 1974; Boicourt and Hacker; 1976, Beardsley et al., 1976).

## Geostrophic storm currents versus storm surge

It is noteworthy that in the course of events followed by the two storms described above, there was no significant "storm surge ebb" in the sense of offshore water flows.

The term "storm surge" refers to the vertical rise of water associated with a storm; it is commonly the consequence of wind setup, and as such rarely exceeds 60 cm sec$^{-1}$ on open coasts, due to the self-regulating mechanism of geostrophic currents described earlier.

In major storms, there is also the "inverted barometer effect" as the sea surface bulges up in response to the atmospheric drop, and wave setup due to long period surface waves. In hurricanes these effects may combined to induce several meters of setup, but only in the immediate vicinity of the eye as it crosses the coast.

When the sea surface is setup against the coast, water must flow in from elsewhere to fill the setup wedge that lies above mean water level, hence a horizontal current is associated with storm surge. Similarly, a relaxation current occurs as the storm wanes and setup dissipates. However, the initial or final discharge associated with the setup ("storm surge ebb") is trivial when compared to the alongshelf discharge of a geostrophic flow of shelf width and depth, which may be prolonged for several days. Powerful relaxation flows did occur in the first storm described above, but these were directed alongshore. They were shelf waves, not storm surge ebb currents in the sense of gravity-driven, seaward return flow. As such long waves propagate alongshore, their energy is slowly dissapated by the frictional drag of the bottom, and their fluid leaks offshore, but the associated offshore component of flow is negligable.

Seasonal variations in flow

The seasonal cycle of stratification of the shelf water column results in important differences between summer and winter currents. Winter storms are more efficient than summer storms in creating strong bottom flows because their winds are more intense. During November, as strong storm winds become more frequent and colder wind-induced mixing of the water column becomes more vigorous, and eventually, the temperature and salinity-induced stratification of the Atlantic Shelf water column is broken down (Fig. 13). Pressure-induced momentum passes downward through the water column unimpeded until, with the onset of spring, winds weaken, and the water column is warmed and re-stratified. During the summer, surface water is still dragged along by the wind, but the resulting surface current is partially decoupled from lower layers by the pycnocline. In both summer and winter, the inner shelf floor experiences more frequent bottom flows than does the outer shelf because even short-lived wind events can entrain the entire shallow water column. However, the pressure field generated by storm setup is independent of depth for as far seaward as the sea surface slope extends. It can generate strong bottom currents as far seaward as the shelf edge, if the winds flow long enough to build a surface slope that far seaward.

Mid-latitude storms versus hurricanes

Hurricanes are storm systems which originate in the tropics, and which have winds in excess of 35 m sec$^{-1}$ (74 mph). Atlantic hurricanes begin near the African Coast and move across the equatorial Atlantic. They may pass into the Caribbean, or curve to the north, to pass north along the Atlantic Coast. As they move into cooler maritime air (or cooler and dryer continental air), they are no longer able to extract energy from the surrounding air and they dissipate.

Atlantic hurricanes are very different in character from the mid-latitude low pressure systems whose tracks they cross. Their winds are stronger by definition. They are much smaller, generally being less than 100 km in diameter (often less than 50 km) in contrast to the 200-500 km diameters that are characteristic of mid-latitude storms. Hurricanes also move much more rapidly, attaining forward speeds of 10 to 15 m sec$^{-1}$ (18 to 27 mph) compared to the 2-5 m sec$^{-1}$ of mid-latitude storms. Paradoxically, hurricanes are often less efficient in coupling with the water column than are the mid-latitude storm systems, because of their high intensities and velocities. They are intense because their energy is concentrated into a small area. With such restricted spatial scales, their winds cannot parallel the shelf surface for any great distance, as do the those of the "scale-matching" mid-latitude storms. Because they travel so quickly, there is generally not enough time for the turbulent downward transfer of momentum, from the wind-stressed surface layer all the toward the bottom. Furthermore, they occur most frequently in the late summer and early autumn, a time when the shelf water column is thermally stratified and the downward transfer of momentum by turbulence is inhibited by the buoyancy of the warmer, lighter, surface water.

Hurricanes are extremely effective in raising the coastal water level. Sea level may rise tens of centimeters before the advent of storm winds, as the longer, faster waves outrun the storm and break against the coast, causing setup between the surf and the shoreline. Setup increases to a meter or more as hurricane winds drive water toward the coast, and may locally reach five to ten meters as the eye of the hurricane, with its enormous atmospheric pressure drop, crosses the shoreline. The maximum setup, however, is limited to a few hours duration and a few tens of km of beach (Nummedal et al., 1980). As a consequence, hurricanes are able to generate short-lived intense coastal flows, but not the vast, sustained geostrophic flows characteristic of mid-latitude storms. However, while the surge lasts, it may lift the surf zone up into the dunes for one or more high tides, and do enormous damage to the marine marginal environments. Beaches are eroded and barriers extensively breached.

During the morning of August 10, 1976, Hurricane Belle, a moderate-sized hurricane, traversed from south to north across the New York shelf. Wind speeds at sea were estimated to be as much as 45 m sec$^{-1}$ (100 mph), moderating to 22 m sec$^{-1}$ (50 mph) or less on land (Young, 1975). The center of the storm passed over the continental shelf in 6 to 8 hours. The effects of the storm were monitored by an extensive array of tide gages and current meters (Mayer et al., 1981). Mean current speeds at 1 m off the bottom exceeded 10 cm sec$^{-1}$ for 12 hours on the central shelf in 30 m of water and briefly attained a speed of 50 cm sec$^{-1}$. However, mean speeds of 30 cm sec$^{-1}$ lasted only 4 or 5 hours. By contrast, surface flow speeds of 40 cm sec$^{-1}$ lasted more than 16 hours. Prior to Belle, the density structure of the shelf water column was characterized by a strong thermal stratification. Three days after the hurricane a strong thermocline persisted, separating well-mixed surface and near-bottom layers. Surface temperatures were about 22° C, while the bottom temperatures ranged from 7° to 10° C. Although temperatures in the apex were somewhat cooler than in previous years, the thickness, position, and form of the thermocline three days after Belle did not differ greatly from measurements made during August of previous years (Young, 1975). Distributions of suspended matter were found to be similar to those present during previous periods of cooler summer weather.

The most striking velocity signal associated with the hurricane was a series of oscillations of the water column (Mayer et al., 1981). Near the center of the storm, where the wind velocities were most intense, the surface mixed layer above the thermocline was blown out of the storm area. The heavier stratified fluid beneath rose to replace it so abruptly that internal waves

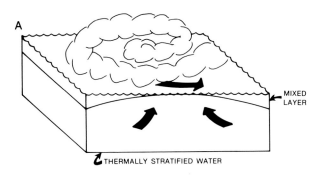

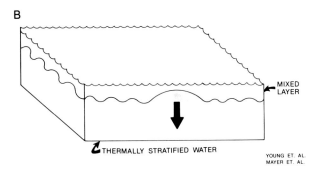

Figure 14. Diagram illustrating "Ekman pumping" of shelf water column by Hurricane Belle, August, 1976. (A) Hurricane winds blow surface mixed water out of area, and raise pycnocline to surface; (B) Internal waves are generated as pycnocline subsides.

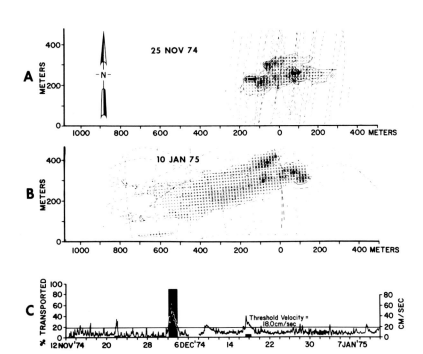

Figure 15. Radioisotope tracer dispersal patterns on the Long Island inner shelf on 25 November, 1974 (A) and on 10 January, 1974 (B). (C) record from Aanderaa current meter mounted 100 cm off bottom during the same time interval. See text for explanation. From Lavelle et al., 1976.

were generated. The water column was in effect struck an impulsive blow, and the resulting oscillations, of the sort described by physical oceanographers as "ringing," moved along the thermocline like ripples spreading from a pebble thrown into a pond (Fig. 14). Such oscillations are characteristic of storms that move more rapidly than the "baroclinic long wave speed"; that is, the hurricane created an internal shelf wave just by raising the thermocline beneath it, but because the hurricane was traveling faster than the depth-determined speed of the shelf wave, it could not continue to pump energy into the wave. Instead, the wave was left behind and subsided into series of oscillations, which had little effect on sediment transport.

Had Hurricane Belle been coincidentally moving as fast as the "baroclinic long wave" created when it passed over the shelf, or had it been intense enough to destroy stratification altogether, then its effect on the shelf floor record might have been very different. But Belle was probably a typical Atlantic Shelf hurricane in this regard, as well as in terms of its other characteristics.

Response of the Shelf Floor to Flow

Storm versus fair-weather regimes

Mid-latitude low pressure systems cross the Middle Atlantic Shelf approximately once every two to four days. About one-fourth of these are accompanied by winds sufficiently intense and prolonged to entrain water, and generate sustained flows in excess of the threshold of grain motion. In most equations for sediment transport, the transport rate varies with the cube of the velocity, and the great bulk of the transport is accomplished by three to five major storms per year. During the winter of 1974, radioisotope tracers released on the Long Island inner shelf were smeared out t the extent of 200 m by flow events which exceeded 18 cm sec$^{-1}$ on November 15, 17 and 22. On December 3-4, however, the flow associated with a major northeaster storm carried the tracer for 1000 m along coast to the southwest (Fig. 15). While the 18 cm sec$^{-1}$ threshold was again exceeded in December 12, 18 and 28, and on Jan. 5 and 12, over 95 percent of the sediment transported during 60-day period was transported by the more intense currents of the December 3 storm (Fig. 15c; Lavelle et al., 1976, 1978a).

Atlantic Shelf storms are relatively short-lived affairs. Winds in excess of 5 m sec$^{-1}$ (10 knots) rarely last more than 36 hours, and sediment suspended by storm generally settles out withi a few days. As light returns to the shelf surface and sediment ceases to move, an algal film spreads across the sand bottom. Vas sectors of the inner shelf become "zebra bottoms"; the active crests of oscillation ripples retain the white hue of clean sand while the intervening trough becomes covered with brownish algae. Echinoids dig themselves out of the sand, plow through the rippled surfaces, or resume their stations on megaripple crests. Ripple patterns are made, erased, and remade during the days and weeks between storms, but this activity involves only the upper several centimeters of the shelf surface. Seaward of the ten meter isobath there is no "fair-weather" sediment transport regime on most storm-dominated shelves; there is only the peak storm regime, and the waning storm regime.

Fair-weather shelf currents, like storm currents, flow along shelf. Tidal flows on the open continental shelf are rotary; that is, the current direction rotates steadily to the right or left through the tidal cycle. The ebb is weaker than the flood but does not drop to zero speed. Likewise, waning storm currents tend to rotate in direction, due to the shifting balance between the Coriolis and pressure terms in the equation of motion, as the coastal setup relaxes and the water particles decelerate (Fig. 7c). Waning, rotating storm flows are called inertial currents (Fig. 3); the inertial period is a function of latitude (Neumann and Pierson, 1966). Inertial currents appear in the spectrum of Fig. as the peak labeled I.

## Sediment transport by combined wind and wave currents: a high efficiency process

Storms are important agents of sediment transport not only because of the wind-driven currents that they cause but also because of the surface waves generated at the same time, and the resulting high-frequency wave orbital currents that stir the shelf floor. Combined-flow currents, in which a wave orbital current is superimposed on a wind driven mean flow component, have unusual properties which make them high-efficiency transporters of sediment.

The relationship between these two components in a typical Atlantic shelf storm is analyzed in Figure 16. A tidal sinusoid dominates the velocity record for a two-week interval on the 10 m isobath of the Long Island coast. The maximum tidal current of 11 cm sec$^{-1}$ is insufficient to entrain the fine to medium sand that floors the area. During a two-day period, however, alongcoast flows generated by the winds of a "northeaster" storm displace the tidal signal from its base line; here the tidal current modulates a storm flow, that varies in velocity from 5 to 40 cm sec$^{-1}$. Wave energy is significant only during this storm period. The combined effects of the wave orbital component and the mean flow component raise sediment concentration, recorded 100 cm off the bottom, from 10 mg $\ell^{-1}$ to ten times that value.

The inserts in Figure 18 show expanded sections of the record from storm and fair-weather periods. During the fair-weather period, the short-term velocity record is dominated by reversing wave orbital currents, as the time-averaged velocity record in the main diagram is dominated by a tidal signal. During the storm

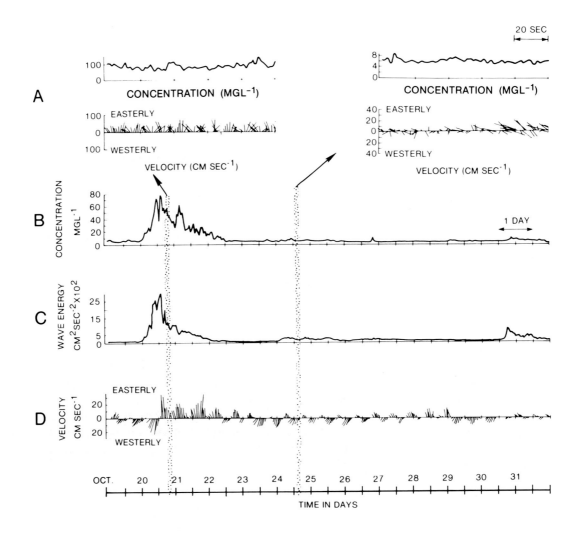

Figure 16. Time series of current velocity, sediment concentration, and wave energy from the Inner Long Island Shelf. Parameters were sampled once per second for five minutes every hour. Data is presented as five minute averages. Note the very different scales in the right and left band panels of row A. Modified from Lavelle et al., 1978a.

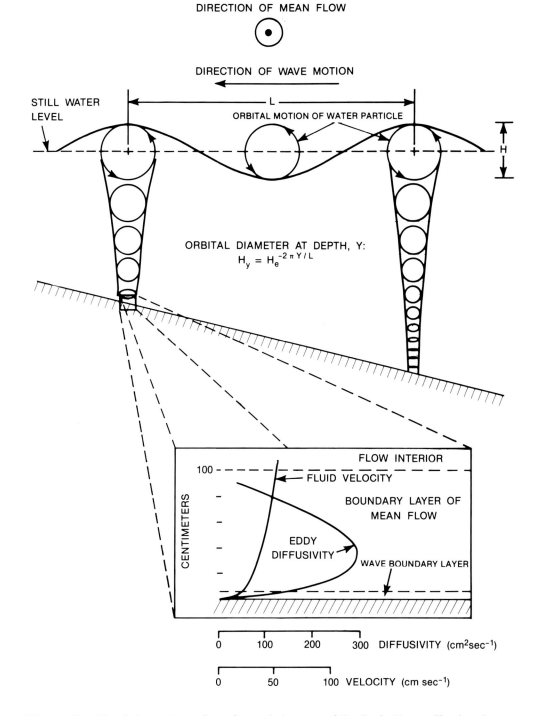

Figure 17. The interaction of surface wind waves with the bottom. The inset reveals the structure of the bottom boundary layer under combined flow conditions. After Komar, 1976, and Vincent et al., 1982.

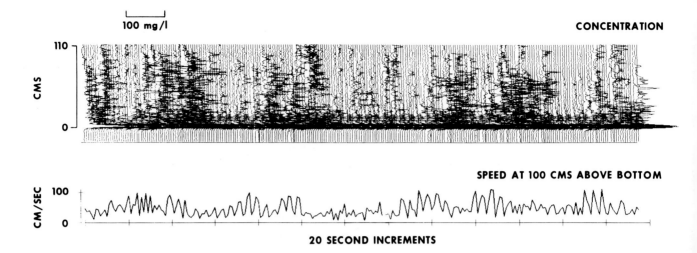

Figure 18. Time series of sediment concentration profiles and current speed from the Inner Long Island Shelf. Parameters were sampled once per second. From Clarke et al., 1982.

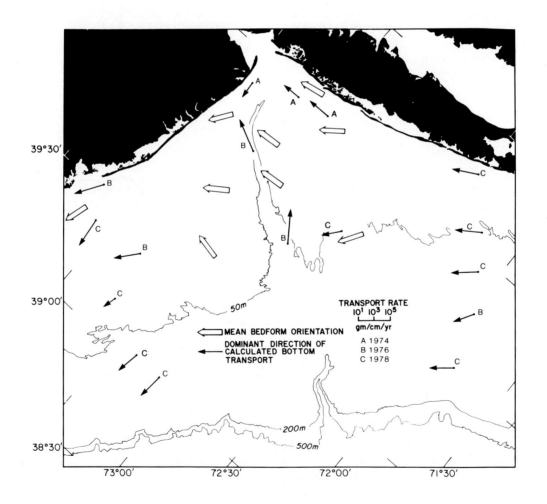

Figure 19. Near-bottom sediment flux on the New York--New Jersey Shelf, and direction of bedform orientations. From Vincent et al., 1981, and Freeland et al., 1981.

period, the mean flow component increases and the wave orbital currents modulate it to form a pulsing flow. Note the different velocity and concentration scales in panel C.

The wave orbital current component of a storm flow plays a critical role in sediment entrainment and transport (Fig. 17 . Wave orbital diameter at first decreases rapidly with depth then more slowly, until at a depth equal to 1/4 of the surface wavelength, wave-induced water motion is negligable. In water sufficiently shallow for wave motion to impinge on the bottom, the orbits become flatter as the bottom is approached, and just above the bottom, exist as a simple back and forth motion.

Wave orbital currents create thin, transient boundary layers which are more effective than the mature, thick boundary layer of the mean flow in entraining the sediment (Figure 17). A fluid boundary layer is that portion of the flow near a fixed boundary (such as the sea floor), where there is appreciable retardation of the flow due to frictional energy loss to the sea bed. Boundary layer thickness is a function of flow duration. A wave boundary layers can only grow to a thickness of several centimeters before the wave orbital current slows and reverses; the boundary layer then decays and must grow anew. The boundary layer associated with the mean flow, however, is able to maintain its equilibrium height which is on the order of 1 m. The vertical velocity gradient, and hence the vertical shear stress gradient, is gentle in such a thick boundary layer.

During a storm the near-bottom wave orbital current component may approach or exceed peak values of the mean flow component. Its velocity gradient, confined to a boundary layer several centimeters thick, is very steep, hence the shear stresses associated with wave orbital currents are greater than those induced by the mean flow component. Furthermore, when a wave orbital current component and a mean flow component coexist near the bottom, they interact in a nonlinear fashion because of the nature of the turbulence generated by the combined flow. The resulting boundary shear stresses are greater than the sum of the stresses that would be developed by either the wave orbital component or the mean flow component by themselves (Grant and Madsen, 1979). In effect, the mean flow boundary layer experiences the thinner wave boundary layer as an additional degree of turbulence-generating bottom roughness.

There is a second important and related effect that results from the combination of the mean flow component and the wave orbital flow component. The repeated accelerations and decelerations cause eddies to spin upwards from the rippled sea floor with greater frequency and intensity than they would from either flow component alone (Fig. 17). Eddy diffusion values as high as 290 $cm^2$ $sec^{-1}$ have been observed within 50 cm from the bottom (Vincent et al., 1982). Such highly turbulent bottom flows may be able to support a higher ratio of suspended load to bedload, and consequently resemble the "upper flow regime" flows that in laboratory flumes occur only at markedly higher velocities (Engelund and Fredsoe, 1974). Sediment transport by combined wave-orbital and storm-wind driven flow components is therefore a highly efficient process.

In Figure 18, an acoustic concentration meter has been used to create an acoustic record of the bottom boundary layer during a storm, in a manner analogous to a seismic record. The associated fluid speed record reveals a storm-driven flow modulated by periods of wave orbital activity. The periods occur because two wave trains of similar frequency are "beating," or experiencing alternating periods of constructive and destructive interference. The acoustic record shows that the periods of intense wave orbital velocity are suspending clouds of bottom sand, which are being swept down the coast as they rise off the bottom. Simple boundary layer theory does not account for turbulence of this intensity at this distance from the bottom, and it is probable that the phenomenon known as "boundary layer bursting" is occurring during periods of peak wave orbital velocity (Clarke et al., 1982.), further increasing the efficiency of the combined flow boundary layer in transporting sediment.

## Sediment transport patterns

Figure 19 shows the effects of geostrophic storm flows on the Atlantic Shelf floor over a period of ten months. The black arrows represent computed average sediment transport rates normalized to a mean annual value. They are based on records from a current meter moored within 2 m of the sea floor. Assumptions entering these estimates are described by Vincent et al. (1981). The open arrows provide "ground truth" in the form of current directions inferred from bedform orientations. Several important points emerge from this diagram. The first is that transport direction is uniform, and is oriented down shelf and slightly offshore in the direction of prevailing geostrophic storm flows. This parallelism is the consequence of the mechanisms described on previous pages (continuity, geostrophy, conservation of vorticity), which constrain fluid transport and therefore sediment transport to be largely parallel to depth contours.

The large northeasterly cross-shelf transports in the center of Figure 21 are not anomalous, but are the result of conservation of vorticity, as the flows enter the Hudson shelf valley, and swing northeast, parallel to the contours. The constant orientation of the transport vectors, in the sense that the barbs on the vectors are almost all on the south end, is a consequence of a second set of mechanisms. Storm flows are superimposed on a steady southern drift of shelf water on the order of 5-7 cm sec.$^{-1}$. This background drift is due in part to an essentially constant setup of the shelf surface by river runoff (Bumpus, 1973), and more importantly to a constant set-down of the sea sur-

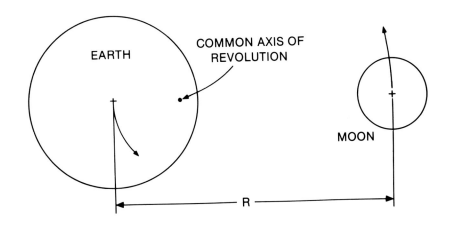

Figure 20. The earth-moon system. From Komar, 1976.

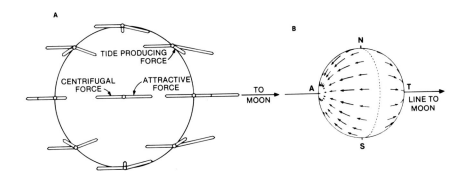

Figure 21. A: Analysis of the tidal forces. B: Distribution of the tide producing forces. From Komar, 1976.

face at the shelf edge with respect to the shoreline, as a consequence of the dynamics of the Gulf Stream (Chase, 1979; Beardsley and Wenant, 1979; Hopkins, 1982). This drift velocity enhances southerly storm flows and retards northerly flows. There is a more fundamental cause, however, in addition to the southernly drift. This is the wave-like behavior of coastal water masses that results when they are set up against the coast; the shelf wave must propagate counter-clockwise in the northern hemisphere (see discussion of time-dependent flow behavior).

## TIDAL CURRENTS ON CONTINENTAL SHELVES

### The tide-producing forces

Tidal currents are sufficiently strong to play an important role in sediment transport in perhaps 15 percent of the world's continental shelves. These have been described as "macrotidal" coasts by Davies (1964), or zones in which the spring tidal range is in excess of 4 meters. In economic terms however, the deposits of macrotidal shelf environments are more important than their frequency would indicate, since these deposits are frequently coarse, permeable sandstones, suitable for petroleum reservoirs.

Tidal currents are horizontal water motions induced by the vertical rise and fall of the astronomical tide. The tide is not simply the response of the water column to the moon's gravitational attraction, but rather to the gravitational interaction of the earth-moon system. In this system the earth and moon revolve around a common center (Fig. 20), without rotation. The center of the system lies several thousand kilometers beneath the earth's surface. To visualize this complex behavior, imagine that the earth in Fig. 20 is a nickel, with a groove in its underside about one third of the way in from the rim. The nickel lies on a flat surface. A pin projects from the surface and is engaged in the groove. The trick is to revolve, (not rotate) the nickel so that the pin passes along the groove, without tipping Jefferson's portrait! Because of this gravitational interaction between the earth and the moon, the earths orbit about sun is not a true ellipse, but a sort of wobbling path which at time brings the earth ahead of the path it would occupy if there were no moon; at other times it is behind that position.

The earth in this system can be thought as constantly moving through the common point at constant velocity as a consequence of its kinetic energy, and at the same time, dropping toward the common point in response to the attractive force between the earth and moon; the combination of these two motions through time is the orbital motion of the earth about the common point. The total attractive force between the earth and the moon is applied to the earth's center. Because the earth is a rigid body, all parts must respond as though the same force were being applied to every part,

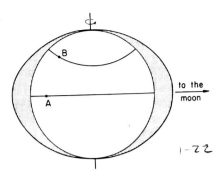

Figure 22. The earth and its tidal bulges. From Komar, 1976.

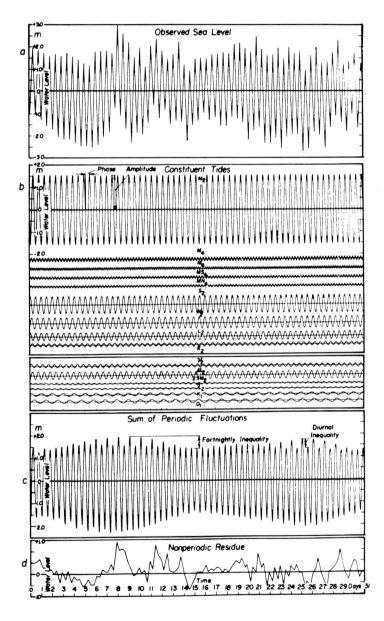

Figure 23. Fourier analysis of the tidal curve showing the numerous components which make up the observed tide, from Komar, 1976, after Defont, 1958.

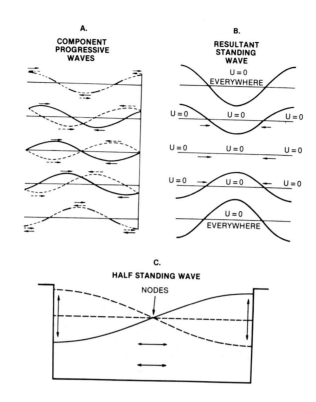

Figure 24. A, B: The interaction of a progressive wave with a reflected progressive wave to form a standing wave. From King, 1972. C: A half standing wave. From Komar, 1976.

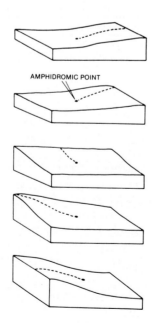

Figure 25. Behavior of an amphidromic system. Tide level does not charge at amphicromic point, while tidal edge wave sweeps around margin of system. Dashed line connects points of highest tide level for a given stage, defining rest of tidal wave. From Komar, 1976.

in fact, parts near the moon feel a stronger attractive force than do parts further from the moon. The solid earth responds only slightly to this force differential, but its fluid envelopes can respond more fully. Thus it is the vector difference between the mean attractive force and the local attractive force that is the tide producing force, or more precisely the component of that force that is attractive to the earth's surface (Fig. 21). The result of these forces acting on the hydrosphere is a water bulge that develops on the moonward side of the earth, where the water is, so to speak, pulled away from the earth, and a second bulge on the opposite side, where the earth, so to speak, is being pulled away from the water. As the earth rotates on its axis over a 24 hour period, an observer at a given point (on an earth without continents) would pass through 2 water bulges daily (Fig. 22).

A plot of sea level against time at any point in the ocean, however, is not a simple sinusoidal curve. Because the earth is tilted 23° with respect to the plane of the ecliptic, most places on earth lie closer to the trajectory of one of the two water bulges than the other (Fig. 22), hence one of the two daily tides is higher than the other at that spot. The sun as well as the moon effects the tide, although more weakly. Twice a month, when the earth, moon and sun are in line, the tides are higher than normal (Spring tides, from the old English word springan; rising water). The distance between the earth and the moon varies through the month, as does the distance between the earth and the sun through the year. As a consequence the tidal curve is a complex polyharmonic curve (Fig. 23). The most important tide producing components, as resolved by Fourier analysis, are shown in Table 1.

Tidal dynamics

Tidal behavior is complicated further because the tidal bulges that move around the earth behave as waves. As waves, they are reflected from the continental margin as they sweep across the ocean basin. The reflected wave moves back through the next incident wave, and the interaction of the two progressive waves results in a standing wave (Fig. 24a,b). A standing wave does not have a horizontally propagating crest libo?? a progressive wave; instead, its crest grows vertically (in place) to full height, sinks back to become a trough, then rises to become a crest again. Since the tidal wave is very large with respect to the ocean basin, the standing wave is in fact usually a half standing wave rather than a whole standing wave (Figure 24c). At the scale of an ocean basin, Coriolis force is an important term in the equation of motion (see discussion of geostrophic currents in the preceding section). Consequently, as tidal currents flow from one side of an ocean basin to the other, the water piles up on the side of the basin that is to the right, looking down current. As the water flows back, it piles up on the other side (Figure 25). This sort of interaction between the Coriolis force and a standing

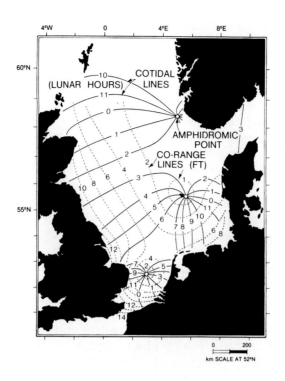

Figure 26.  The amphidromic system of the North Sea and large arrows.  From Komar, 1976.

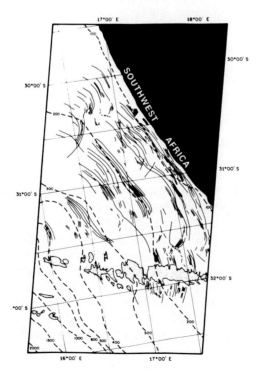

Figure 27.  Internal wave packets crossing the southwest African shelf.  Drawing based on an ERTS 1 image.  Irregular horizontal band (stippled) is cloud cover.  Contours (dashed lines) in feet.  Internal wave motions at depth cause surface upwelling, which changes small-scale wave patterns on surface, outlining internal wave crests.  From Apel and others, 1975.

wave results in an <u>amphidromic system</u> in which a tidal edge wave sweeps around the margin of the basin in a counterclockwise direction, while a central amphidromic point experiences strong tidal currents but a zero tidal range (Figure 26).

As the oceanic tide propagates up onto the continental shelf, its behavior changes markedly. The tidal wave is such a large-scale wave that it is technically a shallow water wave; in other words, even in the deep ocean, the tidal wave "feels" the bottom, and loses energy due to frictional drag. Because of its shallow water character, much of the reflection of the tidal wave occurs on the continental slope; some, however, propagates onto the shelf. During the summer, when the shelf water column is well stratified, tidal energy leaks onto the shelf in the form of internal waves. As the packet of interval waves generated by a shelf-edge, tidal reflection event moves up into the thin shelf water column, it thins, slows, and a second packet from the next reflection event crowds in behind it (Figure 27). During the summer, when the shelf water column is stratified all the way across the shelf, internal waves stir the entire shelf floor. During the winter, the wind-stirred mixed layer at the top of the water column thickens until it rests first on the inner shelf floor then on the outer shelf floor. As a consequence, a sort of internal wave "surf zone" expands across the shelf during the spring and summer, and contracts back to the shelf edge during the fall and winter. Sea floor currents up to 20 cm $sec^{-1}$ have been reported as a result of breaking internal waves (Butman et al., 1979).

The tidal wave also propagates onto the shelf as a surface wave. On macrotidal coasts this is by far the most important aspect of its behavior, since on these coasts the strong tidal currents stir the water column so effectively that density stratification generally cannot form.

The tidal range on the open sea is generally less than a meter, and the near-bottom, mid-tide velocities of tidal currents are generally less than ten $cm^{-1}$. On continental shelves, however, the tide range locally increases to as much as 15 m and tidal currents on the order of 200 cm $sec^{-1}$ (4 knots) are observed (for instance, Ungava Bay, Canada; Bay of Fundy, Canada). Amplification of the tidal wave on the continental shelf occurs because the tidal wave is modified by the shelf topography that it passes over, and because the tidal wave on some portions of continental shelves are in resonance with the oceanic tide.

Topographic effects of shelf topography on the tidal wave are twofold. As the tide moves into the thin shelf water column, its kinetic energy is steadily concentrated into a smaller cross sectional area. The tidal range and tidal currents are enhanced. However, as the tidal wave continues to propagate across the shallow continental shelf, it is damped by a frictional loss of energy into the bottom, and the rate of loss increases as the water col-

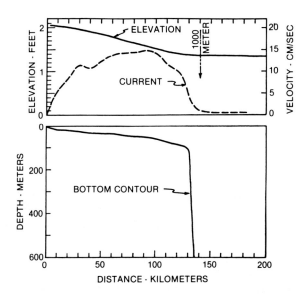

Figure 28. Diagram showing relationship between sea surface elevation (tidal amplitude), tidal current velocity (above) and shelf topography (below). From Redfield, 1958.

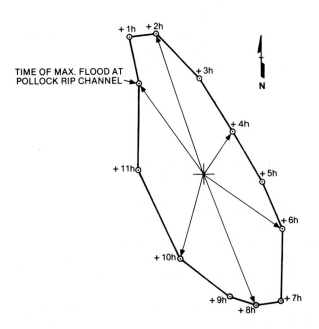

Figure 29. A rotary tidal current showing current vectors at successive times (arrows) and trajectory of vector tips (heavy lines). Modified from Stewart and Jordan, 1964.

umn shallow. The competition between increasing energy density on one hand, and frictional loss on the other can be described by the equation

$$u_{max} = \frac{2\pi CX}{TL}$$

where $u_{max}$ is the maximum tidal current at distance X from the coast, C is the amplitude of the tide (assumed to be constant), L is the width of the shelf, and T is the tidal period. The strongest currents are to be expected when X/L is at a maximum; generally on the central or outer shelf (Fig. 28).

If the shelf tidal wave is not entirely dissipated in crossing the shelf, it tends to be reflected by the shoreline, just as the oceanic tidal wave is largely reflected at the continental slope. The shelf tidal wave is therefore in many areas a standing wave rather than a progressive wave (Fig. 24) and is said to co-oscillate with the oceanic tidal wave. Every basin has a natural period of oscillation for standing waves. A child taking a bath soon learns that if he (or she) moves his (her) body back and forth at the appropriate speed, the water in the tub will easily slosh back and forth. The period of oscillation of a basin, T, is given by

$$T = \frac{2L}{\sqrt{gd}}$$

where L is the length of the basin d is its average depth, and g is the acceleration of gravity. If the length L is comparable to one fourth of the semidaily or daily tidal wavelengths, then the period, T, will approximate that of the oceanic tide, and resonance will occur. In resonance the interaction of the oceanic and the shelf tide is analogous to pushing a child on a swing. Transfer of momentum from oceanic to the shelf water mass is at maximum efficiency under resonant conditions, and the tide builds up to a much higher range before frictional energy loss balances the kinetic energy input. Macrotidal shelves gain most of their excessive tidal range by means of resonance.

## Tidal currents in shallow water

On the open continental shelf, Coriolis force interacts with the pressure force during the passage of the tidal wave, so that a water particle follows an elliptical trajectory (Fig. 29) with a clockwise sense of motion (Poincare wave). In embayments such as the North Sea, with sides whose coasts are parallel to the direction of tidal wave propagation, the sea surface is set up against the coast, and a rotary tidal flow develops with a counterclockwise sense of rotation (Kelvin wave; Fig. 30). As the shoreline is approached, the tidal ellipses become narrower, and within a few kilometers of the beach, only shore-parallel reversing tidal flow is possible.

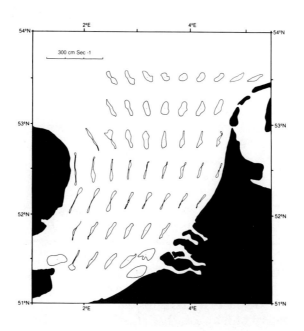

Figure 30. Rotary tidal currents in the southern Bight of the North Sea. From McCave, 1971.

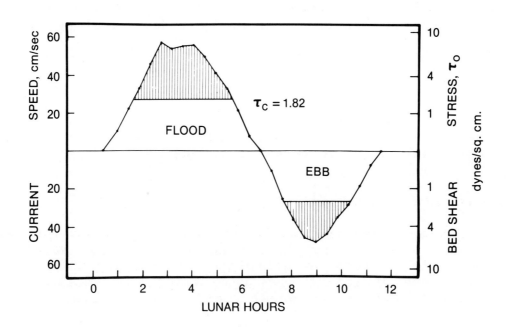

Figure 31. A flood dominated tidal cycle from a station in Chesapeake Bay (from Ludwick, 1970).

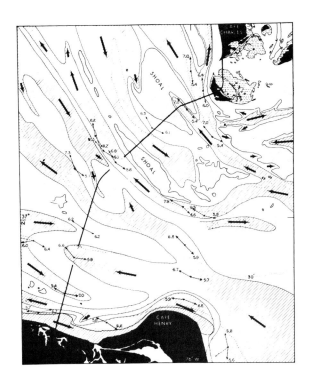

Figure 32. Flow pattern in the mouth of Chesapeake Bay, averaged over the tidal cycle. The pattern shows interfingering ebb-flood channel patterns. Small arrows who measured direction of near bottom currents. Numbers are measured ebb and flood velocities at the bottom in knots. Stippled areas are shoaler than 18 ft. Ruled areas and large arrows show where there is an ebb or flood predominance. From Ludwick, 1970.

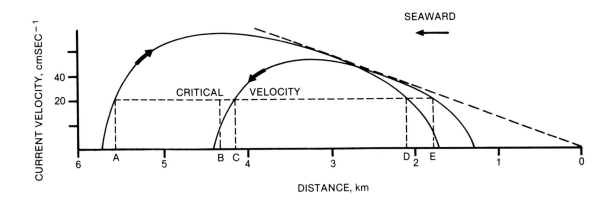

Figure 33. Diagram showing the velocities with which different water masses move with the tides at each point along a section through a tidal area from the inlet to the shore. The curves apply only to averaged, ideal conditions. See text for explanation. From Postma, 1967.

When the tidal range is a significant fraction of the depth, the crest of the tidal wave will be in deeper water than the trough and the wave undergoes a "shoaling transformation." Since the wave's velocity is proportional to $\sqrt{gh}$, and since it is deeper water, the crest will travel faster than the trough, and will develop a steeper forward side. The crest will extend higher above the still water level than the associated trough will extend below it, and the crest will be narrower than the trough. As a consequence of this characteristic deformation, crestal (flood) currents are more intense but shorter in duration than trough (ebb) currents. These changes do not fully compensate for each other, and when the distortion of the tidal wave is large, then significantly more landward water discharge is associated with the crest of the tidal wave than there is seaward discharge associated with its trough (Fig. 31). A portion of the sea floor experiencin a tidal regime of this sort is said to be subjected to a flood-directed residual current (residual to the semidiurnal tidal cycle).

The sensitivity of the tidal wave to water depth greatly complicates the pattern of shallow water currents in shelf regions of strong tidal flows. For one thing, the tide, in a particularly shallow embayment or estuary, tends to get out of phase with the tide elsewhere on the shelf. The velocity of the tidal wave in th estuary is so reduced by frictional interaction with the bottom, that it may be still ebbing, when the tide on the rest of the shel has already begun to flood. This phase lag frequently occurs at estuary mouths and in tidal inlets between barrier islands. For a period of some minutes during each tidal cycle, ebb tidal flow wil occur in the main channel, while flood tide currents will develop on either side. If the sea floor is cohesionless (made of readily erodible sand) a large scale interaction may start between the tidal current pattern on one hand, and the topography of the shelf floor in the vicinity of the estuary or tidal inlet, on the other. Sand will accumulate in the zones of shear, where for part of the cycle, tidal currents are opposed. In such shallow sandy regions, any slight bottom irregularity will tend to accelerate the flow where the water is deeper and retard it where it is shoaler. As a consequence shallow sea floors are inherently unstable in zones of steep horizontal gradients in the tidal phase. In such areas broa zones of interdigitating ebb and flood channel systems eventually develop (Fig. 32).

## Tidal sediment transport

The transport of sediment on continental shelves by tidal cur rents is in most respects similar to transport by storm currents. In fact tidal currents are homologs of storm currents (have simila properties), although not analogs (have similar cause). Both classes of currents are episodic, and tend to flow along shelf, first in one direction, then in the other.

Tidal currents do this because they are very long-period waves, characterized by a reversing flow with a specific period. Tidal currents, like storm currents, are constrained by interaction with the shoreline shelf surface to flow along isobaths. A system of strong tidal currents on the continental shelf may be viewed as storm flow event that repeats every 6 hours. Likewise, storm current events can be viewed as a stochastic (random) tides, whose period and intensity are variable. However, strong tidal currents are in general more intense than strong storm currents and some aspects of sediment transport by shelf currents, which are barely noticeable in storm currents become pronounced in tidal currents. These are lag effects, in which the trajectories of sediment particles in suspension differ from the trajectories of parcels of the water with which they are associated.

These effects were first described from the Dutch Wadden sea, the lagoonal area between the Dutch barrier islands and the mainland (Postma, 1967), but they occur also in shelf areas of strong tidal currents and high fine-sediment concentrations such as the German Bight of the North Sea (Eisma, 1981). In such areas, fine sediment tends to migrate from source areas towards the shallow subtidal zone of the inner shelf and the intertidal mud flats that fringe it. This movement is not a simple diffusion effect because the suspended particles move against the concentration gradient, from areas of low suspended sediment concentration to areas of high suspended sediment concentration.

One important cause of this flux is the scour lag phenomenon. This effect results from the fact that it takes a lower rate of fluid power expenditure to transport sediment than it does to erode it. This is because once deposited, a depositional surface composed of fine sediment is protected by the laminar sublayer of the fluid boundary layer. Particles cannot be resuspended until eddys penetrate this layer. Therefore, as fine sediment is moved by flood tidal currents onto the inner shelf or intertidal zone, it will be deposited at a lower velocity and later in the tidal cycle than the corresponding velocity and time of erosion. The excursion of the particle in the landward direction during the flood tide will be greater than the seaward excursion during the ebb tide, hence fine sediment will tend to move landward, to accumulate on the inner shelf floor, and ultimately in tide flat and marsh deposits. This effect is accentuated by the settling lag; during a period of waning flow, suspended particles require a finite time interval to settle out of the water column after velocity has decreased below the threshold value required for transport (settling lag).

The scour lag phenomenon is enhanced by a related mechanism; settling lag. Figure 33 presents in schematic fashion water parcel trajectories in shallow tidal waters in which both scour lag and settling lag are operating. A fine sediment particle is entrained into the flow at point A, when the water flowing over it attains the appropriate threshold velocity. At point D, velocity drops below the threshold value again, and the particle begins to settle out. As it does so, the water is still moving landward, and the particle finally comes to rest on the bottom near point E (settling lag effect). When the tide begins to ebb, the particle is entrained later in the cycle and at a higher speed because of the scour lag phenomenon. It moves only as far as point C, before it settles out again. The particle thus moves alternately landward and seaward, but a little further landward than seaward in each cycle.

Settling lag and scour lag are facets of a more general phenomenon of distance-velocity asymmetry of sediment transport by tidal currents. A second group of effects, are due to the time-velocity asymmetry of sediment transport by tidal currents. In an area where tidal currents decrease from the open sea to the coast, the period of low tidal velocities is larger around high tide than it is around low tide. Thus suspension is more efficient at low tide when the particles are further offshore, while particles settle out in greater numbers during high tide, near the shore.

## Transport by combined storm and tidal currents

As noted on a preceding page, tide and storm currents are in some respects similar agents of sediment transport; both are episodic alongshelf flows. Peak tidal flows occur twice a day in most areas, and are easily measured. Consequently tidal currents tend to be considered in isolation, and to be contrasted with storm currents. However shelf sectors with macrotidal regimes are as likely to be subjected to intense storms as any other shelf sector. Caston (1976) has shown that during the December 8, 1967 storm in the Southern Bight of the North Sea, peak tidal current velocities increased by as much as 50%, from 60 to 90 cm sec$^{-1}$. The increase was not distributed uniformly over the northerly and southerly (ebb and flood) tidal streams, but was associated primarily with the flood, so that the prevailing southerly transport of water and sediment was greatly intensified.

Pattiaratchi and Collins (in press) have shown that storm effects can actually reverse the sense of sand transport in the macrotidal Bristol Channel, not so much because of intensification of the mean bottomflow but because of the intensification of the wave orbital current component, with the result that sand is entrained not just during a small portion of the tidal cycle, but over the entire cycle. Tryggestad, Selanger, Mathisen and Johanson (1983) observed a two to three-fold increase in near-bottom current velocity near the Ekofisk Teeside pipeline, from 30 to as high as 90 cm sec$^{-1}$ during the February 10-15 storm of 1978. They attribute the increase to an increase in eddy visosity as a result of the increase in the wave orbital velocity component (see Fig. 17 and discussion). The heightened exchange of slow-moving near-bottom water parcels with more rapidly moving water parcels lighter in the water column "stiffened" the flow, and cause the friction-retarded near-bottom

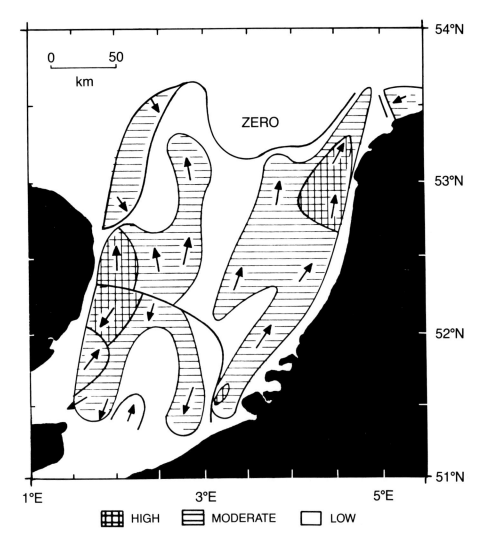

Figure 34. Computed tidal sediment transport system in the southern Bight of the North Sea. Patterns indicate rates of sediment transport. From McCave, 1971.

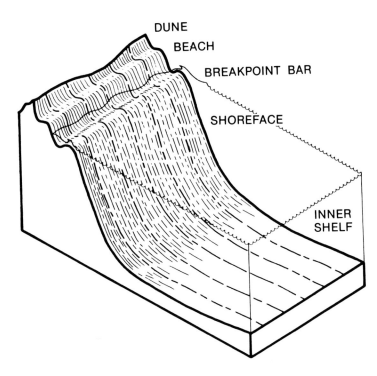

Figure 35. Geomorphology of a straight (two-dimensional) shoreface. From Swift, 1975.

velocity boundary layer to become thinner, so that fast-flowing water moved nearer to the bottom.

## Sediment transport patterns on tidal shelves

Sediment transport patterns on macrotidal shelves are controlled by current components residual to the semidiurnal tidal cycles that are induced by the distortion of the tidal wave in shallow water (see previous discussion). In such areas, pathways of flood-dominated tidal transport interfinger with areas of ebb-dominated tidal transport. McCave (1971) has analysed the tidal current data from the Southern Bight of the North Sea in order to compute bedload transport. His map (Fig. 34) indicates transport parallel to the shoreline and to the contours of the shelf floor. While transport zones are parallel, the transport directions are not uniform over the entire area; transport in ebb-directed in some portions, and flood-directed in others. Transport diverges from a "line of bedload parting" and moves towards lines of bedload convergence. one of which appears in the upper right hand corner. Transport increases in intensity from the axis of the Southern Bight landward towards the shore on each side. The pattern is probably more complex than that reported from storm dominated shelves (compare Fig. 34 with Fig. 19).

## FLUID AND SEDIMENT DYNAMICS ON THE SHOREFACE

The proceding sections have discussed fluid and sediment dynamics on storm and tide-dominated shelves respectively. This section will consider fluid and sediment dynamics on the shoreface in both storm- and tide-dominated regimes. Flow in this narrow inner shelf zone is complex and highly structured compared to the generally uniform flow field on the rest of the shelf, and it has an importance in shelf sedimentation out of proportion to its areal extent; the zone of shoreface flow is a gateway through which all shelf sediment must pass.

### The Shoreface and its Hydraulic Regime

The shoreface (Barrell, 1912) is the relatively steeply dipping, innermost portion of the continental shelf. The slope at the crest of the breakpoint bar may be as steep as 1:20; but it decreases to 1:200 a few meters further seaward, and at a distance of 3 to 5 km seaward of the shoreline, it flattens to 1:2000 or lower, where the shoreface merges with the inner shelf floor. This break in slope may take place at depths of 15 to 20 m, the depth being generally greater as the rigor of the wave and current climate increases.

On unconsolidated coasts exposed to the full force of marine processes, shorefaces are surfaces curved about an axis parallel to the shoreline, and exhibit little change in the alongshore direction (Fig. 35). On rocky coasts however, the time required for the profile to be incised into the substrate is long relative to the rate of sea-level change, hence rocky shorefaces are poorly developed. Local sand accumulations on rocky coasts, such as spits, barriers and tombolos develop well-defined shorefaces, but these shoreface fragments are irregularly distributed in plan view.

Since the days of classical geomorphology (Fenneman, 1902; Johnson, 1919), it has been assumed that the shoreface is an equilibrium response of an unconsolidated coast to the coastal regime of waves and currents. Attempts to quantify the complex, nonlinear relationships between process and response variables have not been entirely successful (see, for instance, the review of Harrison and Morales-Alamo, 1964). In general, we know that the shoreface slope decreases with decreasing grain size (Langford-Smith and Thom, 1969; Wright and Coleman, 1972); with increasing sediment input (Wright and Coleman, 1972) and also with increasing fluid power (Wright and Coleman, 1972). Repeated measurements of coastal profiles show that on the modern, retreating barrier coasts, the profile oscillates about its mean value with a time scale on the order of several tens of years (Moody, 1964).

The lower shoreface regime has been the victim of a sort of conceptual gap. Until recently, Coastal engineers and dynamicists concerned with behavior of the surf have relegated the somewhat deeper shoreface to the realm of oceanography. Oceanographers cannot bring their deep-draft ships in sufficiently closely to examine it, and as a consequence our ignorance of the transition between wave-dominated and oceanic flows has been extreme. The new material presented in this chapter is the consequence of modern concern with coastal environmental management and pollution control, which has led to financial support for a generation of oceanographers who find the same intellectual challenge in coastal flows that their elders found in the large-scale flows of deep ocean basins.

It is helpful to divide shoreface processes into two broad subregimes, those of the upper shoreface and those of the lower shoreface, respectively. The upper shoreface regime is dominated by shoaling and breaking waves. The orbital paths described by water parcels beneath waves decrease exponentially with depth; for every depth unit equal to 0.11 times the wavelength, the orbit is halved, and orbital velocity decreases linearly with the decrease in orbital diameter (Fig. 17). Therefore, the rigor of wave agitation decreases markedly from the upper to lower shoreface. On the other hand, marine currents tend to be less intense over the upper shoreface where they are inhibited by greater bottom friction than over the lower shoreface.

On the lower shoreface, the reverse relationship prevails. Wave orbital currents are important in agitating the bottom, but generally will not produce net sediment transport unless a mean flow component is also present. The boundary between these two

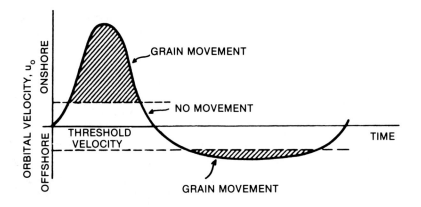

Figure 36. Generalized plot of bottom orbital velocity against time for a shallow water wave. Cross-hatched areas indicate velocities and durations in excess of threshold for sediment transport; note that forward stroke beneath wave crest can carry more sediment and coarser sediment then reverse stroke beneath trough. From Komar, 1976.

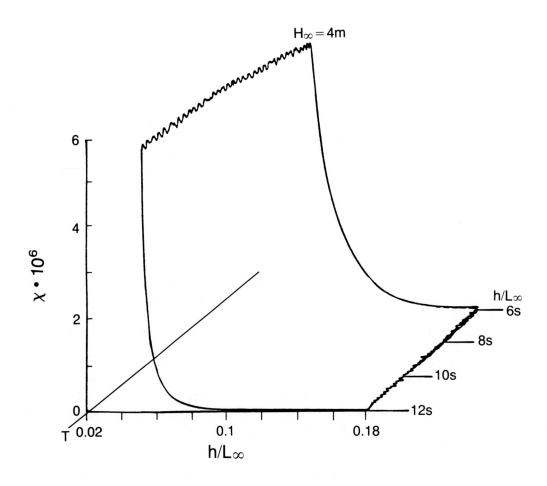

Figure 37. Three-dimensional plot of the sediment transport parameter ($\chi$) against the depth to wavelength ratio $h/L$ and the wave period (T) for 4 m waves. The horizontal projection of the zone of maximum inflection is plotted in $h/L$ versus T space. From Niedoroda and Swift, 1981.

zones is not absolute, depending on the intensity of the wave and current climate, but in terms of effect on the bottom, depths of 10 to 15 m seem to mark a valid generalized division between the upper and lower shoreface. The upper shoreface regime includes the wave-powered domain of the surf zone, where wave-driven longshore currents and rip currents are generated. Surf zone processes have undergone prolonged and intensive study, and elegant fluid dynamical models have been devised to explain them (for instance Bowen, 1969). This paper is concerned with the surf zone only as it interacts with the shoreface. The problem is to relate the upper and lower shoreface regimes to each other and to the larger scale coastal flow regime, and to determine their role in the coastal sediment budget.

### Wave Orbital Currents and Onshore Sand Transport

The boundary between the upper and lower shoreface regimes can be partly understood as a consequence of the shoaling transformation and its effect on wave orbitals.

As waves impinge upon the bottom they undergo a shoaling transformation (Fig. 36). Orbital paths very near the bottom degenerate to forward and reverse horizontal strokes (Fig. 19). In nonlinear wave theory, the forward stroke in waves deformed by shoaling is shorter and faster than the succeeding backstroke. As a consequence, sediment grains may be transported more efficiently forward than backward, since their threshold velocity is exceeded for longer on the forward than on the reverse stroke. In addition to this velocity asymmetry, there is also a discharge asymmetry a greater volume of water is moved forward than backward, and with it, a greater volume of sediment.

In a recent analysis of sediment transport on the shoreface in response to wave orbital currents, Niedoroda and Swift (1981) have substituted a simple Bagnold relationship (Bagnold, 1963) for sediment transport by wave orbital currents into a second order Stokes equation for wave orbital velocity (Komar, 1976). As a result, it is possible to solve the equation for sediment transport, X, in gm m$^{-1}$ sec$^{-1}$ (average mass flux of bedload per unit width of wave crest per unit time), as a function of wavelength, wave height and local depth.

Figure 37 explores this relationship for a range of wave periods (6-12 sec) and wave heights (1-4 m) and the ratio of local depth to open-water wavelength ($h/L^a$) that are typical of shoreface conditions. The model shows that at depths greater than 15 meters, the asymmetric wave orbital currents are ineffective in moving sediment, while at depths less than 3 meters they are intensely effective. The existence of a distinct zone of asymmetric wave orbital motion has an important implication. As first noted by Ippen and Eagleson (1955), the steeper slope of the upper shoreface must be at least in part a response to the tendency for waves to shift bedload landward. For equilibrium to be approached

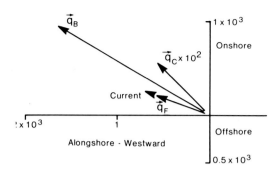

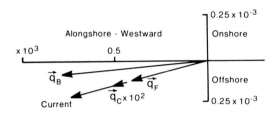

Figure 38. Average transport vectors for 2 storm events on the Long Island coast. Units for q, bedload transport, are gm cm$^{-1}$ hr$^{-1}$. The units for the current are x10$^{-2}$ cm sec$^{-1}$. From Vincent et al., 1983.

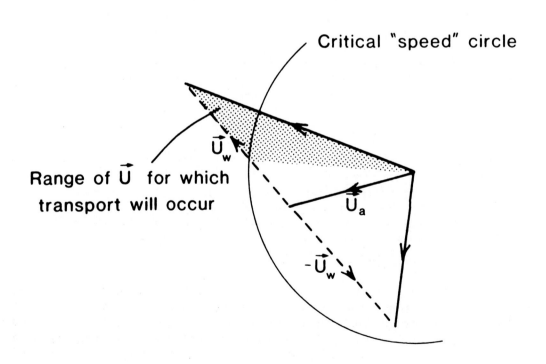

Figure 39. Diagram showing the addition of a steady current vector, $u_a$ and the time-varying wave oscillatory current vector $u_w(t)$, and the resulting variation in speed and direction of the combined current. A critical speed circle, determined from bed and flow characteristics of the site, defines the threshold speed for sediment transport. From Vincent et al., 1983.

or maintained, this onshore drift tendency must be offset by a greater offshore gravitational force component due to the upper shoreface slope.

Only the bedload transport due to the fluid shear stress caused by the wave orbitals has been considered to this point. The downslope component of the gravitation force acting in the the agitated sediment should also be considered. Although we do not have a complete expression at this time, it appears likely that the greater slope of the upper shoreface results from the greater tendency for onshore bedload transport due to pronounced orbital asymmetry on the upper shoreface.

### Wave-Current Interactions and Onshore Sand Transport

Observations of fluid motions on the shoreface indicate that the importance of wave orbital currents in sediment transport is by no means confined to the upper shoreface. The orbital asymmetry model (Fig. 36) demonstrates that orbital asymmetry becomes unimportant below 10 to 15 m. However, recent studies of the Long Island shoreface (Vincent et al., 1983) have shown that the interaction of symmetrical wave orbital currents with slowly-varying wind-driven currents plays a very significant role in determining the movement of sand across the inner shelf and up the lower shoreface.

Vector averages of fluid and sediment transport on the Long Island shoreface during 2 storms (Fig. 38) show that the mean bedload transport vectors are deviated markedly onshore from the mean current direction. This onshore transport component is caused by the combination of the wave orbital current vector $\vec{u}_w$, and the steady current vector $\vec{u}_a$. The combined current vector, $\vec{u}$, which is responsible for the friction imposed on the particles at the sea bed, varies in magnitude and direction as shown in Fig. 39. Transport will occur only when the critical Shields number is exceeded (critical speed). A critical speed circle has been drawn in Fig. 41 to show the range of $\vec{u}$ vectors. The coarser the bed materials, the larger the critical speed circle, and the more the transport direction (integrated over a wave period) approaches the extreme value of $\vec{u}$.

Vincent et al., 1983, have calculated sediment transport in response to waves and currents for two typical sets of observations on the Long Island barrier coast in order to better appreciate the role of wave-current interactions onshore sediment transport. In the first situation, easterly winds produce a surface wind stress which enhances the west-southwesterly coastal flow (about 5 cm sec$^{-1}$) that is characteristic of the regime (Han and Mayer, 1981). A 1 dyne cm$^{-2}$ easterly wind stress produces an

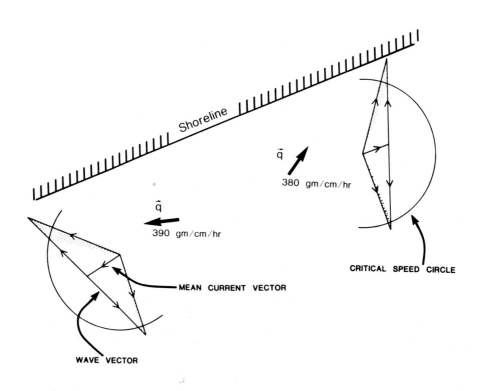

Figure 40. Two typical wave and current conditions on the Long Island inner shelf. a) Easterly or southeasterly winds produce an onshore transport component. b) Southerly or southwesterly winds also produce an onshore transport component. From Vincent et al., 1983.

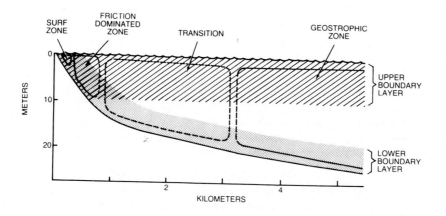

Figure 41. Dynamic zones of the coastal ocean. Thicknesses of the two boundary layers and positions of the zone boundaries varies with the intensity of flow. From Swift et al., in press.

additional 10 cm sec$^{-1}$ flow in the alongshore direction (Han and Mayer, 1981), giving a total of 15 cm sec$^{-1}$ towards 240° at an elevation of 1.1 m above the sea floor. Easterly winds will induce coastal downwelling and an offshore component of the bottom transport (see analysis of Han and Mayer 1981). The resultant flow, southwestward and offshore, can be simulated by a steady flow of 15 cm sec$^{-1}$ towards 240°. Surface waves due to the easterly wind will have a long fetch, and an approach direction of around 325°. Despite the offshore component of steady flow, bottom transport is westward and markedly onshore; $q_v$ is 340 gm cm$^{-1}$ hr$^{-1}$ towards 265° (Fig. 40, Table 2).

In the second situation, winds from the southwest reverse the prevailing southwestward flow. A 1 dyne cm$^{-2}$ wind stress now results in a steady alongshore current of 5 cm sec$^{-1}$ towards 070°. Waves caused by winds from the southwesterly quadrant will have have limited fetch and will generally be smaller than those from the southeast. The current will be downwelling, with an offshore Ekman component of transport near the bed.

These two situations, granted ± 25% variation in parameters, are representative of perphaps 80% of the Long Island coastal flow regime between April and October 60% of the November through March regime (Lettau et al., 1976). They show the onshore bedload transport to be between 20 and 50% of the total transport, even though an offshore component of steady flow is present. On the Long Island coast, wave-current interaction is clearly a significant mechanism inducing onshore sediment transport well seaward of the nearshore zone of pronounced wave orbital velocity asymmetry.

Coastal Wind-Driven Currents and Offshore Sand Transport

Coastal ocean currents

The preceeding section describes shoreface bedload transport processes whose rates and directions are largely determined by near-bottom wave orbital velocities and resulting bottom fluid shear stresses. This is the hydrodynamic mechanism that Allen (1970) has called the "coastal energy fence" wherein the dominant bedload transport is onshore. However, the survey does not provide a complete description of the bedload transporting processes. Although fluid shear stress resulting from vigorous wave orbital motion is needed to entrain bottom sediments, the direction and rate of the resulting sediment transport is commonly controlled by the steady or slowly-varying current flowing over the sea floor. For the first several hundred meters seaward of the beach, these currents are themselves driven by wave orbital motions (the wave-driven alongshore current) through the mechanism of radiation stress (Bowen, 1969). Over the rest of the shoreface, however, wind-driven and tidal currents become dominant agents of sediment bedload transport (Niedoroda 1980; Niedoroda and Swift, 1981) and their complex behavior must be considered here.

## Coastal ocean zones

The shoreface and inner shelf consists of distinct dynamic zones (Fig. 41). The zones are defined by dynamical considerations, which means that they are not fixed in space. As the intensity of the forcing mechanisms changes, the shore-normal length scale of these zones expands or contracts. During peak flows, the characteristic length scales of the dynamic zones do tend to coincide with morphologic zones of the shoreface and inner shelf. It is during these extreme conditions rather than during "typical" flow conditions that most of the transport occurs.

The innermost dynamic zone of the coastal ocean is the surf zone, an area intimately coupled to the beach and dominated by processes caused by breaking waves. Under calm conditions the surf zone may be reduced to a narrow swash zone while under extreme storm conditions it may extend offshore to the shoreface where the waves first begin to break. Characteristic shore-normal scales of the surf zone along open Atlantic shores of the U.S. are widths of 300 to 600 m and offshore depths of 3 to 5 m. This zone has been studied extensively and will not be further discussed.

Seaward of the surf zone lies the friction-dominated zone. This is a relatively turbulent zone where the effect of bottom friction strongly influences all flows. As a consequence, the coefficient of eddy viscosity is relatively high. Turbulent overturn causes masses of relatively slowly moving, friction-retarded bottom water to rise and be replaced by sinking masses of rapidly moving surface water. The flow is consequently "stiffened"; it does not shear as easily (is more viscous) than it would be without the turbulence.

Water masses responding to horizontal forces in the northern hemisphere tend to veer to the right (Coriolis acceleration; see previous discussion). In the friction-dominated zone, however, the effect of the Coriolis force is reduced. The relatively shallow depths and strong turbulence cause the currents to flow parallel to the direction of the forcing mechanism. Thus, surface currents in the friction-dominated zone tend to flow parallel to the surface wind stress, or along the horizontal (shore-normal) pressure gradient. In the Newtonian equation of motion, a wind stress or pressure term, or both, are balanced against a friction term.

On the Long Island coast, the friction-dominated zone characteristically extends to a depth on the order of 10 m. This depth is commonly encountered about a kilometer from the shoreline, making the characteristic shore-normal length scale of the friction-dominated zone similar to that of the surf zone.

The outer portions of the friction-dominated dynamic zone grade smoothly into the geostrophic zone. Locally forced currents in the geostrophic zone arise mainly from the effects of winds.

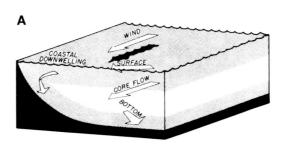

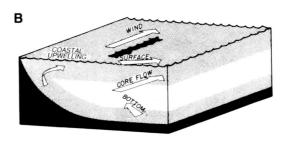

Figure 42. Consequences of Ekman veering in the geostrophic zone. A: north wind case. B: south wind case. From Swift et al., in press.

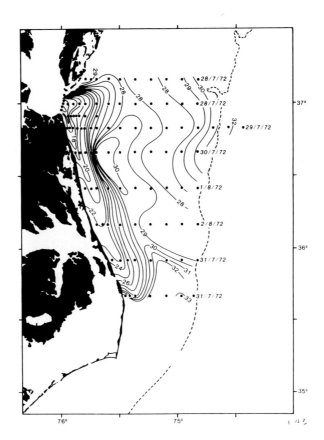

Figure 43. Baroclinic coastal jet on the North Carolina Coast, August, 1972. Tongue of less saline water issues from mouth of Chesapeake Bay and streams southward along the coast. From Schubel, 1975.

The principal differences in the dynamics of the geostrophic zone, as compared to the friction-dominated zone, arise simply as a result of the greater depths. While the geostrophic zone is still relatively turbulent, the greater depth causes bottom friction to have less influence on the overall flow. The effect of Coriolis force is proportionately greater, hence the name "geostrophic" (literally, earth-turning). In this zone, currents tend to flow normal to the horizontal pressure gradient, or shore-parallel. In the equation of motion for an alongshore geostrophic current, an offshore-directed pressure gradient term is balanced by an onshore-directed Coriolis term. Geostrophic currents have been discussed in detail in an earlier section. Momentum imparted to the flow by surface wind stress is not directly expended against bottom friction in this zone as it is in the friction-dominated zone. The shoreward margin of the geostrophic zone begins around a depth on the order of 20 to 40 m.

There is really no distinct dynamic coastal ocean zone between the friction-dominated zone and the geostrophic zone. However, true frictionally-dominated flow is restricted to depths less than 10 m while the true geostrophic dynamic zone does not extend shoreward of 30 to 50 m depths. It is expedient to refer to a transition zone which has characteristics of both adjoining zones. In this transition zone, the relative importance of friction-dominated and geostrophic processes changes according to the intensity of the forcing mechanism and the depth.

## Comparison of coastal ocean zones

The shoreward margin of the geostrophic zone begins at 20 to 40 m. In the geostrophic zone, the water column at any point can be thought of as having three dynamic layers (see previous discussion of the Ekman 3-layer system). These are an upper boundary layer, an interior geostrophic layer and a lower boundary layer (Fig. 41). In each, Coriolis force turns the flow from the direction of the principal causative force.

The overall consequences of this Coriolis turning or veering in the different layers of the geostrophic zone are important in understanding coastal ocean dynamics and the resulting sediment transport. Consider a steady north wind blowing along an east-facing north-south coast in the northern hemisphere (Fig. 42a). The surface transport due to the wind will be onshore. The shore-normal slope of the mean sea surface will be offshore causing a southward interior current. Transport in the bottom frictional layer will be rotated back to the east and offshore. Under similar conditions a south wind yields an offshore surface component, northward interior flow and an onshore bottom flow component (Fig. 42b). Note that an alongshore wind stress is more effective in yielding along-shore surface currents and hence in yielding offshore-veering bottom currents in the geostrophic zone. Because the sea surface slope tends to be steeper in the nearshore than further seaward, flow in these zones may take the form of a coastal jet (Csanady, 1977). Thus, the alongshore component of current velocity may increase seaward, a maximum value several kilometers off the beach, and decrease seaward from that zone.

The dynamics of currents in the friction-dominated zone differ considerably from those of the geostrophic zone. The principal differences arise from the shallower depth, greater proximity of the shoreline and the tendency for enhanced turbulence. There is no room for the interior geostrophic layer. The surface and bottom boundary layers overlap so that the right-hand and left-hand rotations cancel. The direction of the principal driving forces (e.g., surface wind stress, horizontal pressure gradients) and that of the resulting currents are parallel. Wind-induced sea surface slopes are steeper and develop more rapidly in the friction-dominated zone than in the geostraphic zone.

An alongshore wind stress in the friction-dominated zone does not cause set-up or set-down of the mean sea surface against the shoreline. Alongshore currents result in which the surface wind stress is balanced directly against the bottom friction stress.

Wind stresses in the shore-normal direction, or those with shore-normal components, do cause set-up or set-down in the friction-dominated zone. In the presence of such a sloping mean sea surface, the surface current is aligned with the direction of the wind stress and the bottom current has a directional component aligned with the horizontal pressure-gradient force produced by the shore-normal mean sea surface slope. Thus, winds with an onshore component yield surface currents with an onshore component, a set-up of the mean sea surface against the shoreline, a consequent offshore horizontal pressure gradient throughout the water column, downwelling near the shore, and a bottom current with an offshore component. Winds with an offshore component yield the reverse, an upwelling circulation.

The mass of water which is subjected to surface wind stresses is smaller in the shallower friction-dominated zone than in the adjoining geostrophic zone. Thus, the response of the water column in the friction zone to wind stresses is faster. The set-up or set-down of the mean sea surface develops faster and with greater magnitude in the friction-dominated zone. The currents in this zone rapidly expend energy derived from the wind as frictional "losses" against the bottom.

## Effect of stratification on coastal ocean zones

Density stratification is common in the coastal ocean water column during the warmer seasons. The surface mixed layer lies above a strong vertical density gradient called the pycnocline. The underlying water is generally colder and may have higher

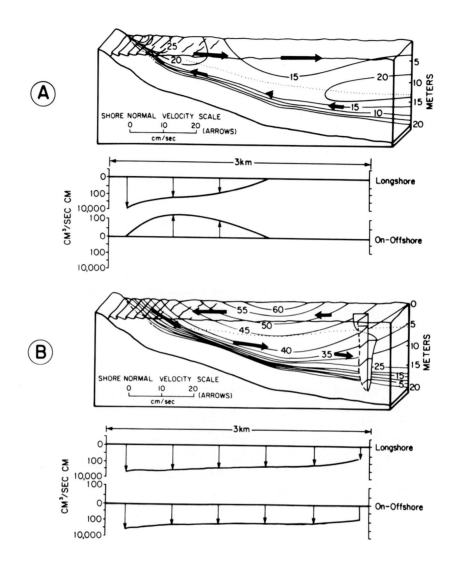

Figure 44. a) Alongshore current component at Tiana Beach on August 26, 1976 (contours), and the onshore-offshore component (vectors). Computed sediment transport rate is plotted below. b) Sediment transport computed for currents of August 24 combined with waves of August 26. From Niedoroda and Swift, 1981.

salinity. Under calm conditions the pycnocline commonly intersects the bottom in the transition zone. The various processes which yield upwelling or downwelling in unstratified conditions are effective in stratified conditions as well. The position of the pycnocline is affected by these flows in much the same way that the mean sea surface is when the coast has been set up by Ekman transport in response to winds. The landward flow of warmer, less salty surface water pushes the landward edge of the pyconocline down, just as the landward edge of sea surface is pushed up. Conversely, when the sea is set down against the coast, the upwelling of cool, salty bottom water pushes the landward edge of the pycnocline up.

Fluid flow in the presence of stratification is complex. It can be divided into a baroclinic flow component (The component of flow driven by the slope of internal density surfaces) and a barotropic flow component (The component of flow driven bv the slope of the sea surface; see previous discussion and Fig. 8). Tilted pycnoclines thus produce internal horizontal pressure gradients and baroclinic flow components which in general oppose the pressure gradients and barotropic flow component associated with the sea surface slope. The presence of stratification thus acts to reduce the bottom stress magnitude over the geostrophic zone. The friction-dominated zone is little affected by stratification since it is strongly mixed by surface winds, and by bottom-induced eddies. Therefore when stratification occurs in the geostrophic zone, it serves to exaggerate the shore-normal gradient in bottom layer flow.

While the primarily barotropic coasts of the winter season are important in moving sediment off the shoreface and onto the sea floor, coastal jets which are primarily baroclinic in nature are also important in this respect. Periods of excessive runoff may create baroclinic coast jets that are driven mainly by the density contrast. When Hurricane Agnes dissipated over west Virginia in 1972, a vast slug of nearly fresh water began to issue from the mouth of Chesapeake Bay after a lag of several days, and ran down the Viriginia - North Carolina coast as a baroclinic jet (Fig. 45). The winter storm of March 3, 1974 did not result in excessive precipitation, but during the storm, offshore winds blew vast masses of brackish bay water out of Chesapeake Bay mouth. The water also turned to the right and ran down the Virginia-North Carolina Coast as a coastal jet with a mid-depth velocity of 52 cm $sec^{-1}$ (Fig.62).

This behavior pattern has important implications for deltaic coasts. The buoyant, inertial jets of fresh water detach from the bottom as they pass over the crest of the distributary mouth bar, and ride out over water that is flowing alongshore (Wright and Coleman, 1974). But this is only part of the story. The freshwater jet, expanding over the sea surface, is subject to the Coriolis force and in the northern hemisphere veers to the right to run along the delta front as a ribbon of brackish water held against the coast in a manner similar to the baroclinic jets of the Atlantic Coast (Fig. 43). Prior to the advent of the Aswan Dam the east-flowing coastal jet that developed during floods at the mouth of the Nile attained speeds of 200 cm sec$^{-1}$, or 4 knots; a flow fast enough to knock people off their feet in thigh-deep water (Sharaf El Din, 1977). This flow was often still traceable, with velocities up to 25 cm sec$^{-1}$, as far east and north as Mt. Carmel on the Northern Israel coast. To talk about deltaic shoreface deposits as "frontal splays" analogous to splay deposits in the tranquil supra-delta basins is to do violence to principles of fluid dynamics. Delta front deposits are shoreface deposits and their textures and structures reflect strong open marine, along-coast flows, rather than the shore-normal flow of the river channel.

## Coastal wind-driven currents and offshore sand transport

The role of wind-driven coastal flows in shoreface sediment transport has been illustrated by a second series of observations from Tiana Beach on the Long Island coast (Fig. 47). The observations have been reported in detail by Niedoroda (1980) and Niedoroda and Swift (1981) and will be summarized here.

Figures 44a and 44b present the velocity field synthesized from boat station, surf zone tripod and Shelton Spar measurements on two representative occasions during the study period. The contours on the back planes of these figures define the alongshore flow component, which was westward in both cases. The arrows indicate shore-normal currents. At the right end of Figure 10a, vertical profiles are drawn for the two flow components measured by the Shelton Spar. The solid profile represents the shore-normal component, while the dashed curve is drawn in perspective and represents the profile of the longshore current component.

Figure 44a shows flow conditions on August 26, 1976. It is typical of the fair-weather regime in which sediment transport is driven mainly by fluid motions associated with shoaling and breaking waves, rather than by the mean flow component. Similar conditions occurred during 34 of the 36 study days. The wave regime on August 26 was characterized by a 13 sec period, a 150 cm significant wave height, and an 8 degree breaker angle opening to the east. Winds were less than 2 m sec$^{-1}$. An offshore, upwelling current was evident. The sediment flux computations shows that during August 26, the combination of the shore-normal wave orbital flow component and the wave-driven mean alongshore flow component created stresses sufficient to entrain and transport sediment on the shoreface. Within the 10 m isobath movement is onshore, into the surf zone; within the surf zone there is intense alongshore transport (not shown).

The situation on August 24 (Fig. 44b) appears to represent the wind-driven flow pattern characteristic of storms. Although waves

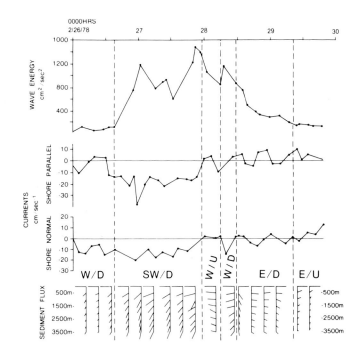

Figure 45. Wave energy (a) and current velocities (b) during the March, 1978 storm. Flows are classed as to the west (w), east (e), downwelling, (d) or upwelling (u). Bottom figures show horizontal vectors of bedload flux. From Niedoroda et al., 1984a.

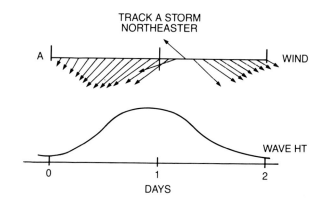

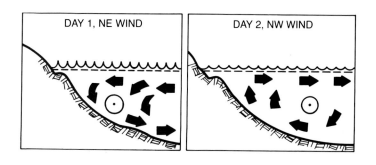

Figure 46. Typical winds, wave heights and coastal ocean (shoreface) currents for a northeaster storm that passes seaward of the study area. From Niedoroda et al., 1984a.

were low and wind stresses were small compared to the major storm events, the coastal flow appears well forced and organized. This regime was confined to two consecutive days of the 36 day study period. The wave regime was characterized by a 9 sec period and a 65 cm significant wave height. The wind was 5.5 m sec$^{-1}$ from the east and later from the southeast. The breaker angle was essentially zero indicating that most of the waves were not locally generated. Data (not given in 44b) show that early in the day a strong jet-like, westward flow appeared at the surface near the upper margin of the shoreface. As the wind shifted to the southwest, downwelling began, and the jet spread seaward and occupied the entire shoreface zone. The maximum surface current velocity was in excess of 60 cm sec$^{-1}$. However, the combined action of both waves and currents was not sufficent to entrain bedload at depths greater than 4 meters.

Because the waves were too small to cause significant sediment transport on the shoreface at the time when the most interesting flow event was measured, it was decided to synthetically combine the wave regime of August 26 with the current regime of August 24 and to use these conditions to model the bedload sediment transport (Figure 44b). The large swells of August 26 originated at great distance from the study area and could have arrived several days earlier without significantly changing the wind-driven flow over the shoreface. By combining the mean flow induced by the mild summer northeaster of August 24 with the wave regime of August 26, we are more nearly approximating the conditions of a major winter northeaster in which the coast may simultaneously experience up to 60 cm of set-up, an intense, downwelling, geostrophic flow (Beardsley and Butman, 1974) and an intense wave regime. Under these synthetic conditions, computed bedload sediment transport extends down the entire shoreface (Figure 44c). Transport is alongshore and offshore. These two components are of almost equal intensity.

Niedoroda et al. (in press) have modeled the effects of storm flow on the shoreface. They compared the intensities and directions of current profiles measured at boat stations for 3 days during 1976, with a much larger near-bottom velocity time series from an electromagnetic current meter located further offshore. Comparison of these data allowed hindcasting the probable conditions along the entire 3 km transect for storm flows that were recorded at 1.5 km station during the winters of 1977 and 1978.

Figure 45 is a model of coastal flow during the storm of March 26-29, 1978. As the storm began, a low pressure center migrated from the Ohio Valley across southeastern New York and New England. The early phases of strong winds along the coast were from the northeast. This drove a western and downwelling flow on the Long Island coast. As the storm center approached, the wind veered more to the east and strengthened. The change accelerated the westerly current along the shoreface and intensified the downwelling. The early phase of the storm was also characterized

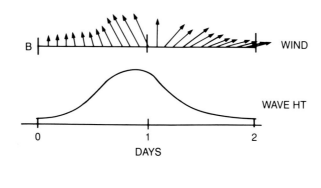
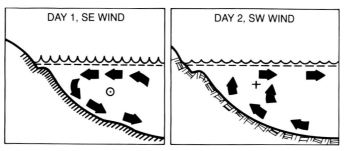

Figure 47. Typical winds, wave heights and coastal ocean (shoreface) currents for a northeaster passing landward of the study area. From Niedoroda et al., 1984a.

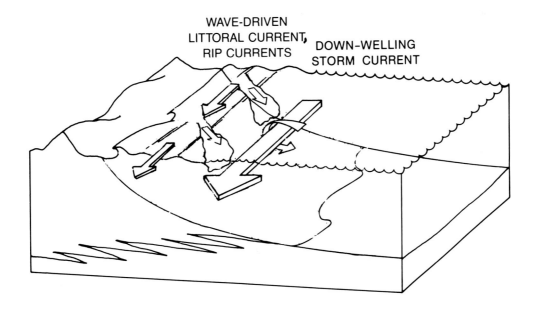

Figure 48. Model for sedimentation by a coastal storm flow.

by rapidly building waves. As the storm center passed north of the coastal area the wind backed to the northwest and began to weaken. The northwesterly wind rapidly diminished the wave heights and caused an easterly upwelling current. Figure 50 shows a time series of sediment transport at seven stations located at 500 m intervals across the shoreface. These sediment transport data were determined by calculating bedload transport due to the combined effects of both waves and currents according to the method given by Madsen and Grant (1976).

The results show that the majority of sediment transport occurred during the strongest northeasterly winds. During this period of the storm the maximum wave heights developed. These were coupled with a strong westerly and downwelling coastal current. The relatively brief period of easterly flow was accompanied by reduced wave heights. Thus the bedload sediment was not shifted eastward to compensate for the large westward transport which occurred earlier in the storm. In the very last portions of the storm an upwelling easterly current developed. However, at this stage the waves were relatively low and little bedload sediment was returned up the shoreface.

The net effect of this coastal storm was to transport sediment westward and down the shoreface. The transport rate decreased in the offshore direction, indicating that the shoreface underwent aggradation during the storm event. It therefore served as a reservoir for a portion of the sand removed from the surface zone and beach. However, seaward transport continued out across the shelf floor.

An obvious question is whether this one storm studied in detail was a typical one in terms of response of the water column and the resulting pattern of sediment transport. This issue has been examined by (Niedoroda et al. (in press a). Storm tracks over the northeastern United States cross the Middle Atlantic Shelf at a low angle, with the storm centers moving from southwest to northeast. If the storm passes seaward of the coastal sector in question, the sector experiences northeast winds which blow alongshore and slightly onshore (Fig. 46). Heavy seas are generated rapidly, and downwelling ensues in both the friction-dominated and geostrophic zones. As the storm center migrates past the coastal point, the winds back rapidly and blow offshore from the northwest. With progressive movement of the storm along the offshore track, these northwesterly winds diminish in intensity.

If the storm center passes to the landward, winds will develop from the southeast and will strengthen as the storm center passes over (Fig. 47). Again, heavy seas will be rapidly generated. With an onshore wind component, downwelling will occur in the friction-dominated zone, although in this case, with the coast to the left of the wind, downwelling will not extend into the geostrophic zone. As the storm center progresses past the coastal point, the winds will veer rapidly to the southwest and diminish in intensity. In both of these cases, initial winds have an onshore component leading to the rapid generation of heavy seas, downwelling, and offshore sediment transport; upwelling and onshore transport occurs late in the cycle if at all, and is accompanied by reduced wave heights and decreased transport rates.

   Turbidity current versus geostrophic current deposition.--Studies of storm sedimentation on modern coasts have revealed non-evidence for deposition by high velocity, spasmodic turbidity currents. Hayes (1967) observed an inner shelf sand deposit on the west Texas shelf in the aftermath of Hurricane Carla, and suggested that it was deposited by turbidity current action during the hurricane landfall. However, a more recent analysis by Morton (1981) suggests that the Carla bed was deposited by alongshelf geostrophic flows.

   High velocity turbidity currents are driven by a feedback process known as autosuspension (Bagnold, 1962). In this process, the turbidity current becomes self-sustaining when it attains a velocity sufficient that the eddies created by the flow are intense enough to suspend the particles, and the density of the particles thus suspended is inturn sufficient to drive the flow. Autosuspension is thus a feedback mechanism. Turbidity currents achieve "ignition" when they travel fast enough to exceed the critical value for autosuspension. Studies by Pantin (1979, 1983) and Parker (1982) suggest that turbidity currents are unlikely to achieve autosuspension under shelf conditions. Muddy shelves are so flat (less than 1°) that most measured shelf suspensions cannot achieve critical velocity. Sandy shelves are slightly steeper but the stirring velocities required to maintain sand in suspension are so high that yet higher slopes are required. Sandy shorefaces have relatively big slopes, but they are continuously wave graded. Submarine slope failure has been reported from the shoreface of an active Mississippi Delta distributary (Linsay et al., 1984) but is not an important process on most shorefaces. A study of the San Francisco inner shelf shortly after an earthquake indicated that zones of sea floor liquification did occur on this muddy shelf, but that the autosuspension criterion was not satisfied; lateral movement was confined to less than 100 m extent (Field and Hall, 1982).

   Theoretical considerations suggest that a rather different sort of density-driven sediment transport may indeed occur on shorefaces (Fig. 48). Sediment loading of the bottom boundary layer during storms will add an additional driving force to the balance of forces governing the current. A flow driven in this fashion by a vertical density gradient is called a baroclinic flow, and its behavior can be computed. The suspended sediment concentrations observed on modern shelves are capable of significantly affecting the velocity of storm created geostrophic flows by increasing their spped, and causing them to veer more directly offshore. However, the baroclinic component of the flow will not be spasmodic, but will last as long as the storm waves resuspend sediment, and the flow will not achieve ignition, but will decay, as it moves offshore and its sediment rains out.

### Coastal tidal currents and offshore sand transport

Tidal fluid and sediment dynamics on open coasts is much more poorly understood than is wind-driven fluid and sediment dynamics. Studies of fluid dynamics and sedimentation on macrotidal coasts have concentrated mainly on such macrotidal estuaries as the Bay of Fundy (dalrymple et al., 1978); Bristol Channel (Collins et al. 1981) or the Wash (Collins et al., 1979). Probably the most studied macrotidal open coast is the eastern coast of Great Britai (McCave, 1978, Carr, 1981, Jago, 1981, Lees, 1983). Tidal current are essentially rectilinear along this coast, with the main tidal axis parallel to the shoreline. The tidal range is large; 4.60 m for mean spring tides; 2.23 m for mean neap tides. The most commo waves are those with a significant height ($H_s$; average of the highest two-thirds of the distribution) of 0.5-0.6 m and a period of 3.5-4.5 sec. Maximum wave heights and periods during 1969-1970 were 7.7 m and 12 sec. (Draper, 1971). Jago (1981) has shown that the direction of surface fluid transport depended on wind direction. Bottom drifters revealed a largely fluid transport during the period, but this may have been an upwelling phenomenon due to the prevailing westerly (offshore) winds.

The alongshore component of transport is toward the south in this area (Jago, 1981, Lees, 1983), consistent with the movement o the tidal wave southward along the coast. The tidal water bulge that sweeps southward along the coast as it rotates about its amphidromic point is identical in its dynamics to the long shelf waves (coastally trapped waves; Kelvin waves) that constitute the "relaxation currents" subsequent to Atlantic shelf storms. Flood tide along the British coast, like storm set up along the Atlantic coast, should result in downwelling and offshore sediment transport, especially when flood tide and storm setup coincides, as in the storms described by Steers (1971) and Caston (1976). It seems probable that a comprehensive observation scheme would show that tide and storm-driven offshore sediment tranpsort would dominate over onshore transport over the course of the year. The geologic setting certainly suggests this; the western coast of Great Britai is a cliffed coast, cut into mesozoic and paleozoic sediments in the north, and into Quaternary unconsolidated sediments in the south. Retreat is rapid (up to 10 m yr$^{-1}$ in the south). An ephemeral beach prism is separated from the inner shelf by a broad sediment starved, wave cut platform extending from the foreshore t 20-40 m water depth; only seaward of this surface does the sand debris created by erosional shoreface retreat come to rest.

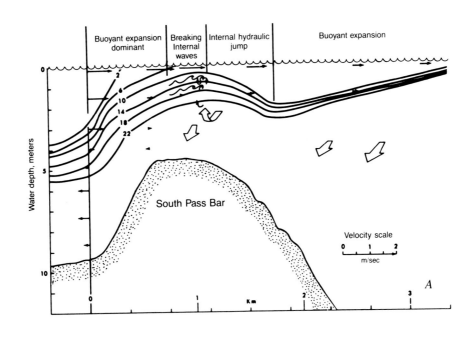

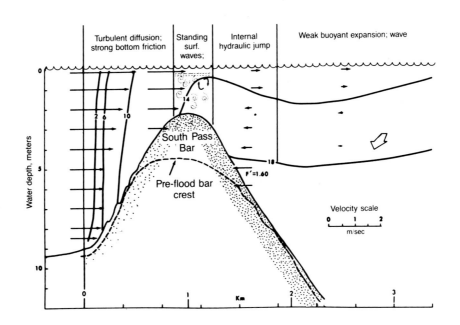

Figure 49. Velocity structure at the mouth of south pass, Mississippi delta, during flood. From Wright and Coleman, 1974.

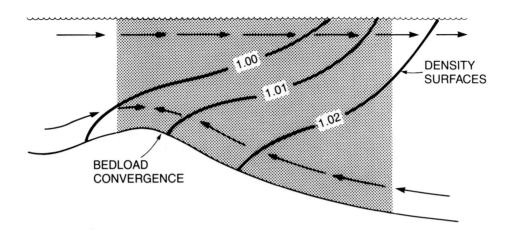

Figure 50. Density stratification, bedload convergence and the turbidity maximum in an estuary.

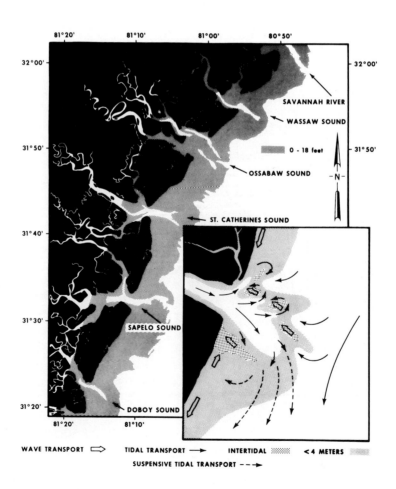

Figure 51. Estuarine coast of Georgia generalized sediment transport pattern on an estuary mouth shoal, showing interfingering ebb flood channel pattern.

## Flow and Sedimentation at River Mouths

On coasts undergoing active sedimentation, stretches of relatively straight shorefaces alternate with coastal inlets. These may be river mouths, tidal creeks, or tidal inlets servicing lagoons. Since these openings are gateways through which all sediment entering the shelf must pass, these areas are much more important than their areal extent would indicate. Coastal inlets tend to be dominated by baroclinic (stratified) flow, because the intracoastal water bodies that they service (rivers, lagoons) are hyposaline (brackish) or in a small percent of cases they (sabkhas, some lagoons) are hypersaline.

We are concerned here with the flow patterns seen in two basic hyposaline cases; that of estuaries, and that of deltas. In dynamical terms, a delta is a river mouth whose fresh water discharge "fits" its mouth, and (at least during flood stage) pushes salt water out; an estuary in this sense is a river mouth that is oversized with respect to its flow; salt water intrudes beneath the fresh water outflow most of the time.

In Figure 49a, Southwest Pass, a distributary of the Mississippi Delta, is larger than required for the shrunken low-stage discharge. Saltier water pushes into the floor of the channel from the sea (salt wedge); lighter river water flows out at the surface. Internal waves develop at the interface between the salt wedge and the river water. The internal waves break, and release gobs of salt water into the fresh; as salt water is thus consumed, more must flow into the channel, which at low stage has an upstream bottom current.

At high stage however, the salt wedge is pushed completely out of the channel (Fig. 49b). Flood water detaches from the channel floor at the crest of the delta mouth bar and flows seaward over the salt water, undergoing buoyant expansion. Flow in the salt water beneath the surface outflow is <u>alongshore</u> shelf flow.

Estuaries are like low stage delta mouths, in that the ratio of freshwater to salt water discharge ranges from 1 to 10 to 1:100. Their circulation patterns are very similar (Fig. 50). If there is an appreciable tide range however, a sharp interface between a salt wedge and the overlying river flow does not occur. Instead there is a zone of tidal mixing and maximum stratification. However, if salinity contours are averaged over a tidal cycle, the salt wedge reappears. The zone of maximum stratification is typically also a turbidity maximum. Suspended fine sediment particles enter the estuary near the surface with the river outflow, and also from the sea, with the saline underflow. Muddy nearbottom water rises in the zone of mzximum stratification as it mixes into the brackish surface outflow, and as this outflow moves seaward, it rains particles back into the saline underflow. Sediment particles thus get caught in a closed loop. The sediment

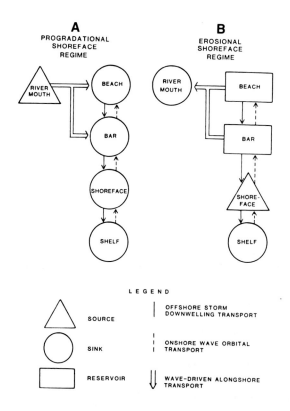

Figure 52. Schematic representation of general coastal sand budgets in a) prograding and b) eroding shoreface regimes. From Niedoroda et al, in press.

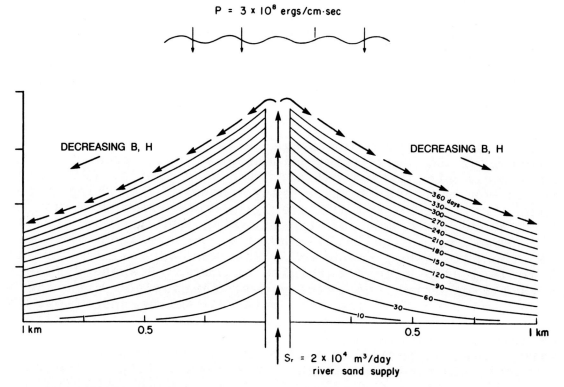

Figure 53. Computer model of the growth of a cuspate distributary mouth. From Komar, 1977. Sand is deposited on the flanks of the delta by the wave-driven littoral current. Horizontal discharge gradient dq/dx is created by the decrease in wave height and breaker angle downdrift.

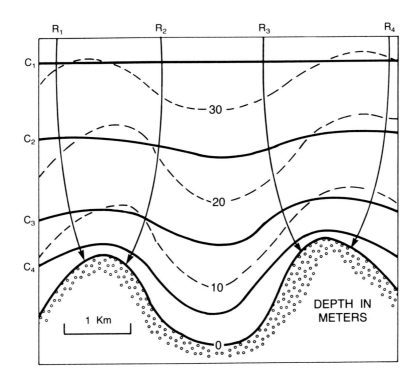

Figure 54. Diagram showing relationship between wave crests ($C_1$-$C_9$), wave rays ($R_1$-$R_4$) and submarine contours (dashed lines) in wave refraction pattern on unirregular coast.

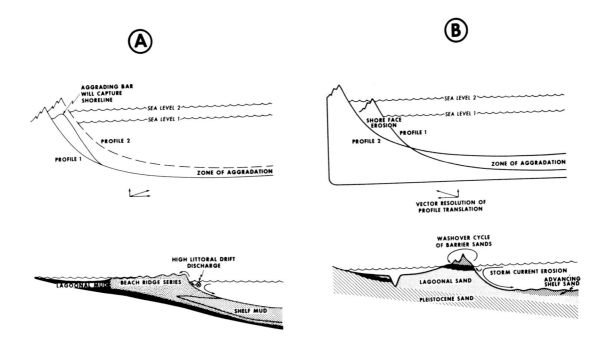

Figure 55. Models for A) prograding coast and B) a coast undergoing erosional shoreface retreat. Upper diagrams in A and B are dynamical models, lower diagrams are stratigraphic models. From Swift, 1976.

concentration in the turbidity maximum builds up until the loss to the open ocean at ebb tide equals the amount gained during a tidal cycle. The suspended sediment build up in the turbidity maximum results in high rates of deposition in adjacent lagoons, tidal flats and marshes.

Coarse sediment transport in estuaries is dominated by the bottom flow convergence between entering salt water and the overriding river flow. Bedload can go no further than this point and must be deposited. The result in microtidal and low mesotidal river mouths is a river mouth bar, whose forward face is moulded by wave forces as a shoreface. In high mesotidal and macrotidal estuaries however, interactions occur between the tidal regime and the mobile sea floor. The tide in the shallow estuary is retarded because the speed of the tidal wave depends on water depth. It continues to ebb for up to an hour as a central jet after the shelf tide has already begun to flood. During this time, the flood tide penetrates the estuary on either side of the central jet. Eventually, sand shoals grow up in areas where zones of shear between the interfingering tidal flows occur duirng part of the tidal cycle. The sand ridge-tidal channel morphology enhances the tidal phase lag that causes it because the tidal wave travels faster in the deeper channel areas, and eventually the estuary mouth bar may develop quite complex patterns of interdigitaling ebb- and flood-dominated channels (Fig. 51).

## Coastal Sediment Budgets

The observations presented in this section show that sand moves landward across the shoreface most of the time. This movement occurs landward of the 10 m isobath at rates up to $1 \times 10^4$ gm $cm^{-1}$ $sec^{-1}$ in response to asymmetric wave orbital currents. On the lower shoreface and inner shelf floor of storm-dominated coasts, rates are an order of magnitude lower; computed values are on the order of $10^2$ gm $cm^{-1}$ $sec^{-1}$. However the observations described above also show that during storms, complex wind-driven circulations develop, which in conjunction with intensified wave orbital currents, reverse the direction of transport and drive large volumes of sediment down across the shoreface.

Numerous studies have shown that most of sand transferred by storms from the beach to the upper shoreface will eventually be returned to storage in the beach prism by fair-weather landward asymmetry of wave orbital currents (summary in Komar, 1976). However, during storms such as the one previously analyzed (Fig. 49), the upper sediment was transported across the shoreface and lost to permanent deposition on the inner continental shelf.

The storm-driven and tide-driven coastal processes described in this section can be time integrated over tens or hundreds of years to establish a time-averaged rate of fluid power expenditure. This long term process variable interacts with two other variables, the rate of sediment input and the rate of relative sea level change, to define one of two basic coastal sedimentary regimes (Fig. 52a).

On coasts down drift from river mouths with high sand discharge, aggradation of the shoreface by the fairweather regime dominates over storm erosion of the shoreface, and the shoreface progrades.

On such coasts, sand moves in the high-intensity sand stream, associated with the nearshore littoral current, driven by shoaling and breaking waves, mainly inshore of the 2 m isobath (Fig. 55). The mean annual sediment transport rate, $q$, in this zone is a function of two parameters of the wave climate; the time-averaged breaker angle, $b$, and the time-averaged wave height, $h$. Deposition is controlled by the alongshore discharge gradient, $dq/dx$; where the rate of sediment transport decreases in a down-current direction sand is deposited. The discharge gradient controlled by the wave refraction pattern. As a wave train moves towards an embayed coast, the waves feel the bottom first on submarine extensions of the headlands, and are retarded over these extensions, so that the wave crests begin to conform to the submarine contours (Fig. 54). As a consequence, wave rays converge on headlands. Wave heights decrease from headlands towards bays, breaker angles open towards the bay, and become smaller in a bayward direction, and littoral sand transport is therefore driven from headlands towards bays. On coasts with high sand discharge, the headlands are generally cuspate (wave-dominated) deltas which are sand sources. The wave-driven littoral current redistributes sand from the delta mouth along the coast (Fig. 54a), and intermittent, downwelling coastal storm flows sweep sand down the shoreface onto the shelf floor.

Coasts undergoing high-intensity sand nourishment prograde by means of beach ridge capture (Fig. 55a). Breakpoint bars form during storms. They expand in volume as they migrate onshore. As they weld to the berm, they are captured by Aeolian processes, and develop dune ridges. In this fashion, a shoreface sandstone lithosome (rock body) progrades seaward beneath the seaward advancing shoreface lithotope (depositional surface; Fig. 54a). The delta mouth is the sand source, the upper shoreface lithosomes (beach prism, breakpoint bar) are reservoirs, and the shoreface is a sediment sink (Fig. 55a).

The rate at which such a sediment distribution system works depends on the relationship between the stratigraphic process variables of fluid power expenditure, the sediment supply rate and the behavior of sea level. As long as sand is supplied more rapidly than downwelling storm flows can spread it over the inner shelf floor, the coast will prograde. When this situation reverses the coastal sediment budget is thrown into reverse. Very often the balance shifts not because the storm climate has intensified, but because relative sea level is reising. River mouths flood with water faster than they can fill with sediment; in morphodynamic terms they cease to be deltas and become estuaries. As such they are sediment traps rather than sediment sources. Storm erosion of the shoreface now provides sediment, both for the littoral drift system and the shoreface (Fig. 52b). Sediment moves from the shoreface into the surf zone and thence into estuary mouths which are filled from the sea. Storms also sweep sediment from the eroding shoreface onto the inner shelf floor (Swift, 1976; Niedoroda and Swift, 1984, 1985, Fig. 55b, this paper).

## FLUID AND SEDIMENT DYNAMICS AT THE SHELF EDGE

### Shelf edge currents

The shelf edge is a dynamically active and complex environment, in which sediment deposition and bypassing occurs. This depositional environment has much in common with the shoreface. Like the shoreface, it is a sharply defined morphological zone in which there are steep gradients of fluid power expenditure, and of frictional energy dissipation into the sea floor, and like the shoreface, the shelf edge is a gateway through which sediment must pass in order to reach deep marine environments.

Figure 56 presents kinetic energy spectra from North American shelf breaks. In this diagram, kinetic energy content is plotted against the frequency of the fluid movement in cycles per day (compare with Fig. 3). In oceanographic terms, these are "red" spectra in that the energy increases with increasing period of motion as one gets towards the "red" or low frequency end of the spectrum. The most energetic events are frontal waves associated with oceanic currents, wind events including wind-generated shelf waves, and tidal events.

Most shelf edges are characterized by oceanic fronts (Fig. 56). Oceanic fronts are more or less vertical boundaries between oceanic water masses, characterized by steep horizontal gradients in physical water properties, mainly temperature and salinity, and also dynamic water properties (momentum). They occur at the shelf break because of the control by water depth on the speed and direction of water movement; shelf water is decoupled from deep ocean water, and the oceanic front is the zone of shear. Shelf edge fronts are dynamically analogous to atmospheric fronts, except that they are restricted in their movements by their genetic relationship to the shelf edge. Shelf edge fronts are <u>retrograde</u> if they dip landward (Fig. 57a), and <u>prograde</u> if they dip seaward (Fig. 57b,c).

The most dynamic shelf breaks are those bounded by major oceanic currents, off the southern Atlantic Bight of the U. S.

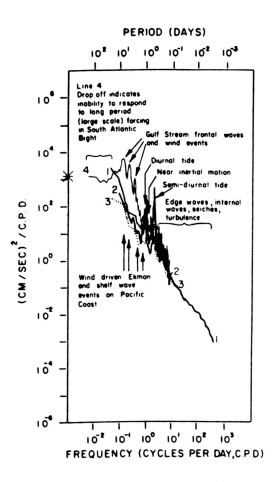

Figure 56. Kinetic energy spectra from currents at shelf break of 1) California, 2) New England, 3) Oregon, 4) North Carolina. Sigma T is a dimensionless measure of fluid density computed from salinity and temperature measurements. From Pietrafesa, 1983.

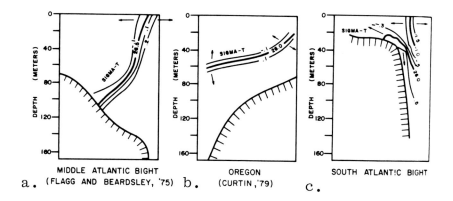

Figure 57. Shelf break fronts. From Pietrafesa, 1983.

East Coast, where the oceanic front is the landward wall of the Gulf Stream (Beardsley and Flagg, 1975) the Gulf Stream is a meandering current, whose meanders tend to undergo cut off. When they do, they spin off from the front as large eddies with diameters of about 100 km, and depths of 200 m and an anticyclonic sense of rotation. These "warm core" eddies (their Gulf stream water is warmer than the shelf water) effect the outer shelf floor as submarine storms. At Cape Hatteras, the west wall of the Gulf stream may ride up onto the shelf for several days at a time (Hunt et al., 1976; Fig. 58).

Slope water may invade the shelf at the bottom (upwelling) or at the surface (downwelling). Flagg (1977) showed that the currents at the shelf break of the middle Atlantic Bight respond to the alongshore component of wind stress in a manner consistent with time-dependent Ekman concepts (as in Fig. 7). The slope of the front increases with northerly winds as shelf downwelling and offshore bottom flow pushed the base of the front (Fig. 57a) seaward and decreases with southerly winds, as upwelling shelf conditions resulted in the invasion of slope water into the lower part of the shelf water column.

On the Oregon and Washington shelf break, a substantial front exists in the summertime, associated with shelf edge upwelling. Northerly winds drive surface waters offshore to the west, which requires a compensatory flow from the upper slope onto the shelf floor (Curtin, 1979).

The shelf edge is critical to the generation of the shelf tide, since the extent to which the shelf tide can co-oscillate with the oceanic tide depends on the steepness of the upper slope, and the depth of the shelf break. The extent to which shelf tide is a standing wave, also depends on the sharpness of the shelf edge; the shelf edge is one of the two main reflectors of the shelf standing tidal wave (the other is the shoreface).

Topographic variations in the shelf break control the intensity and distribution of fluid motions. Upwelling occurs downcurrent from shoals that extend toward the shelf break (Pietrafasa, 1983). Submarine canyons also exert a major effect. They focus and intensify downwelling and upwelling. Cascading events occur, particularly in the winter time when outer shelf waters are cold and dense, and the longshore wind blows with the coast to its right, so that downwelling occurs; cold shelf water thus spills down the canyon as density currents (Pietrafasa, 1983). Internal waves are also focused by submarine canyons (Shepard et al., 1979). Features on computer enhanced Landsat images have been interpreted as groups of internal waves generated by the reflection of the oceanic tide at the shelf break around the head of the Hudson Canyon by the semi-diurnal and diurnal tides (Apel et al., 1975).

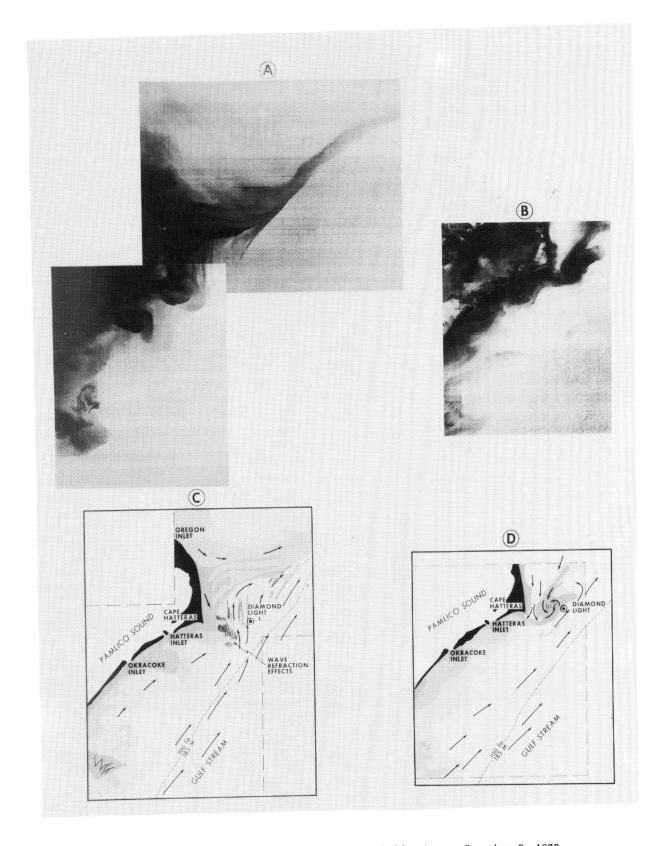

Figure 58. Interaction of the Gulf Stream and shelf water on December 2, 1972. (a,c) and January 25, 1973. From Hunt et al., 1977.

## Sedimentation at the shelf break

The dominant dynamical characteristic of the shelf break in terms of sediment transport is the sudden cessation of storm-induced wave orbital motion at or near the shelf break. Recent detailed studies of the Washington--Oregon shelf break show that over a two-year period, wave-orbital motion as well as the unidirectional wind-driven current component of flow ceased at 200 m the upper slope (Hickey et al., in press). The mud line, which separates shelf sands from shelf edge or slope muds on modern shelves (Wear and Stanley, 1974), is determined mainly by the intensity of the wave climate on muddy shelves. The depth of the shelf break is a response to, as well as a cause of, the intensity of near-bottom wave-orbital motion, and the shelf break marks a zone of sudden seaward increase in the sedimentation rate, as fine sediment bypassed across the shelf begins to accumulate.

The role of submarine canyons in sediment accumulation has been long recognized (Karl et al., 1983). Canyon heads intercept alongshelf sediment flux. Wave-orbital current motion is reduced in their deep floors, and sediment accumulates there many times more rapidly than it does on the adjacent shelf. Even where strong along-axis currents exist, more sediment spills over the canyon walls than can be flushed from the Canyon by the currents (Freeland et al., 1981). Because of the higher rate of fine sediment accumulation, the sediment cannot easily expel its water; its strength is low, and the resulting Canyon fill is liable to fail and slump Canyons thus serve as point sources for redistributing shelf sediments to deeper water.

## REFERENCES CITED

Allen, J. R., R. C. Beardsley, J. O. Blanton, W. C. Boicourt, Bradford, Butman, L. K. Coachman, Adriana Heryer, T. H. Kinder, T. C. Royer, J. D. Schumacher, R. L. Smith, W. Sturges, and C. D. Winant, 1982, Physical oceanography of continental shelves: Reviews of Geophysics and Space Physics, v. 21 p. 1149-1181.

Allen, J. R. L., 1970, Physical processes of sedimentation: New York, American Elsevier, 433 p.

Amos, C. L., and B. F. N. Long, 1980, The sedimentary character of the Minas Basin, Bay of Fundy, in S. B. McCann, (ed.), The coastline of Canada: Geol. Surv. Canada paper 80-10, p. 153-180.

Apel, J. R., H. M. Byrne, J. R. Proni, and R. L. Charnell, 1975. Observations of oceanic internal and surface waves from the earth resources technology satellite: Jour. Geophys. Res. v. 80, p. 865-881.

Asquith, D. O., 1970, Depositional topography and major marine environments, Late Cretaceous, Wyoming: Am. Assoc. Petrol. Geol. Bull., v. 54, p. 1184-1224.

Bagnold, R. A., 1963. Mechanics of marine sedimentation, p. 507-582 in Hill, M. N., (ed.), The Sea: New York, Wiley Interscience, 963 p.

Bagnold, R. A., 1962, Autosuspension of transported sediment; turbidity currents: Roy. Soc. London Proc., v. A265, p. 315-?.

Bally, A. W., 1980, Realms of subsidence, in Miall, A., Ed., Facts and principles of world petroleum occurance: Canadian Soc. Petroleum Geologists, Calgary, Ala., p. 9-94.

Barrell, J., 1912, Criteria for the recognition of ancient delta deposits: Geol. Soc. America Bull., v. 23, p. 377-446.

Beardsley, R. C., W. C. Boicourt and D. V. Hansen, 1976, Physical oceanography of the Middle Atlantic Bight: Am. Soc. Limnol. Oceanogr. Spec. Symp. 2, p. 20-33.

_____, and B. Butman, 1974, Circulation on the New England continental shelf: response to strong winter storms: Geo. Res. Let., v. 1, p. 181-184.

_____, and C. D. Wenant, 1979, On the mean circulation in the Mid Atlantic Bight: J. Phys. Oceanog., v. 9, p. 613-619.

Bennett, J. R., and B. A. Magnell, 1979, A dynamical analysis of currents near the New Jersey coast: J. Geophys. Res., v. 84, p. 1165-1175.

Boicourt, W. C. and P. W. Hacker, 1976, Circulation on the Atlantic continental shelf of the United States, Cape May to Cape Hatteras: Mem. Soc. Royale des Sciences de Liege, 6 serie, tome X, p. 187-200.

Bowen, A. J., 1969, Rip currents. 1. Theoretical investigations: Jour. Geophys. Res., v. 74, p. 5467-5478.

Bumpus, D. F., 1973, A description of circulation on the continental shelf of the east coast of the United States: progress in Oceanography, v. 6, p. 117-157.

Butman, B., M. Noble, and D. W. Folger, 1979, Long-term observations of bottom current and bottom sediment movement on the Mid-Atlantic Continental Shelf: J. of Geophys. Res., v. 84, p. 1187-1205.

Carr, A. P., 1981, Evidence for the sediment circulation along the coast of east Angela: Mar. Geol., v. 40, p. M9-M22.

Caston, V. N. D., 1976, A wind-driven near-bottom current in the Southern North Sea: Estuarine and Marine Science, v. 4, p. 23-32.

Charlesworth, L. J., 1968, Bay, inlet and nearshore marine sedimentation: Beach Haven--Little Egg inlet region, New Jersey: Doctoral Dissertation, Univ. Michigan, Dept. Geol., 23 p.

Chase, R. R. P., 1979, The coastal longshore pressure gradient: temporal variations and driving mechanisms: J. Geophys. Res., v. 84, p. 4898-4904.

Clarke, T. L., B. Lesht, R. A. Young, D. J. P. Swift, and G. L. Freeland, 1982, Sediment resuspension by surface-wave action: an examination of possible mechanisms: Mar. Geol., v. 49, p. 43-59.

Collins, M. F., C. L. Amos, and G. Evans, 1981, Observations of some recent sediment-transport processes over intertidal flats, the wash, in S.-D. Nio, R.T.E. Shuttenhelm, Tj.C. E. Van Weering, Holocene marine sedimentation in the North Sea Basin: Int. Assoc. Sedimentologists Spec. publ. 5, p. 81-98.

_____, G. Feretinos and F. T. Bonner, 1979, The hydrodynamics and sedimentology of a high (tidal and wave) energy embayment (Swansea Bay, Northern Bristol Channel): Estuarine and Coastal Marine Science, v. 6, p. 49-77.

Csanady, G. T., 1977, The Coastal jet conceptual model in the dynamics of shallow seas, in Goldberg, E. D., I. N. McCave, J. J. Obrien, and J. H. Steele, eds., The Sea: v. 6, p. 1045-1061.

Csanady, G. T., 1982, Circulation in the coastal ocean: Reidel, Boston, 280 p.

Curray, J. R., F. J. Emmel, and P. J. S. Crampton, 1969, Lagunas costeras, un simposio, in Mem. Simp. Int. Lagunas costecas: UNAM-UNESCO Nov. 28-30, 1967, Mexico, p. 63-100.

Cutchin, D. L., and R. L. Smith, 1973, Continental shelf waves: low frequency variation is sea level and currents over the Oregon shelf: J. Phys. Oceanogr., v. 3, p. 73-82.

Dalrymple, R. W., R. J. Knight and J. J. Lambiase, 1978, Bedforms and their hydraulic stability relationships in a tidal environment, Bay of Fundy, Canada: Nature, v. 275, No. 5676, p. 100-104.

Davies, J. L., 1964, A morphologic approach to world shoreline: Feet f. Geomorph, v. 8, p. 127-142.

Defant, A., 1958, Ebb and flow. Univ. Michigan press, Annarbor, 121 p.

Draper, L., 1971, Waves at North Carr Light Vessel, off Fife Ness:Nat. Inst. Oceanog. Internal Report Ago, 5 p.

Eisma, D., 1981, Supply and deposition of suspended matter in the North Sea, in. Nio, S. D., R. T. E. Schuttenhelm and T.C.E. Van Weering, Holocene marine sedimentation in the North Sea Basin: Int. Assoc. Sedimentologists Spec. Publ. 5, p. 415-428.

Ekman, V. W., 1905, On the influence of the earth's rotation on ocean currents: Ark. fur Math. Astron Och Fysik., v. 2, p. 1-53.

Emery, K. O., 1968, Shallow structure of continental shelves and slopes: Southeastern Geology, v. 9, p. 178-194.

Engelund, F., and J. Fredsoe, 1974, Transition from dunes to plane bed in alluvial channels: Tech. Univ. Denmark, Inst. Hydrodynamics Hydraulic Engineering Series, paper 4, 32 p.

Fenneman, N. M., 1902, Development of the profile of equilibrium of the subaqueous shore terrace: Geol., v. 10, p. 1-32.

Field, M. E. and R. K. Hall, 1982, Sonographs of submarine sediment failure caused by the 1980 earthquake off northern California: Geo-Marine Letters, v. 2, p. 135-141.

Flemming, B. W., 1980, Sand transport and bedform patterns on the continental shelf between Durban and Port Elizabeth.

Freeland, G. F., D. J. Stanley, and D. J. P. Swift, 1981, The Hudson Shelf Valley: it's role in shelf sediment transport, in Nittrouer, C. E., Ed., Sedimentary Dynamics of Continental Shelves: Mar. Geol. v. 42, p. 399-427.

Grant, W. D., and O. S. Madsen, 1979, Combined wave and current interaction with a rough bottom: J. Geophys. Res. v. 84, p. 1797-1808.

Han, G. C. and T. Niedrouer, 1981, Hydrographic observations and mixing processes in the New York Bight, 1975-1977: Limnol. Oceanogr., v. 26, p. 1126-1141.

_____, and D. A. Mayer, 1981, Current structure on the Long Island Dunes Shelf: J. Geophs. Res., p. 4205-4214.

Harbaugh, J. W. and G. Bonham-Carter, 1977, Computer simulation of continental margin sedimentation, in The Sea: New York, John Wiley & Sons, 625 pp.

Harms, J. C., J. B. Southard, and R. G. Walker, 1982, Structures and sequences in clastic rocks: Soc. Econ. Paleon. Mineral Short Course 9, not sequentually paged.

Harrison, W., and R. Morales-Almo, 1964, Dynamic properties of immersed sand at Virginia Beach, Virginia: Coastal Engineering Res. Ctr. Tech. Memo 9, 52 p.

Harrison, W., and K. A. Wagner, 1964, Beach changes Virginia Beach, Virginia: U.S. Army Coastal Eng. Res. Center Misc. Paper No. 6-64, 25 p.

Homewood, P., and P. Allen, 1981, Wave-, tide-and current-controlled sandbodies of Miocene molasse, western Switzerland: The American Association of Petroleum Geologists Bulletin, v. 65, p. 2534-2545.

Hopkins, T. S., 1982, On the sea level forcing of the Mid-Atlantic Bight: J. Geophys. Res. v. 87, p. 1997-2006.

Hopkins, T. S. and L. A. Slatest, in press, Vertical momentum exhange in coastal waters: J. Geophys. Res.

Hopkins. T. S. and A. L. Swoboda, in press, The nearshore circulation off Long Island, August, 1978: Continental Shelf Res.

Inman, D. L., and C. E. Nordstorm, 1971, On the tectonic and morphologic classification of coasts: Jour. Geology, v. 79, p. 1-21.

Ippen, A. T. and P. S. Eagleson, 1955, A study of sediment sorting by waves shoaling on a plane beach: Beach Erosion Board Tech. Memo. 63, 81 p.

Jago, C. F., 1981, Sediment response to waves and currents, north Yorkshire shelf, North Sea, in S-D. Nio, R. O. Shuttenhelm and JJ. C.E. Van Weering, Eds., Holocene marine sedimentation in the North Sea Basin. Int. Assoc. Sedimentologists Spec. publ. 5, p. 283-301.

Jordan, T. E., 1982, Thrust loads and foreland basin evolution, Cretaceous, Western United States: Am. Assoc. Petrol. Geol. Bull., v. 65, p. 2506-2520.

King, C. A. M., 1972, Beaches and Coasts: New York, St. Martin Press, 570 p.

Komar, P. D., 1976, Beach Processes and Sedimentation: Englewood Cliffs, NJ, Prentice-Hall, 429 p.

Komar, P. D., 1977, Modeling of sand transportation beaches and th resulting Shoreline Evolution. p. 499-513, in Goldberg, E. D. Ed., The Sea: ideas and observations on progress in the Study of the Seas. New York, John Wiley and Sons, 725 pp.

Langford-Smith, T., and B. G. Thom, 1969, New South Wales coastal morphology: Geol. Soc. Australia, v. 16, p. 572-580.

Lavelle, J. W., P. E. Gadd, G. C. Han, D. R. Meyer, W. L. Stubblefield, and D.J.P. Swift, 1976, Preliminary results of coincident current vector and sediment transport observations for winter-time conditions on the Long Island inner shelf: Geophys. Res. Letters, v. 3, p. 47-100.

_____, D.J.P. Swift, P. E. Gadd, W. L. Stubblefield, F. N.Case, H. R. Brashear, and K. W. Huff, 1978a, Fair weather and storm transport on the Long Island, New York, inner shelf. Sedimentology, v. 25, p. 823-842.

_____, R. A. Young, D. J. P. Swift, and T. L. Clarke, 1978b, Near bottom sediment concentration and fluid velocity measurements on the inner continental shelf, New York: J. Geophys. Res., v. 83, p. 6052-6062.

Lees, Barbara J., 1981, Sediment transport measurements in the Sizewell-Dunwich Banks Area, East Anglia, U.K. Sediment v. 5, p. 269-281.

Lettau, B., W. A. Brower, and R. G. Quayle, 1976, Marine Climatology: National Oceanic and Atmospheric Administration MESA New York Bight Atlas Monograph 7, 72 p.

Lindsay, J. F., D. B. Prior and James M. Coleman, 1984, Distributary bar development and the role of submarine landslides in delta growth, South Pass, Mississippi Delta: Am. Assoc. Petrol. Geol. Bull., v. 68, p. 1732-1743.

Ludwick, J. C., 1970, Sand waves and tidal channels, entrance to Chesapeake Bay: The Virginia Journal of Science, v. 21, p. 178-184.

Madsen, O. S., and W. P. Grant, 1976. Quantitative description of sediment transport by waves: Proc. 15th Coastal Eng. Conf. Am. Soc. Civ. Eng. p. 1093-1112.

Mayer, D. A., H. O. Mofjeld, and K. D. Leaman, 1981. Near inertial internal waves observed on the outer shelf in the Middle Atlantic Bight in the wake of Hurricane Belle: J. Physical Oceanography, v. 11, p. 87-106.

McCave, J. N., 1971, Sand waves in the North Sea off the coast of Holland: Mar. Geol., v.10, p.199-225.

McCave, I. N., 1978, Grain-size trends and transport along beaches Example from eastern England: Mar. Geol., v. 28, p. M43-M51.

McGrail, D. W. and R. Rezak, 1977, Internal waves and the nepheloid layer on the continental shelf in the Gulf of Mexico: Trans. Gulf Coast Assoc. Geol. Soc., v. 27, p. 123-124.

Meade, R. H., 1969, Landward transport of bottom sediments in estuaries of the Atlantic coastal plain: J. Sed. Petrology, v. 34, p. 144-122.

Moody, D. W., 1964, Coastal morphology and processes in relation to the development of submarine sand ridges off Bethany Beach, Delaware: Ph.D. thesis, Johns Hopkins Univ., 167 p.

Mooers, C. N. K., 1976, Wind-driven currents on the continental margin, in Stanley, D. J., and Swift, D. J. P., Marine Sediment Transport and Environmental Management: New York, John Wiley, p. 29-52.

_____, J. Fernandez-Portagos, and J. F. Price, 1976, Metorological forcing fields of the New York Bight: Tech. Rep. 76-8, Univ. Miami, Fla, 122 p.

Neumann, G., and W. J. Pierson, Jr., 1966, Principles of physical oceanography: Englewood Cliffs, N.J. Prentice-Hall, 545 p.

Niedoroda, A. W., 1980, Shoreface surf-zone sediment exchange processes and shoreface dynamics: NOAA Tech. Memo, OMPA-1, 89 p

_____, and D. J. P. Swift, 1981, Maintenance of the shoreface by wave orbital currents and mean flow: observations from th Long Island coast: Geophys. Res. Letters, v. 8, p. 337-340.

_____, D. J. P. Swift, T. S. Hopkins, and M. Chen-Mean, in press a, Shoreface morphodynamics on wave dominated coasts: Sedimentary Geology.

_____, D. J. P. Swift, A. G. Figueiredo, and G. L. Freeland, in press b, Barrier island evolution Middle Atlantic Shelf, U.S.A., Part II: Evidence from the shelf floor: Mar. Geol.

Nummedal, D., S. Penland, R. Gerdes, W. Schramm, J. Kahn, and H. Roberts, 1980, Geologic response to hurricane impact on low profile Gulf Coast Barriers: Trans. Gulf Coast Assoc. Geol. Soc. V. XXX, p. 183-195.

Pantin, H. M., 1979, Interaction between velocity and effective density in turbidity flow: Phase plume analysis, with criteria for autosuspension: Mar. Geol., v. 31, p. 59-99.

Pantin, H. M., 1983, Conditions for the ignition of catastrophically erosive turbidity currents-Comment: Mar. Geol., v. 52, p. 281-290.

Parker, G., 1982, Conditions for the ignition of catastrophically erosive turbidity currents: Mar. Geol., v. 46, p. 307-327.

Pitman, W. C. III, 1978, Relationship between eustasy and stratigraphic sequences of passive margins: Geol. Soc. of Amer. Bull., v. 89, p. 1389-1403.

Postma, H., 1967, Sediment transport and sedimentation in the estuarine environment, in Lauff, G. H. Ed., Estuaries, Washington D.C.: Am. Assoc. Adv. Sci., p. 158-179.

Pattiaratchi, C. B., and M. B. Collins, in press, Sediment transport under waves and tidal currrents: a case study from the Northern Bristol Channel: U.K. Mar. Geol.

Redfield, A. C., 1958, The influence of the continental shelves on the tides of the Atlantic coast of the United States: J. Mar. Res., v. 17, p. 432-448.

Schubel, J. R., 1975, Distribution and transportation of suspended sediment, in Manowitz, B., ed., Effects of energy related activities on the Continental shelf: Brookhaven National Laboratories, p. 207-230.

Sharaf El Din, S. H., 1977, Effect of the Aswan High Dam on the Nile flood and on the estuarine and coastal circulation pattern along the Mediterranean Egyptian coast: Limnology and Oceanography, p. 194-207.

Shaw, A. B., 1964, Time in Stratigraphy: New York, McGraw Hill, 223 p.

Shepard, F. P., 1963, Submarine Geology: New York, Harper and Row, 517 p.

Sheridan, R. E., 1974, Atlantic Continental Margin of NorthAmerica in C. A. Burke, and C. L. Drake, The Geology of Continental Margins: New York, Springer-Verlag, p. 391-408.

Steers, J. A., 1971, The east coast floods 31 January - 1 February 1953, in Steers, J. A., (ed.), Applied Coastal Geomorphology: The MIT press, Cambridge, Mass., p. 198-224.

Stewart, H. B. Jr. and G. F. Jordan, 1964, Underwater sand ridges on Georges Shoal, in Miller, R. L. Ed., Papers in Marine Geology, Shepard Commenorative volume: MacMillan, New York, p. 102-114.

Strahler, A., 1963, The Earth Sciences: New York, Harper and Row, 681 p.

Swift, D. J. P., 1975, Barrier Island Genesis: Evidence from thecentral Atlantic shelf, Eastern USA: Sedimentary Geology, v. 14, p. 1-43.

Swift, D. J. P., 1976a, Coastal sedimentation, in D. J. Stanley an D. J. P. Swift, (eds.), Marine Sediment Transport and Environmental Management: New York, John Wiley and Sons, Inc., 602 pp.

Swift, D. J. P., 1976. Shelf sedimentation, p. 311-350, in Stanley, D. J., and D. J. P. Swift (eds.), Marine Sediment Transport and Environmental Management. New York: John Wiley and Sons, Inc., 602 p.

_____, Niedoroda, A. W, Vincent, C. E., and Hopkins, T. S., in press, Barrier Island evolution, middle Atlantic shelf, USA, Part 1: shoreface dynamics: Mar. Geol.

_____, W. L. Stubblefield, F. L. Clarke, R. A. Young, G. L. Freeland, G. Harvey, and B. Hillard, in press a, Sediment Budget in the vicinity of the New York Bight dumpsites: implications for pollutant dispersal, in I. Duedall, (Ed.), wastes in the ocean: John Wiley and Son, N.Y., June 1984.

_____, R. A. Young, T. Clarke, and C. E. Vincent, 1981, Sediment transport in the Middle Atlantic Bight of North America: synopsis of recent observations, in Nio, S. D., R. T. E. Schuttenhelm, and T. C. E. Van Weering, Holocene Marine Sedimentation in the North Sea Basin: Int. Assoc. Sedimentologists Special. Publ. 5, p. 361-383.

_____, R. W. Tillman, J. M. Rine, and C. T. Seimers, in press b, Storm flows and storm deposits on a modern shelf: implications for models of ancient shelf sedimentation. Sedimentology.

Tryggestad, S., K. A. Selanger, J. P. Mathesen, and O. Fobansen, 1983, Extreme bottom currents in the North Sea, in J. Sundeunan and W. Leng, North Sea dynamics: New York, Springer-Verlag, p. 148-158.

Vincent, C. E., R. A Young, and D. J. P. Swift, 1983, Sediment transport rates on the Long Island shelf, North American Atlantic shelf: Role of wave-currents interactions in shoreface maintenance: Continental Shelf Research, v. 2, p. 163-181.

_____, D.J.P. Swift, and B. Hillard, 1981, Sediment transport in the New York Bight, North American Atlantic Shelf: Mar. Geol., v. 42, p. 369-398.

_____, in press, Modelling sediment transport on the New York Bight continental shelf: NOAA Tech. Memo, Boulder, Co.

_____, R. A. Young, and D.J.P. Swift, 1982, On the relationship between bedload and suspended sand transport on the inner shelf, Long Island, New York: J. Geophys. Res. v. 87, p. 4163-4170.

Woodrow, D. L., and A. M. Isley, 1983, Facies, topography and sedimentary processes in the Catskill Sea (Devonian), New York and Pennsylvania: Geol. Soc. Amer. Bull., v. 94, p. 459-470.

Wright, L. D., and J. M. Coleman., 1972, River delta morphology: wave climate and the role of the subaqueous profile: Science, v. 176, p. 282-284.

Wright, W. D., and J. M. Coleman, 1974, Mississippi River mouth processes: Effluent dynamics and morphologic development J. Geol., v. 82, p. 751-778.

Young, R. A, 1975, Suspended matter distribution in the New York Bight apex related to Hurricane Belle: Geology, v. 6, p. 301-304.

# RESPONSE OF THE SHELF FLOOR TO FLOW

Donald J. P. Swift, ARCO Exploration Technology
2300 W. Plano Parkway, Plano, TX 75075.

## INTRODUCTION

The storm and tidal currents that sweep the surfaces of continental shelves imprint a variety of morphologic and textural patterns on these surfaces. As the surfaces aggrade, the grain size gradients and bedform arrays become the textures, structures and stratification patterns of the resulting sedimentary sequences. This paper describes textural gradients and bedform arrays characteristic of shelf surfaces, and the process of strata formation.

## GRAIN SORTING BY MARINE CURRENTS

### Grain size sorting

The tidal- and storm-driven flows that disperse sediments across continental shelves leave characteristic patterns of textures and primary structures in the sedimentary sequences that they build. These patterns provide clues to depositional environment and paleogeography. Shelf floor grain size gradients are one of the most important patterns.

Early petrographers (Wentworth, 1919, Wadell, 1932) noted that most sediments show a progressive decrease in mean grain size in the direction of sediment transport. The decrease was initially attributed to particle abrasion, but in 1939, Russell pointed out that progressive sorting was the more likely cause. This process has been described as the fine particles outrunning the coarse ones, but the description is fallacious. The process is not a continuous one in which finer particles overtake the coarser ones because they have higher velocities, but an inherently episodic and probabilistic one, during which coarser particles have a higher probability of failing to be re-entrained after a period of quiescence than do fine ones. In this model, sediment transport is seen as a diffusion process, in which a large series of separate movements that are random in direction and duration are summed over time. In the case of continental shelf sediment transport, random movements occur at several scales; at the scale of fluid

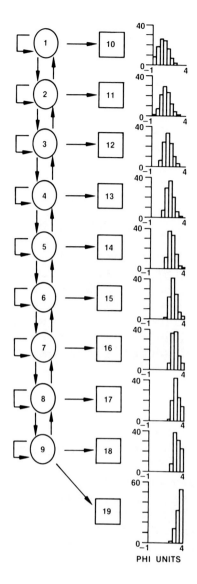

Transition Probability Matrix for Size Class 5. Distribution Exponential with Size and Distance

| From station | To station | | | | | | | | | | | | | | | | | | |
|---|---|---|---|---|---|---|---|---|---|---|---|---|---|---|---|---|---|---|---|
| | 1 | 2 | 3 | 4 | 5 | 6 | 7 | 8 | 9 | 10 | 11 | 12 | 13 | 14 | 15 | 16 | 17 | 18 | 19 |
| 1 | 0.500 | 0.441 | | | | | | | 0.059 | | | | | | | | | | |
| 2 | 0.303 | 0.197 | 0.303 | | | | | | 0.197 | | | | | | | | | | |
| 3 | | 0.162 | 0.338 | 0.162 | | | | | | | | 0.338 | | | | | | | |
| 4 | | | 0.068 | 0.432 | 0.068 | | | | | | | | 0.432 | | | | | | |
| 5 | | | | 0.022 | 0.478 | 0.022 | | | | | | | | 0.478 | | | | | |
| 6 | | | | | 0.006 | 0.494 | 0.006 | | | | | | | | 0.494 | | | | |
| 7 | | | | | | 0.001 | 0.499 | 0.001 | | | | | | | | 0.499 | | | |
| 8 | | | | | | | | 0.500 | | | | | | | | | 0.500 | | |
| 9 | | | | | | | | | 0.500 | | | | | | | | | 0.500 | |
| 10 | | | | | | | | | | 0.100 | | | | | | | | | |
| 11 | | | | | | | | | | | 0.100 | | | | | | | | |
| 12 | | | | | | | | | | | | 0.100 | | | | | | | |
| 13 | | | | | | | | | | | | | 0.100 | | | | | | |
| 14 | | | | | | | | | | | | | | 0.100 | | | | | |
| 15 | | | | | | | | | | | | | | | 0.100 | | | | |
| 16 | | | | | | | | | | | | | | | | 0.100 | | | |
| 17 | | | | | | | | | | | | | | | | | 0.100 | | |
| 18 | | | | | | | | | | | | | | | | | | 0.100 | |
| 19 | | | | | | | | | | | | | | | | | | | 0.100 |

**Figure 1.** Structure of a Markov process model for the offshore component of dispersal of sediment on a continental shelf. The size frequency distributions created by the model demonstrate progressive sorting during cross-shelf transport. From Swift et al., 1972.

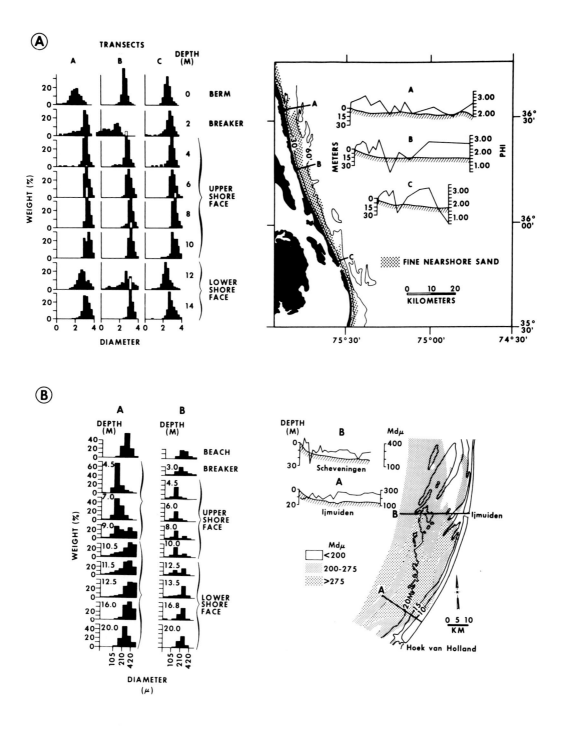

Figure 2. Grain size gradients on the shoreface of A: the North Carolina Shelf and B: the North Sea. From Swift, 1976, after Swift et al., 1971 and Van Straaten, 1965.

turbulence in the near-bottom boundary layer (0.2-5.0 sec) at the scale of wave orbital currents (5-20 sec); at the scale of tidal currents (6 hrs) and at the scale of storm events (6-10 days). It is the periodicity of movement at large time scales that drives the progressive sorting process. The process can be envisaged as a Markov model, in which a series of Markov states represent a transect through a shelf, from shoreline to shelf edge (Fig. 1). In this model, a grain has four options during each transport event; to move to the next station further seaward, to move to the next station further landward, to stay where it is, or to pass to an associated trapping state. In a continental shelf setting, the probability of each of these responses to a typical transport event would depend on the sediment input rate and wave and current regimes characteristics for the shelf. In figure 1, size frequency distributions have been constructed for each trapping station on this simiplified shelf sediment transport system, by assuming a set of probabilities for transitions in which the values decrease with increasing depths and decreasing grain size. The system was run through 1,000 transitions for each grain size and the results combined, in order to obtain the grain size distributions. In the distributions, mean diameter becomes increasingly finer in a seaward direction as a consequence of progressive sorting. The probabilities selected in this model were arbitrarily assigned, but models like this can be calibrated against the wave current, tidal current, and storm current regimes of modern shelves (Clarke et al., 1982, 1983). The mean diameter of a sedimentary rock sample is thus more than a record of the grain size of the parent material as modified by the fluid power expenditure characteristic of the depositional site. It is a <u>cumulative</u> record of grain size filtering at successive positions along the sediment transport pathway.

A variety of grain size gradients are apparent on modern continental shelves. Alongshore grain size gradients are apparent on most beaches, with grain decreasing away from headlands where wave energy is concentrated, and toward recesses (Swift 1975, McCave, 1978). Strong, seaward-fining grain size gradients occur on most shorefaces (Fig. 2). Grain size becomes finer in a seaward direction on the shoreface as a consequence of several hydrodynamic mechanisms. One is the null point mechanism, in which grains migrate towards positions of equilibrium, in which the net landward fluid force, averaged over a wave cycle, is balanced by the downslope component of gravitational force. In shallow water, frictional interaction of wave orbital currents with the sea floor distorts the wave cycle so that the landward stroke is more intense than the seaward stroke (Ch. 1, Fig. 38); thus each grain will tend to roll up the shoreface. The grain will come to rest at a point where the slope is sufficiently steep so that the downslope of component gravitional force will balance the net fluid force averaged over a wave cycle (Ippen and Eagleson, 1965, Bowen, 1980). However, shoreface sands are largely deposited as rip current fallout (Cook, 1970, Cook and Gorsline, 1972) and progressive sorting also occurs during the fallout process.

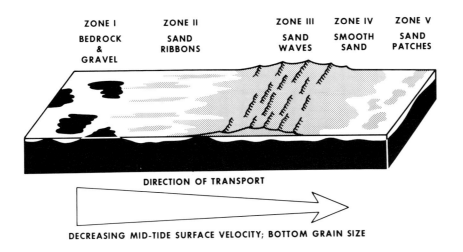

Figure 3. A sediment transport pathway in the North Sea. After Belderson and Stride, 1966.

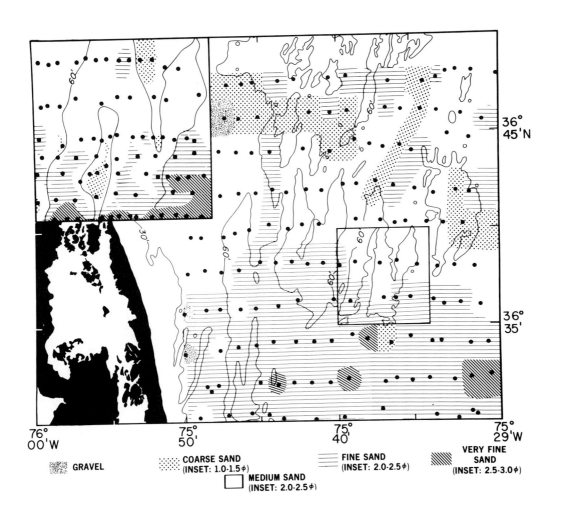

Figure 4. Grain size gradients on the Virginia Beach Massif, North American Atlantic Shelf. The Massif was built by the landward retreat of a littoral drift depositional center during Holocene sealevel rise. Southward storm currents are eroding the upcurrent side and aggrading the downcurrent side. Note east-west belts of coarse sand (top of map), medium sand (center), fine sand (bottom). Inset shows small-scale east-west grain size variation due to ridge topography incised into massif. From Swift et al., 1977.

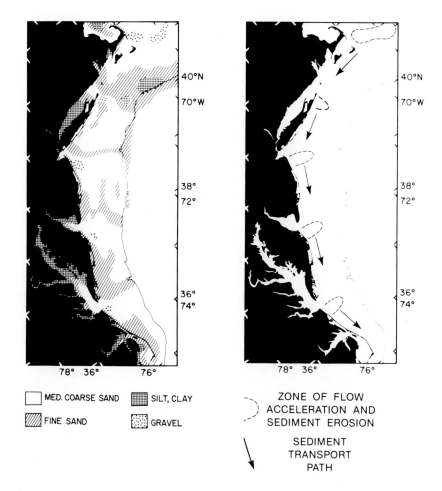

Figure 5. A: Grain size distribution on the North American Atlantic Shelf. B: Transport paths inferred from sediment distribution pattern. Sources of data listed in Swift et al., 1981.

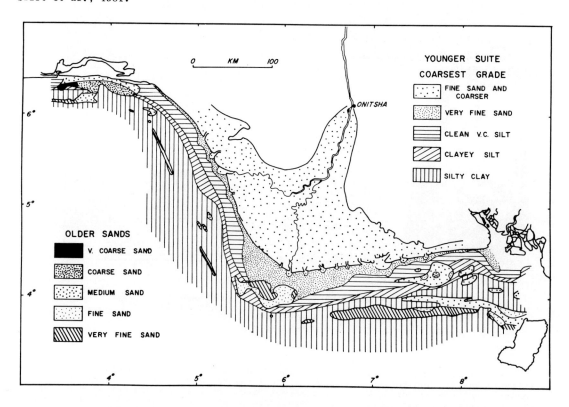

Figure 6. Grain size distribution on the Niger Shelf. From Allen, 1964.

Continental shelves floored by relatively coarse grained sediments (sand and gravel) tend to develop alongshelf grain-size gradients in which the role of alongshore flow components of storm or tidal origin are more important than are high-frequency wave orbital currents. Belderson et al. (1966) for instance, have outlined sediment transport pathways in the tide-dominated continental shelves around the British Isles, which are controlled by the direction of the residual tidal current, and by horizontal gradients in mid-tide velocity (Fig. 3). Along these pathways, grain size becomes finer in a downcurrent direction. However, sorting of the bottom sand at a given locality is controlled by the strength of ambient, wave-induced currents, not by the strength of tidal flows (Channon and Hamilton, 1976). On storm-dominated continental shelves such as the Atlantic continental shelf, along-shelf size gradients are also apparent. Where bottom highs partially restrict alongshelf flows, the upcurrent flanks of the highs experience flow acceleration and tend to undergo erosion, resulting in lag deposits (Fig. 4). Friction retards the flows in the thinner water column over the rest of the feature. Sands that accumulate on the crest and on the downcurrent flank become finer in a downcurrent direction.

Similarly, zones in which the shelf narrows in a downcurrent direction experience flow acceleration. Where the coastline curves sharply into the flow path, zones of flow acceleration must also occur. The regional pattern of grain size distribution on the Middle Atlantic shelf of the North American margin can be best understood in terms of the alongshelf pattern of storm flow. This shelf sector is divided into a series of compartments by major estuaries; the Hudson estuary, Delaware Bay and Chesapeake Bay. Storm flows are oriented alongshelf to the southwest. The coast of each successive compartment, from northeast to southwest, has a more nearly north-south orientation, so that the coastline curves through a series of steps into the flow path in a downcurrent direction. A repeating pattern of grain size appears in each shelf compartment, whereby gravels occur in the zone of flow acceleration on the upcurrent end of each shelf compartment, followed by coarse, medium-, and then fine-grained sand on the downcurrent side of each compartment (Fig. 5).

Seaward-fining grain-size gradients are very prominent on shelves floored by fine-grained sediments. As noted in the preceding chapter, wave orbital current velocities decay exponentially with depth, and as a consequence the threshold velocity is less frequently exceeded by the high-frequency wave orbital current component on the outer margin of the shelf than it is on the inner shelf (Shubel and Okubo, 1972; Clarke et al., 1982, 1983). As a consequence, fine-grained shelves tend to be size-graded; to become increasingly finer-grained in a seaward direction (Fig. 6). It should be noted, however, that in such cases, sediment transport is primarily alongshore, and only secondarily offshore, with transport vectors crossing grain-size isopleths at a low angle.

While alongshore transport is primarily advective (due to wholesale alongshore movement of grains), the offshore component of transport is primarily diffusive; that is, fine particles in shallow water are continuously resuspended and may move in various directions. However, when they happen to move offshore, they tend to undergo permanent deposition. Onshore-offshore grain size gradients are less apparent on sandy shelves, because sand particles are less easily put into suspension, and offshore diffusion is less effective.

Sediment transport on continental shelves, and the resulting textural gradients can be explained by viewing as a diffusion process. Clarke et al. (1983) have presented a stochastic model for shelf sediment transport in which transport on the shelf is assumed to result from a series of statistically independent movements of individual particles. Sediment transport thus becomes a random walk or diffusion process but one in which the spatial scale of diffusion is on the order of tens of kilometers, and the characteristic time is measured in years or centuries. Clarke et al. (1982, 1983), have undertaken to model fine sediment transport by tidal, wind-driven and wave orbital currents as a diffusion process, using random walk theory. In a model of this type, the semidiurnal tidal component of flow is seen as a deterministic transporting agent, which systematically moves suspended sediment several kilometers away from its starting point, then back again. Randomly occurring wind-wave events constitute a probabilistic element of the model. The back-and-forth motion of the tidal currents, randomly loaded with sediment by storm-wave induced currents, results in diffusion transport at large time and space scales. Clarke et al. (1983) have determined the coefficients of the diffusion equation by calculating the probability of entrainment of a sediment particle at each point in a grid. Wave statistics and wind and tidal current data for the middle Atlantic shelf are available for this purpose. Tidal currents and wave orbital currents are statistically independent so that their joint distribution can be obtained by multiplying their individual distributions. Wind-driven water motions are not statistically independent of wave-generated motions, and correlation coefficients must be determined before the effects of wind-driven flow can be included in the diffusion coefficients.

Clarke et al. 1983 have computed diffusion and advection coefficients for transport by tidal, wind, and wave orbital currents, according to the equation

$$\frac{\partial c}{\partial t} = - b_i \frac{\partial c}{\partial x_i} + 1/2 \frac{\partial}{\partial x_j} a_{ij} \frac{\partial c}{\partial x_i}$$

where c is sediment concentration, t is time, $x_i$ and $x_j$ are along-shelf and across shelf directions, $a_{ij}$ is the diffusion coefficient and $b_i$ is the advection coefficient. The sediment

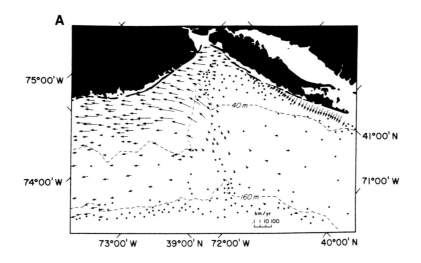

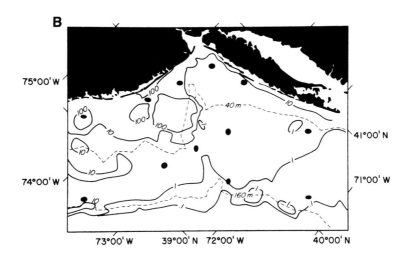

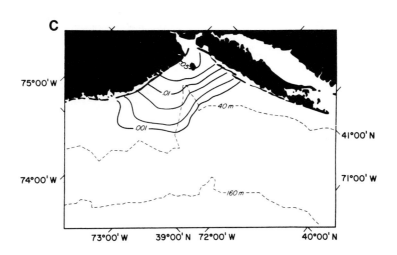

Figure 7. Advection-diffusion model of fine sediment transport in the New York Bight, North American Atlantic Shelf. a: Advection coefficient estimate; units are km $yr^{-1}$. b: Diffusion coefficient estimate, same units. Lengths of axes of black ellipses are proportional to intensities shelf diffusion. c: Contour map of sediment distribution 100 years after a dump at dredge spoil dumpsite. Units are volume percent in sediment in upper 1 cm of bed resulting from dump. From Clarke et al., 1982.

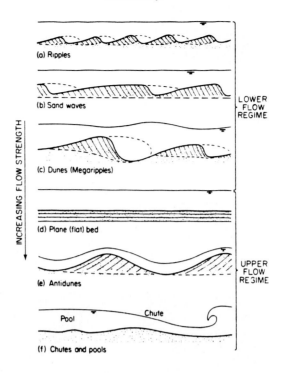

Figure 8. Types of bedforms in quasi-equilibrium, unidirectional flows. Dashed lines indicate zones of flow separation. Not to scale. From Blatt et al., 1980.

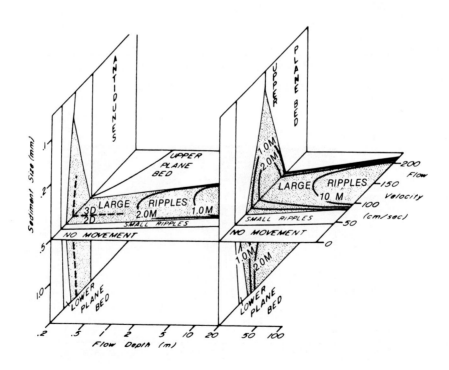

Figure 9. Bedform stability as a function of grain size, flow depth, and fluid velocity. From Harms et al., 1982.

entrainment probability function used in the model is based on observations (Clarke et al., 1983) which reveal that the amount of sediment suspended during a storm event depends on: (1) the grain size of the bottom sediment, which determines how much fine sediment can be pumped out of the bottom by wave-pressure currents; (2) the thickness of the muddy sand substrate entrained as bedload; and (3) the percentage of mud in the bottom sediment.

In the computations, the contours of equal diffusivity (in $km^2\ y^{-1}$) tend to parallel isobaths except where there are marked changes in bottom sediment grain size and permeability (Fig. 7a). A vector map of values of the advection coefficient (Fig. 7b) reveals a southwesterly bias in sediment transport revealed in Fig. 6, as a consequence of the preferred southwesterly trend of storm flows. Clarke et al. (1983) have used the model to compute the dispersal of material dumped as dredge spoil in the New York Bight apex, near New York harbor. The results are shown in Fig. 7c as a contour map of the volume percent of fine sediment in the upper centimeter of the bed at 1 year and 100 years after the fine soil is dumped.

## BEDFORM PATTERNS ON THE CONTINENTAL SHELF

### Bedforms and flow regimes

The tidal and storm-driven flows of the continental shelf create, in addition to grain-size gradients, arrays of bedforms. Shelf bedforms are of interest because they are clues to the character of the depositional regime as a whole, and because they are intimately involved in the process of strata formation, by which a shelf sedimentary sequence is developed.

Bedforms are perturbations of the general level of the bed, either bumps or hollows, which result from the interaction of the flow with the bed. They commonly (but not always) occur in a repeating pattern, and at a vast range of physical scales; spacings vary from a few millimeters to several kilometers. Harms et al. (1982) distinguish between one individual <u>bedform</u> and a <u>bed configuration</u> made up of many such elements and constituting the <u>overall bed</u> geometry; (Allen (1982a) speaks of a pattern of repeating bedforms as a <u>bedform array</u>.

Bedforms have been studied primarily by creating them in laboratory flumes, under controlled conditions. This approach largely stems from the work of Simons and Richardson (1963), who showed that for sands finer then 1 phi (0.5 mm; coarse-medium boundary) there is a sequence of bed configurations as water speed increases from plane bed with no grain motion, to rippled bed, to a bed with large ripples (dunes), to a plane bed with grains in sheet flow, t large ripples which migrated upstream (antidunes; Fig. 8). These last two bed configurations only occurred under conditions of supercritical flow, during which, among other things, water speed is greater than the speed of propagation of a shallow water wave

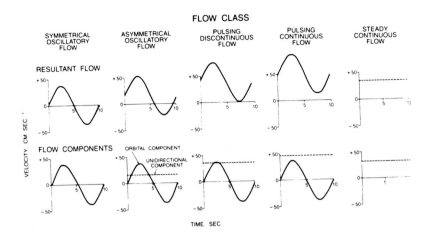

Figure 10. Varieties of flow in which an oscillatory component and a mean flow component occur together. From Swift et al., 1983.

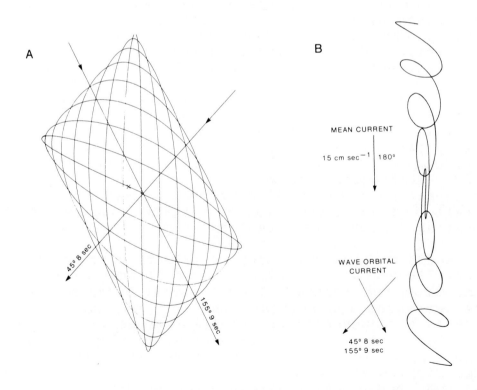

Figure 11. Near-bottom wave orbital trajectories for a: movement of a near bottom-water particle, at 20 m depth in response to the passage of an 8 sec wave train from 450° and a sec wave train from 155°. b: the same trajectory with a 15 sec southward current superimposed. From Swift et al., 1983.

($\sqrt{gh}$). Under these conditions the Froude number, a dimensionless ratio, is greater than unity. The Froude number is defined as f = u/gh, where u is the depth-averaged velocity, g is the acceleration of gravity and h is water depth. The last two bed configurations have been described as upper flow regime configurations, while those occurring under subcritical flow are called lower flow regime configurations. The upper flow regime can occur only in shallow flows (less than 1 m); otherwise unrealistically high fluid speeds are required. Further studies (summary in Harms et al., 1982) have shown that grain size as well as flow speed and flow depth are important controls of bed configurations (Fig. 9).

Flume studies of bedforms are very helpful in understanding bedform arrays on the continental shelf. Unfortunately it is possible to explore only a small part of the range of flow conditions in a flume that actually occurs on continental shelves. Obviously, continental shelves extend to much greater depths and can develop bedforms of greater areal extent than can be conveniently studied in flumes. Futhermore, there is a problem of flow character. Flows may be classified in terms of steadiness and uniformity; a _steady_ flow does not change with time, and a _uniform_ flow does not change along the flow direction. Shelf flows are, to the first approximation, uniform, but they are characteristically unsteady. As noted in the proceeding paper, shelf flows are typically combined flows, in which a wind- or tide-forced, slowly-varying component and a wave orbital component are of subequal value. In fact, a spectrum of flows can be defined, ranging from purely oscillatory flow through asymmetrical oscillatory and pulsating flow to steady flow (Fig. 10). Further complexity is introduced by the fact that if the waves are of sufficiently long wave length with respect to water depth, the time-velocity record of orbital motion becomes distorted as the wave undergoes the shoaling transformation. This process generally occurs under oceanic conditions in less than 10 m of water. Finally, wave orbital currents may be produced by waves of several different wavelengths whose rays may be oriented at angles to each other and to the mean flow component, resulting in very complex water particle trajectories (Fig. 11).

Nevertheless, the bedform patterns observed on continental shelf surfaces do resemble those observed in laboratory flumes, although they sometimes occur at much greater spatial scales. Three broad classes of current-induced bedforms are important on continental shelves. They are, flow transverse bedforms (ripples, megaripples and sandwaves), flow-parallel bedforms (small scale lineations, sand ribbons), and flow-oblique bedforms (sand ridges). These features will be discussed in turn.

## TRANSVERSE BEDFORM DYNAMICS

INITIATION

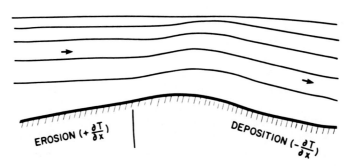

MATURITY

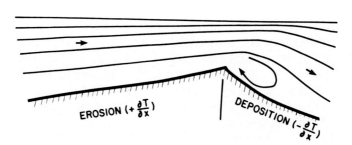

Figure 12. Above: Initiation of a flow-transverse bedform. Bump on the bottom causes near-bottom flow to converge and accelerate on the upcurrent side of the bump, resulting in a positive horizontal gradient of bottom shear stress, and erosion. Deposition prevails downcurrent from point of maximum shear stress. Below: Maturity. Point of maximum shear stress moves toward crest. Crest of growing bedform migrates faster than trough; downcurrent side steepens; boundary layer separation results.

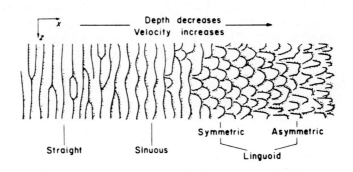

Figure 13. Continuum of ripple types from straight to linguoid patterns. From Allen, 1968.

Flow transverse bedforms

Ripples

Ripples are flow-transverse bedforms whose spacings vary between 10 cm and 1 m. This estimated range of spacings is a generalization; spacing depends on grain size and flow parameters, and will vary from situation to situation. Ripple formation is attributed to streaky velocity variations in the laminar boundary layer of the near bottom velocity boundary layer. Such streakiness creates local irregularities in the bed of some few millimeters in height. Once formed, these bumps are self-propagating because of a phase lag that develops between flow properties and bottom topography (Richards, 1980; review in Allen, 1982a, p. 284). The bottom flow is compressed and must accelerate over the upstream side of the bump, and expand and decelerate over the downstream side; sediment is eroded on the upstream side and deposited on the downstream side (Fig. 12).

Because of momentum effects in near-bottom flow, the peak velocity and shear stress is not on the crest of the bump but on its forward side (Smith, 1970; Richards, 1980). This is because at this spatial scale, the bottom layers do most of their squeezing down and accelerating on the up-current side of the bump. The faster moving, high-momentum, upper layers run into the slower moving lower layers that are rising with the bottom. Maximum scour is forward of the crest where the flow accelerates. Flow expands and slows over the crest, so the crest <u>aggrades</u> as does the down current slope. There is thus a feedback mechanism; a bump in a sandy bottom will tend to grow as currents wave sand over it. The mechanism causes any bottom bump, no matter how small, to grow and to migrate. Flow expansion and deceleration around the sides of the bump causes the bump to propagate laterally and become a ripple.

Because the crest moves faster than the base, the downcurrent side eventually oversteepens and becomes an avalanche face (Smith, 1970). Boundary layer separation in the lee of the growing feature and results in the development of a roller eddy (Fig. 12).

Where the boundary layer reattaches to the bed, it is no longer of equilibrium thickness for the degree of bed roughness present; it has become thinner in its transit over the roller eddy. Consequently, there is intense turbulence at the point of reattachment and for some short distance down stream, until the boundary layer attains an equilibrium thickness once more. The intense turbulence at the point of boundary layer reattachment scours the bed, and initiates the growth of a second ripple down stream from the first. Observations of flumes indicates that as flow over a sandy bed approaches the threshold velocity for grain movement, a shell dropped into the flume will develop a scour collar, which will in turn develop into a ripple, and within minutes trigger a ripple train down the length of the flume

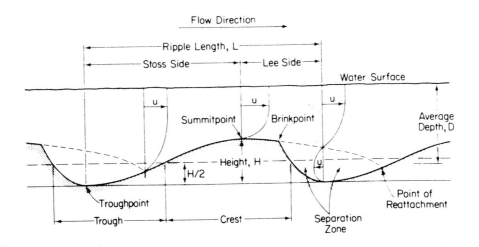

Figure 14. Schematic cross-section of a ripple illustrating ripple nomenclature. From Blatt et al., 1980, after Allen, 1968.

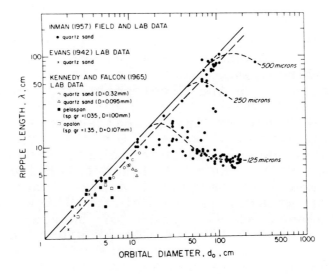

Figure 15. Oscillatory ripplemark spacing related to near-bottom wave orbital diameter for a range of sediment grain sizes. From Komar, 1976, after Inman, 1957, and Komar, 1974.

(Southard and Bugachwal, 1973). As the ripple grows up through the basal part of the boundary layer, the point of maximum bottom velocity and bottom shear stress shifts towards the crest. When at the crest, the mature ripple can still migrate but can no longer increase in amplitude.

Current ripples formed in flumes are commonly highly three-dimensional in nature; that is, their profile varies in the along-crest direction as well as in cross section. Current ripples commonly develop a pattern known as "linguoid" (tongue-like; Fig. 13). Ripples generated by wave orbital currents on the contrary are symmetrical and straight, with "tuning fork" bifurcations; combined flow ripples are intermediate in nature. The geometry of the resulting ripple is described in Fig. 14.

The few available measurements of ripple spacings in coarser sands indicate that sand grain size is the single most important factor in determining ripple spacings (Komar, 1976, p. 122). For a given wave orbital diameter, ripple spacing in 500 μ sand could be more than a factor of 10 greater than ripples formed in fine sand (Cook and Gorsline, 1972). The wave length of surface waves is a second important consideration. As bottom orbital diameter (and therefore maximum orbital velocity) increases, ripple spacing increases, up to a maximum value which is a function of sand grain size (Fig. 15).

When wave ripples first form, they are low <u>rolling-grain ripples</u> (see summary of Harms et al., 1982), but they rapidly grow into <u>vortex ripples</u> (Harms et al., 1982). During each forward and reverse wave stroke, vortex ripples develop a characteristic lee-side vortex that at the height of the stroke, rises from the trough and dissipates. The suspended sand is swept forward or backward by the current to the next ripple. As ripple spacing increases, so does the ripple index (ratio of ripple height to ripple length). As wave orbital diameter (and maximum orbital velocity) increase to the value for maximum ripple spacing, the ripples are increasingly remade during each stroke. At larger wave orbital diameters, ripple asymmetry reverses through the wave cycle and ripple index and spacing decrease until the ripples disappear and sheet flow of sand (upper flow regime) occurs during maximum orbital velocity.

Komar (1976) notes that this ripple sequence can be observed by swimming shoreward toward an ocean beach. In deep water the ripples are sharp-crested with pronounced heights. As the breaker zone is approached the ripple heights progressively decrease and finally disappear within the intense shear under the breaking waves. As the ripple heights decrease there is less development of lee vortices during the wave orbital motions and therefore less suspended sand thrown upward off the bottom. When the ripples disappear under the intense shear, the sand moves as a thin carpet of high concentration close to the bottom rather than as clouds of sand thrown up into the water.

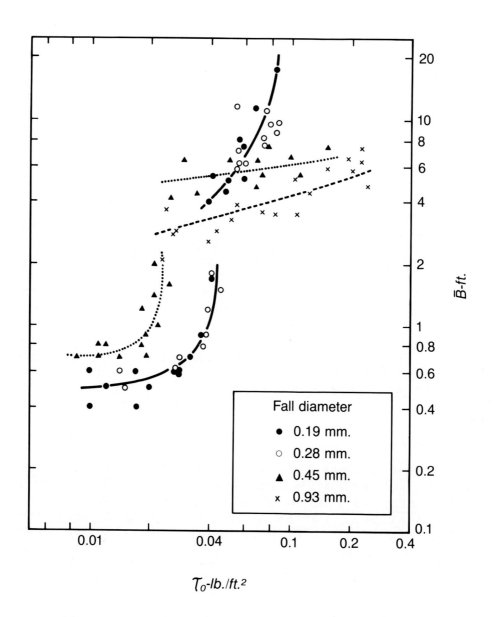

Figure 16. Mean ripple spacing (chord) as a function of boundary shear stress. From Allen, 1968, based on data of Guy et al., 1966.

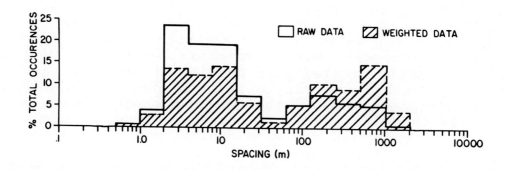

Figure 17. Bimodality of large-scale flow-transverse bedforms on Nantucket Shoals. Weighted data has been corrected for greater abundance of small-scale forms per unit area. From Mann et al., 1981.

Methods for attempting to estimate wave parameters and water depth from fossil ripple marks have been described by Allen (1966, 1979) and Miller and Komar (1980). For a more pessimistic analysis, see Harms et al., 1982.

## Megaripples

Flow transverse bedforms with spacings between 2 and 20 m are called megaripples by field workers, and dunes by flume workers. Megaripples appear to form in response to the same sort of flow-substrate interaction that creates current ripples (Richards, 1980), but megaripples are responses to flow conditions in the outer portion of the turbulent boundary layer, while ripples are governed by conditions in the inner zone (Jackson, 1975).

Experiments in laboratory flumes seem to show that there is a fundemental genetic difference between current ripples on one hand and megaripples on the other (Allen, 1968; Fig. 16). It is generally not possible to create the next larger class of flow transverse bedform (sandwaves) in laboratory flumes, and experimental workers have tended to assume that megaripples and sandwave together form a continuous spectrum of bedforms (the "large ripples" of Harms et al., 1982). Field workers, however, commonly en counter a spectral gap between sandwaves and megaripples (Fig. 17) as well as between megaripples and ripples.(Jackson, 1975; McCave and Gieser, 1978; Swift et al., 1979; Dalrymple et al., 1978; Fig. 24, this chapter. For a field worker with an opposing view, see Flemming, 1978, 1980). However, studies by Allen (Summary in Allen, 1982a) suggest that the spectral gap is a secondary phenomenon, due to the behavior of megaripples in unsteady flows (see discussion in next section) rather than an intrinsic phenomenon, a the difference between ripples and megaripples apparently is.

## Megaripple population in unsteady flows: Hummocky megaripples

Continental shelf currents are unsteady at several time scales, the most important being the wave period (5-20 sec), the semidiurnal and diurnal tidal periods (12 and 24 hrs) and the stor event period (6-10 days). Flow variability is an important contro of bedform patterns. There is generally a "tuning" relationship between the spatial scale of the bedform and the temporal scale of the flow variability. In this relationship, there is a flow reversal frequency, for each spacing of flow-transverse bedforms. Above the limiting frequency, the bedform will be completely remade with each flow reversal, so that thebedform becomes an equilibrium response to undirectional flow.

As noted by Wilson (1972), the migration rates and response times of bedforms are a functions of their volumes, since the spee with which bed slope and elevation can change in response to changing conditions depends on how much sand has to be moved in

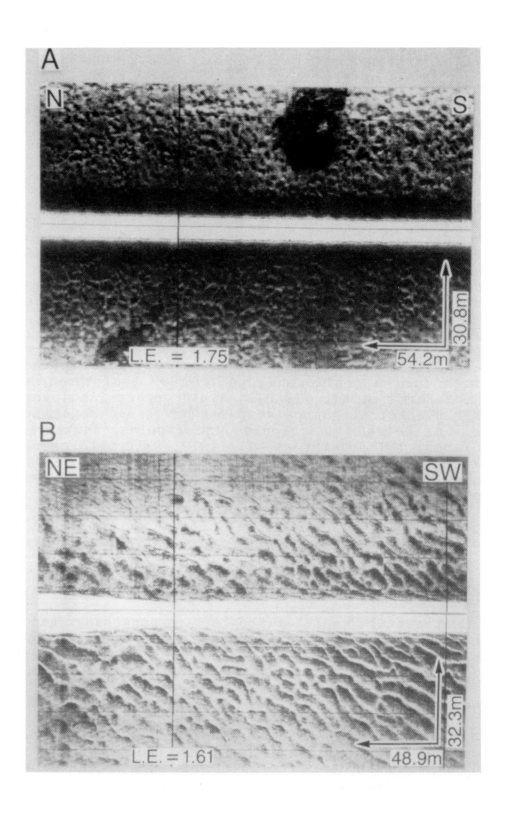

**Figure 18.** Side scan sonar records of submarine megaripples. The two horizontal **bands are oblique** views of the sea floor to the right and to the left of a towed **transducer.** A: Storm generated megaripples in the New York Bight Apex, North American Atlantic Shelf. B: Tidal current megripples from Nantucket shoal, area described by **Mann et al.,** 1980. From Swift et al., 1983.

order to make the changes. For instance, current and oscillation ripples, being small, are remade completely by such relatively high frequency events such as long period waves.

Megaripples, at somewhat larger spatial scales are modified rather than remade by the high frequency reversals of flow beneath surface gravity waves. If the bed is sufficiently fine grained, hummocky megaripples result. However, megaripples are "tuned" however, to somewhat longer period variations in flow. Rebuilding (rather than modifying) of megaripples in response to such long period variations has been mainly observed in rivers of variable discharge, butundoubtedly occur on the shelf floor as well, during storm flow episodes. Megaripple modification in response to surface waves andrebuilding in response to long period flow variations will be described in detail.

Hummocky megaripples, attributed to storm-wave modification o fine-grained flow-transverse, sharp-crested megaripples have been recently described from the Atlantic shelf of North America (Swift et al., 1983; Fig. 18, this PAPER ). Their belated recognition as a unique bedform has come about in response to a deliberate search for modern analogs of the primary structure known from ancient strata as hummocky cross stratification (Swift et al., 1983). Hummocky megaripples are elliptical to circular mounds of fine orvery fine sand beds on the continental shelf. Center to center spacings are on the order of 1-10 m. Side slopes have not been accurately measured, but avalanche slopes do not appear, and resolution on side scan sonar records is due to grain-size reflectivity differences, rather than to acoustic shadowing, so the slopes are probably 15° or less.

Hummocky megaripples have been observed on the continental shelf of Virginia, New Jersey, and New York at depts of 15-40 m. They are inferred to responses of the sea bed to combined-flow storm currents, in which the wave orbital current component is equal to or stronger than the mean flow component (Swift et al., 1983).

The mound-like nature of hummmocky megaripples is believed to be due to the complex velocity structure of storm flows. In combined flow storm currents, the directions of mean flow and the wave orbital currents generally do not correspond. As storms cross the Atlantic Continental Shelf, winds blow in a more or less circular pattern several hundred kilometers in diameter, and generate waves at every point. The waves are not characterized by a single frequency but are rather by a broad frequency band. Each point within the storm pattern is subject to the passage of waves from other points, and the propagation direction varies continuously from point to point. Storm waves are, therefore, generally described by both directional and frequency spectra (Kinsman 1965), and the trajectory of a water particle on the sea floor in response to these motions is correspondingly complex. Current directions are further modified near the shoreface; mean flow

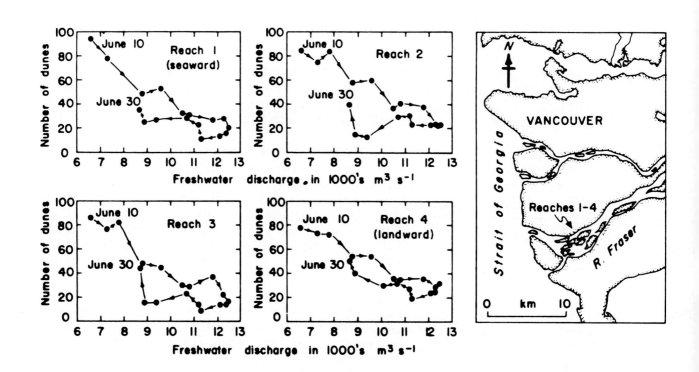

Figure 19. Phase diagrams illustrating change in a population of dunes as a function of fluid discharge, Fraser River, British Columbia, Canada. From Allen, 1982, after Pretious and Blench, 1951.

intensifies and becomes aligned with the shoreline, while waves begin to refract so that these orbital currents move along rays normal to the shoreline and to the mean flow.

Storm wave trajectories are similar in some respects to the Lissajous figures that electrical engineers create on their oscilloscopes (J. Ludwick, personal communication; Ruiter 1959, p. 137). The figures repeat after a period that is determined by the characteristics of the two wave spectra, and is generally many times longer than the largest component frequency (Fig. 12). When a unidirectional flow component is superimposed on such a trajectory, the figure becomes a "braided" one, elongated indefinitely in the flow direction. Scrubbing of the sea floor by such a complex water particle trajectory would presumably have a modifying effect on the geometry of bedforms developing in response to the flow.

## Megaripple populations in unsteady flows: response to long-period variability

Allen (1978b) believes that polymodal bedform assemblages, in which megaripples occur on the rocks of sandwaves, occur in response to flows which are unsteady at storm current or tidal frequencies. He cites the work of Pretious and Blench (1951) who measured the characteristics of 4 distinct megaripple populations in a relatively uniform reach of the Fraser River in British Colombia.

Daily measurements were made for a 20 day period which spanned the flood of June, 1950. During this period, the number of megaripples decreased (wavelength increased) as the river rose, then the number of megaripples increased as the river fell. The response was not symmetrical however; megaripple wavelength is a double-valued function of discharge, with longer wavelengths for a given discharge on a falling river stage than during a rising river stage (Fig. 19). In other words, dune wavelength lags (displays hysteresis) with respect to discharge on both rising and falling stages. As the river falls, small megaripples rather abruptly appear on the back of larger forms which, presumably are then slowly degraded until the bed has an appearance similar to that at the beginnning of the survey. A similar relationship has been noted in tidal estuaries between megaripple spacing and the spring tide-neap tide cycle (Allen et al., 1969).

Allen (1976a, 1976b, 1976c, 1978a, 1978b) has developed an numerical model of megaripple time-lag. In Allen's model, the lag between mean wavelength or height on one hand, and discharge on the other, increases as the time-averaged life span of the megaripples grows relative to the flow period. Increase of life span relative to flow period also changes population structure for large discharge ranges; assemblages are unimodal at all times when the life span is small, bimodal for some of the time at intermedi-

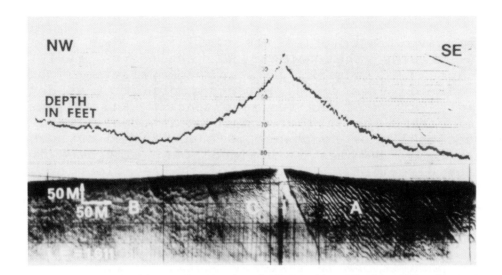

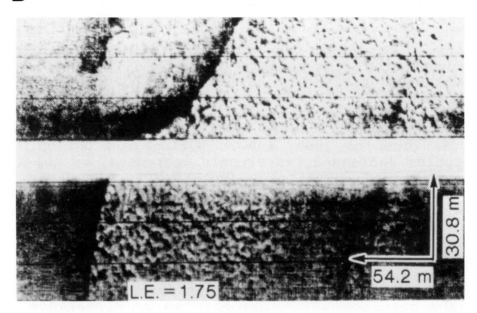

Figure 20. Above: Side scan sonar and fathometer records of tidal sandwave on Nantucket Shoals, North American Atlantic Shelf. Note variation in superimposed megaripple patterns at A, B, and C. From Mann et al., 1981. Below: Low amplitude sandwave with superimposed hummocky megaripples, Long Island sector, North American Atlantic Shelf. From Swift et al., 1983.

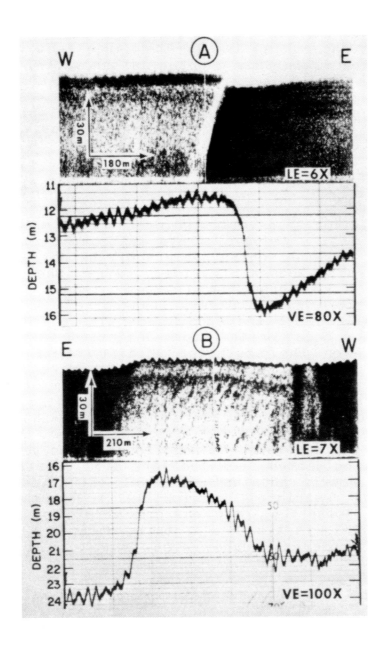

Figure 21. Side scan sonar and fathometer records records of of sand waves, North Carolina sector, North American Atlantic Shelf. From Swift et al., 1978.

ate life spans, and permanent polymodal for large life spans. The polymodality occurs because large bedforms created at high stages persist through the subsequent recession, to be joined during low discharge by a number of megaripples with a wavelength appropriate to the now much reduced discharge. The large bedforms occur at wavelengths referred to in this chapter as <u>sandwaves</u>. Polymodality is probably also "locked in" to a certain extent once large-scale bedforms form; the boundary layer must reform on the back of each successive sandwaves, and the megaripples on their backs can not grow very large before they spill over the avalanche face and lose their identity.

## Sandwaves

Sandwaves are flow-transverse bedforms with spacings of 40 to 200 m (Jackson, 1975). Sandwave arrays are common in such macrotidal shelf sectors as the southern bight of the North Sea (McCave 1971), the Yorkshire coast (Stride, 1970) the Celtic Sea (Cartwright, 1959, the Irish Sea (Harvey, 1966), Georges Bank (Twichell, 1984), the Thames estuary (Langhorne, 1982) Nantucket Shoals (Mann et al., 1981) and Cook Inlet Alaska (Bouma et al., 1977) Sandwaves also occur on the storm-dominated Atlantic Shelf o North America and on the southwest African shelf beneath the Aguilhas current (Flemming, 1978, 1980). Sandwaves in zones of intense currents commonly grow to amplitudes of 7 m or more. Sand wave thickness probably relates to thickness of the whole flow, rather than to boundary layer thickness (Allen, 1982a). In general, sandwaves appear to be able to attain heights of 0.1 to 0.5 of the water depth. Stride (1970) has challenged this relationship o the basis of data collected from the Yorkshire coast. The relationship is most nearly obtained in tidal estuaries where wave fetch is relatively restricted (Ludwick, 1972). In the the southern bight of the North Sea, the eastern margin of the Hook of Holland sandwave field is the 15 m isobath; McCave (1971) suggests that wave activity prevents sand waves from forming in shoaler water.

On locally the storm-dominated North American Atlantic shelf, sand waves occur at scales comparable to the sandwave arrays in tidal seas. Fields of high amplitude, storm built sandwaves occur where the shelf narrows or shoals in the downcurrent direction, (Fig. 20, 21). Here also, sandwave height is modified by wave activity. However, elsewhere in the North American Altantic shelf sandwaves occur at similar spacings, but rarely at amplitudes greater than a meter. In these areas, sandwave heights are surpressed not only by wave activity, but the infrequency of stron flows; they are reactivated perhaps 3 to 5 times a year for a duration of hours. Construction during these short periods is balanced by destruction by the activities of burrowing invertebrates and by wave orbital currents during the rest of the year.

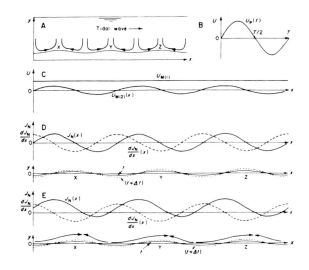

Figure 22. **Control of sediment** transport in the velocity structure of a tide-generated oscillatory **boundary** layer. X, Y, and Z are successive flow cells. T is period, t is time, **and u is** velocity, $J_N$ is sand wave amplitude. From Allen, 1980.

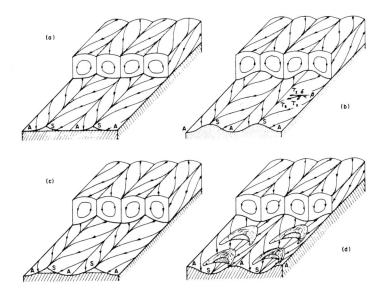

Figure 23. A boundary layer with secondary flow structure. Note variation in velocity gradients between upwelling and downwelling limbs of horizontal vortices. A is bottom current divergence, S is bottom current convergence; T is bottom shear stress. From Allen, 1982b.

161

Sandwaves in unsteady flows

Sandwaves, like other flow-transverse bedforms, may occur in several geometric configurations (Van Veen, 1935); as progressive sandwaves with pronounced asymmetry, as asymmetrical-trochoidal sandwaves with less pronounced asymmetry, and as trochoidal (symmetrical) sandwaves. A fourth kind, catback sandwaves, occurs in tidal settings and are due to reactivation of the sandwave crest by a subordinate phase of the tidal cycle. Reactivation creates a narrow ridge of sand that rests on the brink point (in profile, the cat's ear).

Allen (1980) has pointed out that the growth of a symmetrical sand wave in response to the oscillation of a flow at tidal frequency is analogous to the growth of a symmetrical ripple in response to a flow oscillating at wave orbital frequency. He agrues (Allen 1980, a, b; 1981 a, b, c) that tidal sandwaves are fundamentally different in genesis from the equally large bedforms formed by unidirectional flows in rivers. He cites experimental and theoretical studies (Sleath, 1975, 1976) indicating that in the short-period laminar oscillatory boundary layer over wave-formed bottom ripples, there is a mass transport current that is due solely to the waviness of the bed. A mass transport current in this sense means that when the oscillatory flow is averaged over many oscillations, a residual unidirectional current is apparent. This mass transport current is spatially periodic; that is, it forms flow cells of half the wave length of the bedforms (Fig. 23). Allen argues that because there is no qualitative difference between the oscillatory boundary layers due to short-period gravity waves (wind waves) and long period gravity waves (tidal waves), it is reasonable to postulate that the tidal wave acting on a wavey bed also creates curvature-related mass transport currents of the same kind as those known to be generated by the short period waves.

Allen's analysis illuminates the formation of sandwave scale bedforms in response to tidal flows. However, his argument that these bedforms are fundamentally different from sand waves seems largely semantic in nature. The spatially periodic mass transport currents that he describes are due the thickening and thinning of the boundary layer over the bed waviness, and ultimately, to the same phase lag between bottom shear stress and topography cited by Smith (1970) and Richards (1980) as the generating mechanism for transverse bedforms in unidirectional flows (Fig. 13).

Flow transverse bedforms and the dynamical environment

Studies of flows in laboratory flumes (Southard, 1982, p. 2-14), and in the intertidal zone (Dalrymple, 1978) suggest that megaripples form at higher flow velocities than do sandwaves. However, on modern continental shelves, large sandwaves (7-15 m high) consistently occur in the highest velocity sectors in which thick sand deposits are accumulating (McCave, 1971, Terwindt, 1971, Mann et al., 1980, Johnson et al., 1981). These are commonly surrounded by halo zones in which small sand waves or megaripples only occur. Most examples occur macrotidal shelves, or else in zones of flow constriction and acceleration on storm-dominated shelves (Swift et al., 1978; Hunt et al., 1977).

The apparent contradiction can be resolved by considering the difference between the concepts of "equilibrium" and "climax" in the context of periodic flows. Among uniform, steady flows those with higher velocity, will, when in equilibrium with a cohesionless substrate, produce megaripples rather than sandwaves as indicated by Fig. 9b. However, flows on the shelf are generally unsteady, over a scale of hours to days, in response to tides, or to storm flows. Areas where currents are the most intense during these episodes are areas where the full hierarchy of small-, medium-, and large-scale flow transverse bedforms are most likely to come to equilibrium; such areas are areas of climax development of bedforms.

## Flow-parallel bedforms

A second class of bedforms is oriented with long axes parallel to the direction of flow. Bedforms of this class arise in response to secondary currents; that is, motion in which a transverse circulation is superimposed on the primary longitudinal flow component of the fluid, so that the streamlines form a corkscrew pattern (Fig. 24). In this type of current, the flow is compartmentalized into flow cells whose axes are parallel to the main flow direction; adjacent cells rotate with the opposite sense. Unlike flow transverse bedforms, where deformation of the substrate causes deformation of the flow and vice versa, flow parallel bedforms are, at smaller scales, a passive reflection of water column dynamics on the sea floor. Larger scale flow parallel bedforms interact with their associated currents not as a consequence of their relief, but as a consequence of their grain size patterns (in fluid dynamical terms, their bottom roughness patterns).

### Small scale lineations

Henry Clifton Sorby (1859, 1908; in Allen, 1982a) was the first to report that currents transporting sand over a sandy bed cause a streaky pattern on the bed. He also noticed sandstones composed of extensive, flat-lying parallel laminae arranged like the leaves of a book. The surfaces of these laminae have the same kind of streaking (parting lineation; Crowell, 1955). Allen (1964) showed that grain shape fabrics associated with parting lineation are symmetrically bimodel about the parting lineation trend and lie 20-40° apart. The grains are imbricated upcurrent by 8-12° (Potter and Mast, 1963; Allen, 1964).

Allen (1982a) has reviewed recent studies of boundary layer dynamics, and has been able to show that this bedform and its

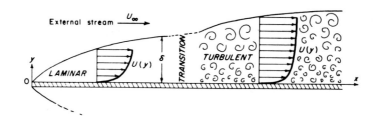

Figure 24. Schematic representation of the boundary layers developed on a flat plate in parallel flow. U is fluid velocity. From Allen, 1982a.

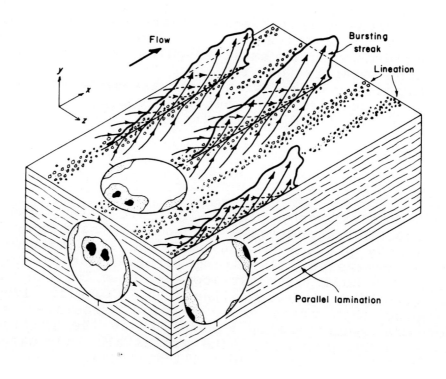

Figure 25. A model for the origin of parting lineations by boundary layer streakiness. Contoured polar plots indicate orientations of long axes of sand grains. From Allen, 1982a.

associated primary structure are responses of the bed to boundary layer processes. The complex boundary layer associated with combined wave and wind driven storm flow has been described in the proceding paper, where a fluid boundary layer has been defined as that part of the flow near the bed whose velocity is appreciably diminished by the proximity of the bed. Small-scale lineations may be observed beneath combined flow boundary layers, but also in the much simpler boundary layers beneath steady uniform flow in flumes. Fig. 24 shows a boundary layer developing on the upper surface (and the lower surface) of a flat plate, suspended in such flow. As the fluid passed over the leading edge of the plate, a boundary layer is initiated. It is a laminar boundary layer for some small distance downstream (fluid sheets slip uniformly over each other) but as the boundary becomes increasing mature, and attains its equilibrium thickness, instabilities develop. Downstream of a transition zone the boundary layer is fully turbulent.

Recent studies (Summary in Allen, 1982a) have shown that in turbulent boundary layers, there exist eddies over a wide range of scales, and tend to increase in size with increasing height above the bed. The eddies are persistant, traveling downstream a distance of several times their diameter before decaying. The eddys control the downward flux of momentum through the boundary layer, and thereby determine its structure, as indicated by the characteristic parabolic velocity curve. Very near the bed (within a few cm or less) however, eddies are damped out. Momentum is transfered downwards through this "viscous sublayer" mainly by intermolecular forces; here velocity profile is linear. We now know a great deal about the velocity structure very near the bed, as a consequence of experiments in which this structure is made visible by releasing dye, or by generating trains of fine hydrogen bubbles (Summary in Allen, 1982a). The "viscous sublayer" turns out to exist only in time-averaged measurements. The near-bed boundary layer has a streaky structure organized into streamwise zones (zones parallel to flow) of alternately high and low velocity. In flume experiments, the dye or other marker is concentrated into zones of slow moving (low momentum) fluid. The intervening zones are clear because they consist of faster moving or high momemtum fluid that has penetrated the lower boundary layer from higher in the flow. The streaks are shifting and wavy in appearance and form randomly in space and time. Many of us have seen this structure made visible in the atmospheric boundary layer, when we have watched streamwise wisps of sand blowing over a beach, or wisps of snow flowing over a road.

Allen (1982a) summarizes many recent studies to provide a model of near-bed structure in fully turbulent boundary layers and its effect on the bed. The near bottom streaks, as they are transported by the flow undergo three stages of behavior; 1) a period of quiescence (tens of seconds), 2) periodic eruption ("bursting", or "ejection") of the low momentum fluid away from the bed, into the upper boundary layer, and 3) simultaneous inrushes of fluid (sweeps) from the laterally adjacent high speed

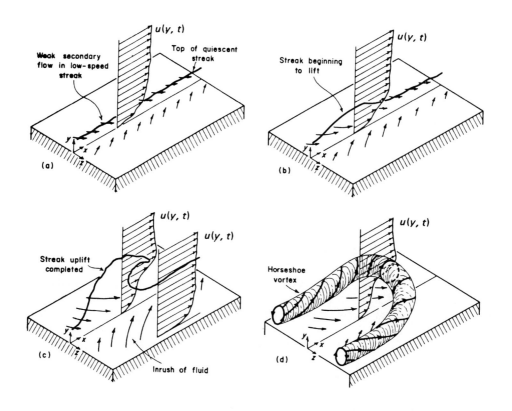

Figure 26. Behavior of sublayer streaks. U is bottom velocity as a function. a-c: to completion of lift-up; d. Horseshoe vortex stage. From Allen, 1982a.

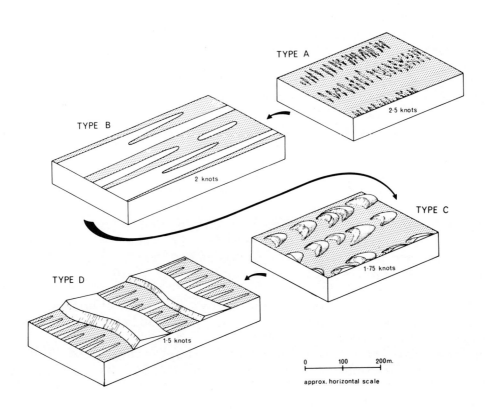

Figure 27. Sand ribbon patterns in the North Sea, as a function of mid-tide velocity. One knot is approximately 50 cm sec$^{-1}$. From Kenyon, 1970.

zones (Fig. 25). As the streak lifts from the bed, it becomes wave-shaped in profile, then develops a downstream overhang like a breaking wave; as it does so, a horseshoe-shaped eddy develops beneath overhang, with ends pointing upstream. Streak-bursting coupled with inrush is a continuous and repeating cycle of events which appears to be controlled by the movement of the larger turbulent eddies past the more retarded fluid in the immediate vicinity of the bed.

Sediment beds may be expected to respond in some semi-permanent manner under the action of the fluctuating forces associated with the sweeps and bursting streaks found in the inner parts of flows. Allen (1982a) argues convincingly that the small-scale streamwise bed lineation seen in flumes, and the correlative primary structure (parting lineation) in sedimentary rocks such responses to turbulent boundary layer flow (Fig. 26). The ridges of the lineation form beneath zones of converging flow in the bed. Coarser grains are concentrated here, since smaller particles are lifted off the bed in suspension. The bimodal shape fabrics reflect the convergence of the flow beneath the rising streak. Allen (1982a) concludes that small scale lineation, combined with parallel lamination, should arise wherever turbulent currents attain or exceed a specific and definable strength. The transverse scale of the lineation may therefore be useful for estimating flow parameters. The lineation is a useful indicator of current path, and the shape fabric associated with it gives the current direction.

## Sand ribbons

Sand ribbons are a relatively common mesoscale, flow-parallel bedform on tide-dominated continental shelves; horizontal spacings range from 2 to 200 m. They are especially abundant on shallow, tide-swept seas (Belderson, 1964; Kenyon, 1970, Werner and Newton (1975). They are most common and most visible where small amounts of sand exist over a coarser substrate; the sand is swept into flow-parallel ribbons between gravel "streets" (Fig. 27). Low, flow-parallel ridges also occur on continuous sand sheets.

McLean (1981) has shown that sand ribbons form in response to horizontal helical vortices in the flow (Fig. 23). The fluid in each cell follows a spiral path; any two adjacent to cells rotate with the opposite sense. Unlike boundary layer bursting cells, the helical flow cells are regularly spaced, and are relatively long lived.

The angle of convergence of bottom flow with the mean flow direction is generally no greater than 10-15°, and often much less. Along separation lines, where the fluid ascends from the bed, the downcurrent velocity increases only slowly with height (Fig. 23). The bottom shear stress along zones of bottom flow convergence is therefore less than the average on the bed. Where the fluid descends towards reattachment lines, the velocity gradient is steep, and the bottom shear stress exceeds the average.

McLean (1981) demonstrates by means of mathmatical analysis (a perturbation expansion) coupled with a flume study that the helical circulation pattern is linked to differences in bottom roughness, which both causes, and is caused by, the circulation. McLean (1981) created downstream variations in bottom grain size in his flume. He found that bottom shear stress was twice as high over gravel zones than sand zones. The gravel strips were more efficient in shedding eddys into the flow. The intensified vertical exchange of water "stiffened" the flow, and caused a steeper vertical velocity gradient over the gravel, so that high speeds were maintained close to the bottom, relative to those over sand strips. The horizontal momemtum gradient, from high-speed water over the gravel to low-speed water over sand, drove a secondary component of circulation as in Fig. 23.

McLean (1981) notes that the very small secondary circulation creates a sizeable difference in the velocity and stress fields over the strips, with boundary shear stress over the smooth region being as much as 50% less than the mean. Sands ribbons are thus regions of much slower sand transport than are the gravel streets that occur between them, which have been swept free of sand.

While sand ribbons might be triggered by random turbulent fluctuations in near-bottom flow, this would be a slow process. However, McLean (1981) suggests that the comet marks, seen on coarse bottoms, may initiate sand ribbon patterns. On gravel beds veneered with sand, comet tails occur behind projecting pebbles and shells. Roller eddies form behind such projections during periods of flow, and the boundary layer separates from the peak of the projection, passes over the eddy, and reattaches to the bed a sent distance downstream. For some distance beyond that, before it grows to equilibrium thickness, there exists a greater than normal stress, hence the bed is swept free of sand in a long streak behind the projection (comet mark). Once the tail of coarser sediment is exposed, the same mechanism responsible for the sand ribbons can cause the feature to extend downstream. If several obstacles are present, their comet marks will create a ramdom pattern of alternating smooth and rough bands like sand ribbons. The tendency toward a preferred spacing would cause some comet marks to disappear or adjust downstream. The resulting sand ribbon field can continue to grow downstream and eventually may appear unrelated to the originating comet marks.

## Flow Oblique Bedforms

Bedforms that trend obliquely with respect to the prevailing flow direction have only slowly been recognized as a discrete class. Stewart and Jordan in 1964, Off in 1963, and Houboult in 1968 described tidal sand ridges, although the critical flow-

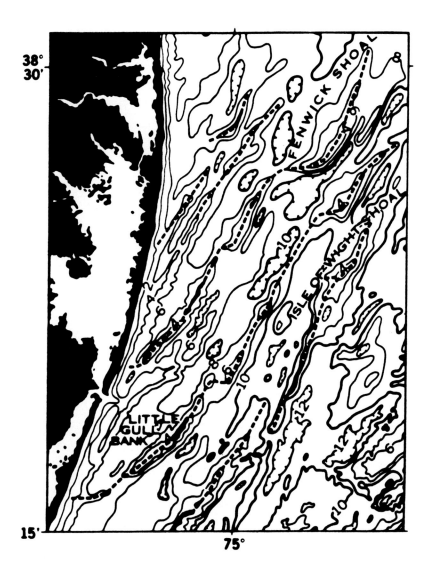

Figure 28. Storm-built sand ridges on the Delaware sector of the North American Atlantic Shelf. From Swift, 1976.

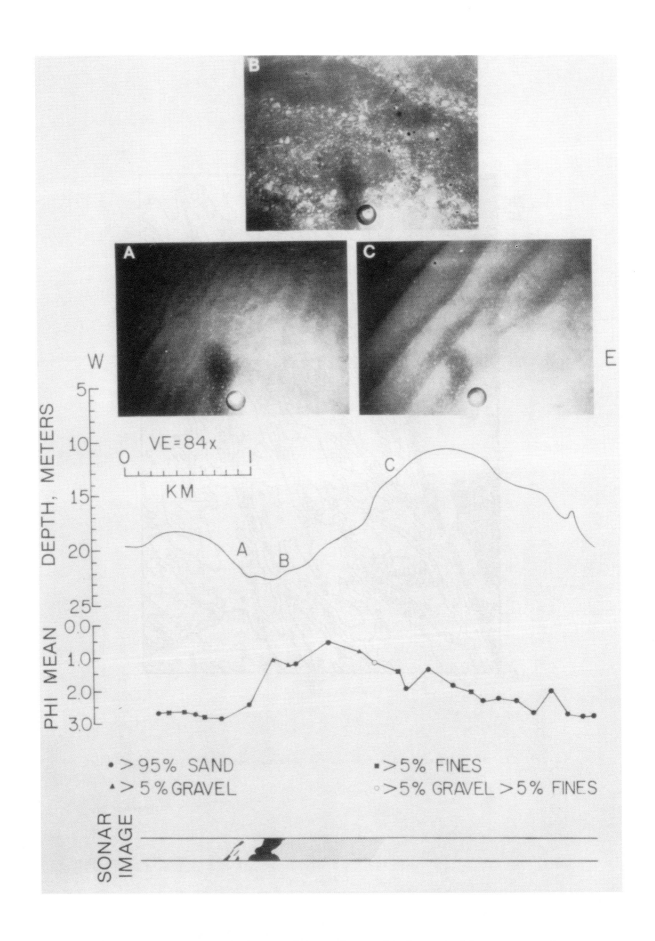

Figure 29. Relationship between topography, grain size and side scan sonar image on storm-built sand ridge, Maryland sector, U.S. North Atlantic Shelf.

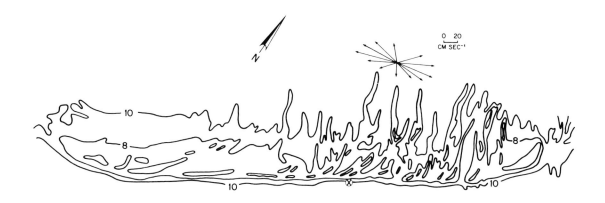

Figure 30. Tidal sand ridge on Nantucket Shoals, U. S. North Atlantic shelf. Contours outline sandwaves as well as ridge crest. Depth in fathoms. Tidal current ellipse based on 12.5 hour record from an electromagnetic current meter mounted 1 m off bottom at point marked "X." From Mann et al., 1981.

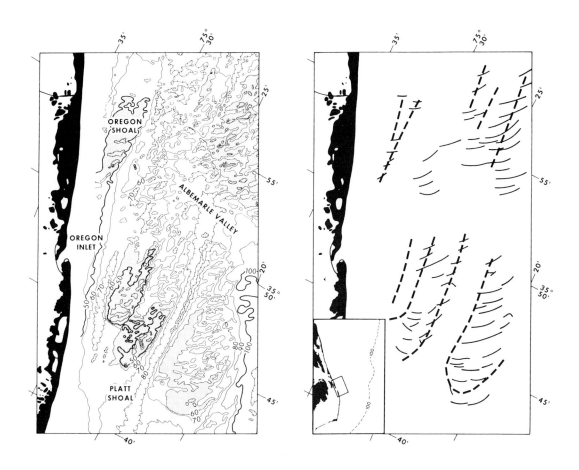

Figure 31. Bathymetric map of the inner shelf, North Carolina sector, U. S. Atlantic shelf. Contour interval 10 ft. Pattern of sand ridges (dashed lines) and sand waves (solid) is indicated in the right hand panel. From Swift et al., 1978.

oblique relationship came to be recognized by degrees, (Caston, 1970, Caston and Stride, 1979; Huthnance, 1972) and was only fully articulated by Kenyon et al. in 1981. Storm-built sand ridges were first described by Swift et al., (1972, 1973) and Duane et al. (1972) from the North American Atlantic Shelf, and later from the Argentine Shelf and East Frisian Shelf of the North Sea (Swift et al., 1978). Here, too, the critical flow-oblique relationships was only slowly recognized (Swift et al., 1978, Figueiredo et al., 1981, Parker et al., 1982).

Sand ridges on continental shelves share the following characteristics. They are large-scale bedforms, with crestal spacings of 1 to 5 km (Fig. 28). Storm-built sand ridges are commonly 10 m high at their peaks (highest points along their crests). Side slopes are low; commonly less than a degree, although where the maintaining processes are vigorous, slopes of up to 7 degrees may occur (Swift and Field, 1981). Sand ridges tend to be asymmetrical. Free-standing sand ridges tend to have steeper downcurrent flanks, but shoreface-attached sand ridges may exhibit reverse asymmetry near the zone of attachment. Sand ridges are texturally zoned, with the coarsest sediment occuring on the upcurrent flank, and the finest on the downcurrent flank (Stubblefield and Swift, 1976, Swift and Field, 1981, Fig. 29). On many ancient shelves, sand ridges developed on muddy surfaces, and when an appreciable portion of the ridge itself was built of less than sand-sized material, this material collected on the downcurrent flank Swift and Rice, 1983. Fine-grained sediment has a lower effective submarine angle of repose than coarse-grained sand, and such sand ridges commonly have reverse asymmetry, with the upcurrent flank steeper than the downcurrent flank.

On open shelves of simple morphology, the obliquity of the ridges with respect to the prevailing direction of flow appears as an obliquity with respect to the shoreline, the shelf edge, and the regional trend of the contours. This is because of the topographic steering of shelf flows, which themselves trend parallel to the contours. The obliquity of sand ridges with respect to the regional trend of the contours is revealed by the fact that the basal contours of shelf sand ridges almost never close. The ridges are instead defined by an echelon kinks in the shelf contours. Where the prevailing direction of flow has been established on modern shelves, the angle of the sand ridge with respect to the shoreline opens into the prevailing direction (Swift et al., 1978). This does not appear to have been the case for ancient muddy shelves, such as the Campanian shelf of the Cretaceous Western Interior Basin (Swift and Rice, 1983; Parrish et al., 1984) where the relationship was reversed.

The relationship of sand ridges to the prevailing flow is sometimes revealed with great clarity, in areas where sand ridges occur in conjunction with sand waves. This is a common phenomena in tide-dominated shelves (Fig. 30), but also occurs on storm-dominated shelves, where the shelf cross-sectional area decreases

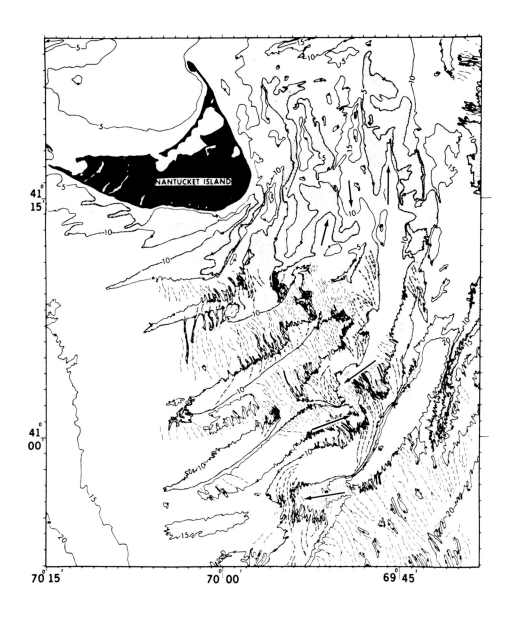

Figure 32. Nantucket Shoals tidal sand ridge field. Arrows indicate sense of residual flow in major ebb-flood channel pairs. Dashed lines are position of sand wave crests. Stippled areas are major highs. Shoreface-attached ridges are presently being formed by process of erosional shoreface retreat. Older ridges to southwest were similarly formed during submergence of ancestral Nantucket land mass. From Swift, 1975, after C&GS map 0708N-53.

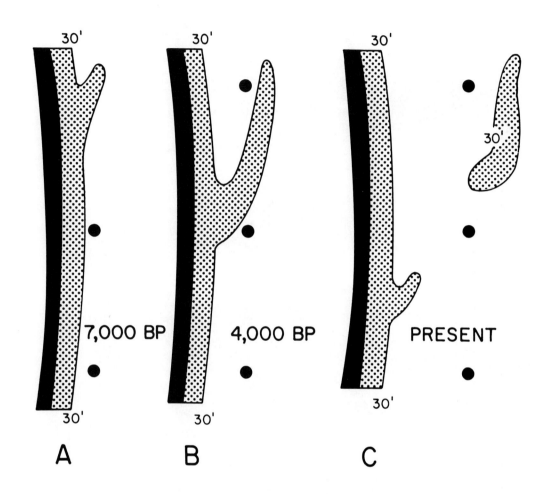

Figure 33. Schematic diagram illustrating shoreface detachment mode of sand ridge formation. Dots are fixed points with which to contrast respective positions of retreating shoreline and migrating ridges. From Swift, 1976.

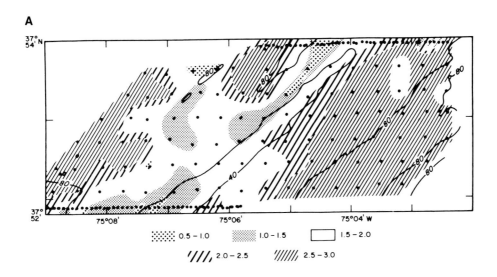
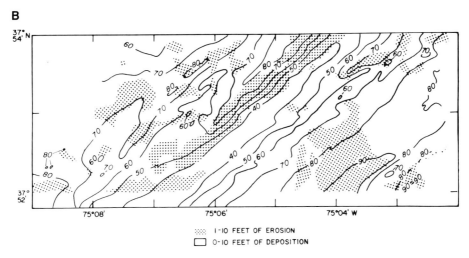
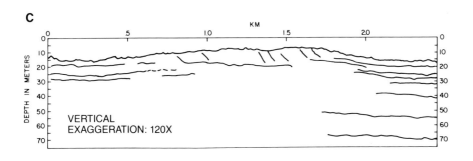

Figure 34. A. Mean grain size in phi units on a sand ridge, Maryland Sector, North American Atlantic Shelf. B. Morphologic change over a 33 year period. Contours in feet. C. Seismic profile, revealing internal structure. Vertical exaggeration is 10:1; Strata dip 1-2° to the south. From Swift and Field, 1981.

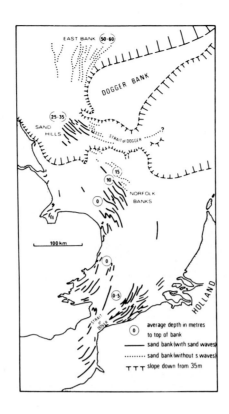

Figure 35. Tidal sand ridge fields in the North Sea. Dash lines indicate moribund ridges. From Kenyon et al., 1981.

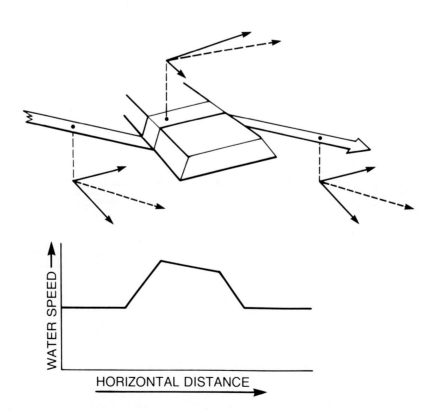

Figure 36. Schematic diagram indicating pattern of flow over a sand ridge. From Parker et al., 1982.

in a downstream direction, so that the flow must accelerate (Fig. 31). On tidal ridges, the sandwave orientation rotates as the crest is traced from the ridge flank up to the ridge crest. On the lower flank, the sandwave crest is almost orthogonal to the ridge, but as the sandwave crest is traced towards the ridge crest it curves into parallelism with the crest (Fig. 30). On such shelves the shoreline, sandwave crests, and sand ridge crests form a topographic system in which no trends are parallel or perpendicular; all angles are obtuse or acute (Fig. 31). In both the tidal and storm-dominated cases, the sandwaves are basically obliquely oriented with respect to the ridge crest. Sand waves are flow-transverse bedforms and therefore the sand ridges must be oriented obliquely with respect to flow. The sandwaves are oriented obliquely with respect to the shoreline because the formative flows are down welling geostrophic storm flows and the near-bottom water is moving obliquely offshore.

Sand ridges on modern shelves appear to be responses to a transgressive sedimentary regime. As sea level has risen through the Holocene, the coastal shoreface profile has undergone erosional retreat and the ridges are composed of debris from the retreat process. Sand ridges form on retreating shorefaces in response to both tidal and storm-dominated regimes (Figs. 28, 32). They migrate offshore and downcurrent as they form (Fig. 33), as indicated by their internal structures (Fig. 34c). As the shoreline becomes more distant and the water column deepens, the ridges continue their activity. Topographic time series (comparison of successive bathymetric surveys) show that ridges continue to grow upward (Sallinger et al., 1975), and they continue to migrate, with erosion on the upcurrent side, and deposition on the downcurrent side (Fig. 34b). Radiocarbon dates from storm-built sand ridges on the Atlantic continental shelf indicate that ridges with crests at 10 m and bases at 20 m began to form 4,000 years ago (Swift et al., 1972b), while ridges at 50 m began to form 10,000 years ago. In general, continental shelf ridge fields have evolved over the course of the Holocene transgression.

Kenyon et al. (1981) have shown that the deeper fields of tidal sand ridges in the central portion of the North Sea (Fig. 35) no longer bear sandwaves on their backs. These authors suggest that such sand ridges are moribund, or no longer subject to vigorous formative processes. McClennen (in Swift et al., 1973) and Stubblefield and Swift (1976) showed that storm built sand ridges on the central New Jersey shelf in many cases have internal strata steeper than their surface slopes, suggesting that formative processes for these ridges also were most intense during the early, shallow water stage. Sand ridges on ancient shelves are also related to transgressions, but the sequence of events was rather different; see the discussion on later pages.

A variety of explanation have been proposed for the origin of sand ridges on continental shelves, including helical flow cells

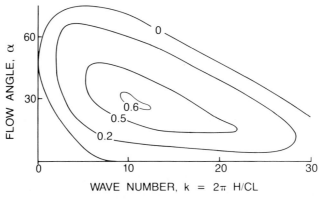

Figure 37. Growth rate of sand ridges, as a function of angle of ridge with respect to flow, α, and wave number, K. After Huthnance, 1982a.

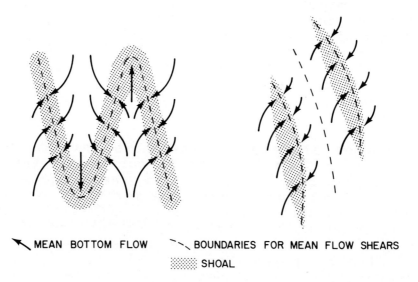

Figure 38. Patterns of mean bottom flow for (left) an ebb-flood channel pair and partition ridges on the inner shelf, and (right) sand ridges dominated by rotary tides on the outer shelf. From Swift, 1976.

(Houbolt, 1968), the interaction between a long ridge and cross ridge components of flow (Smith, 1969), and differential sediment transport rates (Swift and Field, 1981). Sand ridges on the Bering Continental shelf have been considered to be relict littoral bedforms (Field et al., 1981).

The problem seems to have been solved by Huthnance (1972, 1982a, and 1982b; who alone of the above workers has proposed a theory of formation which accounts for all of the aspects of sand ridges behavior described above. In Huthnance's approach, a sand ridge must form in an alignment that is oblique with respect to the prevailing flow direction. Ridges cannot form parallel to flow because no sand would be transported from their flanks to their crests. Sand ridges cannot form at their characteristic spacings in an orientation that is transverse with respect to flow because the flow would accelerate across their crests and the crests would experience erosion, not deposition.

Should flow shift until it were at right angles to sand ridge crests, the large scale sand ridges (2-4 km spacing) would slowly break up into a series of sand waves (100 m spacing). At sandwaves scales there is a sufficient phase shift between bottom topography and the bottom shear stress profile to aggrade sand crests. At sand ridge scales, however, the phase shift is no longer large enough for such aggradation to occur.

Interaction between bedform and flow in such an alignment is illustrated in Fig. 36. In this illustration the flow, obliquely approaching the ridge, can be resolved into an across-ridge component and a ridge-parallel component. As the flow moves up over the upcurrent flank of the ridge, the cross-section area decreases and the cross-ridge component of flow must accelerate in order to satisfy continuity. Consequently flow veers towards the ridge crest and the resultant speed increases (Fig. 36b). However, on the crest, the along ridge component of flow is reduced by frictional drag in the shallow water column and the resultant speed of (Fig. 36b) the flow wanes across the crest. As the flow passes off the ridge crest, it expands, decelerates, and veers towards its original orientation. But because the flow has decelerated on the crest, it can deposit there the sand entrained during acceleration up the upcurrent side, even though the crestal velocity is at all times higher than the velocity of the flow off the ridge.

As well as an optimum angle with respect to flow, there is also an optimum spacing. At small spacings (large k), slopes for a given bedform height are relatively steep, and transport of sand upon to the bank by the mechanism described is inhibited, although flow transverse bedforms (sandwaves) may form. "Down hill" sediment transport predominates. At very large spacings (small wave number) bottom slopes, and velocity changes over the slopes are so gradual, that the flow deceleration over the crest is negligable.

Huthnance (1982a) has shown by means of a stability analysis, that the two critical dependent variables in sand ridges formation are the anlge of the ridge axis with respect to flow (α), and ridge spacing, formulated as a dimensionless wave number, k given by

$$k = \frac{2\pi}{CL} H$$

where π is 3.14, H is water depth, C is the drag coefficient, and L is the distance between sand ridge crest. These variables are in turn functions of the Coriolis term, a friction term, the ratio of the axes of the tidal ellipse, and the exponents of the drag coefficient in the governing equation. Huthnance (1982a) shows that for given values of these parameters, the field of ridge angle and ridge spacing (Fig. 37), can be contoured for the growth rate σ(α,k). Fig. 37, and related figures for different values of the controlling parameters describe the growth of sand ridge fields. Fig. 37 indicates that if periodic flows occur across a sandy shelf floor, and if these flows are in excess of the threshold of sediment transport for the size class of sand present, then whatever initial topography is present on the sea floor will be enhanced. These initial bumps may range from centimeters to many kilometers wide, and may have various shapes. They may be only centimeters high. However, if ample time is allowed (thousands of years), bumps of the appropriate spacing and orientation will grow faster than other bumps, and eventually an ordered sand ridge field will emerge. Thus the characteristic large spatial scales and the oblique orientations with respect to flow are not superficial characteristics of shelf sand ridges; they are innate.

Huthnance's model for sand ridge formation was designed to provide a fluid dynamical basis for V.N.D. Caston's (1972) description of tide-built sand ridges. However, as noted in the previous chapter, along-shelf geostrophic flows can be considered as stochastic analogs of shelf tidal currents (or tidal currents can be considered to be rigorously periodic "storm" currents). Analysis of the storm-built sand ridge topography of the Atlantic continental shelf (Figueiredo et al., 1981) suggests that the Huthnance model is applicable here also.

Within the broad divisions of storm-built and tide-built sand ridges, further subcatagories may be established (table I). The storm-built sand ridges of the North American Atlantic continental shelf are being formed by the process of erosional shoreface retreat and rest on an unconformity cut into a subaerial surface (Swift et al., 1976). Sand ridges on the Campanian shelf of Wyoming were also formed during a marine transgression, but they formed seaward of the maximum seaward position of the shoreline during the preceeding regression; they rest on a marine surface of non-deposition (Swift and Rile, 1984). Field et al. (1981) have described sand ridges on the Bering Shelf, for which they infer a relict origin. They have suggested that these are littoral sand

TABLE I.  SAND RIDGE TYPES

1. Storm-built sand ridges

    a. Storm Sand ridges formed by erosional shoreface retreat

    b. Storm Sand ridges formed on marine erosional surfaces

2. Tide-built sand ridges

    a. Tidal sand ridges formed by erosional shoreface retreat

        1. Sand ridges formed by ebb channel--flood channel interactions

        2. Sand ridges formed by rotary tidal currents in the lee of promontories (banner ridges)

        3. Sand ridges formed by rotary tidal currents on the open shelf

    b. Tidal sand ridges formed on marine erosion surfaces

bodies (barrier spits or islands) that were overstepped during the Holocene transgression. However, the relationship between the axes of the ridges and the orientation of superimposed bedforms suggests that these features are oriented obliquely with respect to flow and may instead be responses to the Huthnance mechanism. The velocity field in the Bering sea consists of a northward flow driven by differences in dynamic height between the Pacific and Arctic ocean, with episodic northerly intensifications of flow resulting from storm winds.

Tidal sand ridges in estuaries and on the shallow inner portion <30m) of tide dominated shelves form primarily as partition shoals separating ebb- and flood-dominated channels. Paired ebb- and flood dominated channles occur in areas where there is a steep horizontal gradient in the tidal phase; for instance at an estuary mouth. The velocity of the tidal wave varies with the square root of water depth; the shelf tidal wave therefore becomes retarded as it enters the shallow estuary and is common still ebbing from the estuary mouth for some time (typically 15 minutes) after the shelf tide has turned and has began to flood. At such periods, Ebb and flood tidal currents interpenetrate. A topography of interfingering ebb and flood channels develops through constructive feed back between the flow and the cohesionless substrate; the greater depth of the channel enhances the tidal phase shift and the intensity of the flow, which in turn causes scour on the channel floor. However, the phase lag mechanism operates in conjunction with the huthance mechanism, and tidal flow ellipses are oriented obliquely with respect to the axes of the partition ridges (Ludwick 1970). Tidal sand ridges may also form as banner ridges, extending along coast from rocky Headlands (Pingree and Matlock, 1979).

While table I lists a possible catagory of tidal sand ridges that form on marine erosional surfaces, this catagory has not yet been described in the literature.

## STRATIFICATION

### Physical Processes of Strata Formation

Stratification in sedimentary sequences has long been understood be caused in some way by changing fluid dynamical conditions during the period of deposition. The deposition of sediment from moving water must be a response to the loss of capacity of the flow; that is, with time or in space (down the transport path), the fluid is exerting less stress per unit area of the bed, and hence can suspend or otherwise transport less sediment.

There are two basic ways in which a current can lose capacity and deposit a bed. In the temporal acceleration model for the formation of a bed, a current, (for instance a storm current),

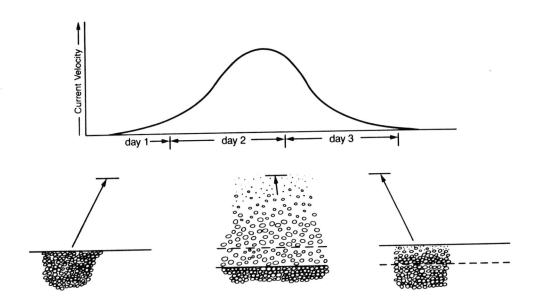

Figure 39. Temporal acceleration model for the formation of a storm bed. From Niedoroda and Swift, in preparation.

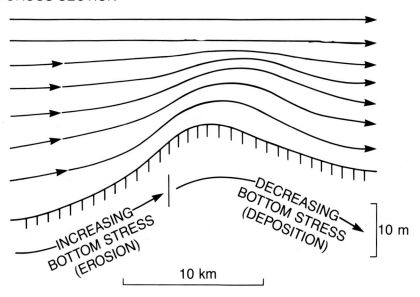
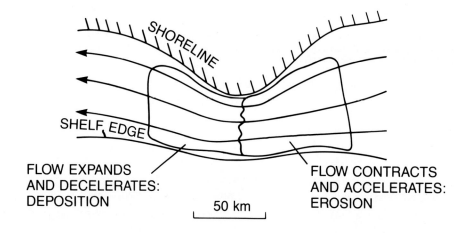

Figure 40. Spatial acceleration models. A: Flow expansion and acceleration over crest and in lee of shelf high results in deposition. B: Flow expansion and deceleration results in deposition in a zone of downshelf widening.

is assumed to accelerate to some value in excess of the threshold criterion for a finite period of time (perhaps several days, for a storm over the continental shelf; Fig. 39). The time history of water speed during a storm event on the shelf is usually more or less Gaussian in nature; that is, the speed usually climbs to a peak, then decelerates to the ambient value. For convenience however, the time history of flow may be thought of as a square wave; that is, the storm is switched on and the current jumps instantaneously, from a low ambient value to the storm value. As it does so, it increases in capacity (ability to carry sediment, measured as total load concentration per unit area); sediment moves from the bed into the flow, to be transported as bed load or suspended load, depending on the grain size frequency distribution of the bottom sediment and the bottom shear stress being exerted by the flow. Very shortly after the increase in velocity however, the boundary layer becomes as loaded with sediment as it can be at that level of fluid power (attains capacity). The bottom is now dynamically armored. Particles are still hopping up off the bed into the flow, but now as many settle out as are entrained per unit time. The storm may continue for days, but if fluid velocity and the resultant bed shear stress do not change no further erosion can occur. If the flow shut off as abruptly as it began, bed load, and then suspended load must settle out, according to their respective settling velocities. The result will be a graded (upward fining) sand bed overlain by a mud bed.

Beds that are mainly entirely as a consequence of temporal deceleration of flow in this manner are relatively common in nature. In some cases, however, flow acceleration and deceleration in space dominates the strata formation process. Erosion and deposition under these conditions is described by the sediment continuity equation;

$$\frac{\partial \eta}{\partial t} = - \varepsilon \frac{\partial q}{\partial x}$$

where $\eta$ is the sediment-water interface, t is time, $\varepsilon$ is a dimensional constant related to porosity, of sediment discharge, and x is horizontal distance. The equation states that the time rate of charge of the level of the water sediment interface, whether undergoing erosion or sedimentation, is equal to the product of a constant and the horizontal discharge gradient. If the flow is contracting in a downstream direction, and therefore accelerating, the sediment discharge gradient is negative (increasing downstream). The near-bottom water arriving at any point along the gradient is constantly under-saturatured with respect to its sediment load and sediment is being eroded (Fig. 40). If the flow is expand (decelerating) the gradient is positive, and sediment is being decreased. Contrary to the case of a square wave storm (temporal deceleration model), the bottom does not become dynamically armored, and will continue to Aggarde throughout the duration of the flow event.

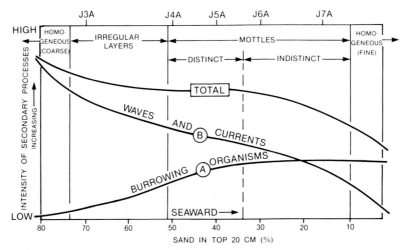

Figure 41. Schematic diagram describing primary structures in continental shelf cores as a function of waves, currents and burrowing organisms. From Moore and Scruton, 1957.

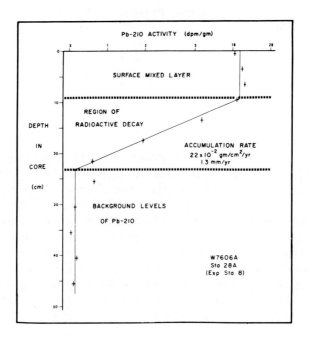

Figure 42. A typical profile of Pb-210 activity for a box core on a muddy shelf. From Nittrouer et al., 1974.

## Biological process of strata formation

Marine invertebrates play a major role in the formation of shelf strata (Nittrouer and Sternberg, 1981; the following discussion is summarized from this source). Benthic organisms displace sediment particles during feeding and locomotion (Rhoads, 1974). Biologic activity has an important effect on strata when particles are moved vertically, because they cross compositional and textural gradients left by physical processes during deposition. Deposit feeding infauna usually dominate sediment reworking (Aller, 1977) and "Conveyor-belt species" which feed at depth and defecate at the surface are especially important group for strata alteration (Rhoads, 1974). Activity of the benthic community decreases rapidly with depth below the sea bed; Myers (1977) estimates the turnover time for a lagoonal sediment to differ from about two days to 2 years between depths of 1 and 10 cm below the surface. On continental shelves, relatively few organisms operate below the upper 10 cm of the sea bed (Mare, 1942), but even slow reworking of sediment will have significant geological effects over a 100 year time scale. Although studies have estimated sediment processing rates for individual organisms which can be found on shelves (Rhoads, 1963, 1967; Gordon, 1966; Nichols, 1974), the rate of community reworking is more difficult to evaluate, especially for deep sediment (below 10 cm where rates are slow).

In the simplest case, sediment reworking can be assumed to resemble a random walk, diffusion process. In fact on time scales of feeding by benthic organisms, particles are moved advectively; conveyor belt species transport particles from depth to the surface. Many other species displace particles laterally, and some may even work sediment downward (Powell, 1977). Aller and Dodge (1974), however suggest that deposit feeding resembles a diffusive process when summed over a time scale that is greater by a factor of 3 or more than the turnover time of the sediment. Another simplifying assumption for a diffusion model of strata formation is that organisms are non-selective in their choice of particles. Recent studies have shown that many deposit feeders are selective, choosing particles according to size, (Whitlach, 1974) Fenchel et al., 1975), density or surface textura (Self and Jumars, 1978). Some organisms have been shown to use specific sizes and shapes of particles to line tubes and burrows (Fager, 1964, Aller, 1977). In the absence of physical processes, selectivity tends to increase the residence time within the zone of biological reworking for selected particles, and to cause the preferential accumulation of non-selected particles (Jumars et al., 1981). For the case of grain size selection, the net result can be biological grading (upward fining) as described by Rhoads and Stanley (1965).

## A model for strata formation

Moore and Scruton (1957) were the first to suggest the formation of sedimentary strata in the marine environment is governed by the rates of biological activity, physical processes, and sediment accumulation (Fig. 41). Nittrouer et al. (1979) and Nittrouer and Sternberg (1981) have devised a framework for the quantification of these relationships by of the examination of profiles of Pb-210 activity. The following discussion has been condensed from this source. Box cores of bottom sediments are collected, and are sampled along a vertical profile at 1 cm intervals. The sample are analysed for the activity of Pb-210, an unstable isotope with a half life of 22.3 years. In most cases Pb-210 values are constant for the first 10 or 20 cm below the surface (Fig. 42). This is the zone of mixing (Fig. 43), where biological and physical processes are very intense relative to the rate of accumulation. Below the zone of mixing, the Pb-210 activity decreases in linear fashion with depth, as a consequence of radioactive decay.

Several models have been developed to explain the vertical distribution of radioisotopes within the seabed (Goldberg and Koide, 1962, Guinasso and Schink, 1975). Radioisotopes such as Pb-210 are irreversibly attached to particle surfaces, and have a known rate of fallout from the atmosphere, and therefore can be used to estimate rates of mixing and accumulation. The steady state profile (ignoring consolidation) for excess activity (above levels supported by the parent isotope) is given by the advection diffusion equation:

$$D \frac{\partial^2 C}{\partial Z^2} - A \frac{\partial C}{\partial Z} - \lambda C = 0$$

Where C is the activity of the radioisotope (disintegrations per minute per gram), D is the particle mixing coefficient ($Cm^2 yr^{-1}$), A is the sediment accumulation rate ($Cm\ yr^{-1}$), $\lambda$ is the decay constant for the radioisotope ($0.693 \cdot half\ life^{-1}$), and Z is the depth below the sediment surface in cm. In general, the surface mixed layer requires the complete equation to describe the profile of a radioisotope, but for relatively low accumulation rates ($A^2 \ll \lambda D$) the accumulation term can be ignored. Below the surface mixed layer where D = 0, the mixing term can be ignored.

Guinasso and Schink (1975) have developed a time-dependent model for biological mixing based on the advection diffusion model presented above. In their model, strata formation is dependent on a non-dimensional parameter G:

$$G = \frac{D_b / L_b}{A}$$

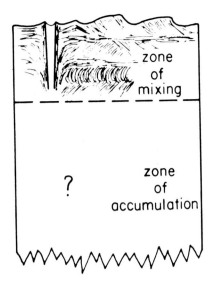

Figure 43. A strata formation model for the sea bed. Particles are actively displaced by biological and physical processes in the zone of mixing. With net accumulation, particles are preserved below, in the zone of accumulation. From Nittrouer and Sternberg, 1981.

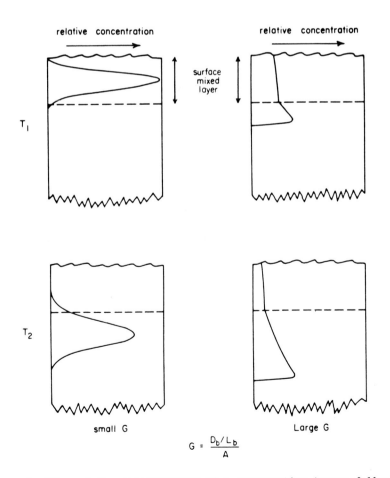

Figure 44. Profiles of the concentration of a conservative tracer following implacement at surface at time $t_o$, for different values of the biological parameter G. $D_b$ is mixing coefficient, $L_b$ is mixing length and A is accumulation rate. From Nittrouer and Sternberg, 1981.

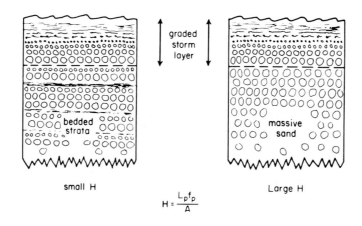

Figure 45. Models of the seabed describing sedimentary structures expected for different values of the parameter H. $L_p$ is depth to which seabed is eroded by storms, $t_p$ is frequency of erosional events, and A is sediment accumulation rate. From Nittrouer and Sternberg, 1981, after Smith, 1977.

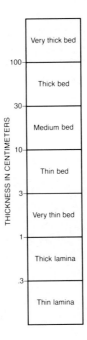

Figure 46. Terminology for strata thickness. From Ingram, 1954.

where $D_b$ is the mixing coefficient and $L_b$ is the mixed layer thickness. The term on the left hand side of the equation is the ratio of the mixing rate to the accumulation rate. Mixing, in the numeration is represented as the volume of sediment reworked per unit area per unit time with units of cm yr$^{-1}$ (the same as the accumulation rate). For small values of G (<0.1; weak mixing relative to accumulation), textural heterogeneities are maintained during their descent through within the mixed layer, and enter the rising zone of accumulation (Fig. 44). For large values of G (>10) the mixed layer is homogenized, and consequently only a very generalized record of strata-forming events is inpressed on the zone of accumulation. The range of G values, from low to high, thus corresponds to the sequence of primary structures ranging from bedded to mottled to homogeneous, as described by Scruton annd Moore (1957; see Fig. 41).

Smith (1977) has suggested a physical mixing parameter, H, whether can defined in a similar fashion:

$$H = \frac{L_p f_p}{A}$$

where $L_p$ is the depth to which the seabed is eroded by storms, and $f_p$ is the frequency of erosional events. The equation does not parameterize the grading of sediment or the size frequency distribution of sediment, and therefore does not allow explict conclusions about selective preservation. However, as Smith indicates, large values of H (intense mixing relative to accumulation) means that small beds would tend to be destroyed as intense events reach deep into the substrate; thick beds would be preferentially preserved. For such thick beds, the top of each bed woulf be stripped off to be reconstituted as the next bed; thus a texturally homogenous sequence of bed bases would result (Fig. 45). Small values of H would allow preservation of much more of the sedimentary record. More of the upper, finer-grained portions of each bed would be preserved, resulting in a well-stratified deposit.

### Stratification terminology

Terminology for stratification has been proposed by McKee and Weir (1953) and has been elabrated on by Ingram (1959) and Campbell (1967). A fundamental distinction occurs between <u>laminae</u> (less than 1 cm thck) and <u>beds</u> (1 cm thick or thicker). The term <u>strata</u> encompasses both classes. Thickness scales for strata have been proposed by both McKee and Weir 1953, and Ingram, 1954; Ingram's scale is used in this paper (Figure 46).

Cross-stratification is the arrangement of layers at one or more angles to the dip of the formation (McKee and Weir, 1953). A cross-stratified unit is one with layers deposited at an angle to the dip of the formation. A cross stratum is single layer of homogeneous or gradient and lithology deposited at an angle to the original dip of the formation and separated from adjacent layers by surfaces of erosion, non deposition, or abrupt change of character (McKee and Weir, 1953). Cross strata may be either cross laminae or cross beds.

A strata set is a group of essentially conformable strata or cross-strata separated from other sedimentary units by surfaces of erosion, non deposition, or abrupt change in character. Cosets consist of more than one set, and composite sets consist of both strata and cross strata sets (McKee and Weir, 1953).

## Classes of Shelf Strata

Einsele and Seilacher (1982) have classified strata as periodites or event strata. Periodites are strata formed by cyclic events such as climatic cycles or astronomical cycles; Einsele, 1982a). Varves belong to this class, as do tidally deposited strata (tidalites; Klein, 1971). Beds formed in response to climatic cycles are beginning to be recognized as the consequence of variations in the earths orbit (Milankovitch cycles; Barron et al. 1985). However, these are commonly fine-grained beds in deep (basinal) settings; on shelves, the climatic signal tends to be obscured as a consequence of the higher depositional rate, and because of overprinting by storm and tidal stratification. Event strata, are formed by random events (Seilacher, 1982a). Important variants of the event class are turbidites, formed by density underflows (Kuenen and Menard, 1952) and tempestites (Ager, 1974), formed by geostrophic storm currents; inundites (Seilacher, 1982a) deposited from terrestrial floods, and seismites; beds formed or altered by by earthquake shock (Seilacher, 1969). Sielacher also describes ashfall beds that result in "tephro-stratigraphy" (Winter, 1977). Should we call them tephrites? At any rate, they are a subclass of event strata.

Event strata may be either single event beds, as most of those described above are, or multiple event beds, (Aigner, 1982). Multiple event beds that are fine grained and are deposited from suspension may be characterized by horizontal "amalgamation surfaces" (Walker, 1965) revealed by sharp, slight changes in grain size, as when two turbidites or two tempestites are stacked on top of each other in such rapid sequence that no intervening clay is deposited. Coarse sand beds are deposited from bed load by migrating bedforms during successive events, are composed of cross-strata set, in which the multiple event origin is indicated by "reactivation surfaces" separating cross-strata sets (Collinson, 1970).

The genetic stratification classes of Einsele and Seilacher (1982) are useful in ordering our thinking. However, a simpler, operational descriptive scheme is more directly applicable to sedimentological problems. In this kind of classification, names are

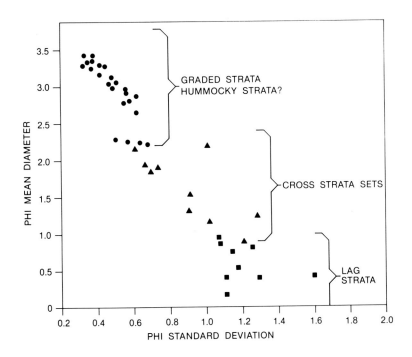

Figure 47. Strata types as a function of mean diameter and standard deviation. From Swift et al., in press b.

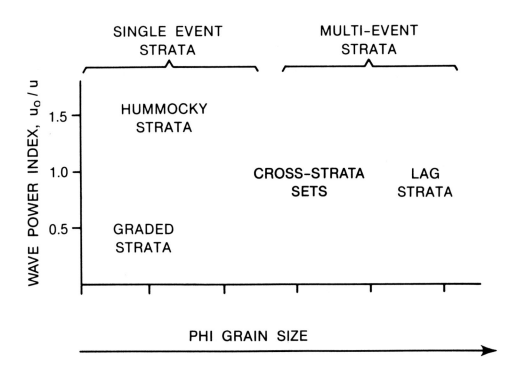

Figure 48. Continental shelf strata types as a function of the strength of wave orbital velocity, Uo/U, and grain size. From Swift et al., in press b.

TABLE II. Strata Classes

(Subclasses important on the shelf are capitalized)

Periodites

Tidalites

    TIDAL CROSS-STRATA SETS

    TIDAL LAG STRATA

    TIDAL ACCUMULATION STRATA

Event Strata

    Single event strata

        Turbidites

        Inundites

        Seismites

        Tephrites

        Tempestites

            HUMMOCKY TEMPESTITES

            GRADED TEMPESTITES

    Multiple event strata

        Multiple event tempestites (amalgamites)

        TEMPESTITE CROSS-STRATA SETS

        TEMPESTITE LAG STRATA

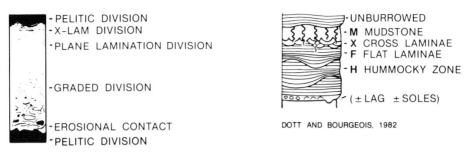

Figure 49. A: Idealized tempestite sequence. From Aigner, 1982. B: Idealized hummock sequence. From Dott and Bourgeois, 1982.

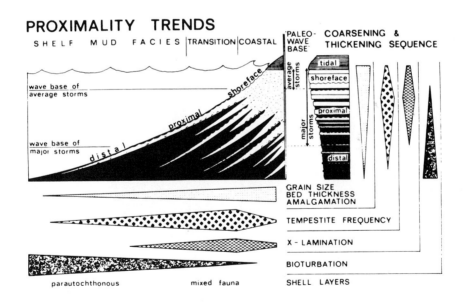

Figure 50. Proximality trends in modern tempestites in the Helgoland Bight. From Aigner and Reineck, 1982.

T. Aigner & H.-E. Reineck: Proximality Trends in Modern Storm Sands from the Helgoland Bight (North Sea) and their Implications for Basin Analysis.

Figure 51. Box core transect across the Helgoland Bight, North Sea, showing basinward (to left) thinning of tempestite sands. From Aigner and Reineck, 1982.

assigned on the basis of observed traits. Genesis of the strata can then be inferred, but by separating observation from inference, circular reasoning is avoided. Two fundamental criteria on which to base on operational descriptive classifiation are grain-size and velocity structure (Swift et al., in press). Grain size determines whether the bed accumulates vertically, from the fallout of a suspension, whether it progrades, as cross strata deposited as a migrating bedform, or whether it forms as an in situ lag accumulate, by the subtraction of finer particles. The importance of grain size is underscored by the separation of primary structures by grain size in shelf beds (Fig. 47); Hummocks cross strata sets and eroded all occur within characteristic grain-size ranges. The term "velocity structure" refers to the relationship between the wave orbital current component $U_o$, and the mean current component U. It is important because it allows us to separate a diagnostic class of shelf strata, namely hummocky strata, from other classes.

Four basic classes of shelf strata can be defined on the basis of these variables usually, graded strata, hummocky strata, cross strata sets and lag strata subscript V(Fig. 48). The ranges of the diagnostic variables are well known from the study of modern shelves, but the boundaries of the stability fields have not yet been defined by laboratory experiments, and have therefore been omitted. Genetic and descriptive strata terms are combined in table 2.

## Graded Strata

Fine- to very fine-grained beds that fine upwards are common in estimated shelf settings. They have been described in the mesozoic limestones of Germany (Aigner, 1982a) in the Miocene sandstones of Morocco (Ager, 1972); in the Cretaceous of the U.S. Western Interior (Hudelson et al., in preparation) and from modern shelf floors, on the Brazilian Shelf (Fiqueredo et al., 1981), in the Bering Sea (Nelson, 1982), in the German Bight of the North Sea (Ager and Reineck, 1982). Graded shelf beds are characterized by a vertical succession of primary structures very similar to the Bouma sequence (Bouma, 1967) that characterizes turbidites (Fig. 49). These strata begin with sharp erosional bases that may exhibit sole marks (flute cast, groove casts or tool marks. Lenses of shell hash or other lay accumuates may sometimes rest on the erosional surface. In the main body of the bed, a massive division (not always present) may be succeeded by a plane laminated division, a ripple cross-laminated division, and a mudstone division. Internal structures includes ball and pillow structure due to intense loading of the sand into the underlying sand, amalgamation surfaces, shale clasts, and The cross-laminated division may have either climbing or horizontally migrating ripples. Internal ripples are usually asymmitical, while ripples on the bed top are more commonly symmetrical. Graded shelf beds are deposited from suspension by either tidal or storm flows (see later discussion). These flows are presumably combined flows, but the absence of hummocks suggests that the ratio of the wave orbital current component to the mean flow component is relatively low.

Graded shelf beds typically occur in an inner shelf setting, where they pinchout seaward into shelf rivers, and thicken landwards into an amalgamated hummocky sequences (Fig. 50, 51).

## Tempestites Versus Turbidites

Even since the recognition of a distinctive deepwater facies characterized by turbidites (beds deposited from turbid density underflows) there have been periodic attempts to equate graded bedsin continental shelf settings with turbidites (Van Straaten, 1959; Kelling and Mullin, 1975, Hamblin and Walker, 1979). In 1967, Hayes described the impact of the Hurricane Carla on the Texas inner shelf, and proposed a model for shelf density underflows in response to coastal storm surge. Hayes (1967) described a graded post-Carla bed on the inner shelf floor. He argued that as Carla moved north across the Texas shoreline, the storm surge filled the coastal lagoon (Laguna Madre) to the bankful stage and beyond; turbid lagoonal water burst out through temporary inlets, generating density underflows that swept down theshoreface and across the shelf, depositing the observed graded bed.

There are some serious problems with this scenario. Morton (1981) in a revisionist analysis, describes the tides in Laguna Madre as being below normal during Hurricane Carla. He notes that barrier washover channels do not occur where the main density underflows were beleived to have transpired. He reports the shelf current during the storm to have been running along coast to the southwest, and describes them as wind-driven geostrophic currents.

Beyond the issue of the Carla event, there are general problems of a theoretical nature concerning the likelyhood of density underflows on continental shelves. Turbidity currents are turbulent sediment transporting systems that owe their existence to potential (gravitational) energy released by the sediment load rather than energy transmitted by the fluid (Middleton, 1966). The term "autosuspension" (Bagnold, 1962) has been suggested to describe the feedback nature of the transporting mechanism. The sediment is suspended by fluid turbulence which itself is generated by the motion of the current. The current in turn is driven by thedensity the suspension. Bagnolds' autosuspension model has become the subject of heated debate (Pantin, 1979, 1983; Southard and MacKintosh, 1981; Parker, 1983), but both its supporters and critics seem to agree on at least the qualitative nature of the process.

A density underflow driven by autosuspension is subject to certain limitations. The threshold concentrations reportedly required to maintain the process are very high, on the order of $1.0 gm\ cm^{-3}$; (Middleton, 1966, a, b, c; Lowe, 1976, 1982) several orders above those commonly reported in the storm boundary layer (Cacchione and Drake, 1980, Clarke et al., 1982; Butman et al., 1979). Furthermore, because grain settling velocity must be overcome, the stirring velocity required (to maintain the suspension) increases rapidly with increasing grain size (Pantin, 1979;). In order for the density underflow to be maintained, "ignition" must occur; The mean value turbulent velocity component, $U'$, which is a fraction of the mean flow, must equal the mean value of the settling velocity for the grain sizes present (Pantin, 1979). Once ignition is achieved, a turbid underflow, having attained high velocities, can travel for long distances on a very slight slope (less than 1°) without mixing with the overlying fluid (Middleton, 1966). However, a significantly steeper slope is required to achieve ignition; the value of the slope depends on the concentration and grain sizes available (Pantin, 1979, table III, this chapter). In general, a partially filled, steep canyon head is an excellent place for a slump to turn into a high velocity, sand-mud density underflow; a shelf is not. Mud-rich shelves, whose storm boundary layers might contain especially dense suspensions, are incredibly flat; gradients are measured in small fractions of a degree. Sandy shelves locally attain slopes of a few degrees, but have less dense storm suspensions, and need very high stirring velocities.

Three possible mechanisms exist for generating turbidity currents on continental shelves; earthquakes, sediment discharge plumes at river mouths, and storm resuspension. The effect of an earthquake on the California continental shelf has been examined by Field and Hall (1982), who observed a series of earthquake-generated slump scars by means of side-scan sonar. Zones of apparent fluidization were in no case more than 100 m long; while the muddy sands of this shelf area may have been mobilized, autosuspension was apparently not achieved. Abundant slump scars are visible on the muddy, steep, outer continent of shelf seaward of the Mississippi Delta, but there is no clear association with earthquakes, and in any case, the movement is typically a slow creep (Lindsay et al., 1984).

Marine deltas cannot directly generate density underflows; suspensions in the muddiest rivers are less dense than salt water, and run out over the sea surface as buoyant spreading plumes. However, indirectly generated density underflows may be possible in such a setting. If they occur, however, they would be very different from the high velocity, impulsive turbidity currents believed to operate in deep marine environments. The rain of river sediment during flood from the surface plume into the underlying salt water would add to the vertical density gradient caused by the salinity contrast, and the resulting baroclinic current would, to a certain extent, be turbidity-driven. However, even if this flow component should become dominate, it is unlikely that the extreme concentrations required for autosuspensions would result; a relatively low velocity, continuous flow would result that would decay as the turbid water moved out from under the plume and

Table III. Characteristics of Geostrophic Flows Versus Initiating Conditions for a Turbidity Current of Finer Sand

|  | Geostrophic flow (observed) | Density underflow (required) |
|---|---|---|
| Mean Speed | 30-60 cm sec$^{-1}$ |  |
| Instantaneous Speed | 50-150 cm sec$^{-1}$ | 100-300 cm sec$^{-1}$ |
| Sed. Conc. | 50-150 mg l$^{-1}$ | 1000 mg l$^{-1}$ |
| Slope | <1° | >1° |
| Source | Drake and Cacchione, 1982. Cacchione et al., 1982, Clarke et al., 1982. | Pantin, 1983 Middleton et al., 1966a, 1966b, 1966c Lowe, 1976, 1982 |

sediment rained out of suspension. To date, no such flows have been observed, not even on the Guiana Coast, downstream from the Amazon River mouth, where the sediment suspensions are locally high enough to achieve "fluid mud" status (Wells et al., 1981). Similarly, a low velocity, continuous turbidity-driven current might occur on a shoreface mantled with fine sand and mud, during peak storm flow, which a downwelling coastal jet might develop. Again, such a current would most likely be subcritical, in the sense that the autosuspension criterion would not be likely to be achieved, and would be continued as long as wave resuspension of bottom sediment contued to load the bottom boundary layer with sediment, rather than spasmodic. Storm flows have triggered failures on the underconsolidated Mississippi Delta shelf by means of rhythmic wave loading Proir and Coleman, 1977, Lindsay et al., 1984), but the resulting movement has been slow creep, rather than a turbidity current.

The preceeding discussion suggests that turbid density underflows due to storm resuspensions are probably not major depositional agents on continental shelves. At our present level of knowledge, however, it seems inadvisable to rule them out altogether; probably almost any sort of submarine phenomenon has happened at some time and place in a shelf setting. Perhaps the most significant objection to the density underflow theory for the origin of graded shelf beds is the scientific "principle of parsimony". We don't need to call on something that we have never observed (turbid density underflows on the shelf) when something that we have observed frequently, geostrophic storm flows, explains the characteristics of these beds equally well.

If tempestite and turbidite deposits tend to be wave ripples and mutually exclusive facies, how do we tell them apart? Hummocky cross strata sets are an excellent criterion, and if hummocks are present, deposition may be presumed to have occurred in response to geostrophic storm flows, combined with wave orbital currents. However, if hummocks are missing, the answer seems to be, "with some difficulty." Many of the criteria cited in the past, such as grading, sole marks, and dewatering and escape structures are common to both facies, as noted by Seilacher (1982b). Seilacher (1982b) attributes a higher degree of alignment of paleocurrent indicators to event deposits and argues that flute casts, current ripples and climbing ripples are uncommon in the cross-laminated division, but these criteria stem from a lack of understanding of the storm hydraulic regime; wave orbital currents do not occur by themselves, but in conjunction with a unidirectional flow component. Climbing ripples, flute casts, and current ripples have all been observed in tempestites elsewhere (Hudelson et al., in preparation). Deep water fossils or trace fossil may be more useful criteria. Seilacher (1982b) notes that tempestite tops tend to be more sharply defined than the tops of turbidites. Seilacher (1982) argues that the sharp tops result from the fact that storm shelf sediments are strongly pre-sorted, so that an interval of non-sedimentation separates the deposition of storm sands and

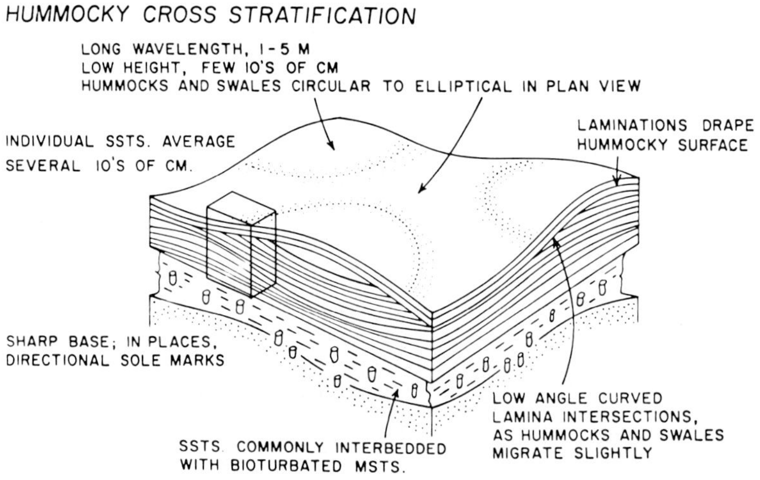

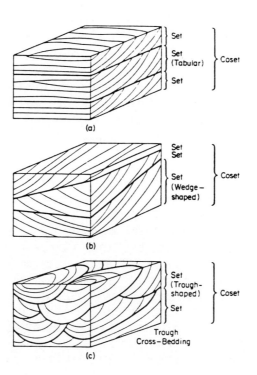

HUMMOCKY CROSS STRATIFICATION

Figure 52. Conceptual model of hummocky cross-stratification. Unpublished diagram of R. Walker. Rectangular volume shows scale of typical box core.

Figure 53. Terminology for cross-stratification. From Blatt et al., 1980, after McKee and Weir, 1953.

storm muds. Turbidity currents, in contrast, carry materials that may have been pre-sorted in their shallower depositories, but have become thoroughly mixed in the slumping event that initiated them so that for turbidity currents, sedimentation goes on without interuption during the waning of the event. Tempestites tend to carry such enigmatic structures as Kenneya ("microripples") and Aristophycus ("dendroid structure") which are probably dewatering structures (Seilacher, 1982b).

Orientation of sole marks may be a valid criterion, but it needs to be used with care. Turbid density underflows, in the absence of geostrophic driving mechanisms, should be oriented down the paleoslope. On the inner shelf, storm downwelling may also drive bottom flows downslope, but on the central and outer shelf, geostrophic flows might be expected to trend more nearly alongshore.

## Hummocky shelf strata

Hummocky strata are fine- to very fine-grained sandstones or course siltstones that are characterized by low-angle (typically less than 15°, wedge-shaped cross-strata sets (Dott and Bourgeois, 1982). The cross-strata sets are organized into antiforms (hummocks) and synforms (swales). The antiforms are defined primarily by the erosional surfaces that truncate the low angle cross-strata sets; lamina within the hummocks tend to be discordant with respect to these surfaces. Lamina sets fan apart from the hummock margins into the intervening synforms (swales); in the swales, the lamina and the erosional surfaces are only slightly discordant to each other or are parallel. Hummocky strata are deposited by storm flow regimes in which the wave orbital current component is high relative to the mean flow component (Swift et al., 1983; Fig. 48; this paper). These are generally assumed to be strom-driven flows, but there is no theoretical reason why wave-modulated tidal flows could not produce hummocks in fine sand. Hummock strata, like graded strata, are vertically aggrading beds that accumulate from suspensive fallout. Like graded strata, they are characterized by a Bouma-like vertical sequence of primary structures. In hummocky strata however, the lower part of the parallel laminated division is hummocky, and this zone is generally very thick relative to the other divisions (Figures 49, 52). Hummock architecture suggests that as the flow began to wane, and hummocks began to grow upwards from the sea floor, hummock growth alternated with hummock erosion so that the antiform structure is defined by erosional surfaces. The conformable fanning lamina sets within sequences indicates that during the latter stages of hummocky bed formation these features aggraded continuously.

## Cross-strata sets

Beds that are somewhat coarser (medium to coarse sand) than hummocky or graded beds tend to form cross-strata sets. In a sense these are lag deposits. As grain size increases, the ratio of bed load to suspended load increases. Finer grains have been winnowed out to form hummocky or graded tempestites further down the transport pathway. The coarse concentrate responds to flow by the deforming of its bed into flow-transverse bedforms. These tend to survive from event to event, hence are multi-event strata, that grow by lateral accretion during the process of bedform migration, rather than by vertical accretion. In this paper the term lag will be reserved for deposits that are winowed in situ but do not migrate.

The mechanism that produces most cross-stratification is avalanching down the lee face of the bedform (Blatt et al., 1980); the following description of the avlanching process is condensed from this source. Avalanching takes place in response to oversteepening of bedload at the brink (Fig. 14) of the bedforms (see Allen, 1982, p. 150). At low rates of bedload movement, avalanching occurs at more or less regular intervals. In the periods between avalanching, fine-grained sediment accumulates by settling on the lower part of the lee slope. In this case, the cross-lamination shows a distinct alternation of grain sizes. Sorting of sediment by size is also produced by the avalanching process itself; many individual cross-lamina are coarser toward the base and show a slight reverse grading across the lamina. Such sorting processes are most marked at low rates of avalanching and become less effective at high rates. At the highest rates of bed load movement, sediment avalanches more or less continuously down the lee slope and the coarsest particles are found at the top, not at the bottom of cross-laminae. The exact mechanism that produces size sorting during avalanching is still not well understood.

The fundamental nomenclature for beds with cross-stratification has been developed by McKee and Weir (1953). A single group of cross-strata, bounded by bedding planes, is called a set. A group of similar sets not separated from each other by any other major discontinuity is called a coset. A set with an upper surface that preserves the shape of the original bedform is called a form set. Cross-stratification may be classified by means of several criteria that includes a) the scale of the sets, (b) the slope and attitude of the cross-strata, and c) the slope and nature of the lower and upper bounding surfaces of sets. Tabular cross-strata sets are bound by plane parallel surfaces; wedge-shaped cross-strata sets are bound by plane, non-parallel surfaces, while trough cross-strata sets have lower bounding surfaces that are trough shaped (Fig. 53). Trough cross-strata sets are characteristic of megaripples; sandwaves, with their lower sinuousity commonly result in planar cross-stratification. Tabular and trough cross-strata sets form a continuum paralleling the continuum between low sinuousity and high sinuousity bedforms (Fig. 54).

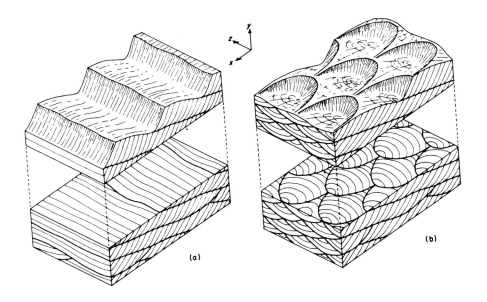

Figure 54. Schematic forms of subcritical cross-stratification in relation to the shape of the parent bedforms. From Allen, 1982a.

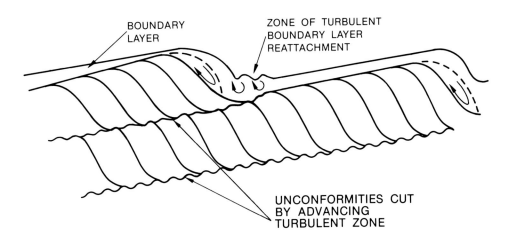

Figure 55. Creation of lower boundary unconformity of migration strata by advancing reattachment zone.

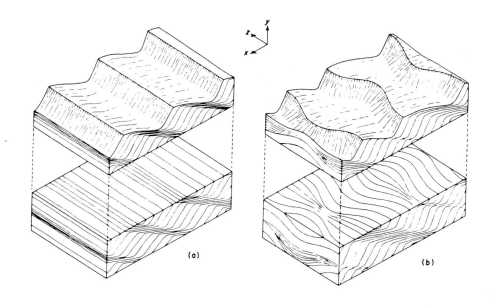

Figure 56. Schematic forms of supercritical cross-stratification in relation to the shape of the parent bedforms. From Allen, 1982a.

"Trough" and "tabular" are terms that are often confused with "planar" and "tangential." The latter terms, however, refer to the relationship of the cross-strata to their lower bounding surfaces. The two end-member relationships are a consequence of the ratio of bedload to suspended load (high for planar, low for tangential, where much sand sweeps over the bedform in suspension, to rain out at the foot of the avalanche face).

A further important characteristic of cross-stratification is its angle of climb. Allen (1982a) describes the geometry of cross-strata sets as dependent on the relationship between the angle of climb, $\zeta$, and the angle of the stoss slope, $\xi$. If the ratio of suspended load to bedload is low, and the downcurrent gradient in fluid power expenditure and sediment discharge is gentle, then the stoss slope will be scoured faster than it is agraded, and the relationship $o<\zeta<\xi$ will prevail (Fig. 54). This is by far the most common condition. In this regime, a portion of each cross-strata set is destroyed by scour at the zone of turbulence associated with the next bedform that migrates over it; this zone of turbulence occurs just downstream of the lee side roller eddy, at the point of boundary layer reattachment (Fig. 55), where the reformed boundary layer has not yet regained its equilibrium thickness, and hence has a steeper velocity profile and a higher bottom shear stress (Fig. 54). Thus the bounding surfaces in this class of cross-stratification (subcritical cross-stratification of Allen, 1982) are erosional surfaces and may be marked by shell lags. As much as 30% of the underlying bed may be sheared off in this fashion. In the special case of $o=\zeta<\xi$ cross-strata are completely destroyed by scour in the troughs, only form sets can occur, and fossilization requires that bedload movement ultimately stop abruptly.

The case where $\zeta=\xi$ represents a critical condition, for stoss slopes neither gain nor lose sediment. However, when the angle of climb $\zeta$, exceeds the slope $\xi$ of the stoss, then strata are preserved on both stoss and lee slopes, and gradational sets accumulate (Fig. 56). Supercritical cross-stratification occurs primarily in cross-strata sets built by current ripples. It is a minor but diagnostic structure of single event beds, and is a consequence of rapid deposition and the high horizontal discharge gradient characteristic of these beds.

The internal geometry of cross-strata sets are much more complicated than indicated by the preceeding discussion. More than one order of moving bed features is usually present at a time. Furthermore, the flow is generally periodic, at the scale of hours, days or both. Allen (1980) has described the consequences of these characteristics in the environment of the developing migration stration. He is particularly concerned with the consequences of migrating tidal bedforms, but his results are sufficiently generalized that they apply to tempestite migration strata as well.

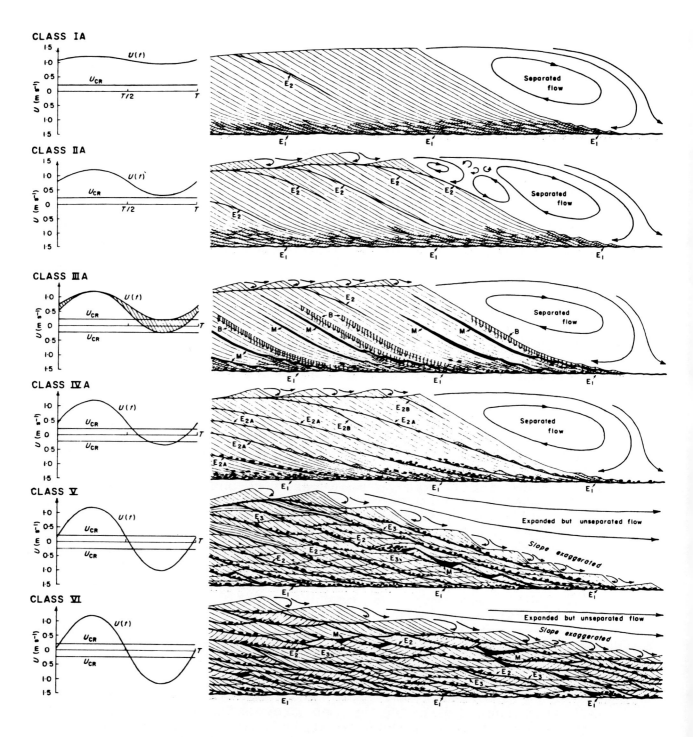

Figure 57. Spectrum of inferred internal structures of sand waves as a function of the velocity asymmetry index. From Allen, 1980. See text for explanation of symbols.

Allen (1980) describes flow-transverse bedforms in unsteady flows as being responses to a time-varying flow, U(t), which is composed of a periodic component which is a function of time $U_{p(t)}$ and a steady component, $U_s$. He defines 2 flow parameters, a velocity strength index;

$$V_1 = \frac{U_{p(m)} + U_s - U_{cr}}{U_{cr}}$$

where $U_{p(m)}$ is the maximum value of the unsteady component of flow, and $U_{cr}$ is the critical velocity (the velocity just able to move bed material. The strength index is thus the ratio of the sum of the flow components (in excess of the critical velocity) to the critical velocity. A second parameter is the velocity asymmetry index,

$$V_2 = \frac{U_s}{U_{p(m)}}$$

On the basis of these parameters, Allen is able to define a spectrum of cross-strata sets (Fig. 57). In class 1A, the periodic component of flow $U_{p(t)}$, is weak and the strength and asymmetry index are both high. The flow is steady, or nearly so, and builds a simple bedform (sand wave) with flow separation and avalanche face. Backset ripples form at the foot of the avalanche face yielding a basal, ripple cross-laminated stratum. A first order erosion surface ($E_1$) is swept out ahead of the bedform by the zone of boundary layer reattachment, advancing in harmony with the bedform. The simplicity of the structure is occasionally complicated by inclined, second order erosion surfaces ($E_2$), resulting from normal random variations of bedform shape and celerity that accompany dune migration (Allen, 1973). Such discontinuities should have a streamwise spacing of several to many times bedform height, and may be lacking altogether.

In class IIA, fluid and sediment again flow wholly in one direction, and the bedform again has separation bubble in its lee. The current, however, is markedly unsteady, but not so unsteady as to temporarily cease transporting bed material. The deposit overlies a first order erosion surface, and is dominated by steep foresets, accompanied by rippled foresets when $V_1$ is comparatively large. Class IIA differs from IA chiefly in the frequency of convex up, second-order erosion surfaces (reactivation surfaces, Collinson, 1970; McCabe and Jones, 1977), which extend the whole or part way down through the unit. Surfaces of this type are due to the overtaking of the bedform in which they are preserved by a smaller, faster moving bedform from behind (smaller bedforms move faster than larger ones, because they have less volume per unit surface area and therefore less sand to turn over). The second order erosion surface is cut by the turbulent zone of flow reattachment that preceeds the smaller bedform (Fig. 55). The greater abundance of second order bedforms (megaripples) migrating over the first order form (sandwave) in this class is attributed by Allen (1980) to the greater unsteadyness of the flow (see discussion of bedform hierarchies.

In Class IIIA the flow is yet more unsteady and transport stops periodically. Clay drapes accumulate, and are ripped up to form clay clasts. If the suspended sediment concentration is not sufficiently high, bioturbated horizons may form instead during the quiet periods.

As the flow becomes yet more unsteady (Class IV), foresets are divided into bundles in the streamwise direction by frequent second-order erosion surfaces of at least 2 kinds. The most frequent emerge out of the bottom sets as gently inclined sigmoidal discontinuities ($E_{2a}$). They record the erosional rounding of the sandwave crest and higher slopes during the reversal of flow. The second kind of discontinuity, usually shorter and steeper ($E_{2b}$) depends on the overtaking of the sandwave crest by superimposed megaripples during the dominant flow.

In Class V, As the flow becomes yet more symmetrical. The first order bedform has in effect, become a traffic jam of second order bedforms, which migrate in a constant direction. The dominant internal structure is a master bedding plane composed of gently inclined, closely spaced, second-order erosion surfaces. Cross-stratification dips in the same general direction as the master bedding and occurs mainly in climbing sets divided by third order erosion surfaces. Finally, in symmetrical unsteady flow (Class 6), cross-strata sets with opposing dips become prevalent.

Cross-strata sets: Tempestites versus tidalites

Allen (1980), distinguishes sharply between tidal sandwaves on one hand, and large-scale dunes which he attributes to unidirectional flow, on the other. The section of this chapter describing bedforms has suggested that this distinction is a semantic one. Storm flows and tidal flows have been compared in Chapter 1. Both are periodic in character. Storm flows can be viewed as stochastic tidal flows, while tidal flows can be viewed as cyclic storm currents. Allen intended Fig. 57 to represent varieties of tidal sandwaves. However because of the broadly similar periodicities of storm and tidal flows, most of the classes of Fig. 57 are probably equally applicable to storm-built sandwaves such as those of the North American Atlantic shelf at Kitty Hawk (Swift et al., 1978), or on Lookout, Frying Pan, and Diamond shoals off the Carolina Capes (Swift et al., 1972; Hunt et al., 1979, Swift et al., 1979). It is fair to say, however, that the low-numbered classes are more probable in storm-dominated settins, and the

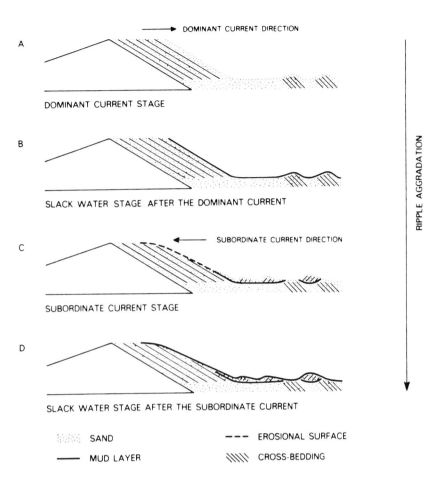

Figure 58. Schematic drawing of a ripple during the four stages of a tidal cycle. From Visser, 1980.

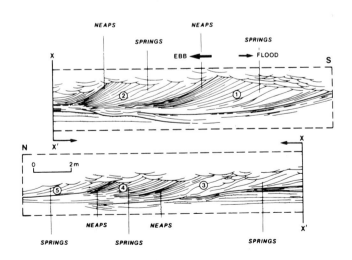

Figure 59. Field sketch of cliff exposure, Miocene marine Molasse, Switzerland. From Allen and Homewood, 1984.

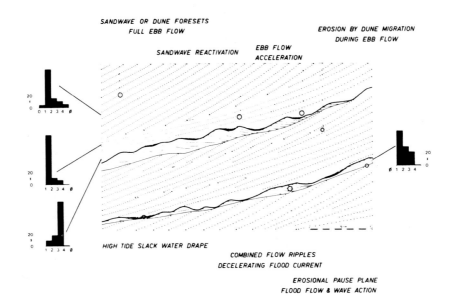

Figure 60. Interpretive sketch of typical spring bundles. From Allen and Homewood, 1984.

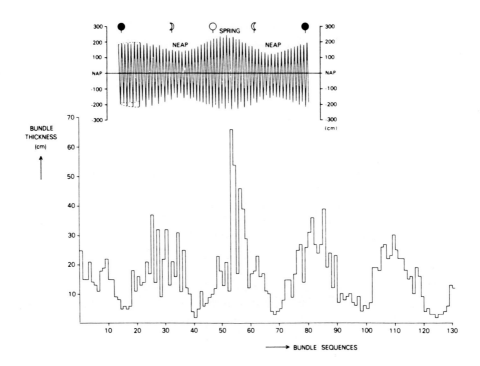

Figure 61. Measurement of the thickness of 131 adjacent strata in one cross-strata set, revealing spring-neap tide pattern. From the OOstuschclde estuary. From Visser, 1980.

high-numbered classes are more probable in tide-dominated settings.

Recent studies of tidal cross strata sets have shown that it is possible to distinguish tidal from storm cross-strata sets on other grounds. Tidal cross-strata sets are in fact periodites, not event strata, and it is possible to detect astronomical rhythms in them. An important recent study of tidal migration strata has been conducted by Visser (1980) and the following discussion is condensed from this source.

Characteristic features of tidal deposits were first described by De Raaf and Boersma (1971). One of these features is the occurrence of unidirectional, large-scale bedform deposits, which, according to Terwindt (1971) reflect a velocity inequality in the tidal cycles with dominance of tidal currents in one direction over tidal currents in the opposite direction. These sequences are also described as successive bundles of cross strata, separated by unconformities (Boersma,1969; Allen and Homewood,1984,Fig. 59).

Visser (1980) has described large-scale tidal migration strata exposed during the emplacement of Caissons for a tidal flood gate in the Oosterschelde Estuary of Holland. In the Oosterschelde, unidirectional, large-scale bedform deposits reflect a predominance of the ebb current (dominant stage, in this area) over the flood current (subordinate stage). The current stages are followed by slack water stages, one after the dominant current, and one after the subordinate current. Fig. 58 presents schematic drawings of a megaripple during these four stages. During the dominant current stage, up to 80 cm of cross-strata are deposited by avalanching on the lee slope of the megaripple (Fig. 60a). During the following slack water stage, a mud layer a few millimeters thick is deposited (Terwindt, 1971). During the sub-ordinate current stage, the current flows up the steep lee side (Fig. 60b). Much of the slackwater deposit is eroded away, to accumulate as clay clasts in a downcurrent trough, and near the crest, foreset laminae were eroded as well. At the same time, several centimeters of sand are deposited in the trough.

During slack water after the subordinate current, a second thin clay layer is deposited. As the dominant current resumes, the second clay drape is buried by avalanching sand, frequently with much less erosion, than occured to the first clay drape. As a result, a cross-strata set consists of a series of bundles of cross-strata, these may be separated by erosion surfaces if subordinate flow is sufficiently intense (Fig. 59). In most areas, however, the bundles are separated by clay drapes that become thin clay-sand-clay sandwiches (clay doublets) near the base of the cross-strata set. Regularly repeated clay doublets are diagnostic of tidal sedimenation. Even when only one of the two slack water periods deposits a clay drape, the tidal inprint is still highly recognizable (Fig. 60).

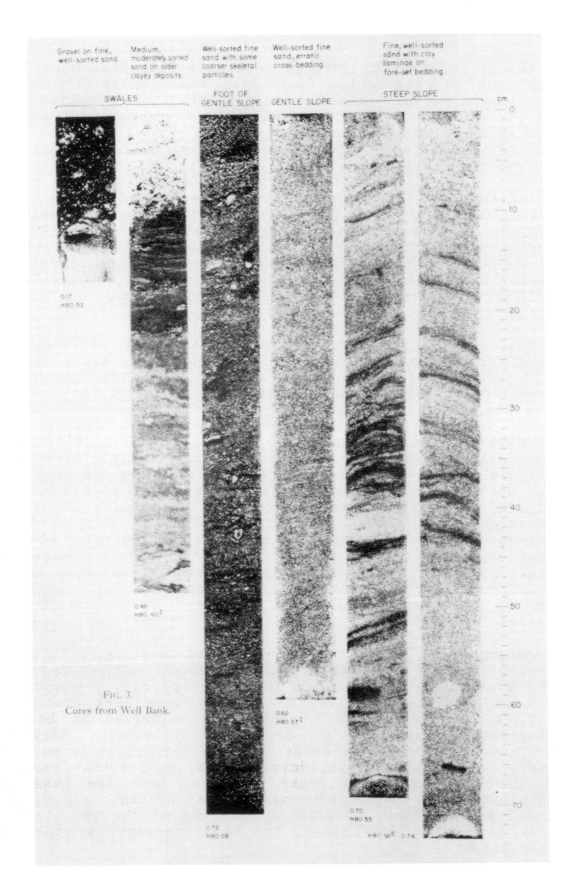

Figure 62. Strata types as a function of sand-fines ratio. After Reineck and Wunderlich, 1968, and Ruby et al., 1981.

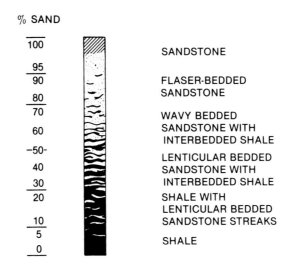

Figure 63. Tidal cross-strata sets in cores through tidal sand ridges in the Southern Bight of the North Sea. From Houbolt, 1968.

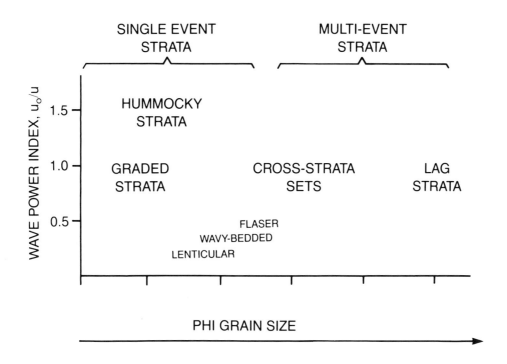

Figure 64. Inferred relationship of shoreface strata assemblage to wave-power index and grain size.

Visser (1980) showed that in the Oosterschelde pits, bundle thickness varied in a clear and consistant cyclic and sinusoidal pattern, with each bundle sequence containing 26 to 30 bundles (Fig. 61). This variation is due to the waxing and waning of the peak velocity of the peak value of the dominant current spring tide-neap tide (14 day) cycle (see Ch. 1, Fig. 2 and accompanying discussion). Spring-neap variation in bundle thickness is a conclusive criterion of tidal sedimentation. Nio et al. (1983) have shown that it is possible to reconstruct such hydraulic parameters as shear velocity, the tidal current velocity, and the tidal range from such measurements. Allen and Homewood, (1984), have analyzed 2 1/2 months in the evolution of a Miocene tidal sandwave in western Switzerland and have estimated sediment transport rates.

Cores through cross-strata sets in the tidal sand ridges of the North Sea are illustrated in Fig. 62.

### Thin-bedded, heterolithic strata

A special terminology has evolved for thin- to very thin-bedded and laminated heterolithic strata. (Strata in which both sandstone and shale beds are present). Thin- to very thin-bedded and laminated stratifiaction is typical of deeper shelf settings where wave-orbital currents are less effective in entraining bottom sediment. Thin-bedded to very thin-bedded and laminated stratification is by no means limited to the outer shelf, however. It is also found in tidal flats (where the terminology first evolved (Reineck and Wunderlich, 1968), and in deep sea fans. The terminology for thin-bedded stratification describes a continuum of stratification types in which the sand becomes finer grained, and the bedding becomes thinner, less contiuous and more silt and clay rich, as a function of decreasing lfuid power expenditure (Fig. 63). Their percent of silt and clay is diagnostic for these stratification types. Massive, laminated, or cross-laminated sandstone contains 0 to 10 percent fine sediment (silt and clay). Flaser-bedded sandstone cotains 10 to 25 percent fines. Flaser bedding is characterized by clay drapes in ripple troughs of bedding surfaces. Wavy-bedded sandstones contain 25 to 50% fines. These thin strata classes are transitional between cross-strata sets and graded strata (Fig. 64). Thin-bedded stratification types are generally more apparent in slabbed cores than they are in outcrop.

### Lag strata and hardgrounds

Lag strata occur on continental shelves in response to both storm and tidal regimes. They tend to occur in zones of higher mean velocity than do cross-strata sets, and consist of coarse sands and gravels. The relationship between graded beds and cross-strata sets parallels that between cross-strata sets and lag strata; in each pair the first type of deposit is an outwinnowed fraction with respect to the second. Both cross-strata sets and

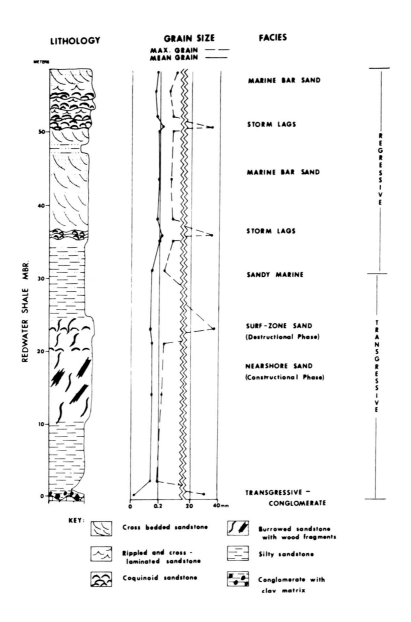

Figure 65. Interbedded lag and cross-strata sets, Sundance Formation Jurassic (Oxfordian) of Wyoming. From Brenner and Davies, 1974.

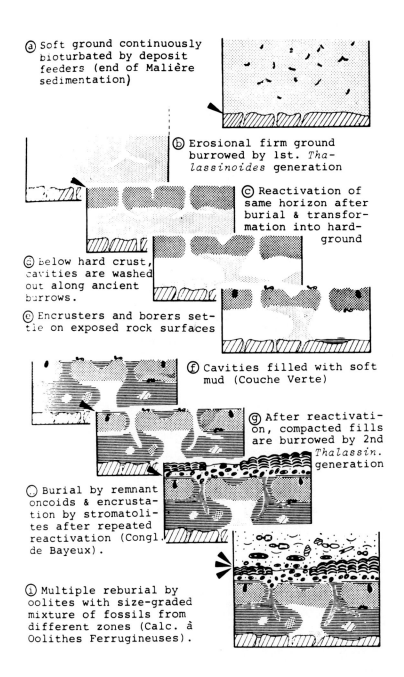

Figure 66. Time-buildup diagram for the Eocene Nummilitic Limestone of Egypt, indicating the effect of multiple erosional events on depositional fabric. From Aigner, 1980b.

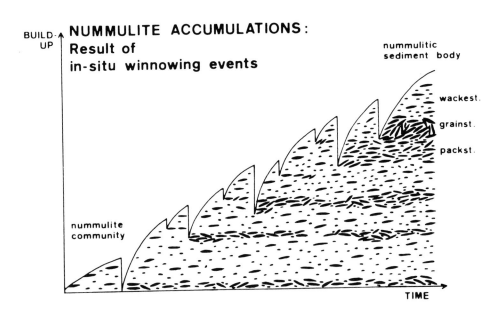

Figure 67. Complex history of a mid-Jurassic condensed section. Arrows indicate burial-erosion cycles. From Seilacher, 1982, after Fursich, 1971.

lag strata tend to be enriched in intraclasts (sharks teeth, shell hash, glauconite, phosphorite) indicating that they are lag concentrates. Lag strata are defined in this paper by these characteristics as containing more than 100% intraclasts. Cross-strata sets and lag strata are often interbedded, with the lag occurring along the erosion surfaces separating cross-strata sets (Fig. 65). Lag strata may also occur at the base of graded or hummocky beds (Kriesa, 1982)

Lag strata have been subjected to a complex winnowing process in which post-depositional vertical grain diffusion within the bed plays an important role. Such "vertical mixing" of sand within the sea floor has been studied primarily in the surf zone, where the mixing depth has been shown to be a strong function of wave period (for large wave heights) and also a function of grain size (Sunamura and Kraus, 1985). The vertical diffusion results because under strong wave-orbital currents and for coarser grades of sand, the contact between the moving bedload and the static bed is not sharp; kinetic energy is transmitted for centimeters or tens of centimeters down into the bed, where grains are "jostled" to an extent that diminishes with depth. The depth and intensity of diffusion increases with the mean grain size of the bed, hence coarse lag strata are the strata class most affected by this process, but the diffusion constant at a given depth is greatest for the finest grains, so that the net effect is to winnow out finer grains from <u>beneath</u> the bedload-bed interface, and reinject them into the moving layer. The bed thus becomes coarser with time, hence more susceptible to the process. In beds exposed to repeated transport events for tens, hundreds or thousands of years, the process may be an important one.

Seilacher (1982a) has described the effects of marine flow on sea floor lags and the following material is condensed from his discussion. He points out that in many cases, the substrata already contains the sedimentologic memory of previous events that may change the erosion behavior from layer to layer. This memory is most important in muddy sediments, in which previous erosion surfaces become further compacted by intermittent overload. The memory becomes most enhanced in carbonate muds, because early diagenesis tends to selectively cement erosional mud surfaces buried under calcarenitic or shelly event deposits, so that these surfaces become more and more resistant <u>reference horizons</u> during repeated re-burial and re-exposure by <u>subsequent</u> events (Fig 66). The processes involved are not yet well understood, but field evidence suggests that they are largely responsible for the formation of hardgrounds and their common association with tempestitic condensed horizons (Fig. 68; Fursich, 1971). Such lag deposits are the result of countless erosional and depositional episodes. Of these, only the strongest can still be recognized because this event tends to wipe out traces of the weaker events (Fig. 67).

The shelly faunas within tempestite lag strata may be quite complex. A soft bottom community, dominated by shallow burrowing

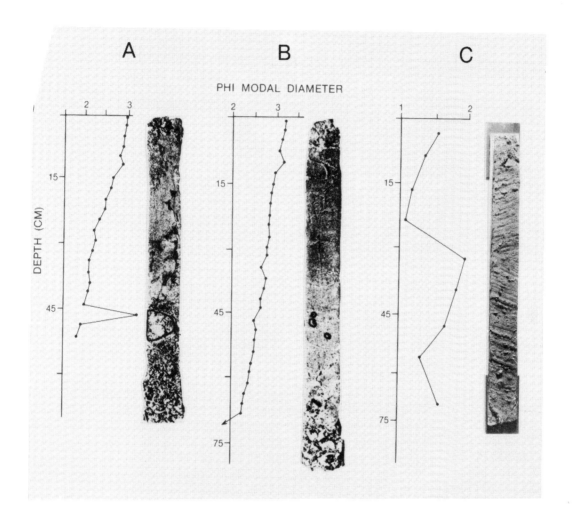

Figure 68. Vibracores through the transgressive sand shelf of the Virginia shelf. A, B: basal lag beds. C: cross-strata sets from a ridge crest.

and epifaunal bivalves, scaphopods and gastropods characterizes some lag deposits. These assemblages are winnowed out of muds that accumulated between major events. Mixed faunas contain soft-bottom forms, plus epifaunal and encrusting forms that grow on the shell ground if burial does not rapidly follow exposure (Aigner, 1982a). In many lag deposits, the fauna indicates either such a multiple provenance, or a complex series of truncation and colonization episodes. Shelly lag deposits composed of in-site accumulations in this manner are described as para-autochthonous; (Aigner, 1982). Shelly lag deposits transported in from other environments are allochthonous (Hagdorn and Mundlos, 1982).

Lag strata and hardgrounds occur extensively as the basal gravels of modern continental shelves, on shelves dominated by storm flows (the Atlantic and Brazilian continental shelves; (Figueiredo et al., 1982; Fig. 68; this paper) and on shelves dominated by tidal flows (Belderson et al., 1966).

Analogous lag strata form sequence boundaries in ancient sequences (Vail et al., 1983). In these settings lag strata and cross-strata sets may be intimately associated, with migration strata forming lenticular sand bodies (sand ridge deposits) over a lag pavement. The lag deposits may be better developed between the sand ridges than beneath them, since here the concentration process has had a longer time to operate.

## STRATA ASSEMBLAGES

The strata types presented in Fig. 48 tend to occur in several characteristic assemblages in continental shelf settings. One assemblage occurs on the shoreface. A second is present in shelf sand ridge sequences. A third occurs in prodelta sandstone deposits. All three are variants of a simple basic pattern in which sediment enters the depositional environment at a point of high fluid power expenditure. Lag beds develop at the place of sediment introduction, while cross-strata sets, hummock beds and graded beds occur further down the sediment transport path, in the more distal portions of the depositional environment.

These variants are _depositional systems_ in the sense of Fisher and McGowen (1967), who defined a depositional system as "an assemblage of process-related facies." In order to understand shelf stratification patterns, it is necessary to think in terms of a second kind of system, a _sediment transport system_, which consists of a network of transport pathways containing fluid and sediment in a state of intermittent motion. As students of ancient rock sequences, we directly observe depositional systems. However, we can only infer the behavior of the transport systems which created them, and we are in this respect, in the position of a paleontologist who observes a fossil and infers the behavior of the living animal.

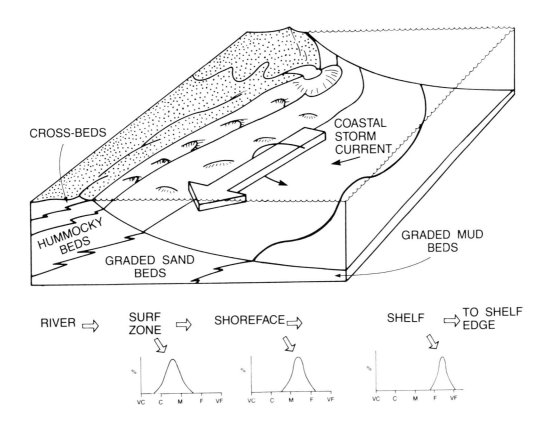

Figure 69. Schematic diagram indicating distribution of strata types on a prograding shoreface.

Strata type changes systematically through other subenvironments of shelf depositional systems, partly because the fluid power level decreases down the transport pathway from sediment source to sediment sink (Swift and Niedoroda Fig. 52, this volume), but also because intense progressive sorting occurs as sediment moves down the transport pathway (see previous discussion) and the sediment available for sediment formation because professionally finer in the distal portions of the depositional environment.

The transport systems associated with shelf strata assemblages will be discussed in detail in a later paper that analyzes shelf facies. One of the transport systems, the shoreface transport system, will be reviewed here, to show how it builds a strata assemblage (Fig. 69). On prograding shorefaces, sand and mud are introduced by the alongshore current driven by shoaling and breaking waves (Swift and Niedoroda, this volume). Interbedded lag beds and cross-strata sets accumulate beneath the surf zone. During storms, rip currents move large quantities of fine sand out of the surf zone, into the zone of coastal storm downwelling. Here storm currents carry sand obliquely down the shoreface into the inner shelf (Swift and Niedoroda, this volume, Figs. 45, 48). The fluid power gradient along this pathway is relatively steep; depth increases rapidly. The wave-orbital current component therefore decreases rapidly, and so does the competence of the flow. In this transport system, the current is too weak relative to the grain size to create high-angle cross-strata sets seaward of the surf. Hummocks form in less than about 10 m, and seaward of the 10 m isobath, graded beds form from the very fine-grained, very well sorted sand that has survived the progressive sorting process.

## REFERENCES

Allen, J. R. L., 1976, Computational models for dune time-lag: an alternative boundary condition. Sed. Geology, v. 16, p. 255-279.

_____, 1976, Computational models for dune time-lag: population structures and the effects of discharge pattern and coefficient of change. Sed. Geol., v. 16, p. 99-130.

_____, 1977, Computational models for dune time-lag: calculations using Stein's Rule for dune height, Sed. Geol., v. 20, p. 165-216.

_____, 1978b, Polymodal dune assemblages: An interpretation in terms of dune creation-destruction in periodic flows. Sedimentary Geol., v. 20, p. 17-28.

_____, 1978c, A model for the interpretation of wave ripple-marks using their wavelength, textural composition and shape. Journal of the Geol. Soc., v. 136, p. 673-682.

Allen, J. R. L., 1980a, Sandwaves: a model of origin and internal structure. Sedimentary Geol., v. 26, p. 281-328.

_____, 1980b, Large transverse bedforms and the character of boundary layers in shallow water environments. Sedimentology, v. 27, p. 317-323.

_____, 1981a, Simple models for the shape and symmetry of tidal sand waves: (1) statically stable equilibrium forms. Mar. Geol., v. 48, p. 31-49.

_____, 1981b, Simple models for the shape and symmetry of tidal sand waves: (2) dynamically stable asymmetrical equilibrium forms. Mar. Geol., v. 48, p. 51-73.

_____, 1981c, Simple models for the shape and symmetry of tidal sand waves: (3) dynamically stable symmetrical equilibrium forms. Mar. Geol., v. 48, p. 321-336.

_____, 1982a, Sedimentary Structures, their Character and Physical Basis. v. I. New York, Elsevier, 663 pp.

_____, 1982b, Sedimentary Structures, their Character and Physical Basis. v. II. New York, Elsevier, 663 pp.

Allen, P. A., and Homewood, P., 1984, Evolution and mechanics of a Miocene tidal sandwave. Sedimentology, v. 31, p. 63-81.

Aller, R. C., 1977, The influence of macrobenthos on chemical diagenesis of marine sediments. M. Sc. thesis, Yale Univ., New Haven, Conn., 600 pp.

Aller, R. C. and Dodge, R. E., 1974, Animal-sediment relationships in a tropical lagoon, Discovery Bay, Jamaeca. J. Mar. Res., v. 32, p. 209-233.

Apel, J. R., Byrne, H. M., Prone, J. R., and Charnell, R. L., 1975, Observations of oceanic internal and surface waves from Earth Resources Technology satellite: Jour. Geophys. Res., v. 80, p. 865-881.

Bagnold, R. A., 1962, Autosuspension of transported sediment; turbidity currents. Roy. Soc. Londen Proc., v. A265, p. 315-319.

Barron, E. J., Arthur, M. A. and Kauffman, Erle G., 1985. Cretaceous rhythmic bedding sequences: a plausible link between orbital variations and climate. Earth and Planetary Science Letters, 72:327-340.

Beardsley, R. C., and Flagg, C. N., 1975, The water structure, mean currents, and shelf-water/slope-water front of the New England continental shelf, Seventh Liege Colloquium on Ocean Hydrodynamics, v. 10, p. 209-226.

Belderson, R. H., 1964, Holocene sedimentation in the western half of the Irish Sea. Mar. Geol., v. 2, p. 147-163.

_____, Kenyon, N. H., and Stride, A. H., 1966, Tidal current fashioning of a Basal Bed. Mar. Geol., v. 4, p. 237-257.

Berryhill, H., 1978, South Texas continental shelf and continental slope: Late Pleistocene/Holocene evolution and sea floor stability. U.S. Geol. Surv. open file report, 78-514, 91 pp.

Blatt, H., Middleton, G., and Murray, R., 1980, Origin of Sedimentary Rocks. Englewood Cliffs, New Jersey, Prentiss Hall, 782 pp.

Boersma, J. R., 1969, Internal structures of some tidal megaripple on a shoal in the Westerschelde Estuary, the Netherlands: Geologie in Mijnbonw, v. 48, p. 409-414.

_____, Terwindt, J. H. J., 1981, Neap-spring tide sequences of intertidal shoal deposits in a mesotidal estuary. Sedimentology, v. 28, p. 151-170.

Bouma, A. H., Hampton, M. A., and Orlando, R. C., 1977, Sandwaves and other bedforms in lower Cook Inlet, Alaska. Marine Geotechnology, v. 2, p. 132-138.

Bowen, A. J., 1980, Simple models of nearshore sedimentation: Beach profiles and alongshore bars. p. 1-11, in: McCann, S. B. ed., The Coastline of Canada. Geol. Surv. Canada paper 80-10, 238 pp.

Boyles, J. M., and Kocurek, G., in press, Depositional model for hummocky stratification based on the special case of hummocky stratified scour fill. Unpublished Manuscript.

Brenner, R. L., and Davies, D. K., 1974, Oxfordian sedimentation in Western interior United States. Am. Assoc. Petroleum Geologists Bull., v. 58, p. 407-428.

Brown, P. J., Ehrlich, R., Colquhoun, D. J., 1980. Origin of patterns of quartz sand types on the southeastern United States Continental Shelf and implications on contemporary shelf sedimentation--Fourier grain shape analysis. Journal of Sedimentary Petrology, v. 50, p. 1095-1100.

Cartwright, D. E., 1939, On reprint submarine sandwaves and tidal-lee waves. Proc. Ray. Soc., v. a253, p. 218-241.

Cacchione, D. A., D. E. Drake, and P. Wiberg, 1982. Velocity and bottom-stress measurement in the bottom boundary layer outer Norton Sound, Alaska. Geologic en Mijnbouw, v. 61, p. 71-780.

Cacchione, D. A. and D. E. Drake, 1982. Measurements of storm-generated bottom stresses on the continental shelf. J. Geophys. Res., 87:1952-1960.

Campbell, C. V., 1979. Model for beach shoreline in Gallup sandstone (Upper Cretaceous) of northwestern New Mexico. New Mexico Bureau of Mines and Mineral Resources Circ., 164, 29 p.

Cartwright, D. E., 1939, On reprint submarine sandwaves and tidal-lee waves, Proc. Ray. Soc. v. a 253, p. 218-241.

Caston, G. F., 1981, Potential gain and loss of sand by some sand banks in the Southern Bight of the North Sea. Marine Geology, v. 41, p. 239-250.

Caston, V. N. D., 1972, Linear sand banks in the southern North Sea. Sedimentology, v. 18, p. 63-78.

_____, 1981, Potential gain and loss of sand by some sand banks in the southern Bight of the North Sea. Mar. Geol., v. 91, p. 239-250.

_____, and Stride, A. H., 1970, Tidal sand movement between some linear sand banks in the North Sea off northeast Norfolk. Mar. Geol., v. 9, p. M38-M42.

Channon, R. D., and Hamilton, D., 1976, Wave and tidal current sorting of shelf sediments southwest of England. Sedimentology, v. 23, p. 17-42.

Clarke, T. L., B. Lesht, R. A. Young, D. J. P. Swift, and G. L. Freeland, 1982. Sediment resuspension by surface-wave action: an examination of possible mechanisms. Mar. Geol., 49:43-59.

_____, Swift, D. J. P., and Young, R. A., 1982, A numerical model of fine sediment transport on the continental shelf. Environmental Geology, v. 4, p. 117-129.

_____, Swift, D. J. P., and Young, R. A., 1983, A stochcastic modeling approach to shelf sediment dispersal. J. Geophys. Res., v. 88, p. 9653-9660.

Collinson, J. D., 1970, Bedforms of the Tana River, Norway. Geogr. Ann., v. 52a, p. 31-56.

Cook, D. O., 1970, The occurrence and geological work of rip currents off southern California. Mar. Geol., v. 9, p. 173-186.

_____, and Gorsline, D. S., 1972, Field observations of sand transport by shoaling waves. Mar. Geol., v. 13, p. 31-55.

Crowell, J. C., 1955, Direction current structures from the pre-Alpine Flysch, Switzerland. Bull. Soc. Am., v. 66, p. 1351-1384.

Curtin, T. B., 1979, Physical dynamics of the coastal upwelling frontal zone off Oregon, Ph.D. Dissertation, Miami, Florida, University of Miami, 317 p.

Dalrymple, R. W., Knight, R. J., and Lambiase, J. J., 1978, Bedforms and their hydraulic stability relationships in a tidal environment, Bay of Fundy, Canada. Nature, v. 275, p. 100-104.

De Raaf, J. F. M., and Boersma, J. R., 1971, Tidal deposits and their sedimentary structures: Geologie en Mijnbouw, v. 30, p. 479-504.

Dott, R. H., Jr., and Bourgeois, J., 1982, Hummocky stratification: significance of its variable bedding sequences. Geol. Soc. America Bull., v. 93, p. 663-680.

Drake, D. E., D. A. Cacchione, R. D. Muench and C. H. Nelson, 1980. Sediment transport in Norton Sound, Alaska. Mar. Geol., 36:97-126.

Duane, D. P., Field, M. E., Miesberger, E. P., Swift, D. J. P., and Williams, S. J., 1972, Linear shoals on the Atlantic Continental Shelf, Florida to Long Island. In Swift, D. J. P., D. B. Duane, and Pilkey. Shelf Sediment Transport, Process and Pattern. Stroudsburg, Pa., Dowden, Hutchinson and Ross, 656 pp.

Dunbar, C. O. and J. Rodgers, 1957. Principles of Stratigraphy. New York, John Wiley & Sons, 356 p.

Einsele, G., 1982, General remarks about the nature, occurrence, and recognition of cyclic sequences (periodites). P. 3-7 in G. Einsele and A. Seilacher, Eds., Cyclic and Event Stratification New York, Springer Verlag, 536 pp.

_____, and Seilacher, A., eds., 1981. Cyclic and Event Stratification. Springer Verglag, 536 p.

Engelund, F., and Fredsoe, J., 1974, Transition from dunes to plane beds in alluvial channels. Tech. Univ. Denmark, Inst. Hydrodynamics Hydraulic Engineering Series paper 4., 25 pp.

Fager, E. W., 1964, Marine Sediments: Effects of a tube building polychaete. Science, v. 143, p. 356-359.

Fenchel, T., Kofoed, L.H., and Lappalainen, A., 1975, Particle size selection of two deposit feeders: the amphipod Corophium volutator and the Prosobranch Hydrobia ulvae, Mar. Biol., v. 30, p. 119-128.

Field, M. E. and R. K. Hall, 1982. Sonographs of submarine sediment failure caused by the 1980 earthquake off northern California. Geo-Marine Letters, v. 2, p. 134-141.

Field, P. M. E., Nelson, C. H., Cacchione, D. A. and Drake, E., 1981, Sandwaves on an epicontinental shelf: northern Bering Sea. Mar. Geol., v. 42, p. 233-258.

Figueiredo, A. G., Swift, D. J. P., Stubblefield, W. L., and Clarke, T. L., 1981a, Sand ridges on the Inner Atlantic Shelf of North America: morphometric comparisons with Huthnance Stability model. Geo. Marine Letters, v. 1, p. 187-191.

Figueiredo, A. G., Sanders, J., and Swift, D. J. P., 1981b, Storm-graded layers on inner continental shelves: Examples from Southern Brazil and the Atlantic Coast of the Central United States. Sedimentary Geology, v. 31, p. 171-190.

Fisher, W. L. and J. H. McGowen, 1967. Depositional system in the Wilcox Group of Texas and their relationship to the occurrence of oil and gas. Gulf Coast Assoc. Geol. Soc. Trans., 17:105-125.

Flagg, C. N., 1977, The kinematics and dynamics of the New England continental shelf and shelf/slope front. Ph.D. Dissertation, Cambridge, Massachusetts, Mass. Inst. Tech., Mass. Inst. Tech.--Woods Hole, Oceanogr. Inst. Joint Program in Oceanogr., 207 p.

Flemming, B. W., 1978, Underwater sand dunes along the southeast African Continental Margin--observations and implications. Marine Geology, v. 26, p. 177-198.

Flemming, B. W., 1980, Sand transport and bedform patterns on the continental shelf between Durban and Port Elizabeth (southeast African Continental Margin). Sedimentary Geology, v. 26, p. 179-205.

Fursich, F., 1971, Hartgrunde und Condensation in Dogger Von Calvados, N. Jb. Geol. Paleont Abh., v. 138, p. 313-342.

Gaynor, G. C. and D. J. P. Swift, 1985. Shannon sandstone depositional model: Sand ridge formation on the Campanian western interior shelf. In preparation.

Goldberg, E. D., and Koide, M., 1962, Geochronological studies of deep sea sediments by the ionium-thorium method. Geochim. Cosmochim. Acta, v. 26, p. 417-456.

Gordon, D., 1966, The effects of the deposit feeding polychaete Pectinaria gouldii on the intertidal sediments of Barnstable Harbor Limnol. Oceanogr., v. 11, p. 327-332.

Guinasso, N. L., and Schink, D. R.,1975, Quantitative estimates of biological mixing rates in abyssal sediments. Geophys. Res., v. 80, p. 3032-3043.

Guy, H. P., Simons, D. B., and Richardson, E. V., 1966, Summary of alluvial channel data from flume experiments, 1956-1961. U.S.G.S. Prop. Paper 462-J, 96 p.

Hagdorn, H., and Mundlos, R., 1982, Allochthonous coquinas in the Upper Muschelkalk--caused by storm events? (Abstract).

Hamblin, D. P., and Walker, R. G., 1979, Storm-dominated shallow marine deposits: The Fernie-Kootenay (Jurassic) transition, Southern Rocky Mountains. Can. J., Earth Sci. v. 16, p. 1673-1689.

Hammond, F. D. C., and Heathershaw, A. D., 1981, A wave theory for sandwaves in shelf-seas. Nature, v. 293, p. 208-210.

Harms, J. C., Southard, J. B. and Walker, R. G., 1982, Structures and sequences in clastic rocks. Soc. Econ. Paleon. Mineral. Short Course 9, Tulsa, Ok., Not sequentially paged.

Harvey, J. G., 1966, Large sandwaves in the Irish Sea: Marine Geology, v. 4, no. 1, p. 49-55.

Hayes M. O., 1967, Hurricanes as geological agents: case studies of Hurricanes Carla, 1961 and Cindy, 1963: Texas Bur. Econ. Geol. Rept. Invest. 61, 54 p.

Houbolt, J. H. C., 1968, Recent sediment in the souther Bight of the North Sea. Geologie en Mijnbouw, v. 47, p. 245-273.

Hudelson, P. M., R. L. Brenner and D. J. P. Swift, in preparation. Storm-dominated prodelta shelf sandstones, Mesaverde Group, Book Cliffs, Utah.

_____, R. Brenner, and Swift, D. J. P., in preparation, Storm stratification on a prodelta shelf, Kenilworth member of the Blackhawk Formation (Mesaverde Group); Book Cliffs, Utah.

Hunt, R. E., Swift, D. J. P., and Palmer, H., 1977, Constructional shelf topography, Diamond Shoals, North Carolina. Geol. Soc. America Bull., v. 88, p. 299-311.

Hunter, R. E., and Clifton, H. E., 1981, Cyclic deposits and hummocky cross-stratification of probable storm origin in upper Cretaceous Rocks of the Cape Sebastian Area, south-western Oregon. J. Sed. Petrol., v. 52, p. 0127-0143.

Huthnance, J. M., 1972, Tidal current asymmetries over the Norfolk sandbanks. Estuarine and Coastal Mar. Sci., v. 1, p. 89-99.

_____, 1982a, On the formation of sandbanks of finite extent. Estuarine, Coastal and Shelf Science, 15:277-299.

_____, 1982b, On one mechanism forming linear sandbanks. Estuarine, Coastal and Shelf Science, v. 14, p. 79-99.

Inman, D. L., 1957, Wave generated ripples in nearshore sands, Beach Erosion Board Technical Memorandum 100, 44 pp.

Ingram, R. L., 1954, Terminology for the thickness of stratification and parting units in sedimentary rocks: Geol. Soc. America Bull., v. 65, p. 937-938.

Ippen, A. T. and Eagleson, P. S., 1955, A study of sediment sorting by waves shoaling on a plane beach: Beach Erosion Board Tech. Memo. 63, 81 pp.

Jackson, R. G., II, 1975, Hierarchical attributes and a unifying model of bedforms composed of cohesionless material and produced by shearing flow. Geol. Soc. of Amer. Bull., v. 86, p. 1523-1533.

Johnson, M. A., Stride, A. H., Belderson, R. H., and Kenyon, N. H., 1981, Predicted sand-wave formation and decay on a large offshore tidal-current sand-sheet. Sedimentology, v. 5, p. 247-256.

Jumars, P. A., Nowell, A. R. M., and Self, R. F. L., 1981, A simple model of flow sediment organism interaction, Mar. Geol., v. 42, p. 155-172.

Karl, H. A., Carlson, P. R., and Cacchione, D. A., 1983, Factors influencing sediment transport at the shelfbreak, p. 219-231, in Stanley, D. J., and Moore, G. T. (eds.), The Shelfbreak: Critical Interface on Continental Margins, Soc. Econ. Paleontol. Mineral. Spec. Pub. 33, 467 p.

Kelling, G. and Mullin, P. R., 1975, Graded limestones and limestone-quartzite couplets: possible ancient storm deposits from the Morroccan Carboniferous. Sedimentary Geology, v. 13, p. 161-190.

Kennedy, S. K., Ehrlich, R., and Kana, T. W., 1980, The non-normal distribution of intermittent suspension sediments below breaking waves. J. Sed. Petrology, v. 51, p. 1981-1985.

Kenyon, N. H., 1970. Sand ribbons of European tidal seas. Mar. Geol., v. 9, p. 25-39.

_____, Belderson, R. H., Stride, A. H., and Johnson, M. A., 1981, Offshore tidal sandbanks as indicators of net sand transport and as potential deposits. Spec. Publs. int. Assoc. Sedimentologist, v. 5, p. 257-268.

Kinsman, B., 1965, Wind Waves. Prentice-Hall, Englewood Cliffs, NJ, 676 pp.

Klein, G. De V., 1971, A sediment model for determining paleotide range. Geol. Soc. Amer. Bull., v. 82, p. 2585-2592.

Komar, P. D., 1974, Oscillatory ripple works and the evaluation of ancient wave conditions and environements. J. Sed. Petrol., v. 44, p. 169-180.

_____, 1976, The transport of cohesionless sediments on continental shelves. p. 107-126, in. Stanley, D. J., and Swift, D. J. P., Marine Sediment Transport and Environmental Management. New York, John Wiley and Sons, 602 pp.

Kreisa, R. P., and Bomback, R. K., 1982, The role of storm processes in generating shell beds in Paleozoic shelf environments. P. 201-207 in Einsele, G., and Seilacher, A., Cyclic and Event Stratification. New York, Springer Velag, 535 pp.

Kuenen, P. H., and Menard, H. W., 1952, Turbidity currents, graded and non graded deposits. J. Sed. Petrology, v. 22, p. 83-97.

Langhorne, D. K., 1982, A study of the dynamics of a marine sandwave. Sedimentology, v. 29, p. 571-594.

Lindsay, J. T., David B. Prior and James M. Coleman, 1984. Distributary mouth bar development and role of submarine landslides in delta growth, South Pass, Mississippi Delta. Am. Assoc. Petrol. Geol. Bull., v. 68, p. 1732-1743.

Lowe, D. R., 1976, Subaqueous liquefied and fluidized sediment flows and their deposits. Sedim., v. 23, p. 285-308.

_____, 1982, Sediment gravity flows II: depositional model with special reference to deposits of high density turbidity currents. J. Sed. Petrology, v. 52, p. 0279-0297.

Ludwick, J. C., 1970, Sand waves and tidal channels in the entrance to Chesapeake Bay. The Virginia Journal of Science, v. 21.

_____, 1972, Migration of tidal sandwaves in Chesapeake Bay Entrance p. 377-410, in. Swift, D. J. P., Duane, D. B., and Pilkey, O. H., Shelf Sediment Transport: Process and Pattern. Dowden, Hutchinson and Ross, Stroudsberg, Pa., 656 p.

_____, 1975, Tidal currents, sediment transport, and sandbanks in Chesapeake Bay Entrance, Va., p. 369-380, In: Cronin, E., ed., Estuarine Research, vol. II, Geol. and Eng., Academic Press, Inc., New York, 587 p.

Mann, R. G., Swift, D. J. P., and Perry, R., 1981, Size classes of flow-transverse bedforms in a subtidal environment, Nantucket shoals, North American Atlantic shelf. Geo-Marine Lett., v. 1, p. 39-43.

Mare, M. F., 1942, A study of a marine benthic community with special reference to micro-organisms. J. Mar. Biol. Assoc. U. K., v. 25, p. 517-554.

Mazzullo, J. M., and Crisp, J., in press, Sources and distribution of relect, palimpsest and modern course silt on the south Texas continental shelf. J. Sed. Petrology.

_____, Ehrlich, R., and Hemming, M. A., in press, Provenance and areal distribution of Late Pleistocene and Holocene quartz sands on the southern New England Continental Shelf. J. Sed. Petrology.

McCave, I. N., 1971, Sandwaves in the North Sea off the coast of Holland. Marine Geology, v. 10, p. 199-225.

_____, 1978, Grain-size trends and transport along beaches: Example from eastern England. Mar. Geol., v. 28, p. M43-M51.

_____, and Geisa, A. C., 1978, Megaripples, ridges and runnels on intertidal flats of the wash, England. Sedimentology, v. 20, p. 353-369.

McCabe, P. J., and Jones, C. M., 1977, Formation of reactivation surfaces within superimposed deltas and bedforms. J. Sed. Petrology, v. 47, p. 707-715.

McLean, S. R., 1981, The role of nonuniform roughness in the formation of sand ribbons. Mar. Geol., v. 42, p. 49-74.

McKee, E. D., and Weir, G. W., 1953, Cross-bedding terminology for stratification and cross-stratification in sedimentary rocks. Geol. Soc. Amer. Bull., v. 64, p. 381-390.

Middleton, G. V., 1966, Experiments on density and turbidity currents I. Motion of the head. Can. J. Earth Sci., v. 3, p. 523-546.

_____, 1966, Experiments on density and turbidity currents II. Uniform flow of density currents. Can. J. Earth Sci., v. 3, p. 627-637.

_____, 1966, Experiments on density and turbidity currents III. Deposition of sediment. Can. J. Earth Sci., v. 4, p. 475-505.

Middleton, G. V., 1966, Small-scale models of turbidity currents and the criterion for auto-suspension. J. Sed. Petrology, v. 36, p. 202-208.

Miller, M. C. and Komar, P. D., 1980, A field investigation of the relationship between oscillation ripple spacing and the near-bottom water orbital motions. J. Sed. Petrol., v. 50, p. 0183-0191.

Moore, P. G., and Scruten, P. C., 1957, Minor internal structures of some recent unconsolidated sediments: Am. Assoc. Petroleum Geol. Bull., v. 41, p. 2723-2751.

Morton, R. A., 1981, Formation of storm deposits by wind-forced currents in the Gulf of Mexico and North Sea. Spec. Publ. Internat. Assoc. Sedimentologists 5, p. 303-396.

Myers, A. C., 1977, Sediment processing in a marine subtidal sandy bottom community. J. Mar. Res., v. 35, p. 609-647.

Nelson, L. H., 1982, Modern shallow water graded sand layers from storm surges, Bering Gulf: A mimic of Bouma sequences and turbidite systems. J. Sed. Petrology, v. 52, p. 0537-0545.

Nichols, F. H., 1974, Sediment turnover by a deposit feeding polychaete. Limnol. Oceanogr., v. 19, p. 945-950.

Nio, S. D., Sigenthaler, C., and Yang, C. S., 1983, Megaripple cross-bedding as a tool for the reconstruction of the paloeo-hydraulics in a Holocene subtidal environment, S.W. Netherlands. Geologie en Mijnbouw, v. 36, p. 499-509.

Nittrouer, C.A., Sternberg, R. W., Carpenter, R., and Bennett, J. J., 1979, The use of Pb-210 geochronology as a sedimentological tool: application to the Washington Continental Shelf. Mar. Geol., v. 21, p. 297-316.

_____, and Sternberg, R. W., 1981, The formation of sedimentary strata in an allochthanous shelf environment: the Washington Continental Shelf. Mar. Geol., v. 42, p. 201-232.

Off, T., 1963, Rhythmic linear sand bodies caused by tidal currents. Am. Assoc. Pet. Geol. Bull., v. 47, p. 324-341.

Pantin, H. M., 1979. Interaction between velocity and effective density in turbidity flow: Phase plume analysis, with criteria for autosuspension. Mar. Geol., v. 31, p. 59-99.

Pantin, H. M., 1983. Conditions for the ignition of catastrophically erosive turbidity currents-comment. Mar. Geol., v. 52, p. 281-290.

Parker, G., 1982. Conditions for the ignition of catastrophically erosive turbidity currents. Mar. Geol., v. 46, p. 307-327.

_____, Lanfredi, N. W., and Swift, D. J. P., 1982, Substrate response flow in a southern hemisphere ridge field: Argentina Inner Shelf. Sedimentary Geology, v. 33, p. 195-216.

Parrish, J. T., Gaynor, G. C., and Swift, D. J. P., in press, Circulation in the Cretaceous Western Interior seaway of North America, a review in D. F. Stott, Mesozoic of Middle North America. Can Soc. Petrol. Geol. Mem.

Pingree, R. D. and Maddock, L., 1979. The tidal physics of headland flows and offshore tidal bank formation. Mar. Geol., 32:259-289.

Potter, P. E., and Mast, R. F., 1963, Sedimentary structures, sand shape fabrics and permeability. J. Geology, v. 71, p. 441-470.

Powell, E. N., 1977, Particle size selection and sediment secondary in a funnel feeder, *Uptosynapta tenuis* (Holothuroidea, Synaptidae). Int. Rev., Ges. Hydrobiol., v. 62, p. 385-408.

Pretious, E. S., and Blench, T., 1951, Final report on special observations of bed movement in the lower Fraser River at Ladner Beach. Rep. Nas. Res. Counc. Canada Fraser River Models Vancouver, 43 pp.

Prior, D. B., and Coleman, J. M., 1977, Disintegrating retrogressive landslides on very low-angle subaqueous slopes, Mississippi Delta. Marine Geotechnology, v. 3, p. 37-60.

Reineck, H. E., and Singh, I. B., 1972, Genesis of laminated sand and graded rhythmites in storm-sand layers of shelf mud; Sedimentology, v. 18, p. 123-128.

_____ and Wunderlich, F., 1968. Classification and origin of flaser and lenticular bedding, Sed., 11:99-104.

Rhoads, D. C., 1963, Rates of sediment reworking by *Yoldia linatula* in Bazzards Bay, Massachassetts, and Long Island Sound. J. Sed. Petrol., v. 33, p. 723-727.

_____, 1967, Biogenic reworking of intertidal and subtidal sediments in Barnstable Harbor and Buzzards Bay, Massachussetts. J. Geol., v. 75, p. 461-476.

_____, 1974, Organism-sediment relations on the muddy sea floor. Oceanogr. Mar. Biol. Sunn. Rev., v. 12, p. 263-300.

_____, and Stanley, D. J., 1965, Biogenic graded bedding. J. Sed. Petrology, v. 35, p. 956-963.

Richards, K. K., 1980, The formation of ripples and dunes on an erodable bed. J. Fluid Mech., v. 99, p. 592-618.

Riester, D. D., Shipp, C., and Ehrlich, R., 1982, Patterns of quartz grain shape variation, Long Island littoral and shelf. J. Sed. Petrology, v. 52, p. 1307-1314.

Ruby, C. H., Horne, J. C., and Reinhart, P. J., 1981. Cretaceous rocks of western North Americas: A guide to terrigenous clastic rock identification. Research Planning Institute, 925 Gervais St., Columbia, SC 29201.

Ruiter, J. H., Jr., 1959, Modern Oscilloscopes and their uses, New York, Reinhart and Co., 346 pp.

Sallenger, A. H. Jr., Goldsmith, V., and Sutton, C. H., 1975, Bathymetric comparisons: A manual of methodology, error criteria and techniques. Virginia Institute Marine Science Spec. Dept. 66, Gloucester Point Va, 23062, 34 p.

Schubel, J. R., and A. Okubo, 1972, Some comments on the despersal of suspended sediment across continental shelves. p. 333-346 in D.J.P. Swift, D.B. Duane, and O.H. Delkey, eds, Shelf Sediment Transport, Process and Pattern. Stroudsburg Pa., Dowden, Hutchenson and Ross, 656 pp.

Seilacher, A., 1982a. General remarks about event deposits. p. 161-174, in G. Einsele and A. Seilacher, Cyclic and Event Stratification. New York, Springer Verlag, 536 p.

_____, 1982b, Distinctive features of sandy tempestites. p. 333-349 in Einsele, G., and Seilacher, A., Cyclic and Event Stratification. New York, Springer Verlag, 536 pp.

Self, R. F. L., and Jumars, P. A., 1978. New resources axes for deposit feeders? J. Mar. Res., v. 36, p. 627-641.

Shepard, F. P., Marshall, N. R., and McLoughlin, P. A., 1974, "Internal waves" advancing along submarine canyons, Science, v. 183, p. 195-198.

Simons, D. B., and Richardson, E. V., 1963. Forms of bed roughness in alluvial channels. Am. Ser. Civil. Eng. Trans., v. 128, p. 284-302.

Sleath, J. F. A., 1974, Mass transport on a rough bed. J. Mar. Res., v. 32, p. 13-24.

Sleath, J. F. A., 1976, On rolling grain ripples. J. Hydrial. Res., v. 14, p. 69-81.

Smith, J. D., 1969, Geomorphology of a sand ridge, J. of Geology, v. 77, p. 39-55.

Smith, J. D., 1970, Stability of a sand bed subjected to a shear flow of low Froude number. J. Geophys Res., v. 75, p. 5928-5940.

_____, 1977, Modeling of sediment transport on continental shelves. p. 539-577 In: E. D. Goldberg, I. N. McCave, J. J. O'Brien and J. H. Steele (Editors), The Sea, v. 6. Wiley, New York, 1084 pp.

Southard, J. B., and Boguchwal, L. A., 1973, Flume experiments on the transition from ripples to lower flat beds with increasing grain size. J. Sed. Petrology, v. 93, p. 1114-1121.

_____, and Mackintosh, M. E., 1981. Experimental test of autosuspension. Earth Surface Processes and Landforms, v. 6, p. 103-111.

Stewart, H. B., and Jordon, G. F., 1964, Underwater sand ridges on Georges Shoal. p. 102-116 in. R. L. Miller, Papers in Marine Geology; Shepard Comemorative Volume New York, McMillan, 531 pp.

Stride, A. H., 1970, Shape and size trends for sand waves in a depositional zone of the North Sea. Geol. Mag., v. 107. p. 469-477.

Stubblefield, W. L. and Swift, D. J. P., 1976, Ridge development as revealed by sub-bottom profiles on the central New Jersey Shelf, Mar. Geol., v. 20, p. 315-334.

_____, and Swift, D. J. P., 1981, Grain size variation across sand ridges, New Jersey Continental Shelf. Geomarine Letters, v. 1, p. 45-48.

Swift, D. J. P., 1975, Tidal sand ridges and shoal retreat massifs. Mar. Geol., v. 18, p. 105-133.

_____, 1975, Barrier island genesis: Evidence from the central Atlantic Shelf, Eastern USA. Sedimentary Geology, v. 14, p. 1-43.

_____, 1976a, Coastal sedimentation. p. 255-310, in: D. J. Stanley and D. J. P. Swift, eds., Marine Sediment Transport and Environmental Management, New York; John Wiley and Sons, Inc., 602 pp.

_____, Duane, D. B., and McKinney, T. F., 1973, Ridge and swale topography of the Middle Atlantic Bight, North America: secular response to the Holocene hydraulic regime. Mar. Geol., v. 15, p. 2272-47.

_____, and Field, M. F., 1981, Evolution of a classic ridge field, Maryland Sector, North American Inner Shelf. Sedimentology, v. 28, p. 461-482.

Swift, D. J. P., Figueiredo, A. G., Jr., Freeland, G., and Oertel, G., 1983, Hummocky cross-stratification and megaripples, a geological double standard? J. Sed. Petrol., v. 53, p. 1295-1317.

_____, Freeland, G. L. and Young, R. A., 1979, Time and space distributions of megaripples and associated bedforms, Middle Atlantic Bight, North American Atlantic Shelf, Sedimentology, v. 26, p. 389-406.

_____, Holliday, B., Avigvane, N., and Shideler, G., 1972b, Anatomy of a shoreface ridge system, False Cape, Virginia, Mar. Geol., v. 12, p. 59-84.

_____, Hudelson, P. M., and Brenner, R. L., in press, Sandstone beds of the Mancos-Mesaverde transition, Book Cliffs, Utah: Turbidites or Tempestites? Geology.

_____, Kofoed, J. W., Saulsburg, F. P., and Sears, P., 1972, Holocene evolution of the shelf surface, central and southern shelf of North American. p. 499-574, in Swift, D. J. P., Duane, D. B., and Pilkey, O. H., Eds., Shelf Sediment Transport, Process and Pattern. Stroudsburg, Pa., Dowden, Hutchinson and Ross, 656 p.

_____, Ludwick, J. C., and Boehmer, W. R., 1972, Shelf sediment transport, a probability model. p. 195-223, In D. J. P. Swift, D. B. Duane, and O. H. Pilkey, Eds., Shelf Sediment transport: Process and Pattern: Stroudsberg, PA.: Dowden, Hutchinson and Ross, 656 pp.

_____, Nelson, T., McHone, J., Holliday, B., Palmer, H. and Sheldon, G., 1977, Holocne Evolution of the Inner Shelf of Southern Virginia. J. Sed., Petrology, v. 47, p. 1454-1474.

_____, Parker, G., Lanfredi, N. W., Perillo, G. and Figge, K. ,in press, 1978, Shoreface-connected sand ridge on American and European Shelves: A comparison. in press. Estuarine and Coastal Marine Research, v. 17, p. 257-273.

_____, and Rice, D. D., 1984. Sand bodies on muddy shelves: A model for sedimentation on the Cretaceous western interior seaway, North America. P. 43-62, in Tillman, R. W., and Siemers, C. T. (eds.), Siliciclastic Shelf Sediments. Soc. Econ. Paleontol. Mineral. Spec. Publ. 34, 268 p.

Swift, D. J. P., and Rice, D. P., Sand bodies on muddy shelves: a model for sedimentation in the Cretaceous Western Interior Seaway, North America. In, Seimers, C. T., and Tillman, R. W., Siliciclastic Sedimentary Sequences. Soc. Econ. Paleon. Mineral. Spec. Publ.

_____, Sanford, R., Dill, C. E., Jr., and Avignone, N., 1971, Textural differentiation on the shoreface during erosional retreat of an unconsolidated coast, Cape Henry to Cape Hatteras western North Atlantic Shelf. Sedimentology, v. 16, p. 221-250.

_____, Sears, P. C., Bohlke, B., and Hunt, R., 1978, Evolution of a shoal retreat massif, North Carolina Shelf: Inferences from areal geology. Mar. Geol., v. 27, p. 19-42.

_____, Stanley, D. J., and Curray, J. R., 1971, Relict sediments on continental shelves: a reconsideration. Jour. Geology, v. 79, p. 322-346.

_____, Tillman, R. W., Rine, J. M. and Seivers, C. T. in press, Storm flows and storm deposits on a modern shelf: implications for models of ancient shelf sedimentation. Sedimentology.

_____, Young, R. A., Clarke, T., and Vincent, C. E., 1981, Sediment transport in the Middle Atlantic Bight of North America: Synopsis of recent observations. p. 361-383. in Nio, S. D., Schuttenhelm, R. T. E., and Van Weering, T. C. E., Holocene Marine Sedimentation in the North Sea Basin, Int. Assoc. Sedimentologists Specl. Publ. 5, 514 pp.

Terwindt, J. H. J., 1971, Sandwaves in the Southern Bight of the North Sea. Marine Geology, v. 10, p. 51-68.

_____, 1971, Lithofacies of inshore estuarine and tidal inlet deposits: Geologie en Mijnbouw, v. 50, p. 515-526.

_____, 1981, Origin and sequences of sedimentary structures in inshore mesotidal deposits of the North Sea. Sedimentology, v. 5, p. 4-26.

Van Andel, T. H., and Poole, D. M., 1960, Sources of recent sediments in the northern Gulf of Mexico, Journal of Sedimentary Petrology, v. 30, p. 91-122.

Van Straaten, L. M. J., 1959, Littoral and submarine morphology of the Rhone Delta in: Proc. Second Coastal Geogr. Conf. Louisiana 1959, p 233-264.

_____, 1965, Coastal barrier deposits in south and north Holland - in particular in the area around Scheveningen and Ijmuden. Meded. Geol. Sticht., NS, v. 17, p. 41-75.

Van Veen, J., 1935, Sandwaves in the North Sea. Hydrogr. Rev., 12, p. 21-28.

Vincent, C. E., Young, R. A., and Swift, D. J. P., 1982, On the relationship between bedload and suspended sand transport on the inner shelf, Long Island, New York. Jour. Geophy. Research, v. 87, p. 4163-4170.

_____, Swift, D. J. P., and Hillard, B., 1981, Sediment transport in the New York Bight, North American Atlantic Shelf. Mar. Geol., v. 42, p. 369-398.

Visser, M. J., 1980, Neap-spring cycles reflected in Holocene subtidal large-scale bedform deposits: a preliminary note. Geology, v. 8, p. 543-546.

Wadell, H., 1932, Volume, shape, and roundness of rock particles. J. Geol., v. 40, p. 443-51.

Walker, R. G., 1965, The origin and significance of internal sedimentary structures of turbidites. Proc. Yorkshire Geol. Soc., v. 33, p. 1-29.

Wear, C. M., Stanley, D. J., and Boula, 1974, Shelfbreak physiography between Wilmington and Norfolk canyons, Jour. Mar. Technol. Soc., v. 8, p. 37-48.

Wells, John T. and James M. Coleman, 1981. Physical processes and fine-grained sediment dynamics, coast of Surinam, South America. J. Sed. Petrol., v. 51, p. 1053-1068.

Wentworth, C. K., 1919, A laboratory and field study of cobble abrasion. J. Geol., v. 27, p. 507-521.

Werner, F. and Newton, R. S., 1975, The pattern of large-scale bedforms in the Langeland Belt (Baltic Sea). Marine Geology, v. 19, p. 29-59.

Whitaker, J. H. M. P., 1973, Gutter casts, a new name for scour and fill structures: with examples from the Llandoverian of Ringerrike and Malmaya, southern Norway. Norsk Geol. Tidskrift, v. 53, p. 403-412.

Whitlach, R. R., 1974, Food-resource partitioning in the deposit feeding polychaete _Pectinaria_ _gouldii_. Biol. Bull., v. 147, p. 227-235.

Wilson, I. G., 1972, Aeolian bedforms - their development and origins. Sedimentology, v. 19, p. 173-210.

Winter, J., 1977, Stabile spurenelemente als leit-indicatoren einer tephrostratigraphischen Korrelation (grenzbereich unter - 1 Mitteldevon, Eifel-Belgien. Newsl. Stratigraphy, v. 6, p. 152-170.

# COMPARISON OF SAND RIDGES ON THE NEW JERSEY CONTINENTAL SHELF, U.S.A.

R. W. Tillman
Cities Service Oil & Gas Corp.
Exploration and Production Research
Tulsa, Oklahoma

J. M. Rine*
Cities Service Oil & Gas Corp.
Exploration and Production Research
Tulsa, Oklahoma

W. L. Stubblefield
National Oceanic and Atmospheric Administration
Miami, Florida

Two linear sand ridges from the nearshore and middle portion of the New Jersey Continental Shelf were sampled using vibracores and box cores. Lithologic descriptions were made of the cores based on epoxy peels, X-ray radiographs, and impregnated core slabs and grain size analysis. Vibracores obtained for the study have an average penetration of 6 m (20 ft.) and 95% recovery. Box cores sampled lithologies and relative abundance of physical and biogenic structures found in the upper 25 to 46 cm (9.8 to 18.1 in.) of the sediment. Bottom topographies were established on the basis of 3.5 kHz seismic data.

The nearshore sand ridge sampled (72°22'W, 39°19'N) exceed 5 km (3 mi) in length and ranges up to 2 km (1.2 mi) in width and has a relief of 6 to 10 m (20 to 33 ft.). The mid-shelf ridge (74°08'W, 39°09'N) is nearly 4 km (2.5 mi) long, up to 1 km (0.6 mi) wide, and has a relief of 10 to 11 m (33 to 36 ft.).

Three to four general lithologic units were recognized; these may be common to both ridges. At the base of many of the cores, nonskeletal mud and poorly sorted sands are present; some of the interlayered sands and muds contain laminations and abundant pebbles. Overlying this unit in the nearshore ridge is a shell-rich mud and sand interval that is relatively massive (bioturbated). This lithology was also recovered in one core from the middle shelf ridge. C-14 dates taken from the shell-rich units indicate that the middle and nearshore ridges differ in age by more than 6000 years.

The top unit in all the cores is a fine to medium-grained sand, here termed the upper ridge sand. This unit is similar in both ridges and consists of laminated stacked beds ranging from 3 to 71 cm (1.2 to 28 in.) in thickness, and generally coarsens upward. This unit in the nearshore ridge system has a slightly coarser mean grain-size range (150 to 400µ) than the mid-shelf ridge (130 to 350µ). Both ridges contain alternating laminated and nonlaminated bed sequences.

---

*ERCO, Houston, Texas

## GEOLOGICAL EVIDENCE FOR STORM TRANSPORTATION AND DEPOSITION ON ANCIENT SHELVES

Roger G. Walker
Dept. of Geology, McMaster University,
Hamilton, Ontario L8S 4M1, Canada

### INTRODUCTION

This paper is designed to review the geological evidence for storm deposits, and will only peripherally discuss modern processes. The papers in this volume by Swift give excellent coverage of the day-by-day and year-by-year processes which operate on modern shelves. However, the rarer events have a low probability of being observed or measured, yet the deposits of such events are probably abundant in the geological record. Thus, the record adds to, as well as compliments the body of knowledge acquired by oceanographers and marine geologists. It will be suggested, for example, that turbidites occur in ancient shallow marine situations, commonly with a periodicity of about 1000-10,000 years. There are no well established examples of modern turbidity currents that have deposited preservable beds in a shelf or shallow marine situation. It must be emphasized, therefore, that the geologist will inevitably have a different perspective on shelf storm deposits from that of a geological oceanographer.

### Storm Deposits - Historical Development of the Idea

This part of the notes is chronologically subdivided as follows:

1. 1899, Silurian storm deposits in New York State; discussion of G. K. Gilbert's deduction of 20 m storm waves in the Medina ocean.

2. 1967, Hurricanes Carla and Cindy. The work of Hayes can be taken as the beginning of "modern" geological studies of storm deposits.

3. 1971-1975, Miscellaneous storm deposits. In this section, several storm interpretations are introduced, emphasizing both the nature of the deposit and the emplacing mechanism.

4. 1975, Hummocky cross stratification. This sedimentary structure, present in both siliciclastic and carbonate rocks, is accepted today by most workers as one of the most characteristic features of storm deposits.

5. 1979, HCS interpretations in the literature. The first journal papers to use HCS in their interpretations appeared in 1979, along with the suggestion that turbidity currents might be important in emplacing shelf sands.

6. 1982, Event stratification. This is part of a book devoted to cyclic and "Event" stratification, the event being mostly storm deposits of one type or another.

7.     The lessons of history - problems of storm deposition.

## 1; 1899 - Silurian Storms in New York State

One of the first interpretations of storm waves in the geological record was that of G. K. Gilbert (1899) writing about the Silurian Medina Formation in the Lockport-Niagara area, New York State. On the floors of some quarries, he observed "cross bedding...the oblique structure is of a peculiarly intricate type, often exhibiting dips toward all points of the compass in the same quarry...there are places where the strike of a dipping layer can be traced through an elliptic arc, like the end of a spoon, for 150 degrees...there are places where oblique partings of opposite dip are seen in section to unite at the top, making angular anticlines and other places where they unite below, making smoothly curved synclines...the anticlines and synclines are crests and troughs of sand ripples". Later, Gilbert (1899, p. 136) notes that "these giant ripples range from 10-30 feet [3.3 to 10 m] in wave length; in height, from 6 inches to 3 feet [0.15 to 1 m]". Unfortunately, Gilbert does not give angles of dip of the cross strata, but his descriptions and photographs make these "ripples" possible candidates for hummocky cross stratification (HCS), described in detail below.

Little was known in 1899 of the relationship between ripple wavelength (presumably the orbital diameter of the waves at the bed for these large features), wave height, and wave length. Gilbert deduced from the limited observations and experimental results available to him that "most of the ripple marks are only half as broad as the waves rolling above them are high". We could write this as

$$\lambda = d_o = 2H$$

where $\lambda$ = wavelength of the ripples, $d_o$ = wave orbital diameter at the bed, and H = wave height. From his information, Gilbert concluded that because the maximum wavelength was 30 feet (10 m), "the Medina ocean was agitated by storm waves sixty feet (20 m) high".

We now know that the relationship between H and $d_o$ is given by

$$d_o = \frac{H}{\sinh\left(\frac{2\pi h}{L}\right)} \quad \text{where } h = \text{water depth and } L = \text{wavelength.}$$

A unique solution is not possible, because although $d_o$ can be estimated from the HCS wavelength, neither H, L nor h can be independently estimated. In general, L/H >20; when waves become steeper than this they begin to break.

## 2; 1967 - Hurricanes Carla and Cindy

Although there have been a scattering of storm interpretations since Gilbert's neglected paper, the current vogue for storms can be taken to start with Hayes' (1967) now classic paper on the effects of Hurricanes Carla (1961) and Cindy (1963) on the Texas coast. Hayes demonstrated that Carla set up waves which threw marine debris high onto Padre Island. More importantly, he showed that after Carla, a graded bed was deposited on the shelf off Padre Island. The bed extended at least 15 km offshore into depths of at least 36 m (Fig. 1). It was sharp based, graded, and in places was thicker than 9 cm.

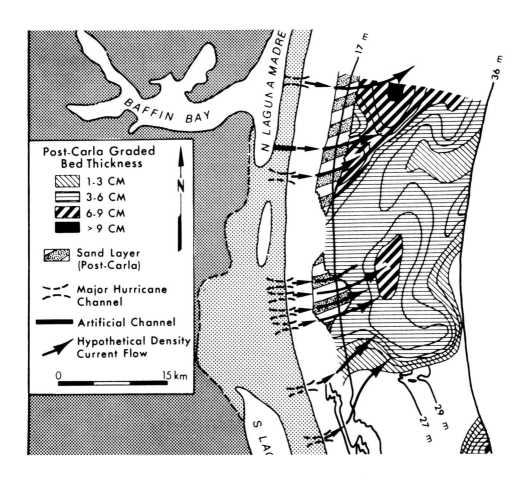

Figure 1. Graded bed resulting from Hurricane Carla (1961) off Padre Island, Texas. From Hayes (1967).

He suggested that the graded bed had been deposited by a "density current", and since the density would have been due to suspended sediment rather than enhanced salinity or cold temperatures, the "density" current would have been a turbidity current. The driving force was believed to be the hydraulic head associated with a 17 foot storm surge tide. Morton (1981) has recently presented evidence that this surge tide occurred at Matagorda Bay (185 km northeast of Padre Island), but was not present at Padre Island. The origin of the Carla graded bed (now destroyed by bioturbation) is therefore in doubt; it may have been introduced by shore-parallel geostrophic currents. Nevertheless, interpretations of storm surge ebb currents in the geological record published during the last 16 years can all be traced back to this work of Hayes (1967).

## 3: 1971-1975 - Miscellaneous Storm Deposits

Some important papers appeared during this time, with the interpretation of storm deposits in both siliciclastic and carbonate rocks. Ball (1971) worked on the Westphalia Limestone (Pennsylvanian, northern midcontinent), and concluded from the presence of mixed faunas, fusilinid concentrations, ripped up clasts, and other criteria that storms had been important in the Westphalia. He emphasizes the onshore transport of debris, and quotes the Hurricane Carla work of Hayes. He begins his paper (Ball, 1971, p. 217) by noting that "the literature suggests a scarcity of ancient storm deposits. This scarcity is possibly one of recognition - not one of existance".

In siliciclastic sediments, our thinking now (1984) is mostly in terms of storm transportation of sand. In the 1971-1975 period under discussion, Hobday and Reading (1972) suggested that storms had been responsible for the erosion of the seaward margins of sand shoals in late Precambrian rocks of north Norway. Accretion of the shoal margins took place during fairweather periods, or possibly during storm waning. The problem of how the eroded sand was transported, and where it went to, was not discussed.

Problems of sand dispersal by storms were reviewed in detail by Goldring and Bridges (1973). The paper discusses several "sublittoral sheet sandstones", with examples from the Upper Devonian Baggy Beds (Devon, England; Goldring, 1971), the Western Interior Seaway (New Mexico, Colorado), and the Devonian Mahantango Formation (Pennsylvania). Goldring and Bridges (1973) anticipated much of our present understanding of hummocky cross stratification (HCS) (see below). They describe sharp based beds with sole marks, with internal cross stratification showing "undulating surfaces of low troughs and swells two to three meters across". This structure had previously been described by Campbell (1966, 1971) as "truncated wave ripple lamination". It is now agreed that truncated wave ripple lamination, and the troughs and swells described by Goldring and Bridges (1973) are HCS. In terms of mechanisms, Goldring and Bridges (1973, p. 743-746) also anticipated much of our present thinking. They suggested five possible processes for the deposition of sublittoral sheet sandstones.

1. Storm waves

2. Storm waves with superimposed tidal ebb or storm surge currents.

3. Turbidity currents: they realized that initiation by slumping was unlikely in the nearshore zone, and suggested "effluent floods highly charged with sediment" or "powerful rip currents channellized at the head of a canyon".

4. Rip currents

5. Tsunamis.

They appear to favor mechanisms 1 and 2, citing Hayes (1967), but cannot be more specific because of "the paucity of detailed description of recent nearshore sediments and of the knowledge of the processes operating in such environments".

## 4; 1975 - Hummocky Cross Stratification (HCS)

The description of hummocky cross stratification (HCS) by Harms (in Harms et al., 1975) is probably the most important geological contribution to storm deposits since Gilbert (1899). It gave a name to a widely distributed sedimentary structure whose significance had not fully been recognized, probably because the multiplicity of earlier names obscured the generality of the structure. Earlier names include "truncated wave ripple lamination" (Campbell, 1966) and "sub-littoral sheet sandstone" (Goldring and Bridges, 1972).

The geometry of HCS is now well known, and is summarized in Figure 2. It occurs

1) in sharp based sandstones generally 5 cm to 1 m thick, interbedded with bioturbated mudstones (Figs. 3, 4), and

2) in amalgamated sandstones several meters thick (Fig. 5) with very thin and discontinuous mudstone partings (or none at all).

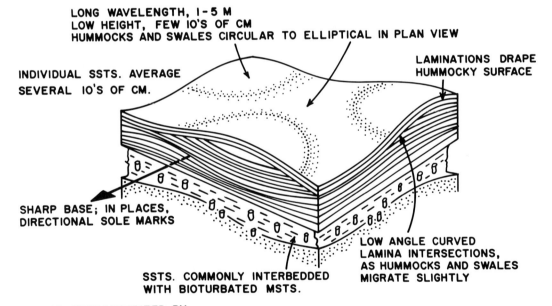

Figure 2. Block diagram showing morphological features of hummocky cross stratification. Diagram is from Walker (1982) and not from Harms et al., 1982 (as mistakenly attributed by Swift et al., 1983, p. 1297).

In plan view, the wavelengths are normally in the 1-5 m range, and heights are a few tens of cm. As the hummocks and adjacent swales shift laterally on the bed during aggradation, they produce a stratification (Fig. 2) characterized by

1) low dips, normally less than about 15°,

2) gently curved laminae, both convex up (hummocks) and concave up (swales), and

3) low angle curved intersections of laminae.

Examples are shown in Figures 3, 4, and 5. Full discussions of the geometry of HCS have been given by Hunter and Clifton (1982), Dott and Bourgeois (1982), Walker (1982) and Walker et al. (1983). In the HCS sandstones alternating with bioturbated mudstones, angle of repose cross bedding is absent or very rare. The implication is that after formation HCS, it is not reworked by "normal" shelf currents (tidal, alongshore, etc.) that form sand waves and dunes, and hence planar tabular or trough cross bedding. If grain sizes are too fine to form medium scale cross bedding, "normal" shelf currents would rework the HCS into current ripples and/or upper flat bed (see Fig. 18). However, in the HCS sandstones alternating with bioturbated mudstones, current ripple cross lamination is rare, and parallel lamination is uncommon.

Figure 3. Interbedded sharp-based sandstones and mudstones within an overall thickening-upward sequence. Sandstones contain abundant hummocky cross stratification. Cardium Formation, Blackstone River, Alberta.

Figure 4. Interbedded sharp-based sandstones and mudstones in Cardium Formation at Blackstone River, Alberta. Stratigraphic top is to left - note person for scale. Lower beds (right) show amalgamated HCS sandstones; arrows mark prominent swale and hummock.

Figure 5  Amalgamated HCS sandstones with almost no preserved mudstone partings. Low angle undulating stratification can be seen best in right half of photo. Cardium Formation at Blackstone River, Alberta, stratigraphic top to right.

---

Harms (1975, p. 87-88) and others have argued that the morphology of the structure suggests formation by waves of long orbital diameter (i.e., storm waves). The interbedding with bioturbated mudstones suggests a normally quiet basin floor, and the absence of reworking suggests depths below normal fairweather wave base. Using these criteria, it is now broadly agreed that HCS is formed by storm waves acting below fairweather wave base, probably combined with a unidirectional flow component (Swift et al., 1983). This raises the problem, addressed later, of how the sand was originally emplaced into such a quiet environment.

There are presently few or no well-authenticated examples of HCS in recent sediments (Swift et al., 1983), and an extremely limited amount of experimental work (Southard, in Harms et al., 1982). The interpretation suggested above is therefore based on evidence from the geological record, and includes the points mentioned above, as well as the characteristic occurrence of HCS sandstones interbedded with mudstones in the lower parts of prograding shoreline sequences. Harms (1975) showed an example from the Gallup Sandstone where bioturbated shelf floor mudstones were overlain by HCS beds, which in turn were overlain by cross bedded shoreface sands and beach sands (Fig. 6). I will discuss other examples later in this paper, and will comment on some of the other implications of HCS. It is clear, however, that the occurrence of HCS is an excellent example of the geological record contributing to our understanding of storm processes in marine geology.

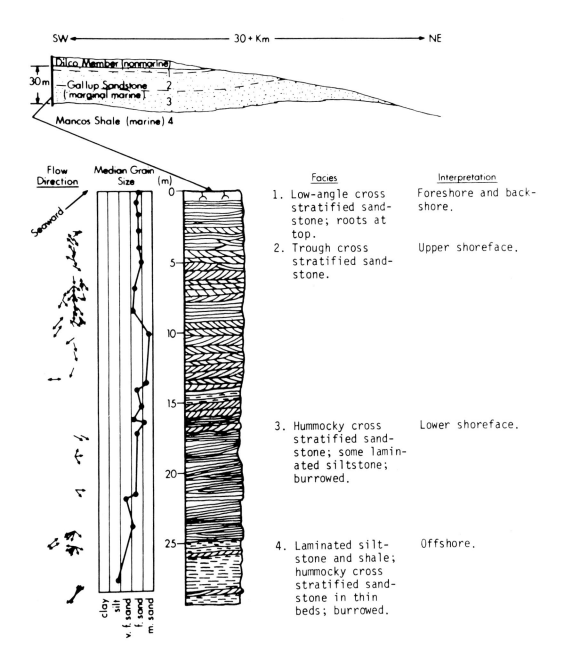

Figure 6. Primary structures and facies sequence of an ancient prograding sandy shoreline (Gallup Sandstone, San Juan Basin, New Mexico). Note beginning of cross-bedded facies at about 17 m below rooted horizon. From Harms et al. (1975).

## 5; 1979 - HCS Interpretations in the Literature

Beginning in about 1979, HCS was used in the interpretation of specific units in the geological record. There are now over 100 examples that either describe HCS by name, or describe the structure sufficiently carefully that it can be identified "between the lines" (Duke, 1985). One of the first papers (Hamblin and Walker, 1979) addressed the problem of how the sand was originally emplaced, and suggested that turbidity currents were the most likely mechanism. This interpretation of the Upper Jurassic section at Banff, Alberta (Fig. 7) was based on the occurrence of sharp based HCS beds immediately overlying classical turbidites. Both the turbidites and HCS beds had essentially identical paleoflow directions (Fig. 7), suggesting emplacement of the HCS beds down the same paleoslope as the turbidites. The most likely process of emplacement for the HCS beds, exclusively <u>down</u> the paleoslope, would also be turbidity currents. Because the local and regional orientation of the shoreline is uncertain, it is not clear whether or not the turbidite-defined paleoslope is perpendicular to the shoreline.

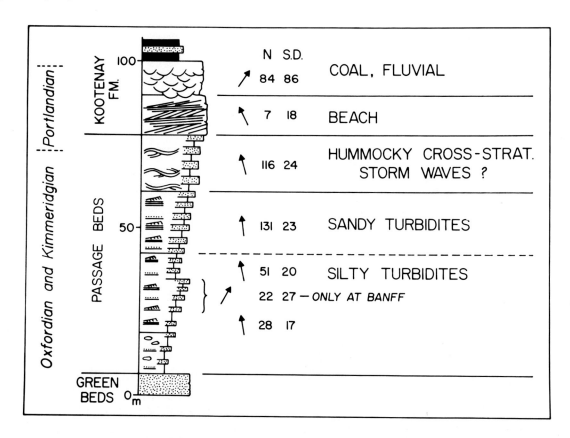

Figure 7. Typical stratigraphic section through the Passage Beds, which are transitional from the Fernie black shales to the coal-bearing Kootenay Formation in southern Alberta. Note consistency of paleoflow direction, with north toward top of page; N = number of readings, S.D. = standard deviation. From Hamblin and Walker, 1979. New highway construction (1983-84) has improved the outcrop of the "beach," which is partly swaley cross stratified, and contains a coal seam overlain by more swaley cross stratification. Thus the sequence is tentatively reinterpreted as shoreface passing into coal, followed by a minor transgression with more shoreface development.

The turbidity current idea was developed into a more general preliminary model by Walker (1979), who combined Hayes' (1967) idea of density currents with storm surge ebb and stratigraphic data from outcrops at Banff (Fig. 8). The model suggested that:

1) deposition below storm wave base resulted in a classical turbidite,

2) deposition between storm and fairweather wave base, under the influence of storm waves, resulted in HCS, and

3) deposition above fairweather wave base might at first result in HCS, but this would subsequently be reworked into cross bedded upper shoreface deposits.

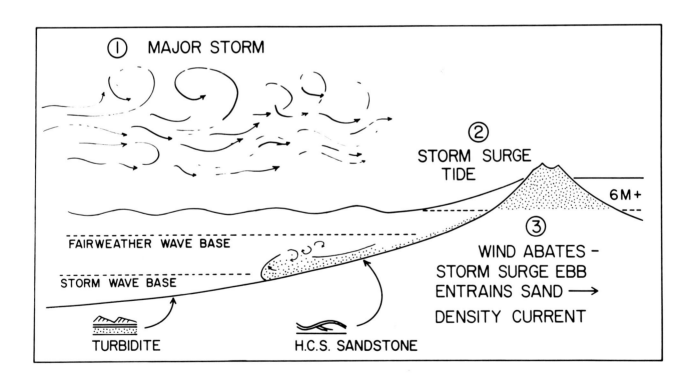

Figure 8. Original model for storm-generation of turbidity currents by storm-surge relaxation, coupled with sediment suspension at the shoreline by storm waves. From Hamblin and Walker (1979) and Walker (1979). A modified version of this model is shown in Figure 33.

A particular problem posed by the Banff section concerns the generation of turbidity currents. Interbeds of HCS sandstones and bioturbated mudstones persist to the upper shoreface, which in turn is dominated by storm-influenced swaley cross stratification (Leckie and Walker, 1982). There appears to be no stratigraphic record of a prograding slope facies in the 200 km long stretch of outcrop between the Crowsnest Pass and Banff (Hamblin and Walker, 1979), nor a pro-delta slope facies, and hence no area where sediment could slump, accelerate, and develop into a turbidity current. Turbidity currents might have been developed directly from rivers in flood sweeping estuarine sand bars seaward, as in the Congo River (Heezen et al., 1964). Alternatively, the model of Figure 8 tentatively suggests that turbidity currents might be developed right at the shoreline by a combination of

1) storm waves suspending sand, and

2) seaward ebb of the storm surge tide, driven by its hydraulic head. If such a suspension could flow initially at about 2 m/sec., the turbulence would continue to suspend sand and the flow might continue seaward for some distance.

A modified and improved version of this model is discussed later in this paper.

There is no data from recent sediments indicating that this mechanism actually operates; however, the recurrence interval of sharp-based sandstones at Banff is of the order of 1000 years, and it may be that marine geologists have not witnessed or recorded the thousand-year event. Swift (this volume) has emphasized that seaward-ebbing surge tides are influenced by Coriolis force, and instead of flowing seaward, they veer to the right (in the northern Hemisphere) to become geostrophic flows following the submarine contours. However, this does not appear likely for the section at Banff, where flows are consistently north-northwestward (Fig. 7), and sediment was supplied from the rising Cordillera. During initial eastward flow of turbidity currents from the mountains, Coriolis force would have deflected flows southward. In the turbidity current interpretation, flows were consistently deflected northward because of the axial dip of the basin between the rising Cordillera and the northern extension of the Sweetgrass Arch (Hamblin and Walker, 1979, p. 1687). The turbidity current model for emplacement of HCS sands is probably only one of several, and I will examine them all in more detail below.

## 6; 1982 - "Event" Stratification

The most recent highlight in this brief review of the history of ideas concerning storm deposition was the publication of "Cyclic and Event Stratification" (Einsele and Seilacher, editors, 1982). It contains 18 papers which discuss event stratification, or the emplacement of calcareous and quartz-sand beds by storms. The deposits are called "tempestites", a Shakespearian-sounding term introduced by Gilbert Kelling (in Ager, 1974).

Despite the current popularity of this term, I strongly recommend that it be abandoned. The deposits of "tempests" are extremely variable, and probably range from small wave ripples to hummocky cross stratified beds one or more meters in thickness. The term also includes winnowed shell lags and graded calcarenites; thus the term can at best be considered as vague! The "Cyclic and Event Stratification" volume is a useful introduction to storm deposits outside North America, and it emphasizes carbonates as well as clastics. I will refer to several specific papers below. There is another very useful review of storm deposits by Marsaglia and Klein (1983). They reviewed the paleogeographic position of Paleozoic and Mesozoic rocks containing storm deposits, trying to relate them to hurricane belts (5-45° latitude) or winter storm belts (latitudes higher than 25°). They tabulate 69 examples of storm deposits, including 16 with HCS (one of these is the same example quoted twice, and one [James, 1980] does not mention HCS), and conclude from their very limited sample of HCS examples that "Mesozoic examples...appear to be restricted to paleo-winter-storm zones, whereas Paleozoic examples include winter storm, one hurricane, and questionable examples" (Marsaglia and Klein, 1983, p. 137). By contrast, the compilation of over 100 examples of HCS by Duke (1985) indicates that over three-quarters of them lie within tropical hurricane belts (for both Paleozoic and Mesozoic examples), and Duke infers that hurricanes are the principal generating agent.

## 7; The Lessons of History - Problems of Storm Deposition

This very brief history enables us to highlight specific problems concerned with storm effects in the geological record. These can be listed:

1) Granted that we can identify the depositional effects of storms, can we reliably identify the erosional effects?

2) To what extent is sediment stirred up and sorted in place, as opposed to being transported for appreciable distances (tens of meters or farther)?

3) How do storms transport sediment?

4) Can turbidity currents be generated by storms, and can they transport sediment across isobaths?

5) Granted that HCS is formed by storm waves combined with unidirectional flows, how is the structure initiated, and how is it modified as storm waves die away?

### FAIRWEATHER WAVE BASE

The concept of "wave base" was introduced by Gulliver (1899) for the ultimate depth of a platform of marine abrasion. Since then, the term has been used by geologists and marine geologists, but it does not appear in the index of oceanography texts.

There are basically two ways of approaching a definition of wave base, one from oceanography, and one from the geological record. It is generally agreed that at depths of $L/2$, there is little or no wave action at the bed, where $L$ is the wavelength of gravity waves at the ocean surface (Fig. 9). However, for geological purposes, a better definition combines

1) depths at which sand moves on a day-by-day basis, and

2) depths at which mud may settle and not subsequently be disturbed.

The wavelength $L$ is a function of wind velocity, and we can therefore distinguish between fairweather and storm waves. Clearly, storm wave base will increase as the intensity of the storm increases, whereas we may be able to define a little more precisely the fairweather wave base.

If we define storm winds as gale force or greater (7 on the Beaufort scale), our maximum fairweather winds can be taken at about 50 km/hour (14m/sec.). These winds produce waves of about 40 m wave length, 6.2 second period. Using the L/2 criterion, the depth at which such waves can just feel bottom is about 20 m. They will not be able to move sand at this depth, so effective fairweather wave base for sand transport must be less than 20 m.

A geological interpretation of fairweather wave base has been suggested by Reineck and Singh (1973), who define a transitional zone between "coastal sand and shelf mud sediments" (Fig. 9). The coastal sand environments include the upper and lower shoreface, and the "lower shoreface [is an environment] where normal waves do not reach and sediment is moved only during storms" (Reineck and Singh, 1973, p. 305). Thus, the lower shoreface/transitional zone boundary is roughly equivalent to or slightly deeper than "normal wave" base, and it varies from "2 to 20 m...on an average it is between 8 and 10 m" (Reineck and Singh, 1973, p. 308). Depths of 10-15 m have been cited from the Atlantic shelf by Dietz (1963), and about 6 m for the outer Jade (North Sea) by Reineck and Singh (1973, p. 318). Swift (personal communication, 1984) suggests depths of 5-15 m for the Atlantic shelf, and 10-20 m for California.

Thus, evidence from modern oceans suggests that fairweather wave base is shallower than 20 m, and in the 5-15, 6, 8-10, 10-15 or 10-20 m range, depending on the wave climate of the shelf in question.

In the geological record there are many prograding sequences that pass from bioturbated shelf mudstones into cross bedded (i.e., current agitated) sandstones and then into non-marine beds. If we assume that for the sandy part of the sequence, stratigraphic thickness roughly approximates depths of deposition (i.e., no complications due to compaction or unusually rapid subsidence compared with rate of supply), then fairweather wave base must approximate the mud-sand contact, or the point at which interbedded HCS sandstones and mudstones pass upward into dominantly cross bedded sandstones. The following list is by no means complete, but gives an idea of where the mud-sand contact occurs below mean sea level, both in ancient and recent examples. The sand thickness given, therefore, approximates fairweather wave base.

GALLUP SANDSTONE (Harms et al., 1975). Cross bedded sandstone above HCS sandstone and mudstone, and shelf mudstones. 17 m (see Fig. 6).

GALLUP SANDSTONE (McCubbin, 1982, p. 262). Nearshore marine sandstone above marine shale and siltstone. 14 m or 27 m.

BRANCH CANYON SS., CALIFORNIA (Harms et al., 1975, quoting Clifton, 1973, unpublished field trip notes). Trough cross bedded sandstones above burrowed fine sandstone and siltstone. 15 m or 21 m.

GALVESTON ISLAND (McCubbin, 1982, p. 257). Barrier island sands overlying shelf silt and clay. 7 m.

GULF OF GAETA (Reineck and Singh, 1973, p. 316). Cross bedded sandstones above very strongly bioturbated fine sand. 6 m.

OXNARD-VENTURA, CALIFORNIA (McCubbin, 1982). Sands overlying highly bioturbated silt or clay. 17 m.

$J_2$ SAND, CRETACEOUS, NEBRASKA (Exum and Harms, 1968, p. 1854). Sandstones overlying shales. 8m.

BELL CREEK, CRETACEOUS, MONTANA (Davies et al., 1971). Sandstones overlying bioturbated siltstone and claystone. 6m.

The average thickness of sandstone in the above examples is about 11-13 m.

Thus, at least two lines of evidence, oceanographic and geologic, converge on numbers in the 5-15 m range for the depth to which waves operate on a day-to-day basis -- i.e. fairweather wave base (Fig. 9). This is an important concept geologically, inasmuch as fairweather wave base separates environments dominated by day-to-day processes from those which are normally quiet (with sand only moving during storms). It is also convenient to define the base of the lower shoreface at the same point where sands pass seaward into muds at fairweather wave base (Fig. 9). However, there is no concensus about this definition, and many workers consider the lower part of the lower shoreface to be storm rather than fairweather influenced.

### Fairweather Wave Base - Conclusion

For this review, I will define the base of the lower shoreface at the point where sands pass seaward into muds. This transition is controlled by fairweather wave base, which is normally shallower than 20 m, and commonly in the 5-15 m range (Swift, personal communication, 1983) Fig. 9).

## RECOGNITION OF STORM DEPOSITS IN THE GEOLOGICAL RECORD

Storm deposits may stand out in some way as "unusual" or "unexpected" beds within the "normal" geological record. An example might be a sharp-based graded bed with a basal shell coquina, surrounded by many meters of monotonous, bioturbated shelf mudstones. Another example might be a sharp-, perhaps erosive-based sandstone with internal upper flat bed parallel laminations, interbedded with flaser-bedded tidal flat mudstones. The sharp-based sandstone could be a storm washover from a barrier island into a lagoonal tidal flat.

Although storms can also give rise to unusual rainfalls, and hence rivers in flood, I will restrict my comments to marine environments. In general, it appears that storm deposits are very hard to recognize in shoreline and nearshore deposits, because the storm damage is commonly repaired by normal, day-to-day processes. These processes would include tidal and alongshore currents, rip currents, and small ephemeral wind-driven currents. We may, therefore, expect to find most recognizable storm deposits below fairweather wave base, in depths where day-to-day reworking is minimal or absent (Fig. 9).

An analysis of the literature suggests five types of evidence that may be used to identify storm effects:

1) Rare and unusual examples of shoreline storm effects, such as storm movement of large boulders at rocky shorelines (Cambrian storm boulder accumulations in Wisconsin, Dott, 1974).

2) In situ winnowing of the sea floor, commonly detected by the formation of shell lags (e.g., Kreisa, 1981).

3) Transported shell accumulations, commonly with the shells in sharp-based, graded calcarenite beds (e.g., Kelling and Mullin, 1975).

4) Storm scouring leading to the development of hardgrounds and condensed horizons (e.g., Aigner, 1982b).

5) Sharp-based sandstone layers, commonly with HCS, interbedded with bioturbated mudstones (e.g., Wright and Walker, 1981).

These various deposits will be discussed in turn below.

---

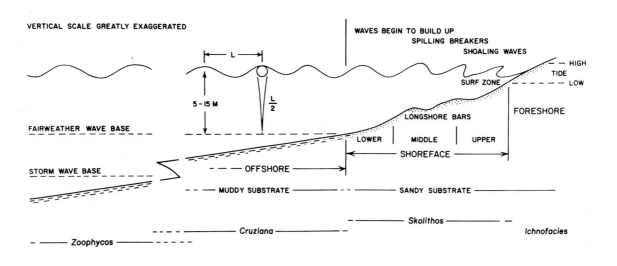

Figure 9. Diagram defining shoreface, fairweather wave base, and effective limit for waves (wavelength L) feeling bottom. The bottom of the lower shoreface is taken at the point where sand no longer moves on a day-to-day basis, and where mud normally settles out.

## Storm Scour Forming Hardgrounds and Condensed Sequences

Interpretations of hardgrounds as storm-scoured surfaces appear to be relatively new. A hardground is a surface "at which early diagenetic cementation has produced a lithified sea floor" (Bromley, 1978, p. 397), and is "thus the product of submarine preburial diagenesis alone". Several authors writing in "Cyclic and Event Stratification" (Einsele and Seilacher, 1982) have suggested that storm scouring of the softer substrate, down to well compacted or semi-lithified levels, may contribute to hardground formation. Also, renewed storm scouring in one area may give rise to condensed sequences.

One of the best descriptions is that of Aigner (1982b), who described burrowed "firmground" horizons from nummulite accumulations (Nummulites gizehensis) in the Eocene of Egypt (Fig. 10). The nummulitic beds are 0.5 to 1.5 m thick, and are interbedded with marls 1 to 50 cm thick. The bases of the nummulitic beds have large scale scours and smaller erosive pockets and potholes (Fig. 10), with burrows of the Glossifungites-facies (Seilacher, 1967) indicating stable coherent substrates. Within beds there are scour and fill structures several tens of cm in diameter. Aigner (1982b, p. 256) suggests that epidosic winnowing by storms is responsible for the removal of the original lime mud substrate, and that scouring down to compacted levels may form "firmgrounds".

Aigner (1982b, p. 257) also describes upper Eocene shell beds from Egypt 0.5 to 1.5 m thick (Figs. 11, 12). They show the following features:

1) Shell bed development on surfaces burrowed by Spongeliomorpha indicating firm substrates (Kennedy, 1975).

2) Colonization of the firmground surface by Carolia (an epibyssate bivalve). There may also be colonizations on internal erosion surfaces within shell beds.

3) Change in fauna, to ostreids or Plicatula, the ostreids being cemented on Carolia shells and in turn encrusted by corals.

4) Erosion of some or all of the proceeding shell beds, producing an in situ shell hash. This hash may be burrowed by Ophiomorpha.

5) Beds containing endobenthonic organisms (living within the soft sediment substrate, e.g., spatangids [echinoids] and Turritella) may be preserved in the sequence. Alternatively, the endobenthonic organisms may be preserved as lags, with the fine sediment winnowed away.

Aigner suggests that a "major storm event" may initiate the firmground substrate which is then colonized by the shelly faunas described above.

Condensed horizons, with many zonal faunas present within an unusually thin sedimentary sequence, have also been attributed to storm scouring of a soft substrate by Hagdorn (1982) and Gebhard (1982 - both very detailed examples). The evidence for storms consists of the absence of the fine sediment in which the shelly faunas normally live (winnowed out and moved into deeper parts of the basin), and the reworking and concentration of the shells themselves. The presence of several stratigraphic zones in a thin sedimentary succession indicates persistant storm activity.

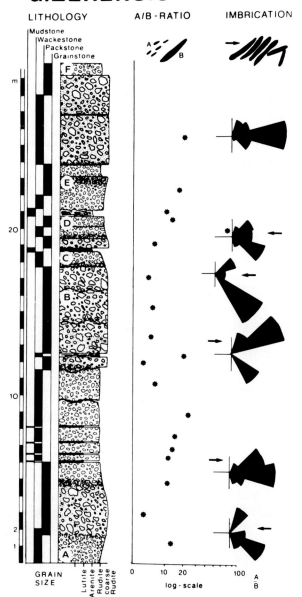
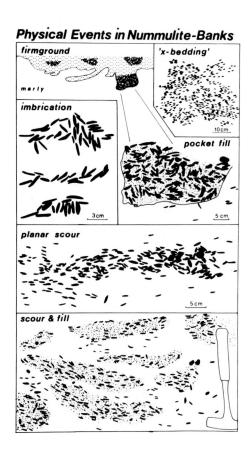

Figure 10. Evidence of scouring and reworking of Middle Eocene Nummulites gizehensis Bed, Egypt. In sketch, note extensively burrowed firmgrounds with erosive relief, erosive pockets filled with biosparite, imbrication, stratification reminiscent of cross-bedding, nummulite concentrations on planar planar scours and small scale scour and fill structures. In stratigraphic section, note bed thickness and variable flow directions suggested by the imbrication. Both diagrams from Aigner, 1982b.

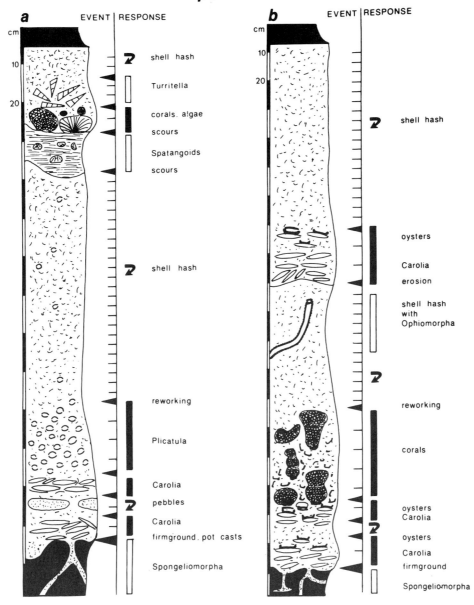

Figure 11. Shell beds (Upper Eocene, Egypt), showing complex series of erosional/depositional events, and specific biological responses. See text for details. Black bar shows epifaunal response, open bar shows endofaunal response. Column a shows "Plicatula-bed", and column b the "Ostrea-bed". Details in text. From From Aigner, 1982b.

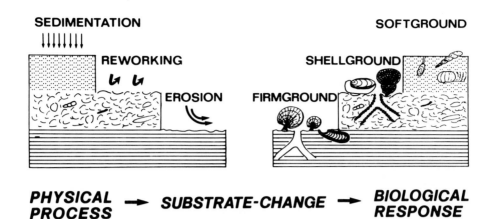

Figure 12. Physical processes of sedimentation, reworking and erosion alter substrate conditions and are followed by specific biological response on the newly created softgrounds, shell pavements, or firmgrounds. Details in text. From Aigner, 1982b.

---

Storm Scour Forming Winnowed, Autochthonous Shell Beds

Accumulations of shells, seen in the geological record as coquinas, have commonly been explained by storm winnowing, as discussed above. Recent descriptions include those of Brenner and Davis (1973), Kreisa (1981), Futterer (1982), Kreisa and Bambach (1982), Fursich (1982) and Aigner (1982b).

The shell accumulations are found interbedded with bioturbated shales and mudstones, or with bioturbated calcilutites. Thus, the "normal" background sedimentation appears to be quiet. The authochthonous shell concentrations tend to have sharp, in places scoured bases, although Brenner and Davies (1973, p. 1693) note that the bases of their "swell lag" deposits "do not appear to be erosional". Kreisa and Bambach (1982, p. 201) note "sharp erosional bases, gutter casts, and tool marks" (Fig. 13), suggesting considerable unidirectional scour, and Aigner (1982b) describes "scour and fill structures" and "erosive pockets and potholes". Aigner (1982b, p. 250) also suggests that "major hydrodynamic events (hurricanes ?) eroding the sea floor down to already compacted levels are inferred to have exposed these laterally fairly persistant firmground surfaces".

Within the shell beds, a variety of textures can be found. Brenner and Davies (1973) describe shells parallel with bedding, convex side up, with beds generally thinner than 20 cm, but rarely up to 3 m. The shells are whole, or nearly whole, thin-shelled bivalves of the genus Camptonectes, which was a free swimmer. Whole shell accumulations only occur in mudrocks, and the concentrations appear to be due to winnowing, the mud being suspended by storm waves.

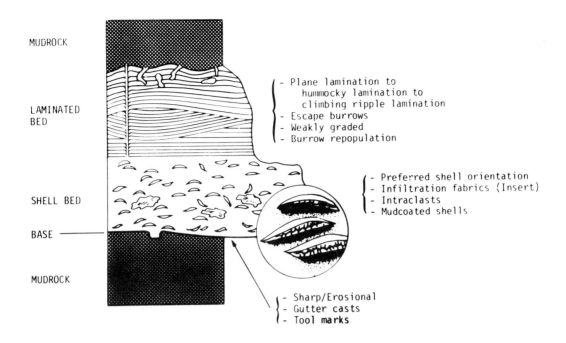

Figure 13. Model for autochthonous storm generated shell bed overlain by laminated fine sand and silt. Ordovician Martinsburg Formation, Virginia. From Kreisa and Bambach, 1982.

---

Kreisa (1981) described shell packstone beds mostly 2-10 cm thick (range 1-40 cm) from the Ordovician Martinsburg Formation of Virginia (Fig. 13). Shells tend to be parallel to bedding (75%) and convex up (2:1). The shells are commonly unbroken and not abraded, and the faunal composition and proportion is the same in the packstones as in the interbedded mudstones (Table 1). The lack of abrasion and faunal similarity of mudstones and packstones is the best evidence of winnowing in situ without appreciable transport. Kreisa also notes that many of the shells are mud-coated or mud-filled, and that the mud is identical to that of the underlying shale. Identical eroded shale intraclasts can also be present within the shell packstone.

Above the packstone there is a laminated unit, composed of carbonate or terrigenous clastic sediment. Terrigenous sediment consists of very fine sand and silt-sized, angular to subangular quartz. Kreisa does not give analyses of the interbedded shales, so it is not clear whether the very fine sand and silt has been winnowed from the shales in situ, or whether it has been transported to the depositional site from elsewhere. Sedimentary structures observed (Fig. 13) include horizontal plane lamination, wave ripple cross lamination, hummocky cross stratification and climbing wave ripple lamination. Horizontal plane lamination is the most abundant, wave ripple cross lamination has wavelengths of 5-15 cm and heights of 1-3 cm, and hummocky cross stratification has wavelengths of one to several meters and heights up to 30 cm.

TABLE 1

Comparison of Relative Abundance of Calcite
Skeletons From Shell Packstones and Interbedded Shales
(from Kreisa and Bambach, 1982, after Plants, 1977).

|  | 10 Collections From Shales | 7 Collections From Shell Packstones |
|---|---|---|
| Sowerbyella | 27% | 30% |
| Zygospira | 28% | 22% |
| Dalmanella | 12% | 17% |
| Ramose Bryozoa | 14% | 14% |
| Rafinesquina | 5% | 5% |
|  | 86% | 88% |

Kreisa (1981, p. 836) notes that "commonly, individual beds consist of only one of the above types of lamination, horizontal plane lamination being the most abundant. However, in many beds several types of lamination are associated in ordered vertical sequences, though no bed contains all types. Vertical transitions from one lamination type to the next are commonly gradational. Plane lamination and/or hummocky stratification typically occur at the base of laminated beds, overlain by climbing wave ripple lamination. This is overlain by wave ripple cross lamination, or a thin matrix-rich laminated cap (<1 cm), or climbing wave ripple lamination may flatten upward into matrix-rich plane lamination". Kreisa (1981) also notes that "evidence of unidirectional traction transport is extremely rare in laminated beds" and suggests that "the hummocky stratification and much of the plane lamination formed by rapid deposition of sediment while wave surge was still relatively strong".

Kreisa (1981) continues his paper with a very interesting and useful discussion of storm deposits, estimating a preserved storm frequency of 1 every 1200 to 3100 years (calculated from the number of observed storm beds and an estimated 8-10 million years for Martinsburg deposition). Other storm frequencies are given later in this paper.

## Autochthonous Shell Beds - Conclusions

The evidence for autochthonous accumulations can be summarized as
1) faunal similarity of shell beds and interbedded shales,

2) shells packed together, commonly parallel to bedding and convex up, not in life position,

3) presence of whole shells, and minimal evidence of abrasion, and

4) mud coatings or fillings of the shells similar to the interbedded muds.

A storm origin for such accumulations can be suggested

1) if the background sediment indicates a normally quiet basin,

2) if there is some evidence of scouring at the bases of the shell beds, and

3) if the shell accumulations grade up into clastic sediments with wave formed (especially HCS) sedimentary structures.

There will probably be a continuum of facies in the geological record from those reworked in situ, to those transported a few tens of meters, to those transported hundreds of meters or kilometers. The allochthonous accumulations, described below, will show more evidence of faunal mixing and shell abrasion, and will have more chance of terrigenous clastic material being mixed in with the shells. Sole marks may show a preferred orientation, whereas erosion in situ might result in less regular orientations.

## Storm Deposits With Transported Shells

There is almost certainly a transitional set of facies between the in situ reworked shells discussed above, and the storm-emplaced shell beds described here. Where the bases of beds are sharp and/or erosional, it is very difficult to know how much transportation (as opposed to winnowing) has taken place. The best clue is probably the degree of mixing of faunas, as discussed above.

Allochthonous storm-emplaced shell beds have been described by several authors, notably Aigner (1982a), Brenner and Davies (1973), Cant (1980), and Kelling and Mullin (1975). In all of these examples, the beds range from a few cm to several tens of cm in thickness. Bases of beds are invariably sharp and commonly erosive, with both tool and scour marks. Scours up to several meters wide and several tens of cm deep are common. Elongate scours known as "gutter casts" also occur, and it is increasingly apparent that gutter casts are associated both with carbonate and siliciclastic storm deposits.

Inside the bed, the shelly component normally occurs at the base, with shells lying in the plane of bedding, convex up. Intraclasts of the substrate that have been ripped up during the storm emplacement are commonly mixed with the shells. The shelly or coquinoid layer grades up into finer calcarenite and calcilutite, or into siliciclastic sand and silt. Most authors describe the storm beds as graded. Above the shell layer, hummocky cross stratification may be present (Aigner, 1982a; Cant, 1980), and Kelling and Mullin (1975) describe a massive to parallel laminated to ripple cross laminated sequence, noting that "such units bear an obvious resemblance to the typical sequence of internal structures found in turbidite beds (Bouma, 1962)".

The tops of the beds are in places rippled (interference or symmetrical ripples; Fig. 14), or are bioturbated. The storm layers are interbedded with "normal" background sedimentation - bioturbated calcareous muds, or bioturbated silty mudstones (Figs. 13, 14).

## IDEAL TEMPESTITE-SEQUENCE +HYDRODYNAMIC INTEPRETATION

| | BEDFORMS | FLOW REGIME | SEDIMENT. RATE |
|---|---|---|---|
| | pelitic division | LAMINAR FLOW | very low |
| | wave ripples | LOWER REGIME | moderate-low |
| | plane lamination | UPPER FLOW REGIME | high |
| | graded bedding | redeposition of suspended detritus | very high |
| | erosional contact | storm erosion | |
| | pelitic background sedimentation | | very low |

Figure 14. Idealized "tempestite" sequence (note similarity to a Bouma sequence) for Upper Muschelkalk Limestones in SW Germany. From Aigner, 1982a.

---

The various features have best been summarized by Aigner (1982a) in an "ideal tempestite sequence" developed primarily from examples in the Upper Muschelkalk Limestones of southwestern Germany (Fig. 14). The best discussions of process are given by Kelling and Mullin (1975) and Aigner (1982a). It appears that storm scouring of the substrate is necessary for winnowing and concentrating the shells (as in the autochthonous beds discussed above), but that a superimposed unidirectional flow is necessary to transport the winnowed shells, mix in the sand and silt, and finally deposit a graded bed (Aigner, personal communication, 1983). This unidirectional flow must be powerful enough to cut channels and gutter casts, and rapid enough to form upper flat bed parallel lamination (Kelling and Mullin, 1975; Aigner, 1982a; Fig. 14). These unidirectional flows are regarded as turbidity currents by some authors, and as "storm surge ebb flows", better termed relaxation flows, (perhaps superimposed on tidal currents) by others. Few authors have presented sufficient paleocurrent data to be able to assess whether dispersal is perpendicular or parallel to isobaths.

## Terrigenous Clastic Storm Deposits

This class of deposits appears to be very common in the geological record. In one "end member" type, sharp-based sandstones typically 5 to 100 cm thick are interbedded with roughly similar thicknesses of bioturbated mudstones (Figs. 3, 4). The typical and characteristic sedimentary structure in the sandstones is hummocky cross stratification (HCS).

This facies poses three main problems.

1) What environment is indicated by the bioturbated mudstones?

2) What is the hydrodynamic implication of the HCS?

3) How was the sand transported into the basin?

### Bioturbated Mudstone Depositional Environment

In most examples of HCS described so far, the bioturbated mudstones are regarded as having been deposited below fairweather wave base. This is based on multiple lines of evidence. First, the lithology itself suggests deposition in quiet water, where the fines can settle out. Second, the mudstones commonly contain a trace fauna indicative of grazing, with members of the Cruziana and Zoophycos ichnofacies (Crimes, 1975; Frey and Seilacher, 1980; Pemberton and Frey, 1983, Frey and Pemberton, 1984). There are many common ichnogenera in units interpreted as storm deposits, tabulated by Pemberton and Frey (1983, p. 111). For example, in the Upper Cretaceous Cardium Formation of Alberta (Fig. 15; see Cardium Formation 4, this volume), the trace fauna includes Chondrites, Cochlichnus, Cylindrichnus, Diplocraterion, Gyrochorte, Muensteria, Ophiomorpha, Paleophycus, ?Phoebichnus, Planolites, Rhizocorallium, Rosselia, Skolithos, Thallassinoides and Zoophycos (Pemberton and Frey, 1983).

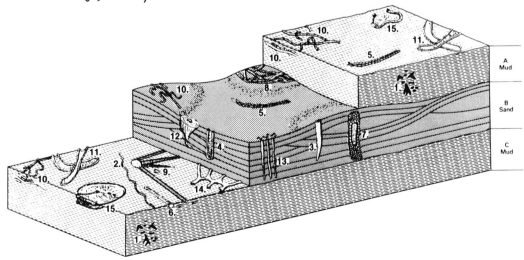

Figure 15. Schematic representation of common trace fossils in the Cardium Formation at Seebe, Alberta. A and C represent background mud sedimentation, and B is a sharp based, hummocky cross stratified sandstone. 1 - Chondrites; 2 - Cochlichnus; 3 - Cylindrichnus; 4 Diplocraterion; 5 - Gyrochorte; 6 - Muensteria; 7 - Ophiomorpha; 8 - Paleophycus; 9 Phoebichnus; 10 - Planolites; 11 - Rhizocorallium; 12 - Rosselia; 13 - Skolithos; 14 - Thalassinoides; 15 - Zoophycos.

Third, there is some sparse information on foraminiferal paleoecology to suggest depths below fairweather wave base. Again using the Cardium example, the arenaceous foraminifera *Haplophragmoides* sp., *H. howardense*, *H. crickmayi*, *Reophax* sp., *Pseudobolivina* sp., *Trochammina* sp., *Dorothia* sp., *Verneuilinoides bearpawensis* and *Ammobaculites* sp., suggest environments below fairweather wave base, and probably deeper than 50 m (C. Mahadeo, Amoco Canada, personal communication, 1982; Walker, 1983b).

Implication of HCS

The sandstones interbedded with the bioturbated mudstones are characterized by HCS. There are two lines of evidence to suggest deposition below fairweather wave base.

1) The nature of the interbedded mudstones, and

2) the fact that in the interbedded HCS-mudstone facies, there is seldom, if ever, any medium-scale (10-50 cm) angle of repose cross bedding and rarely any current ripple cross lamination. This suggests formation of HCS in depths where once formed, it cannot normally be reworked by fairweather processes.

The storm wave interpretation of HCS was first suggested by Harms (in Harms et al., 1975). The interpretation was based on:

1) the morphology of the HCS, with cross sections resembling giant low amplitude symmetrical ripples, and

2) the fact that below fairweather wave base, the only likely candidate for making the symmetrical bedform was storm waves.

This interpretation has not been verified by direct observations in recent sediments (scuba diving in 50 m of water during a hurricane does not seem to be a favorite pastime of students of recent sediments), and is supported by very little experimental work. In situ instrument arrays have recorded flow conditions, but have not photographed or otherwise monitored hummocks and swales on the sea floor. Southard (in Harms et al., 1982) has assembled experimental data suggesting that for wave periods between 2 and 20 seconds (deep water wavelengths of 6 to 625 m), an increase in maximum orbital speed of the waves results in the sequence (Fig. 16) 2-D oscillation ripples $\longrightarrow$ 3-D oscillation ripples $\longrightarrow$ flat bed with sand movement. He notes (1982, p. 2-37) that "wave-molded hummocky bedding surfaces observed in the ancient are strikingly similar in geometry to the bed surfaces associated with experimentally produced three-dimensional vortex ripples (Fig. 17)... it therefore seems a reasonable possibility that hummocky cross stratification is produced by bidirectional oscillatory flow, although some more complicated kind of multidirectional flow might alternatively have been responsible".

Once formed by the storm waves, the HCS (in the HCS-mudstone facies) is only modified by the formation of small scale symmetrical or interference ripples on the top surface.

If we accept the formation of HCS below fairweather wave base by storm waves, the problem of how the sand was first emplaced into this environment becomes important -the various possibilities are discussed below.

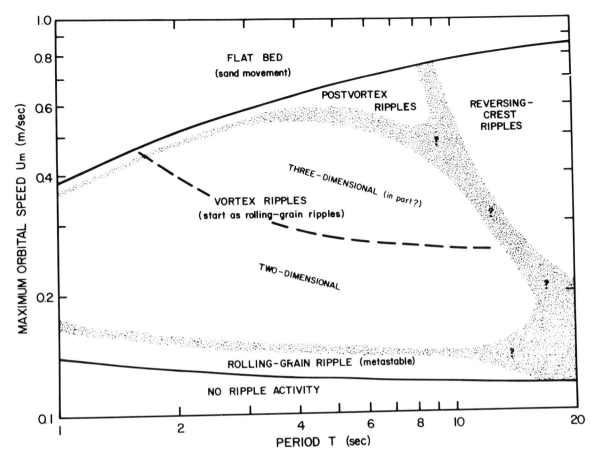

Figure 16. Schematic plot of types of oscillation ripples as a function of maximum orbital speed and oscillation period T for sand sizes of 0.15-0.21 mm. Shaded areas are partly transition zones and partly areas of especially great ignorance. From Southard, in Harms et al., 1982.

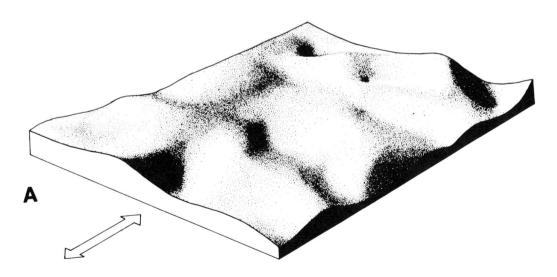

Figure 17. Bed surface formed in 0.19 mm sand by flow in an oscillatory-flow tunnel with maximum orbital velocity 0.56 m/sec and oscillation period 3.5 sec. Arrow shows directions of flow. Width of surface perpendicular to flow 35 cm. Note similarity of surface geometry with (much larger) HCS. From Southard, in Harms et al., 1982.

## HCS in Recent Marine Sediments

The occurrence of bedforms believed to be similar to those of HCS has recently been reviewed by Swift et al. (1983). They note that "the characteristic large scale linguoid pattern of tidal megaripples is sometimes visible in the storm megaripples of deeper areas (>20 m), but the bedforms are more rounded and exhibit less asymmetry than is characteristic of tidal megaripples or no asymmetry at all, and the steepest slopes do not exceed 12 degrees. A distinct slip face is not apparent". Swift et al. (1983) quote the shelf areas of New Jersey, Maryland, Virginia, North Carolina and Argentina as having similar "hummocky megaripples", although they note that the features "were not so designated" in the reports. However, more data is needed on the relationship between specific "hummocky megaripples" and known storm events. Also, internal structures need careful delineation by box coring (preferably a row of box cores across one entire "hummocky megaripple"), and the control of the rounded morphology by combined flow processes needs elucidating. The present data from marine geology can only be described as tantalizing.

## How is Sand Emplaced Below Fairweather Wave Base?

It has been established above that the general depositional environment for sharp-based sandstones with HCS, interbedded with mudstones, is below fairweather wave base. Normal currents which move sand on the bed, day-by-day, tend to form abundant ripples, dunes and sandwaves, which are preserved in the geological record as various forms of cross-lamination and cross-bedding (Fig. 18). These "normal" tidal and alongshore currents do not effectively move sand below fairweather wave base, nor is current ripple cross lamination and cross bedding present in the HCS facies. The problem of what types of episodic currents might be responsible for introducing the sharp-based HCS sands has been tackled by Goldring and Bridges (1973) and Kelling and Mulling (1975), among others.

I believe that the geological record has much to contribute to our understanding of processes in the general area of rare storm events. However, some consideration of modern processes is also appropriate. Following Swift, Stanley and Curray (1971), we can list the _normal_ shelf currents as

1) tidal currents

2) intruding ocean currents

3) density currents (controlled by salinity or temperature)

4) meteorological currents.

The meteorological, or _storm_ currents, can be subdivided into five main types (and here I combine the ideas of many authors summarized in Walker et al., 1983, and Swift et al., 1983).

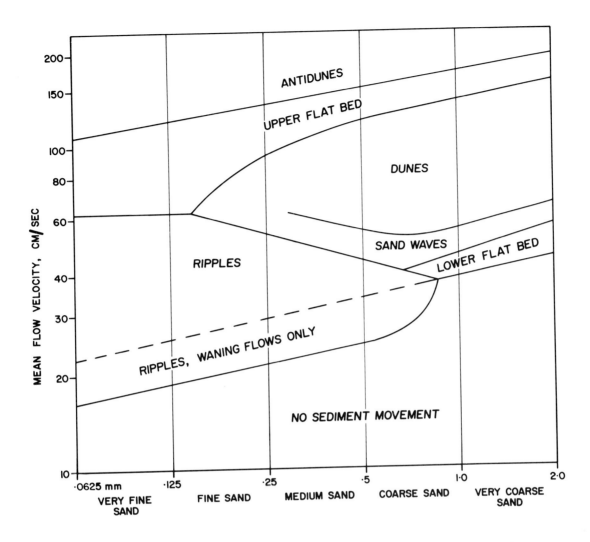

Figure 18. Size velocity diagram for flow depths of 18-22 cm. The original diagram, plotted by Southard (in Harms et al., 1982), shows all data points and sources of data.

1) rip currents

2) incremental storm flows

3) storm surge ebb currents

4) wind-forced currents

5) turbidity currents (the special case of density currents where the excess density is due to suspended sediment).

Let us examine how the geological record contributes to our understanding of these processes.

1. Rip Currents. There is some evidence for rip currents in the geological record. Rip currents return water seaward after it has been forced onshore by winds and waves (Fig. 19). The rips cut seaward from the beach through the breaker zone for up to 500 meters (Swift et al., 1983), and then spread out and die. Rip current deposits can be recognized as graded coarse, commonly conglomeratic, layers which are found a few meters stratigraphically below beach deposits. They have been described by Leckie and Walker (1982), and also occur in cores below Cardium beaches in Kakwa field and below Fahler (Albian) beaches in the Elmworth area, both in Alberta.

Although rip currents may effect local seaward sediment transport, this sediment stands a chance of being reworked by fairweather processes, because the rip currents do not continue below fairweather wave base. They are certainly inadequate to explain sharp-based HCS beds below fairweather wave base.

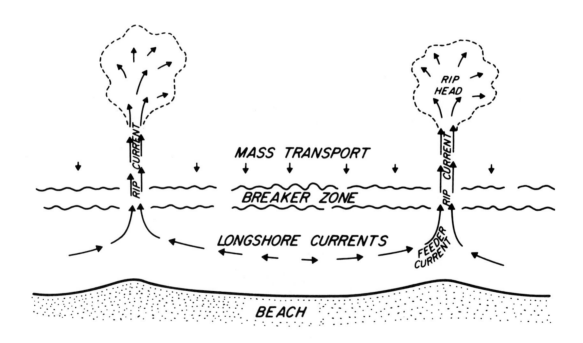

Figure 19. Nearshore cell circulation showing feeder longshore currents, rip currents, and slow mass transport of returning water to the foreshore. From Komar, 1976, after Shepard and Inman.

2. _Incremental Storm Flows_. This category of flows includes wind-forced storm flows with periodicities ranging from several flows per year to one flow every five or ten years. Each event is capable of moving sand a certain distance (increment) toward its final depositional site, and the cumulative effects of many such events would be to move considerable volumes of sand offshore (Swift et al., 1983). Incremental movements may well be capable of moving sand abruptly into areas of former mud deposition, producing sharp-based sand beds, and Swift (personal communication, 1984) has vibracored many examples of sharp-based graded beds on the Atlantic shelf. Combined flows (geostrophic currents plus oscillatory flows produced by the passage of storm waves) could produce hummocky cross stratification. However, if storm waves are not feeling the bottom, unidirectional bed load movement of sand would produce ripple cross lamination and medium scale angle-of-repose cross bedding. Swift (personal communication, 1984) reports horizontal lamination, but it is not clear if this is equivalent to "upper flat bed" of experimental workers. Flow depths are probably too great to produce upper flat bed except in very fine sands.

It has been pointed out many times (e.g., Swift et al., 1983) that fairweather processes cannot and do not move sand from the shoreline out onto the shelf. Swift et al. (1983) state that "no river sand is presently escaping from Atlantic Coast estuaries", but that the problem of where the shelf sand comes from "has been resolved...[it is supplied] by erosional shoreface retreat; submarine contours out to the 15 m isobath are presenting shifting landward at rates of 1-3 m yr.$^{-1}$...a significant proportion of the eroded material moves offshore during storms...during many storms the entire inner 3 km of the coastal water column takes the form of a downwelling coastal jet with mean core velocities in excess of 60 cm sec$^{-1}$, and an offshore component of near-bottom flow on the order of 10 cm sec$^{-1}$...offshore sand transport beneath such coastal storm flows has been calculated as $10^2 - 10^4$ cm$^3$ cm$^{-1}$ sec$^{-1}$ (Niedoroda et al., in press)".

Incremental sand movement is very common on the inner shelf (the "inner 3 km of the coastal water column" for wide shelves), and may continue to the shelf edge "80 km out [where we get] 30, 40 and 50 cm/sec flows" (Swift, personal communication, 1984). At these velocities, one would predict an abundance of current ripples, yet in many geological situations (discussed below), ripple cross lamination is rare or absent. I will, therefore, examine less "incremental" and more "catastrophic" processes below.

3. _Relaxation (Storm Surge Ebb) Currents_. Ever since Hayes' (1967) discussion of the Hurricane Carla graded bed off Padre Island, relaxation (storm surge ebb) currents have been regarded as extremely effective mechanisms for rapidly and catastrophically moving sand seaward.

During the storm, commonly of hurricane proportions, the wind piles water onshore, raising a surge tide (Fig. 8). Heights above normal high tide levels commonly reach 3-4 m, and the 17 foot (6 m) figure cited for Hurricane Carla (Hayes, 1967) is one of the highest on record. Surge tide persists for about 2.5 to 5 hours (Gross, 1982) before the wind changes and the surge tide subsides. During the period of surge tide, there is a constant offshore/alongshore flow of water, which is being driven by the hydraulic head and Coriolis force. However, as Swift (personal communication, 1983, and in press) has emphasized, the flow is dominantly geostrophic and moves parallel to isobaths and hence roughly parallel to the shoreline. Swift (personal communication, 1983) has also estimated that during a three-day storm, the geostrophic discharge is 1000 to 2000 times as great as the relaxation discharge. Since I have put turbidity currents (with sand in suspension) into a separate category, we must regard "storm surge ebb" currents as essentially driving sand as bedload.

The problems with this mechanism concern the effectiveness of such flows in moving sand seaward for long distances - in the case of the Cardium (discussed later), HCS sands were moved tens of km (possibly up to 200 km) from their source at the shoreline. Very few modern studies contribute to this problem. In 1967, Hayes considered that the Carla graded bed (Fig. 1) was the result of a density current (i.e., turbidity current, because the density was due to suspended sediment), generated by the 6 m surge tide. However, Morton (1981) has presented data showing that the 6 m tide occurred at Matagorda Bay (some 185 km northeast of Padre Island), and that tides were only about 2.5 m higher than normal at Padre Island during Hurricane Carla. In Laguna Madre, tides were below normal (Morton, 1981, p. 390). It is clear that the Carla graded bed was not the result purely of storm surge ebb, and the problems are discussed in detail by Morton (1981).

Nelson (1982) has suggested that graded sand layers could be the deposits of storm surges on the Bering Shelf, north of the Yukon Delta (Fig. 20). The evidence consists of graded sand layers encountered in cores (Fig. 21). Nelson (1982, p. 539) comments that

> "Because of the similarity to the vertical sequence of structures defined by Bouma (1962) for turbidite beds, I have chosen to designate shelf structures by substituting the T of Bouma's $T_{a-e}$ designation with a capital S, designating shallow water, graded storm-sand beds. The idealized vertical sequence of sedimentary structures includes basal parallel laminated medium to fine sand ($S_b$), a center section of the sand layer with cross and convolute lamination ($S_c$), upper parallel-laminated very fine sand to coarse silt, commonly containing laminated beds of epiclastic plant fragments ($S_d$), and an upper mud cap ($S_e$) that is absent in many places".

Nelson (1982) (Fig. 21) shows that these graded beds can be found nearly 100 km from the Yukon Delta, but beds farther than 60 km from the delta contain only silt-grade sediment, finer than 4 phi. He proposed (1984, p. 541) three possible mechanisms for deposition of the graded beds. "One, associated with inshore sub-ice channels, may be the sudden high river discharge at the time of spring breakup...the other two proposed processes, wind-forced currents and storm-surge ebb-flow currents, both appear to be responsible on the basis of limited oceanographic data in this remote region". I suggest there are several problems wih these interpretations.

1) Only one of the many sand layers can perhaps be related to a known storm and was deposited "throughout Southern Norton Sound some time after 1970 and probably was deposited during the 1974 storm surge". No data is given for the height of the 1974 surge, but extreme surges during the 1900's have been as high as 5 m (Sallenger et al., 1978). Swift (personal communication, 1983) has suggested that there were no storm surge ebb currents, but that the whole shelf water mass was moving as a geostrophic storm current, driven by a sea surface slope of perhaps 50 cm or more over 200 km.

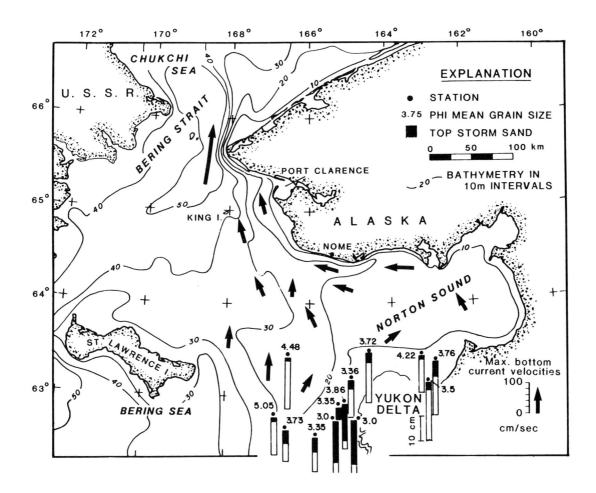

Figure 20. Bottom water currents and bathymetry of the northern Bering Sea. also shown are thickness and distribution of a surface sand layer observed in cores taken after the 1974 storm surge. From Nelson, 1982.

2) If the sand moved <u>incrementally</u> for 100 km, there would be no specific relationship between a given graded bed and a known storm surge.

3) Nelson does not mention or discuss the more likely possibility that the graded sand layers ("a mimic of Bouma sequences and turbidite systems") actually <u>are</u> turbidites, perhaps generated by the initial stages of storm surge but then flowing for considerable distances across Norton sound as sediment-laden density currents (i.e. turbidity currents driven by gravity rather than elevated water surfaces), and depositing <u>real</u> Bouma sequences.

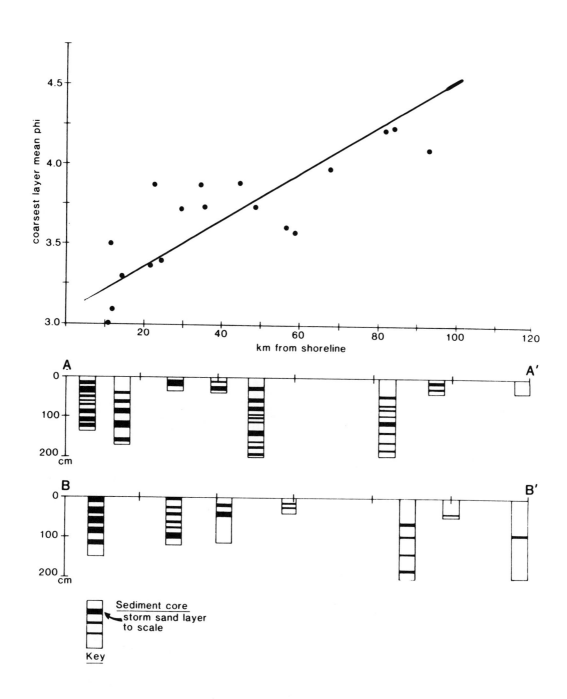

Figure 21. Subsurface distribution and mean grain size change in graded sand layers with distance from the Yukon Delta shoreline. Line AA' showing vertical distribution and thickness of storm sand layers runs northwestward from western margin of delta, and BB' runs north from eastern margin of delta. From Nelson, 1982, with regression line added by R. G. Walker. Note that in cores over 60 km from the delta the coarsest layers are silt (4 ∅) or finer.

The emplacement of these beds from a single storm surge ebb current (i.e., a bottom current) can probably be eliminated. If the storm surge were maintained for as long as 10 hours, the mean flow velocity would have to be about 2.8 m/sec if the flow is to continue for 100 km driven by the hydraulic head above. At this velocity, silt, fine and medium sand would all be in suspension, not moving on the bed, and the flow would closely resemble a turbidity current. Alternatively, the beds cored by Nelson may have been emplaced incrementally, as the result of a series of wind-forced geostrophic flows (Swift, this volume), not one storm surge ebb (Swift, personal communication, 1984). Finally, I suggest that we take the cored Bouma-like sequences at their face value, and consider the possibility that the beds were emplaced by real turbidity currents generated by unknown processes (? slumps) in the area of the Yukon Delta.

4. Wind-Forced Currents. When storm winds blow consistently across the ocean, surface waters are set in motion as a wave drift current. These surface layers may entrain deeper and deeper layers until the wind-forced ocean current is feeling bottom and capable of moving sand. The deflection of the flow to the right in the northern hemisphere with increasing depth is known as the Ekman spiral (Fig. 22), and if the bottom is sloping will result in sand transport paths parallel to isobaths, the sand dispersal being a function both of wind direction and depth (i.e. degree of deflection in the Ekman spiral).

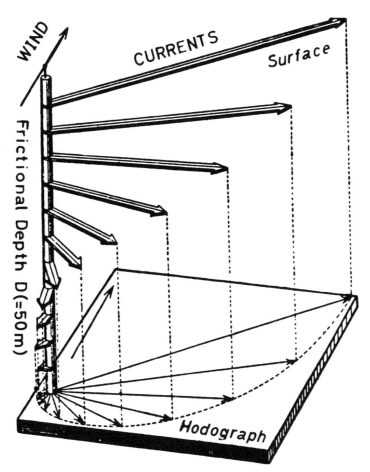

Figure 22. Ekman spiral, showing progressive deflection to the right (northern hemisphere) with depth of surface drift current. From Defant, 1961.

There is some data on wind-forced currents from modern shelves that can be utilized directly in geological interpretations. The three most commonly cited studies are those of Hurricane Camille in 1969 (Murray, 1970), Tropical Storm Delia in 1973 (Forristall et al., 1977) and storms on the Atlantic Shelf (D.J.P. Swift and colleagues).

4A. Hurricane Camille. Murray's current meter and anemometer records are shown in Fig. 23. The meter was installed in a depth of 6.3 m, 300 m offshore. It was about 160 km east of the hurricane landfall at Gulfport, Miss. (Fig. 24). Murray (1970) has divided the records into three phases.

A. No storm winds (Fig. 23A).

Bottom flows about 5-10 cm/sec, toward 270-300° (alongshore).
Breakers on beach near current meter location 1.5 to 2.0 m high.

B. No storm winds, intensified bottom current (Fig. 23B).

Bottom flows increase to about 35 cm/sec, direction more westerly (270°). In absence of storm winds, the "explanation for the observed increase in current speed during phase B" is that the flows appear to have been "a longshore current generated in the surf zone set in motion by lateral friction the water beyond the outer bar" (Murray, 1970, p. 4580).

C. Storm winds (Fig. 23C).

Bottom flows gradually intensifed to about 1.0 m/sec, with pulses to 1.6 m/sec. Current meter impeller jammed at 1900 hours, Sunday August 17. Flow directions gradually changed from westerly (toward 270°) to southerly (offshore, about 160°). This was a counter-clockwise rotation, but the flow was in water too shallow (less than 15 m) to experience much Ekman veering (Swift, personal communication, 1984). A two layer flow was probably generated, with "an upper layer moving onshore with the wind, and a lower layer flowing offshore" (Murray, 1970, p. 4582). Note that the offshore flow velocities were about 1.0 m/sec, peaking to 1.6 m/sec. These would be capable of making dunes (and upper plane beds in shallow flows) in medium sand.

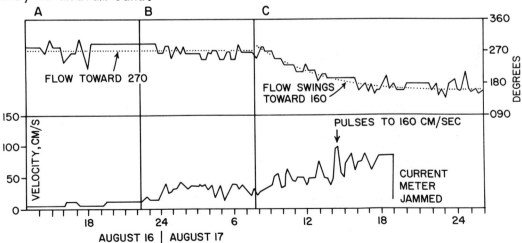

Figure 23. Anemometer and current water data for Hurricane Camille, August 16-18, 1969. Location of instruments shown in Figure 24. Stages A, B, and C discussed above. Redrawn from Murray, 1970.

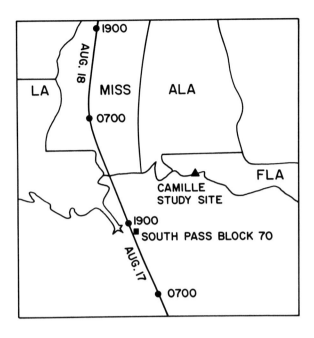

Figure 24. Location of eye of Hurricane Camille, Aug. 16-18, 1969, and location of instruments (black circle). Redrawn from Murray, 1970.

---

From about 1100 hours on Sunday, the wind directions gradually shifted from easterly (from 90°) to southerly (from 180°). This resulted in water being driven onshore, resulting in the offshore flow (storm surge ebb) described above.

It is unfortunate that the impeller of the current meter jammed, because the duration and magnitude of the offshore flow would have been very interesting. Figure 23 shows that it persisted for at least 12 hours.

4B. Tropical Storm Delia. Current meters were installed on three Shell drilling platforms in the Gulf of Mexico, and the data (Fig. 25) were collected 40 km seaward of Galveston Island in a depth of 20 m (Forristall et al., 1977). Velocities measured 3 m above the bed were directed offshore and peaked at 50-75 cm/sec. They remained above 50 cm/sec for about 5 hours. Alongshore velocities reached nearly 2.0 m/sec, and remained above 1.0 m/sec for about 6 hours. As the windspeed died down, the offshore velocities dropped to about 10 cm/sec, and the alongshore velocities dropped to about 40 cm/sec.

Storm surge measurements at Galveston showed a surge of 1.3 m preceeding the maximum velocities in Figure 25. However, maximum surge of 2.0 m was recorded 30 hours after storm landfall (0200 hours, Sept. 6). This is 35 hours after peak alongshore flow, and 38 hours after peak offshore flow. This indicates that the Delia currents were wind-forced currents, not storm surge ebb currents (Morton, 1981, p. 392-3).

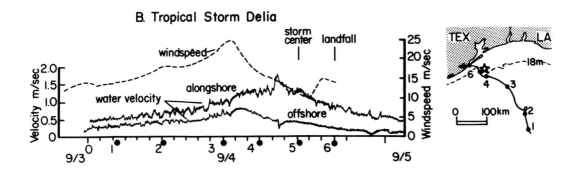

Figure 25. Alongshore and offshore current velocity profiles during Tropical Storm Delia, Sept. 3-5, 1973. Data from Forristall et al. (1977) as drawn by Morton (1981).

---

At the current meter location, the peak flows could have produced dunes and possibly upper flat bed (Fig. 18), but the decreasing flows would have reworked these bedforms into ripples (assuming a fine sand bed; Fig. 18). It is not clear whether new coastal sand was introduced onto the shelf during Camille and Delia, but it is almost certain that there was scouring and/or local transport of bottom sediment. The bedforms made by the storm would probably have been reworked into current ripples as the storm subsided, so that the net geologically preservable results of the storm would be a cross-laminated sand with local <u>in situ</u> winnowing.

4C. <u>Data from Atlantic Shelf</u>. Swift and co-workers have published several records of storm flows on the Atlantic Shelf, for example, off the Maryland Coast (Swift and Field, 1981); off Long Island (Lavelle et al., 1978; Swift et al., 1979; Clark et al., 1982); and in the New York Bight (Vincent et al., 1981). In general, near bottom current meters indicate that "during the winter months, flows sufficiently intense to exceed the threshhold velocity may last for hours or days, and may occur several times a month" (Swift et al., 1979; see Fig. 26 for data from the 10 m isobath off Tobay Beach, Long Island). Off Tobay Beach, tidal current speeds approach the threshhold (18 cm sec$^{-1}$), but efficient sediment suspension was only achieved during the storm of Oct. 19-21 (Fig. 26). The suspended sediment could then be advected by the shore-parallel storm flow.

A similar pattern is shown by data off the Maryland Coast, from a current meter mounted 100 cm above the sea floor, on the 20 m isobath about 6 km offshore (Fig. 27, from Swift and Field, 1981). The thirty-day record shows that "velocities in excess of 30 cm sec$^{-1}$ occurred for periods of hours to days at approximately 10 day intervals, and were accompanied by an intensified wave regime" (Swift and Field, 1981, p. 476). This current meter location is on the landward flank of a linear sand ridge. In the trough adjacent to the ridge there are sand waves which survive from year to year, and which have southward-facing steeper slopes. They are hundreds of meters apart, and have heights of less than 1 m. Swift and Field (1981, p. 477) note that "the bedform pattern... can be interpreted [as a response] to the observed flow regime." Also, "the southward and offshore flows seen during the observation month [Fig. 27] appear to be representative of conditions during sandwave formation".

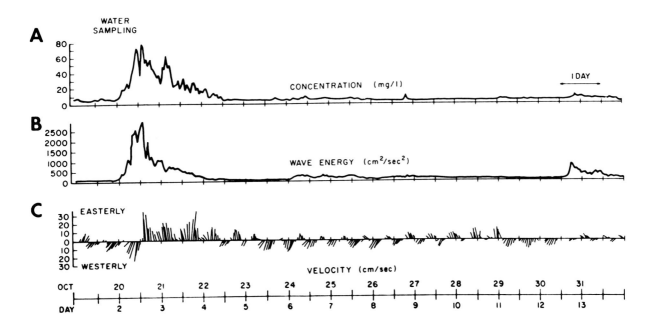

Figure 26. Data on sediment concentration, wave energy and flow velocity (presented as vectors) for the Oct. 19 - Nov. 1 period, 1976. Instruments were at the 10 m isobath off Tobay Beach, Long Island. Note strong tidal influence on current vectors, Oct. 23-30, and storm of Oct. 19-21. Data was sampled once per second for an eight minute burst every hour. From Swift et al., 1979.

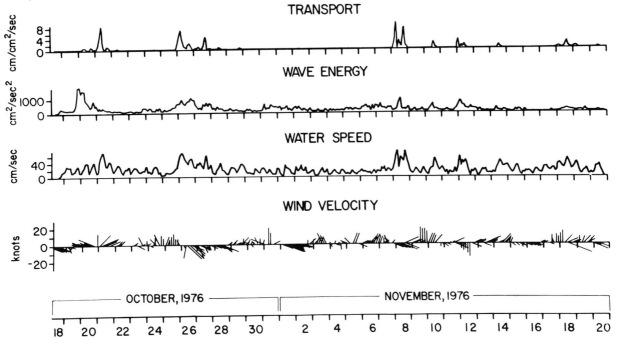

Figure 27. Sediment transport, wave energy, water speed and wind velocity during the period Oct. 18 - Nov. 20, 1976. Instruments were on the landward flank of a linear ridge 6 km off the Maryland coast, on the 20 m isobath. Note that velocities in excess of 30 cm sec$^{-1}$ occurred for periods of hours to days at approximately 10-day intervals. From Swift and Field, 1981.

Thus, they suggest that flows of the order of 30 cm sec$^{-1}$ could be responsible for the formation of the long, low sandwaves. The internal structure of these sandwaves is not known, but may be dominated by ripple cross lamination - velocities closer to 50 cm sec$^{-1}$ are required in unidirectional flow flume experiments to make sand waves in medium sand (Southard, in Harms et al., 1982, p. 2-14). The velocity increases to about 60 cm sec$^{-1}$ for sand waves in fine sand.

These three examples of wind-forced currents in the Gulf of Mexico and on the Atlantic Shelf suggest that this is an important mechanism for incremental movement of sand, and for the formation of current ripples, small sand waves, and sinuous crested dunes. These bedforms show up in the geological record as various forms of ripple cross lamination and medium scale angle-of-repose cross bedding. Wind-forced currents may also be important in producing storm-stirred graded beds and autochthonous graded shell lags.

This category of wind-forced currents overlaps considerably with the category of incremental storm flows, discussed earlier. The geologically preservable results of wind-forced currents may be sharp-based graded beds, with or without hummocky cross stratification. Each preserved bed represents the final stage of what may have been a long series of incremental sand movements. Measured current strengths and pictures of the sea flow suggest that wind-forced currents should leave ripple cross lamination and (in fine sand and coarser sizes) medium scale cross bedding as a record of their passage.

5. Turbidity Currents. The evidence for turbidity currents in modern oceans comes partly from data on the breaking of submarine telegraph cables in relatively deep water. There is little direct evidence for turbidity currents on modern shelves in depths shallower than 100 m, perhaps for three main reasons.

1) In many places, turbidity currents are rare events, with recurrence intervals of hundreds of thousands of years.

2) There are probably few places where turbidity currents could be generated at or near the shoreline, thence to flow seaward.

3) In most shelf depths (<100 m), the deposits of a turbidity current stand a very good chance of being bioturbated and/or reworked by storm wind-forced currents, making turbidites very difficult to recognize in shelf cores.

The generation of turbidity currents at or near the shoreline is a major problem when the process is invoked in basins such as the Cretaceous Western Interior Seaway. Here, there is apparently no "slope" (no Continental Slope, no major delta front slope) into the basin, and hence a limited shoreface zone in which the potential turbidity currents could accelerate. In the Congo River of Africa, however, turbidity currents have been generated within the estuary of the river at times of 1, peak river discharges and 2, during years when the Congo River is changing its path among the estuarine sand bars, and is sweeping sand into the Congo Canyon head (Heezen et al., 1964). The canyon head lies within the estuary of the river. The axial gradient of the first 10 km of the canyon is 0.025, or about 1.43 degrees. Averaged over the 80 km from the canyon head to the edge of the shelf, the gradient is about 0.014, or about about 0.79 degrees. These estimates are from the bathymetric maps published by Shepard and Emery (1973). By contrast, the slope in the epicentral area of the Grand Banks earthquake was about 2.3 degrees (Heezen and Drake, 1964), and one of the largest known turbidity currents was generated (Piper and Normark, 1982, 1983).

If a slope of 0.025 (1.43 degrees) were projected for the margin of the Western Interior Seaway, it need only persist for about 2 km before depths of about 50 m would be reached. The slope could thence flatten gently, and turbidity currents could possibly persist for many tens of kilometers before depositing all of their load. The distance travelled would depend on many factors, particularly the rate at which the slope flattened, the scale of the flow, and its density contrast with the surrounding sea water.

## Geological Evidence for Turbidity Currents

I suggest that this is one case where the geological record contributes in a major way to our understanding of shallow marine processes, because the rare event (turbidite) can be preserved, recognized and interpreted. There are now perhaps five studies which supply evidence for storm-associated turbidites in "shelf" settings.

1) Carboniferous of Morocco, Kelling and Mullin, 1975.

2) Ordovician of Norway, Brenchley et al., 1979.

3) Jurassic of Alberta, Hamblin and Walker, 1979.

4) Triassic of Germany, Aigner, 1982a.

5) Cretaceous of Alberta, Walker, 1983a; and "Cardium Formation 4," this volume; and in press.

In all of these papers, there is independent evidence of a shelf or shallow marine setting. There is also independent evidence of storm action, as well as beds with turbidite characteristics.

1. __Carboniferous of Morocco.__ Here, Kelling and Mullin (1975, p. 179) describe graded limestones "with markedly erosive bases [which] suggest deposition within shallow channels eroded through the adjacent sediments". Also, "within the graded limestones the basal flutes and grooves, the occasional mud clasts, the grading and the large percentage of fine grained matrix all suggest deposition by turbidity currents". Kelling and Mullin (1975, p. 179) consider that these beds "differ from typical deep water turbidites in being strongly lenticular, and in lacking consistently developed upward sequences of internal structure divisions conforming to the Bouma sequence or its recognized variations" (see "Comparison of Shelf Storm Deposits and Deep Water Classical Turbidites", this volume).

The interpretation here centers on the nature of the beds rather than their context (see Hamblin and Walker, 1979; Walker, "Comparison of Shelf Storm Deposits and Deep Water Classical Turbidites", this volume). The same comment can be applied to the studies of Brenchley et al. (1979) and Aigner (1982).

2. Ordovician of Norway. In this study, Brenchley et al. (1979) also describe beds with Bouma sequences (Fig. 28). The beds, 0.5 to 10 cm thick, contain graded bedding, planar lamination and small-scale cross-lamination. If two or more structures are present, they are arranged vertically from base to top in the above sequence. They calculate an index based on these sedimentary structures akin to the ABC index of Walker (1967), and demonstrate an overall upward increase in the index from "open shelf" through "lower shoreface" into "upper shoreface". Brenchley et al. (1979) summarize the basis of their interpretation (p. 208) noting that

1) the sandstones represent very rare events,

2) the events were of short duration,

3) the currents fluctuated in strength but sometimes progressively waned.

Although a turbidity current origin had been suggested by Seilacher and Meischner (1964) for these rocks, Brenchley et al. (1979, p. 209) list some apparent objections.

1) the thin lower Paleozoic cover of the Baltic Shield, with interpreted depths of 0-200 m seemed "inappropriate" for turbidites. I believe this to be an insufficient reason for rejecting turbidites, despite the fact that Brenchley et al. (1979, p. 209) cite Drake et al. (1972), who found in one study of "recent sediment suspensions in a shelf environment [that there was] no evidence to suggest...concentrations sufficient to drive them downslope as turbid layer flows". However, it should be pointed out that the evidence in the geological record may indicate processes that, because of their low recurrence interval, stand little chance of being observed today.

2) the sequence with turbidites "passes up rapidly but transitionally into upper shoreface sandstones...a passage not usually associated with turbidites" (Brenchley et al., 1979, p. 209). I suggest that the interpretation should be based on the nature of the beds. If they can and must be interpreted as turbidites, perhaps we should re-evaluate the occurrence of turbidites in prograding shoreline sequences (see 3. Jurassic of Alberta, below).

3) The sequence of internal laminae, "particularly the interspersed mud laminae commonly differs from that associated with turbidites". It is difficult to assess this particular objection to a turbidity current origin.

The alternatives offered by Brenchley et al. (1979, p. 210-211) seem weaker than the turbidity current suggestion. Their "ebb currents from hurricane surges" (p. 213) could so easily have suspended sand that the flow would in fact be a turbidity current (Fig. 8), driven by gravity acting on the density difference between the flow and seawater, rather than a bottom flow being driven by the hydraulic head of the hurricane surge.

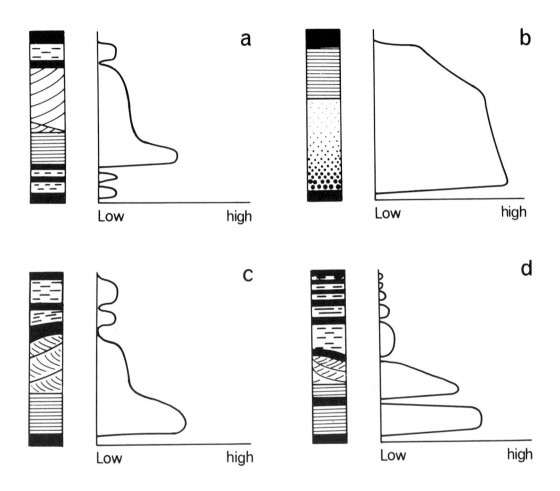

Figure 28. Internal structures of selected sandstone beds (circa 5 cm thick) in Ordovician storm deposits, Norway. a) Mud-sand lamination, overlain by planar lamination, overlain by cross lamination, overlain by mud-sand lamination. b) Graded to planar laminated bed (AB in Bouma terminology).
c) Planar lamination, overlain by cross lamination, overlain by mud-sand lamination (BC [?D] in Bouma terminology). d) Planar lamination with mud interbed, overlain by cross lamination, overlain by mud-sand lamination. From Brenchley et al. (1979).

3. __Jurassic of Alberta__. The sequence (Fig. 7) described by Hamblin and Walker (1979), modified slightly by recent evidence revealed in new highway cuts at Banff, is:

top    4) shoreline.

3) swaley cross stratified sandstone (? storm dominated upper shoreface), originally termed beach in Figure 7.

2) HCS sharp based sandstones alternating with bioturbated mudstones,

base  1) classical turbidites,

The evidence for turbidity current emplacement of the hummocky cross stratified sands lies in their relationship to the classical turbidites below, and the shoreface above. The turbidites below contain Bouma sequences, mostly BC and C types. Division A is absent, probably due to the fine and very fine grain sizes of the sands. These sizes tend to remain in suspension in the turbidity current until it has slowed down to velocities associated with Bouma's division B. However, the Bouma sequences only record the last few minutes or hours of deposition. The beds could be classical turbidites in the sense that the sediment was transported rapidly from the shoreline in one catastrophic event, down the dip of the paleoslope; alternatively, the beds could have been transported a short distance (the final "increment") by a powerful storm-generated geostrophic flow along the strike of the paleoslope.

It is absolutely vital to the following argument that the Banff "turbidites" be established as classical turbidites emplaced down the paleoslope. There are three important lines of evidence.

1. Within the turbidite sequence, there is no evidence of __wave__ ripples, implying transport and deposition below __storm__ wave __base__. If the beds had been emplaced by geostrophic flows, it seems likely that some modification of the bed by wave or combined flows would have taken place.

2. At Banff, the turbidite sequence is at least 220 m thick (Hamblin and Walker, 1979, p. 1675), with more turbidites below the point at which we started measuring. These were omitted because of structural complications. This thickness alone, even without any wave ripples, is more indicative of deep turbidite deposition than geostrophic flows on shelves. The final turbidite lies about 110 m below the newly exposed coal seam (which in turn overlies shoreface sediments).

3. The turbidite sequence contains a relatively restricted trace fauna, but the sinuous fecal filled burrows of __Cosmorhaphe__ are prominent. This trace is accepted as characteristic of the __Nereites__ ichnofacies, which is the deepest of the various ichnofacies (Seilacher, 1967; Crimes, 1975; Ekdale et al., 1984, p. 232 - 253). None of the major "mining" structures such as __Zoophycos__, __Rhizocorallium__ and __Teichichnus__ is present in the turbidite facies.

These points combine to suggest that the Bouma BC and C beds at Banff are truly turbidites, not geostrophically emplaced sands where the deposition from a storm flow onto the bed produced upper plane bed conditions waning into a rippled bed. Sands emplaced by geostrophic storm flows may indeed show some Bouma sequences; however, I would also expect somewhere in a 200 m thickness to encounter many <u>wave</u> ripples and some HCS, and a trace fauna that was <u>not</u> characterized by a <u>Nereites</u> ichnofacies with abundant <u>Cosmorpaphe</u>. The following argument therefore assumes

1. that the lower beds at Banff are classical turbidites, and

2. that the flows travelled <u>down</u> the dip of the paleoslope toward 347 degrees.

It is clear from Figure 7 that the paleoflow directions for the hummocky cross stratified beds are identical with those of the turbidites. Because it has been established that the turbidites flowed down the paleoslope, we may now argue that the sharp-based hummocky cross stratified beds were also emplaced by currents that were <u>constrained to flow down the paleoslope</u>. That is, the HCS beds were emplaced by <u>turbidity currents</u>, <u>not</u> by storm-generated geostrophic flows. It follows that if the turbidity currents deposit their sand below storm wave base, Bouma sequences are generated and preserved. If the flows deposit sand <u>above</u> storm wave base, the storm waves acting on the bed during deposition suppress the Bouma sequence, and HCS is developed in its place.

A final consequence of this model concerns the <u>generation</u> of turbidity currents. In the Banff section, HCS sandstones interbedded with bioturbated mudstones persist to the base of the shoreface. It is therefore necessary to hypothesize how turbidity currents might be generated within or landward of the shoreface, because there is no stratigraphic record at Banff of a slope on which the turbidity currents might have been generated and down which they might have accelerated.

A model for flow generation (Fig. 8) was developed specifically with the Banff section in mind, by Hamblin and Walker (1979) and Walker (1979). It proposed suspension of sand at the shoreline by storms, and a combination of storm surge relaxation and increased density to get the turbidity current moving. Swift (pers. comm., 1984) has criticised this idea, commenting that the density may not be sufficient to cause the flow to move down the paleoslope - instead, the sand would be distributed essentially parallel to isobaths by geostrophic currents. Later in this paper, a modified version of this idea will be put forward.

4. Triassic of Germany. Aigner (1982a) has proposed an idealized "tempestite" sequence (Fig. 14) for the Upper Muschelkalk Limestones of S.W. Germany. The beds strongly resemble turbidites in their sedimentary structure sequence, differing only in the presence of wave ripples rather than current ripples (Fig. 14). Aigner is a little ambiguous as to whether the beds are <u>in situ</u> shelly graded beds, or whether the shells and calcilutite have been transported into the basin. However, he comments (personal communication, 1983) that "the association of structures...suggests to me the complex interaction of <u>in situ</u> reworking with lateral sediment transport by storm flows". His facies model (Fig. 14) and proximal-distal changes (Fig. 29) also suggest flow within the basin, and not simply <u>in situ</u> reworking. He suggests that "stratification similar to the Bouma-sequence suggests that many limestone beds

formed under high energy conditions in a waning flow regime [and that] some of the beds, mainly the thinner and finer grained ones, were rapidly deposited as one-event beds" (Aigner, 1982a, p. 185). He cautiously refrains from describing the beds as turbidites, but offers no alternative explanation for the sedimentary structure sequences and proximal-distal trends, except to relate everything to storms. Storms may indeed be important, but the point surely is <u>how</u> the storms generate sediment-transporting currents.

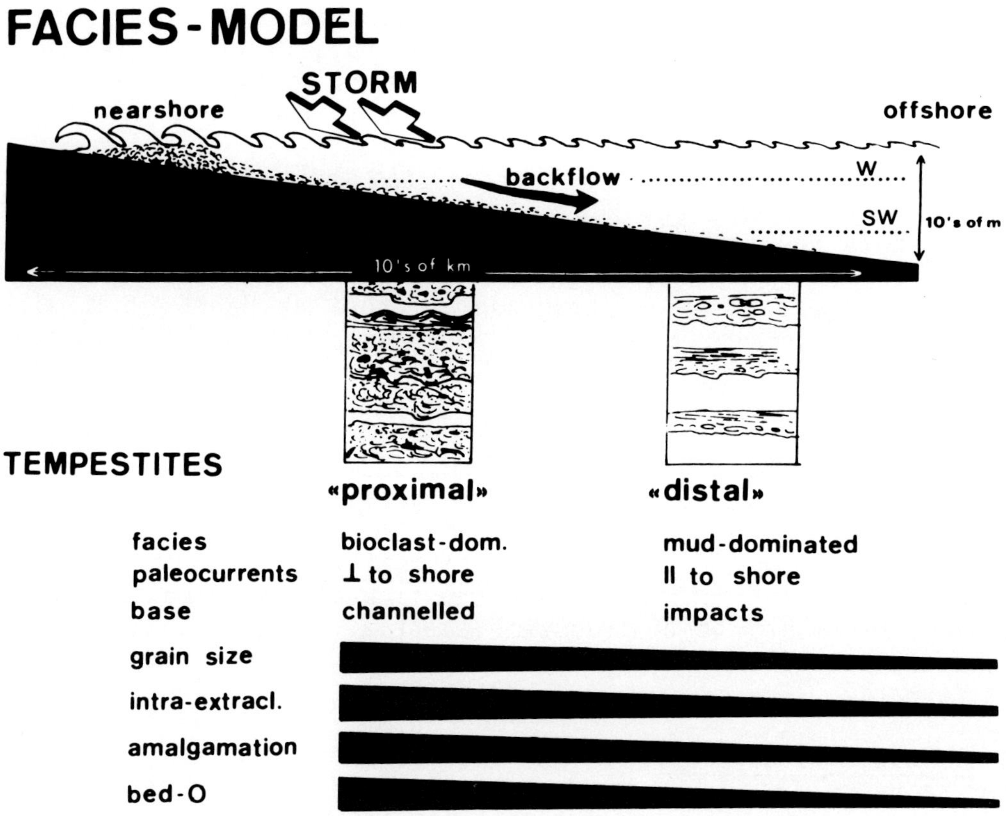

Figure 29. Tentative facies model accounting for the lateral change of tempestite facies in terms of proximity as shown in the Muschelkalk example (W = wave base, SW = storm wave base). From Aigner, 1982a.

5. **Cretaceous of Alberta.** I refer here particularly to my work on the Cardium Formation (Turonian, Upper Cretaceous; Walker, 1983a, b), discussed in detail in "Cardium Formation 4", this volume. The Ricinus oil field is situated in a channellized Cardium sandbody (Fig. 30), the channel being over 45 km long, 5-8 km wide (before palinspastic reconstruction), and 20-40 m deep (Walker, in press). Independent evidence ("Cardium Formation 4", this volume) indicates depths in excess of 50 m, but distances from the closest possible shoreline are difficult to reconstruct due to possible sea level changes.

Within the Ricinus sandbody, individual beds have sharp bases with ripped-up mud clasts, and Bouma BC and (less commonly) ABC sequences of internal structures. Although HCS is ubiquitous in marine members of the Cardium, it is rare at Ricinus. The higher abundance of Bouma sequences relative to HCS beds suggests depths mostly below storm wave base at Ricinus. I suggest that turbidity currents deposited the Bouma sequences, and were responsible for the initial cutting of the channel.

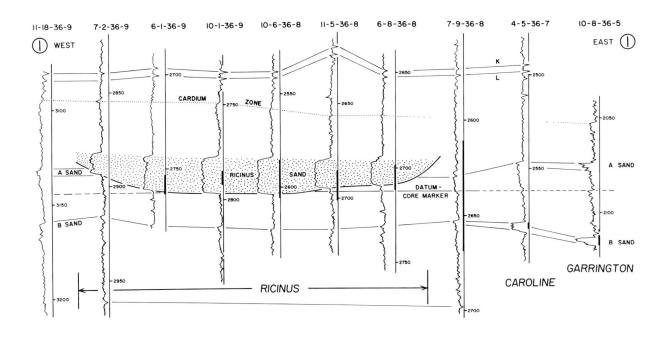

Figure 30. Channelized nature of the Cardium sand at Ricinus, Alberta (Walker, 1983b). Note that A and B sands (Caroline and Garrington) have very different gamma ray profiles from the Ricinus sand. Channelized nature of the sand shows up using a "core marker", and the K and L gamma ray markers as datums (note tectonic thickening in 11-5-36-8). Note blocky gamma ray response of the Ricinus sand. From Walker 1983b.

## GENERATION OF TURBIDITY CURRENTS IN SHALLOW SEAS

In the studies discussed above, all five papers discuss beds which contain sedimentary structure sequences close to, or identical with the Bouma turbidite sequence. All of the authors have independent evidence for relatively shallow seas (less than 300-400 m deep) and storm influences. We may first conclude that turbidites indeed exist in these situations.

However, two main criticisms have been raised concerning turbidity currents in shallow seas:

1) there is no well established mechanism for generating the turbidity currents, and

2) the basins are generally too shallow and too flat-floored for turbidity currents to operate.

Let us consider the second objection first. A turbidity current is a process. It operates in flumes, concrete ditches (Kuenen and Migliorini, 1950), on the shelf, in canyons, on submarine fans, and on abyssal basin plains. The problem is not where the flow will operate -- it is how to recognize the deposits. If waning-flow Bouma sequences are interbedded with quiet-water bioturbated mudstones, recognizing the deposits is normally fairly easy. If the deposits have been modified by storm waves, and HCS is present rather than Bouma sequences, the turbidity current emplacement of sand may not be at all apparent or demonstrable. Consider the following facts concerning basin geometry.

1) The slope at the epicenter of the 1929 Grand Banks earthquake was only 2.3°.

2) The earthquake triggered a turbidity current that broke submarine cables. The turbidity current was moving at over 11 m/sec when it broke the last cable, 13 hours 17 minutes after the quake, on a slope of 0.3° (Uchipi and Austin, 1979). At that velocity, it could maintain low concentrations of clasts up to about 3 cm diameter in suspension by fluid turbulence alone.

3) The Grand Banks turbidity current continued to flow over the Sohm Abyssal Plain for several hundred km, on a slope of less than 1:1500 or 0.04° (Horn et al., 1971).

The Grand Banks flow may have been unusually large, and slopes of 2.3° are abnormally steep compared with most shelf/shallow sea slopes. Yet the gradient of the continental shelf off the Sao Francisco Delta, N.E. Brazil, is 11.2%, or about 6.4° (Coleman and Wright, 1975, p. 141).

The Grand Banks flow is not the only turbidity current known to have travelled a long distance on a very low slope after having accelerated into the basin down a steeper slope (Pilkey et al., 1980). It seems reasonable that turbidity currents could have travelled long distances (hundreds of km) in, say, the Western Interior Seaway, if there were suitable generating mechanisms.

## FLOW GENERATION - SUMMARY

Many factors contribute toward the generation of turbidity currents. The Cordillera was rising during the late Jurassic and Cretaceous, and would have been supplying sediment rapidly to the shorelines of the Western Interior Seaway. The combination of rapid supply, uplift (steep basin margins) and active mountain building (earthquakes) would contribute to slumping and turbidity current formation.

Rapid sediment supply commonly leads to high pore pressures and sediment instability, especially in fine grained sediments (fine, very fine sands and silts). This is well demonstrated by the abundant sediment movements around the Mississippi Delta (Coleman et al., 1983). Some of this sediment appears to move as mudflows spontaneously, but large volumes of sediment can be liquefied or partially liquefied during severe storms and hurricanes. The most dramatic example concerns the loss of Shell's South Pass Block 70 Platform B (Sterling and Strohbeck, 1975), which tipped over and sank during Hurricane Camille (1969; Figs. 31, 32). Comparison of soil tests before and after the hurricane showed a large reduction in cohesive shear strength down to a depth of 80 feet (24 m) following the hurricane. There were associated major changes in sea floor topography, including an area of sediment removal of about $7 \times 10^6$ m$^2$. The average lowering of the sea floor was of the order of a meter, indicating a sediment volume of $7 \times 10^7$ m$^3$ which flowed from this area, and piled up in adjacent areas. The cause of the liquefaction was related to cyclic storm wave loading on the substrate, which being rapidly deposited, had high pore pressures.

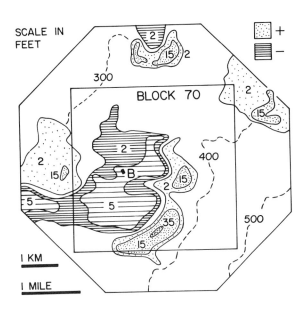

Figure 31. South Pass Block 70, showing area of sea floor lowering (ruled) and elevation (stippled) following Hurricane Camille. Shell's drilling platform is shown by the letter B. Redrawn from Stirling and Strohbeck, 1975.

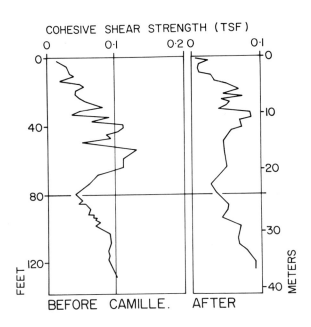

Figure 32. Cohesive shear strength of substrate in area of Shell Platform B (Fig. 31), before and after Hurricane Camille. Note modification of substrate down to a depth of about 80 feet following the Hurricane. Redrawn from Sterling and Strohbeck, 1975.

---

The generation of turbidity currents from slumps which gradually lose their excess pore pressures has been considered by Morgenstern (1967), who concluded that "for a slump to turn into a turbidity current... it is necessary that at failure the strength be reduced sufficiently to permit the acceleration of the mass, and that deeper slumps will transform more readily because, other things being equal, the dissipation of pore pressure will be less".

In Figure 33, I suggest that there is a marginal, dominantly fine-grained delta, with slopes a little steeper than those of the Mississippi. Major hurricanes not only generate coastal set up, but disrupt the substrate to the point where several meters (maybe up to 20 m or so) are liquefied and begin to slump. Storms may not even be necessary -- earthquakes will also liquefy the sediment, as at Valdez, Alaska, during the 1964 earthquake. If the gradient is sufficient, and the pore pressures do not dissipate too fast, the slumped mass may accelerate and develop into a turbidity current (as discussed in detail by Morgenstern, 1967).

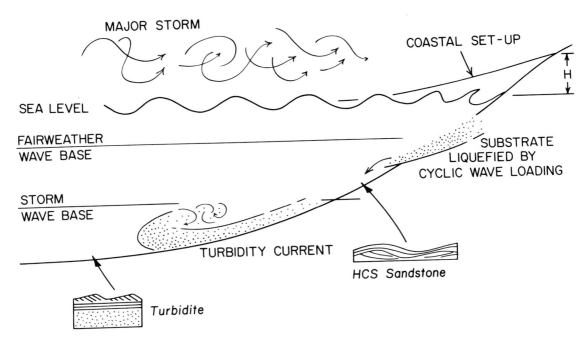

Figure 33. Model for the generation of a turbidity current by a major storm or hurricane. Onshore winds have developed a coastal set-up, H, which may be a few meters high. Storm waves have also liquefied the substrate, which slumped seaward, accelerated, and developed into a turbidity current. This is a modification of an earlier model (Fig. 8) which suggested that the relaxation (storm surge ebb) flow might develop directly into a turbidity current. The problem here is achieving a high enough density (volume of suspended sediment) that the flow would move down the paleoslope driven by gravity, rather than along the paleoslope under the influence of Coriolis force. The liquefaction proposal might overcome this problem. Details in text.

---

The volumes of sediment involved are interesting. About $7 \times 10^7$ m$^3$ may have moved during Hurricane Camille, which is similar to the $7.5 \times 10^7$ m$^3$ quoted by Morgenstern (1967, p. 193) for the Valdez slump. Slumps of this size are believed to have turned into turbidity currents in Orkdals Fjord, Norway ($10^7$ m$^3$) and Suva, Fiji ($1.5 \times 10^8$ m$^3$). The best data, involving slumping and subsequent cable breaking, is from the Magdalena River, Columbia, where before-and-after surveys of the sea floor indicated a missing volume of $3 \times 10^8$ m$^3$.

If the sediment which moved in the South Pass area during Hurricane Camille had turned into a turbidity current, and had been spread on the shelf as a bed averaging 20 cm thick, the bed would cover an area of about 350 km$^2$. By coincidence, this is roughly the area of the Cardium fields at Caroline and Garrington (Walker, this volume).

Finally, if flows can be generated at basin margins by a combination of rapid sediment input, steep basin margins, and earthquakes, we might ask whether storms are even necessary any more in flow genesis. I suggest that in many cases they are, because in many sections (especially the Fernie-Kootenay transition, and the Cardium), every sandstone with preserved sedimentary structures shows either a Bouma sequence (below storm wave base) or hummocky cross stratification. It is the two-birds-with-one-stone argument again -- the storm both generates the flow and imprints the hummocky cross stratification.

### Flow Generation - Conclusions

Rip currents are well known close to the breaker zone, but are inadequate to explain sand transport below fairweather wave base. Storms and wind-driven currents (divided somewhat artificially above into incremental storm flows, relaxation currents and wind-forced currents) are extremely important in moving sand on the shelf, and can form sharp-based, graded sandstones. Nevertheless, the geological evidence for very long transport distances and the close stratigraphic and paleoslope relationships between classical turbidites and HCS sandstone, both suggest that turbidity currents may be important in some shelf situations. In the absence of one major traditional generating process (major slumps on steep slopes within the basin), a combination of rapid sediment input, storm surge tides and wave liquefaction of sediment should be considered.

### Flow Generation - the Dilemma

I have discussed in some detail the geological evidence for the existence of turbidites in shallow marine settings, and have given some criteria by which they can be distinguished from the deposits of the last stages of incremental sand transport by storm-generated geostrophic currents. The evidence from the Jurassic Banff section, and the Cretaceous Ricinus channel, seems compelling. However, I should repeat the dilemma again.

"Is the geological evidence for classical turbidites so compelling that a generating mechanism at or close to the shoreface must be envisaged? Or is the oceanographic evidence for geostrophic dispersal so overwhelming that the beds which resemble Bouma sequences in the geological record must be interpreted in a non-turbidite context?"

The solution of the problem is important in any area where the depositional mechanisms and history are being studied. It is even more important in exploration situations, where a geostrophic interpretation suggests strike-oriented sand bodies, and a turbidite interpretation suggests exploration down-dip for dip-oriented sand bodies.

## RECURRENCE INTERVAL

There is relatively little data on the recurrence interval of storm deposits in the geological record. Most estimates have been made by dividing the absolute duration of an interval by the number of beds. Obviously, estimates of absolute duration and estimates of the number of storm events can be quite variable, and any exercise in dividing absolute time by observed number of sharp-based sandstone beds gives a <u>preservation</u> recurrence interval, not necessarily a <u>depositional</u> recurrence interval.

Hamblin and Walker, 1979, estimated 1240 sharp-based sandstones within the Kimmeridgian at Banff. The Kimmeridgian is now estimated as 4 million years (Palmer, 1983), and hence the preserved recurrence interval is about 3200 years.

Goldring and Langenstrassen, 1979, estimated one preserved event per 400 to 2000 years for Devonian sheet sandstones.

Brenchley et al., 1979, concluded that the preserved periodicity of storm beds in the Ordovician of Norway was about 10,000 to 15,000 years. A full discussion is given on p. 205-6.

Aigner, 1982a, indicates two independently-calculated preserved periodicities for Triassic storm deposits in Germany, either one per 2500 to 5000 years, or one per 5000 to 10000 years.

Kreisa, 1981, suggests a preserved recurrence interval of 1200 to 3100 years for storm events in the Ordovician of Virginia.

These five studies indicate preserved periodicities ranging from 400 to 15,000 years, with an average of a few thousand years. For the Kimmeridgian section at Banff, Alberta, Hamblin and Walker (1979) calculated a recurrence interval of about 4000 years (now estimated as 3200 years), but reasoned that perhaps more flows had been generated, but only 1 in 4 ended up in that part of the basin now exposed at Banff. If so, and if all beds have been preserved, each one can be regarded as the thousand-year event.

## THE GEOLOGICAL VIEWPOINT: ACKNOWLEDGMENTS

Many of the ideas in this review cannot be supported by observations in modern oceans. The ideas are suggested to me by interpretations of the geological record. If correct, these interpretations suggest processes that have not been observed by marine geologists, hence the sometimes contrasting opinions expressed in this paper compared with those of Don Swift (this volume). Both Swift and I agree that in this frontier area of research, there is room for more than one opinion. The reader must appreciate that this is inevitably the case, especially when the probjem is attacked from such contrasting viewpoints. I particularly thank Swift and Rod Tillman for their comments on this review, and I thank my graduate students, especially Tony Hamblin, Bill Duke and Dale Leckie, for helping to sharpen my observations and for their lively discussions over the last eight years. The work has been funded by the Natural Sciences and Engineering Research Council of Canada.

# REFERENCES

Ager, D. V., 1974, Storm deposits in the Jurassic of the Moroccan High Atlas: Paleogeography, Paleoclimatology, Paleoecology, v. 15, p. 83-93.

Aigner, T., 1982a, Calcareous tempestites: Storm-dominated stratification in Upper Muschelkalk Limestones (Middle Trias, SW-Germany), in Einsele, G. and A. Seilacher (eds.), Cyclic and Event Stratification: New York, Springer, p. 180-198.

_____, 1982b, Event-stratification in nummulite accumulations and in shell beds from the Eocene of Egypt, in Einsele, G. and A. Seilacher (eds.), Cyclic and Event Stratification: New York, Springer, p. 248-262.

Ball, S. M., 1971, The Westphalia Limestone of the northern midcontinent: a possible ancient storm deposit: Journal of Sedimentary Petrology, v. 41, p. 217-232.

Bouma, A. H., 1962, Sedimentology of some flysch deposits: Amsterdam, Elsevier, 168 p.

Brenchley, P. J., G. Newall, and I. G. Stanistreet, 1979, A storm surge origin for sandstone beds in an epicontinental platform sequence, Ordovician, Norway: Sedimentary Geology, v. 22, p. 185-217.

Brenner, R. L. and D. K. Davies, 1973, Storm generated coquinoid sandstone: genesis of high energy marine sediments from the Upper Jurassic of Wyoming and Montana: Bulletin of the Geological Society of America, v. 84, p. 1685-1697.

Bromley, R. G., 1978, Trace fossils of omission surfaces, in Frey, R. W. (ed.), The Study of Trace Fossils: New York, Springer, p. 399-428.

Campbell, C. V., 1966, Truncated wave-ripple laminae: Journal of Sedimentary Petrology, v. 36, p. 825-828.

_____, 1971, Depositional model - Upper Cretaceous Gallup beach sandstone, Shiprock area, northwestern New Mexico: Journal of Sedimentary Petrology, v. 41, p. 395-409.

Cant, D. J., 1980, Storm dominated shallow marine sediments of the Arisaig Group (Silurian-Devonian) of Nova Scotia: Canadian Journal of Earth Sciences, v. 17, p. 120-131.

Clark, T. L., B. Lesht, R. A. Young, D.J.P. Swift, and G. L. Freeland, 1982, Sediment resuspension by surface wave action: an examination of possible mechanisms: Marine Geology, v. 49, p. 43-59.

Coleman, J. M., D. B. Prior and J. F. Lindsay, 1983, Deltaic influences on shelfedge instability processes, in Stanley, D. J. and G. T. Moore (eds.), The shelfbreak: critical interface on continental margins: Tulsa, OK., Society of Economic Paleontologists and Mineralogists, Special Publication 33, p. 121-137.

_____ and L. D. Wright, 1975, Modern river deltas: variability of processes and sand bodies, in Broussard, M. L. (ed.), Deltas, Models for Exploration: Houston Geological Society, p. 99-149.

Crimes, T. P., 1975, The stratigraphical significance of trace fossils, in Frey, R. W. (ed.), The Study of Trace Fossils: New York, Springer, p. 109-130.

Davies, D. K., F. G. Ethridge, and R. R. Berg, 1971, Recognition of barrier environments: Bulletin of the American Association of Petroleum Geologists, v. 55, p. 550-565.

Defant, A., 1961, Physical oceanography, Vol. 1: Oxford, Pergamon, 729 p.

Dietz, R. S., 1963, Wave base, marine profile of equilibrium and wave-built terraces - a critical appraisal: Bulletin of the Geological Society of America, v. 74, p. 971-990.

Dott, R. H., Jr., 1974, Cambrian tropical storm waves in Wisconsin: Geology, v. 2, p. 243-246.

_____, and J. Bourgeois, 1982, Hummocky stratification: significance of its variable bedding sequences: Bulletin of the Geological Society of America, v. 93, p. 663-680.

Drake, D. E., R. L. Kolpack, and P. J. Fisher, 1972, Sediment transport on the Santa Barbara - Oxnard shelf, Santa Barbara Channel, California, in Swift, D. J. P. et al. (eds.), Shelf Sediment Transport; Process and Pattern: Stroudsburg, Pa., Dowden, Hutchinson and Ross, p. 307-331.

Duke, W. L., 1985, Hummocky cross stratification, tropical hurricanes and intense winter storms: Sedimentology, v. 32, p. 167-194.

Einsele, G. and A. Seilacher, (eds.), 1982, Cyclic and event stratification: New York, Springer, 536 p.

Ekdale, A. A., Bromley, R. G. and Pemberton, S. G., 1984, Ichnology: Society of Economic Paleontologists and Mineralogists, 317 p.

Exum, F. A. and J. C. Harms, 1968, Comparison of marine bar with valley fill stratigraphic traps, western Nebraska: American Association of Petroleum Geologists Bulletin, v. 52, p. 1851-1868.

Forristall, G. Z., R. C. Hamilton, and V. J. Cardone, 1977, Continental shelf currents in Tropical Storm Delia: observations and theory: Journal of Physical Oceanography, v. 7, p. 532-546.

Frey, R. W. and Pemberton, S. G., 1984, Trace fossil facies models, in Walker, R. G., (ed.), Facies Models, 2nd edition: Geoscience Canada Reprint Series 1, p. 189-207.

Frey, R. W. and A. Seilacher, 1980, Uniformity in marine invertebrate ichnology: Lethaia, v. 13, p. 183-207.

Fursich, F. T., 1982, Rhythmic bedding and shell bed formation in the Upper Jurassic of East Greenland, in Einsele, G. and A. Seilacher, (eds.), Cyclic and Event Stratification: New York, Springer, p. 208-222.

Futterer, E., 1982, Experiments on the distinction of wave and current influenced shell accumulations, in Einsele, G. and A. Seilacher, (eds.), Cyclic and Event Stratification: New York, Springer, p. 175-179.

Gebhard, G., 1982, Glauconitic condensation through high energy events in the Albian near Clars (Escragnolles, Var, SE-France), in Einsele, G. and A. Seilacher (eds.), Cyclic and Event Stratification: New York, Springer, p. 286-298.

Gilbert, G. K., 1899, Ripple marks and cross bedding: Bulletin of the Geological Society of America, v. 10, p. 135-140.

Goldring, R., 1971, Shallow water sedimentation as illustrated in the Upper Devonian Baggy Beds: Geological Society of London, Memoir 5, 80 p.

Goldring, R. and P. Bridges, 1973, Sublittoral sheet sandstones: Journal of Sedimentary Petrology, v. 43, p. 736-747.

_____ and F. Langenstrassen, 1979, Open shelf and nearshore clastic facies in the Devonian: Special Papers in Paleontology, v. 23, p. 81-97.

Gross, M. G., 1982, Oceanography (3rd Ed.): Englewood Cliffs, N. J., Prentice Hall, 498 p.

Gulliver, F. P., 1899, Shoreline topgraphy: Proceedings of the American Academy of Arts and Science, v. 34, p. 149-258.

Hagdorn, H., 1982, The "Bank der Kleinen Terebrateln" (Upper Muschelkalk, Triassic) near Schwabisch Hall (S.W. Germany) - a tempestite condensation horizon, in Einsele, G. and A. Seilacher, (eds.), Cyclic and Event Stratification: New York, Springer, p. 263-285.

Hamblin, A. P. and R. G. Walker, 1979, Storm-dominated shallow marine deposits: the Fernie-Kootenay (Jurassic) transition, southern Rocky Mountains: Canadian Journal of Earth Sciences, v. 16, p. 1673-1690.

Harms, J. C., D. R. Spearing, J. B. Southard, and R. G. Walker, 1975, Depositional environments as interpreted from primary sedimentary structures and stratification sequences: Tulsa, OK., Society of Economic Paleontologists and Mineralogists, Short Course 2, Notes, 161 p.

_____, J. B. Southard, and R. G. Walker, 1982, Structures and sequences in clastic rocks: Tulsa, OK., Society of Economic Paleontologists and Mineralogists, Short Course 9, Notes, 8-51 p.

Hayes, M. O., 1967, Hurricanes as geological agents - case studies of Hurricanes Carla, 1961 and Cindy, 1963: Bureau of Economic Geology, Texas. Report of Investigations 61, 56 p.

Heezen, B. C. and Drake, C. L., 1964, Grand Banks slump: American Association of Petroleum Geologists Bulletin, v. 48, p. 221-225.

Heezen, B. C., R. J. Menzies, E. D. Schneider, W. M. Ewing, and N.C.L. Granelli, 1964, Congo Submarine Canyon: Bulletin of the American Association of Petroleum Geologists, v. 48, p. 1126-1149.

Hobday, D. K. and H. G. Reading, 1972, Fairweather versus storm processes in shallow marine sand bar sequences in the late Precambrian of Finnmark, North Norway: Journal of Sedimentary Petrology, v. 42, p. 318-324.

Horn, D. R., M. Ewing, B. M. Horn, and M. N. Delach, 1971, Turbidites of the Hatteras and Sohm abyssal plains, western North Atlantic: Marine Geology, v. 11, p. 287-323.

Hunter, R. E. and H. E. Clifton, 1982, Cyclic deposits and hummocky cross stratification of probable storm origin in Upper Cretaceous rocks of the Cape Sebastian area, southwestern Oregon: Journal of Sedimentary Petrology, v. 52, p. 127-143.

James, W. C., 1980, Limestone channel storm complex (Lower Cretaceous), Elkhorn Mountain, Montana: Journal of Sedimentary Petrology, v. 50, p. 447-456.

Kelling, G. and P. R. Mullin, 1975, Graded limestones and limestone-quartzite couplets: possible storm-deposits from the Moroccan Carboniferous: Sedimentary Geology, v. 13, p. 161-190.

Kennedy, W. J., 1975, Trace fossils in carbonate rocks, in Frey, R. W. (ed.). The Study of Trace Fossils: New York, Springer, p. 377-398.

Komar, P. D., 1976, Beach processes and sedimentation: Englewood Cliffs, N. J., Prentice-Hall, 429 p.

Kreisa, R. D., 1981, Storm-generated sedimentary structures in subtidal marine facies with examples from the middle and upper Ordovician of southwestern Virginia: Journal of Sedimentary Petrology, v. 51, p. 823-848.

_____ and R. K. Bambach, 1982, The role of storm processes in generating shell beds in Paleozoic shelf environments, in Einsele, G. and A. Seilacher, (eds.), Cyclic and Event Stratification: New York, Springer, p. 200-207.

Kuenen, P. H. and C. I. Migliorini, 1950, Turbidity currents as a cause of graded bedding: Journal of Geology, v. 58, p. 91-127.

Lavelle, J. W., R. A. Young, D.J.P. Swift, and T. L. Clarke, 1978, Near-bottom sediment concentration and fluid velocity measurements on the Inner Continental Shelf, New York: Journal of Geophysical Research, v. 83, p. 6052-6062.

Leckie, D. A. and R. G. Walker, 1982, Storm- and tide-dominated shorelines in Cretaceous Moosebar-Lower Gates interval -- outcrop equivalents of Deep Basin Gas Trap in western Canada: American Association of Petroleum Geologists Bulletin, v. 66, p. 138-157.

Marsaglia, K. M. and G. deV. Klein, 1983, The paleogeography of Paleozoic and Mesozoic storm depositional systems: Journal of Geology, v. 91, p. 117-142.

McCubbin, D. G., 1982, Barrier-island and strand-plain facies, in Scholle, P. A. and D. R. Spearing, (eds.), Sandstone Depositional Environments: Tulsa, OK., American Association of Petroleum Geologists, Memoir 31, p. 247-279.

Morgenstern, N. R., 1967, Submarine slumping and the initiation of turbidity currents, in Richards, A. F. (ed.), Marine Geotechnique: Urbana, University of Illinois Press, p. 189-220.

Morton, R. A., 1981, Formation of storm deposits by wind-forced currents in the Gulf of Mexico and the North Sea, in Nio, S. D., R.T.E. Schuttenheim, and T.C.E. Van Weering, (eds.), Holocene Marine Sedimentation in the North Sea Basin: Oxford, International Association of Sedimentologists, Special Publication 5, p. 385-396.

Murray, S. P., 1970, Bottom currents near the coast during Hurricane Camille: Journal of Geophysical Research, v. 75, p. 4579-4582.

Nelson, C. H., 1982, Modern shallow-water graded sand layers from storm surges, Bering shelf: a mimic of Bouma sequences and turbidite systems: Journal of Sedimentary Petrology, v. 52, p. 537-545.

Niedoroda, A. W., D.J.P. Swift, T.S. Hopkins, and M. Chen-Mean, in press, Shoreface morphodynamics on wave-dominated coasts: Sedimentary Geology.

Palmer, A. R., 1983, The decade of North American geology: 1983 geological time scale: Geology, v. 11, p. 503-504.

Pemberton, S. G. and R. W. Frey, 1983, Biogenic structures in Upper Cretaceous outcrops and cores: Calgary, Alberta, Canadian Society of Petroleum Geologists, Mesozoic of Middle North America Field Trip Guidebook 8, 161 p.

Pilkey, O. H., S. D. Locker, and W. J. Cleary, 1980, Comparison of sand layer geometry and flat floors of ten modern depositional basins: American Association of Petroleum Geologists Bulletin, v. 64, p. 841-856.

Piper, D. J. W. and W. R. Normark, 1982, Effects of the 1929 Grand Banks earthquake on the Continental Slope off eastern Canada: Geological Survey of Canada, Paper 82-1B, p. 147-151.

Piper, D. J. W. and W. R. Normark, 1983, Turbidite depositional patterns and flow characteristics, Navy Submarine Fan, California borderland: Sedimentology, v. 30, p. 681-694.

Reineck, H. E. and I.B. Singh, 1973, Depositional sedimentary environments: New York, Springer, 439 p.

Sallenger, A. H., J. R. Dingler, and R. Hunter, 1978, Coastal processes and morphology of the Bering Sea coast of Alaska, in Environmental Assessment of the Alaskan Continental Shelf: Annual Report of Principal investigators for the Year Ending March, 1978: U. S. Dept. of Commerce, National Oceanic and Atmospheric Administration, Environments Research Laboratory, Boulder, Colorado, v. 12, p. 451-470.

Seilacher, A., 1967, Bathymetry of trace fossils: Marine Geology, v. 5, p. 413-428.

_____ and D. Meischner, 1964, Faziesanalyse im Palaozoikum des Oslo-Gebietes: Geologische Rundschau, v. 54, p. 596-619.

Shepard, F. P., and K. O. Emery, 1973, Congo submarine canyon and fan valley: American Association of Petroleum Geologists Bulletin, v. 57, p. 1679-1691.

Sterling, G. H. and G. E. Strohbeck, 1975, The failure of South Pass 70 Platform B in Hurricane Camille: Journal of Petroleum Technology, v. 27, p. 263-268.

Swift, D.J.P. and M. E. Field, 1981, Evolution of a classic sand ridge field: Maryland sector, North American inner shelf: Sedimentology, v. 28, p. 461-482.

_____, A. G. Figueiredo, Jr., G. L. Freeland, and G. F. Oertel, 1983, Hummocky cross stratification and megaripples: a geological double standard?: Journal of Sedimentary Petrology, v. 53, p. 1295-1317.

_____, G. L. Freeland and R. A. Young, 1979, Time and space distribution of megaripples and associated bedforms, Middle Atlantic Bight, North American Atlantic Shelf: Sedimentology, v. 26, p. 389-406.

_____, T. Nelsen, J. McHone, B. Holiday, H. Palmer and G. Shideler, 1977, Holocene evolution of the inner shelf of southern Virginia: Journal of Sedimentary Petrology, v. 47, p. 1454-1474.

_____, D. J. Stanley, and J. R. Curray, 1971, Relict sediments on continental shelves: a reconsideration: Journal of Geology, v. 79, p. 322-346.

Uchupi, E. and J. A. Austin, 1979, The stratigraphy and structure of the Laurentian Cone region: Canadian Journal of Earth Sciences, v. 16, p. 1726-1752.

Vincent, C. E., D.J.P. Swift, and B. Hillard, 1981, Sediment transport in the New York Bight, North American Atlantic Shelf: Marine Geology, v. 42, p. 369-398.

Walker, R. G., 1967, Turbidite sedimentary structures and their relationship to proximal and distal depositional environments: Journal of Sedimentary Petrology, v. 37, p. 25-43.

_____, 1979, Facies models 7. Shallow marine sands, in Walker, R. G. (ed.), Facies Models: Geoscience Canada Reprint Series 1, Geological Association of Canada, p. 75-89.

_____, 1982, Hummocky and swaley cross stratification, in Walker, R. G. (ed.), Clastic Units of the Front Ranges, Foothills and Plains in the area between Field, B. C. and Drumheller, Alberta: International Association of Sedimentologists, 11th International Congress on Sedimentology (Hamilton, Ontario, August 1982), Field Excursion Guidebook 21A, p. 22-30.

_____, 1983a, Cardium Formation 1. "Cardium a turbidity current deposit" (Beach, 1955): a history of ideas: Bulletin of Canadian Petroleum Geology, v. 31, p. 205-212.

_____, 1983b, Cardium Formation 2. Sand body geometry and stratigraphy in the Garrington-Caroline-Ricinus area: the "ragged blanket" model: Bulletin of Canadian Petroleum Geology, v. 31, p. 14-26.

_____, in press, Cardium Formation at Ricinus Field, Alberta: a channel cut and filled by turbidity currents in the Cretaceous Western Interior Seaway: American Association of Petroleum Geologists Bulletin.

_____, W. L. Duke and D. A. Leckie, 1983, Hummocky stratification: significance of its variable bedding sequence: discussion: Bulletin of the Geological Society of America, v. 94, p. 1245-1249.

Wright, M. E. and R. G. Walker, 1981, Cardium Formation (U. Cretaceous) at Seebe, Alberta - storm-transported sandstones and conglomerates in shallow marine depositional environments below fair-weather wave base: Canadian Journal of Earth Sciences, v. 18, p. 795-809.

# ANCIENT EXAMPLES OF TIDAL SAND BODIES FORMED IN OPEN, SHALLOW SEAS

Roger G. Walker

Department of Geology
McMaster University
Hamilton, Ontario, Canada

## INTRODUCTION

There are many examples of ancient sand bodies which have been interpreted as having a tidal origin. Most of them, however, are intertidal to very shallow subtidal, and represent lagoonal sand flats, tidal channel or estuarine channel and sand flat environments. There are remarkably few well described examples of ancient shelf/shallow marine subtidal sand bodies -- the few examples that exist will be the topic of these notes.

## MODERN TIDAL SEAS AND TIDAL SAND WAVES

The basis for the interpretation of the stratigraphic record lies in studies of recent sediments, and there are now several very good and useful studies of modern sand ridges and sand waves. These studies are reviewed by Swift (this volume), by various authors in Stride (ed., 1982) and by Allen (1980). Sand ridges are the larger features, up to a few tens of meters high, and with spacings of many kilometers (Table 1). Sand waves are up to about 10 m high, and have wavelengths of a few hundred meters.

In the North Sea, the sand was originally emplaced by glacial, fluvial and deltaic processes, and was reworked during the Holocene transgression. Stride (1963) has shown the present dispersal paths for this sand (Fig. 1), and in this figure, Allen (1970) has added the down-transport-path changes in mud/sand/gravel distribution, and the types of sedimentary structures in the sand.

In order to understand ancient tidal deposits, it is necessary to refer briefly to the studies of Houbolt (1968 - the group of tidal sand ridges shown in Figure 1 extending northeastward from the southeast coast of England) and McCave (1971 - the sand waves off the mouth of the Rhine - Meuse estuary).

The ridges studied by Houbolt (1968) are shown in Figure 2, and their average dimensions are shown in Table 1. A sparker profile (Fig. 3) revealed a series of internal reflectors, suggesting to some authors a possible very large scale angle-of-repose cross bedding. However, it is clear from the grossly exaggerated vertical scale that these reflectors cannot be angle-of-repose faces, and the calculated dip is in the 5 to 6 degree range. These surfaces can be referred to as "master bedding surfaces". Smaller bedforms migrate down and up these surfaces to form cross stratification similar to that discussed later in this paper (Figs. 21 and 26). Houbolt (1968, p.257) suggested that sand "seems to go round the ridge", implying a racetrack like circulation with little or no new sand presently being added to the system. This circulation pattern has been confirmed for Haisborough Sand by McCave and Langhorne (1982).

TABLE 1

Comparison of Features between Georges Bank,
the Sand Ridges off the Norfolk Coast,
and the Sand Waves off the Coast of Holland

|  | GEORGES BANK | SAND RIDGES, NORFOLK | SANDWAVES, HOLLAND |
|---|---|---|---|
| Area ($km^2$) | 16-20,000 | 5,000 | 15,000 |
| Average sand thickness (m) | 10-15 | 12 | ? |
| Bedform height (m) | 10-35 | up to 40 | 1-7 |
| Bedform spacing (km) | 15 | 9 | 0.2 - 0.5 |
| Bedform length (km) | 15-90 | up to 60 | ? |
| Inclination of "steep" slope (degrees) | 4 | 5-6 | 5-6 |

Similar large tidal sand ridges cover a large area of Georges Bank, on the Atlantic Shelf off Massachusetts (Twichell, 1983). The dimensions are shown in Table 1. Note the overall nature of the thin veneer of transgressive sand, the scale of the ridges, and the low dips of the "steep" faces of the ridges. The internal structure again probably consists of gently dipping master bedding surfaces, with smaller scale cross bedding marching down and/or up the master bedding surfaces.

The smaller sand waves described by McCave occur off the coast of Holland, and cover an area of about 15,000 $km^2$. Their heights and spacings are shown in Table 1. McCave's echo sounding profiles are shown in Figure 4, and clearly document a series of sand waves with smaller megaripples on their backs. In Figure 5, McCave's interpretation is shown, and in the caption of his Figure 17 he notes that "sand moves as bed load in megaripples up the stoss side and avalanches down the foreset slope". However, the "steep" faces of the sandwaves shown in Figure 5 average only 5 to 6 degrees; I therefore suggest that the "steep" faces again act as master bedding surfaces, with megaripples marching downward (and/or upward). An alternative interpretation is shown in Fig. 6, with possible ancient analogs illustrated in Figs. 21 and 26.

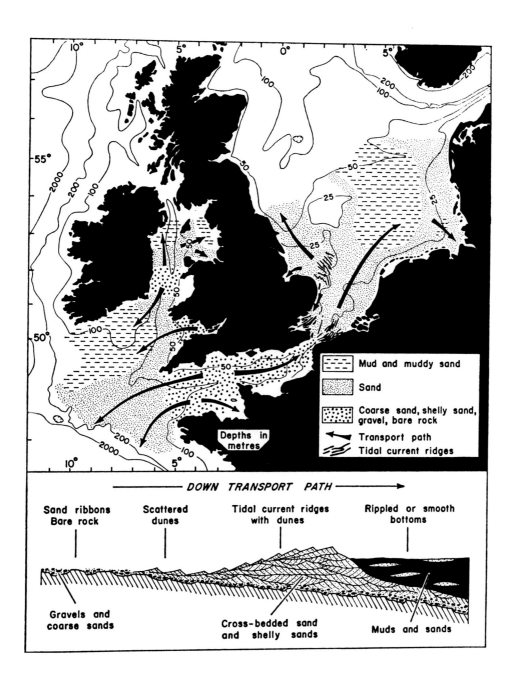

Figure 1. Sediment dispersal around the British Isles, showing location of the sand ridges studied by Houbolt (1968) - these extend as a group northeastward from the S.E. coast of Britain, and the sand waves studied by McCave (1971) - these occur off the Rhine-Meuse estuary of the Netherlands. The lower diagram shows the changes typically encountered downcurrent along a transport path. From Allen, 1970.

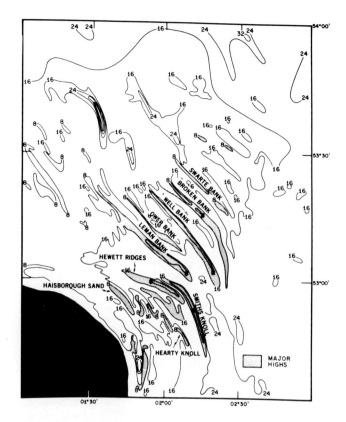

Figure 2. Map of tidal sand ridges in the North Sea, off the Norfolk coast of Britain - see Figure 1 for exact location. From Swift, 1975 after Houbolt 1968.

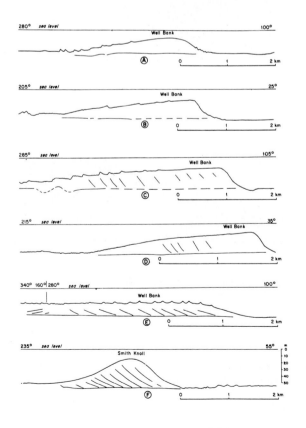

Figure 3. Interpretation of sparker profiles across Well Bank and Smith Knoll (locations in Fig. 2). Note greatly exaggerated vertical scale. From Houbolt, 1968.

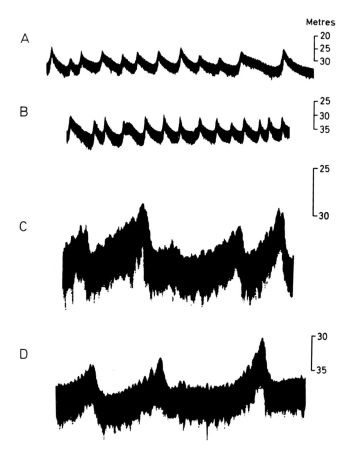

Figure 4. Echo sounder profiles of tidal sand waves off the coast of the Netherlands, from McCave, 1971. Lengths of profiles are A, 3800 m; B, 2800 m; C, 900 m; and D, 1200 m. Again note greatly exaggerated vertical scale.

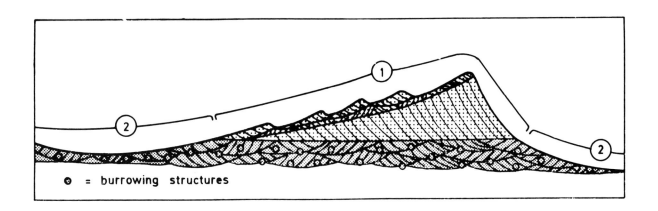

Figure 5. McCave's (1971) interpretation of the internal structure of a sand wave with megaripples. In zone 1, McCave suggested that much of the sand "moves as megaripples up the stoss side and avalanches down the foreset slope" - however, many of the foreset slopes have slopes of only about 5 degrees. See alternative interpretation in Figure 6.

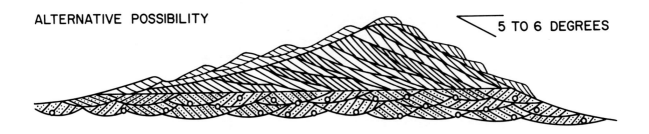

Figure 6. Alternative possible internal structure of McCave's sand waves, based on a 5 degree "foreset" slope, with megaripples acitvely migrating down the slope and forming cross bedding with dipping set boundaries ("master bedding surfaces").

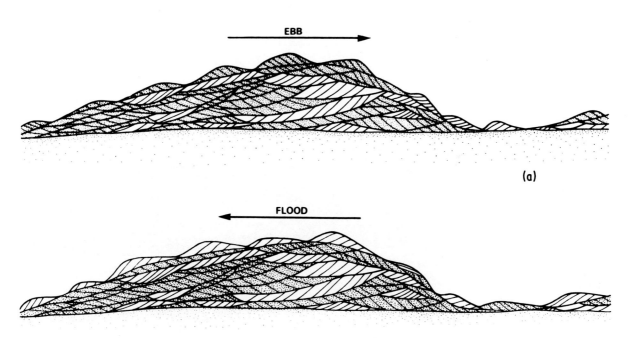

Figure 7. Internal structure of sand waves in the southern North Sea, based on box cores. Note both ebb and flood oriented cross bedding. Height of "giant ripples" varies from 1.7 to 5.5 m, and diagram has a 10 x vertical exaggeration. From Stride, 1982, after Reineck, 1963.

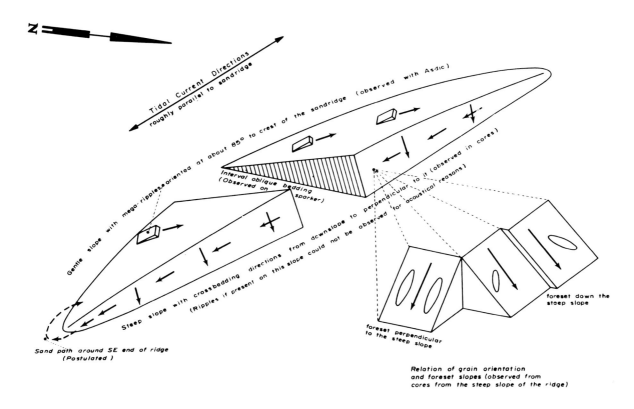

Figure 8. Sedimentary features of Well Bank, North Sea, showing sand circulation pattern and internal oblique bedding. Note that the dip of these internal sparker reflectors, and the dip of the "steep" face, is about 5°. Height of Well Bank is about 25-30 m. From Houbolt, 1968.

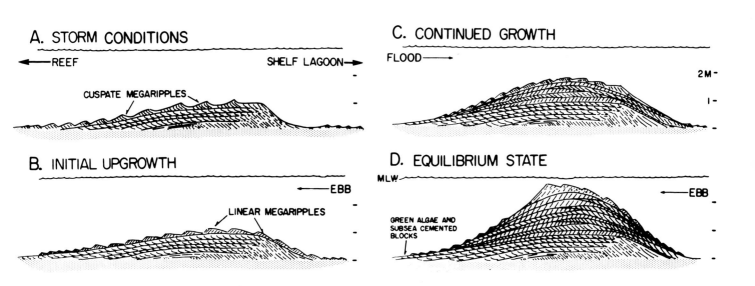

Figure 9. Model for growth and internal structure of oolitic sand waves, Lily Bank, Bahamas, from Hine, 1977. Note both ebb and flood oriented cross bedding, with cuspate megaripples moving up the gentle side, over the crest and down the lee side. Compare with Allen's Classes V and VI, Figure 5.

It is extremely difficult to obtain good continuous cores from these sandy deposits, and Allen (1980, p.282) specifically noted that "of the four workers (Reineck, 1963; Houbolt, 1968; Klein, 1970; Dalrymple et al., 1978) who have sampled acceptably the internal structures of modern sand waves, only Reineck and Houbolt worked subtidally, where [sand waves] are chiefly to be found." Thus, as an introduction to ancient sand waves, I have included the summary or interpretive diagrams of Reineck (1963; Fig. 7, this volume), Houbolt (1968; Fig. 8), McCave (1971; Fig. 5) and Hine (1977; Fig. 9). In his review, Allen (1980, p. 304-305) has combined data from these modern sand wave complexes with measurements and theory concerning tidal currents, and has devised a model (Fig. 10) which relates flow to internal structures. He lists 1) the tidal time-velocity pattern; 2) water depth; and 3) bed-material calibre as the factors controlling the intertidal structure. Class IA (Fig 10) develops from essentially unidirectional currents; Class VI represents asymmetrical sand waves with gentle and steep slopes of about 1° and 3°, respectively, believed to be formed by reversing flows of almost equal velocity and duration.

The scarcity of ancient examples is illustrated by Allen's discussion of his model. He cites no ancient examples of Class I (Fig. 10), one Pleistocene example of Class II, two examples of Class III, one example of Class IV, no examples of Class V, and two examples of Class VI (total, 6 examples).

Figure 10. (opposite). Regime diagram and predicted categories of sand-wave and dune internal structure, from Allen, 1980. U is the velocity, U(t) the velocity at a given time in the tidal cycle, and $U_{CR}$ is the threshold velocity for sand movement. T indicates the duration of a tidal cycle. Refers to Allen (1980) for other symbols. Note that in Classes I and II, flow strength exceeds $U_{CR}$ throughout the tidal cycle. In Class III, there is a period when $U < U_{CR}$, but without flow reversal. In Classes IV, V and VI, there is increasing symmetry between ebb and flood components, resulting in no flow separation over the sand wave crest, and no angle-of-repose lee face. Allen's Classes IV, V and VI describe the North Sea ridges (Houbolt, 1968; Fig. 8, this volume) and sand waves (McCave, 1971; Fig. 5) very well. See also Figures 11-15, 19, 22, 25, 26 and 30.

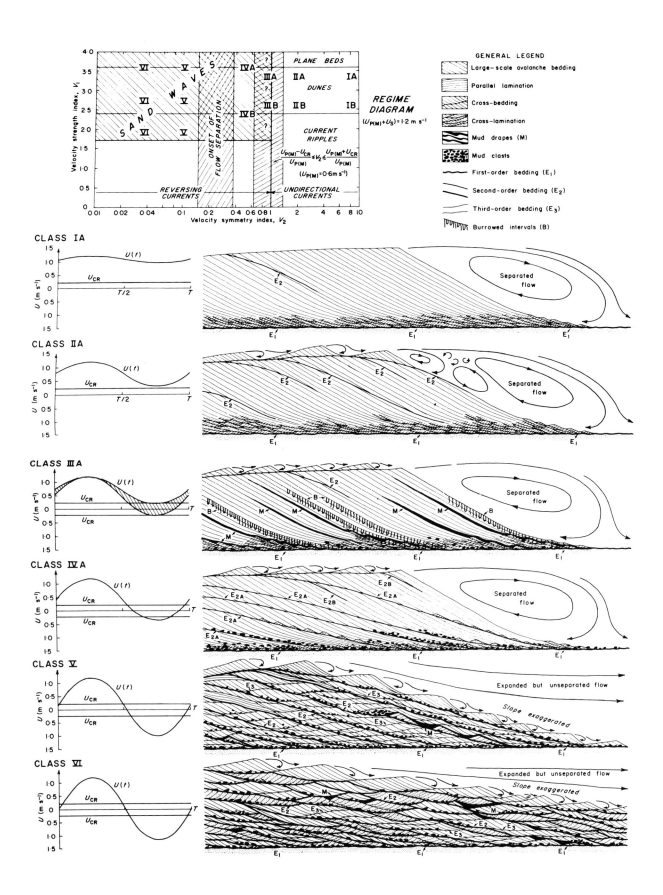

# ANCIENT TIDAL SAND BODIES

With so few examples to compare, it is perhaps premature to attempt any sort of classification or subdivision. For the sake of comparison and discussion, I will separate the examples under three subheadings.

1) transgressive sand-wave complexes

2) very thick quartzites, typical of the late Precambrian

3) early Paleozoic quartzites

## 1. Transgressive Sand-Wave Complexes

This term was introduced by Nio (1976), who described three examples -- the Roda Sandwave Complex of Spain, the Lower Greensand of Britain, and the Burdigalian sandwave complex of the Swiss Molasse.

The Lower Greensand (Aptian-Albian) of southeastern England has been studied by several workers (Allen and Narayan, 1964; Narayan, 1971; Dike, 1972; Dalrymple, unpublished), and a summary has been written by Bridges (1982). The outcrop pattern and paleoflow directions are shown in Figure 11. In the Woburn area, sand deposition began in the Lower Albian, with a total accumulation of about 40 m. The Lower Woburn Sands in the Leighton Buzzard area consist of bioturbated sands with thin clay partings, with angle of repose cross bedding up to a few tens of cm thick (Figs. 12, 13, 14, 15). Mud drapes interlaminated within the cross-bedded sands (Fig. 13, 14) indicate a strong tidal time-velocity asymmetry, probably during neap tides. The part of the sets without mud drapes represents spring tides with stronger tidal currents and, hence, more sand transport (Allen, 1981). The sub-horizontal clusters and single mud drapes suggest flood or ebb tides during which there was only limited sand transport.

The complex of large sand waves occurs in the Upper Woburn Sands, and the increase in scale of the structures perhaps represents a deepening (transgression) from the Lower Woburn Sands. The overall aspect of the large sand waves is shown in Figures 16 and 17, with a sketch of the major bedding surfaces in Figure 18. In Figure 16, the angle of repose cross bedding approaches 5 m in thickness, but the top of the set is truncated by several thinner, horizontal sets of cross bedding (Figs. 16, 18). These appear to cross the crest of the large sandwave, and migrate part way down the lee face (Fig. 19). Careful examination of Figure 19 shows that most of the low angle dark layers are set bounding surfaces, not individual cross beds. Steeper individual cross beds can be seen between these darker bounding surfaces. Many of the bounding surfaces are clearly erosional and hence probably represent reactivation surfaces. In detail (Fig. 20), many of the darker layers are extensively bioturbated, suggesting a close comparison with Allen's (1980) Class III of Figure 10. In Figure 17, the center of the photograph shows a series of prominent reactivation surfaces (arrows), inviting comparison with Allen's Class IV. In the modern sand waves described by Houbolt (1968; Fig. 8) and McCave (1971; Fig. 5), the lee faces dip at about 5-6°. The internal bedding features in Houbolt's (1968) sparker profiles clearly indicate slow migration of the linear ridges in the direction of the "steep" face, probably by smaller bedforms migrating down the "steep" face. An analog of this is perhaps shown in Figure 21, also from the Upper Woburn Sands.

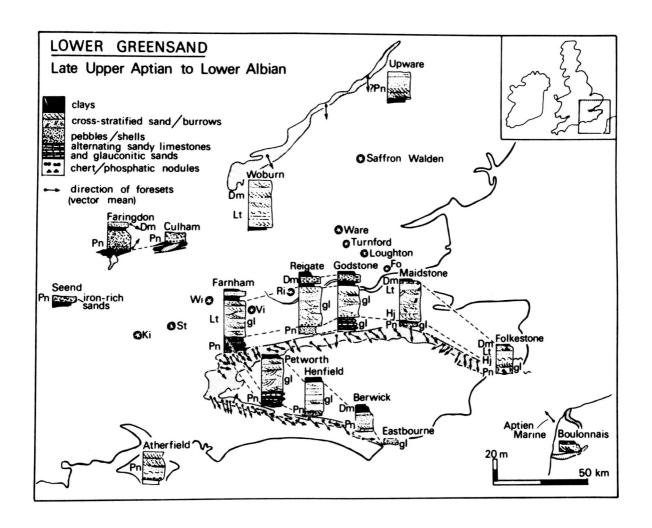

Figure 11. Location and paleocurrent map for Lower Greensand, southeastern England. Figures 12-21 are all from the Woburn area. From Bridges, 1982.

Figure 12. Lower Woburn Sands, Double Arches Pit. See Figure 11 for location. General view of tidal/subtidal sand flats.

Figure 13. Lower Woburn Sands, Double Arches Pit. Photo taken near base of face in Figure 12. Note small scale of sedimentary structures (trowel, right, for scale) with one larger cross bed (Fig. 14) at center.

Figure 14. Lower Woburn Sands, Double Arches Pit. Detail of cross bedding in Figure 13 showing short drapes interlaminated with cross beds and long mud drapes on probable reactivation surfaces. The short mud drapes probably formed at spring tides when drapes were overwhelmed by high rates of sand transport. The long mud drapes formed at times when rates of sand transport were much lower, probably during neap tides ($U < U_{CR}$; see Figure 10).

Figure 15. Lower Woburn Sands, Double Arches Pit. Tidal sand flat facies (intertidal) with cross bedding, ripple cross lamination (near base) and abundant bioturbation (upper part). Local scouring is observed near the trowel.

Figure 16. Upper Woburn Sands, Pratt's Lane. These large cross beds are up to 5 m thick, with some angle of repose cross stratification. They overlie the tidal sand flat facies of Figures 11-15, and are in turn overlain by the marine Gault Clay. This sequence establishes an overall transgressive situation for these large sandwaves. Note truncation of top of sandwave (detail in Fig. 19), and presence of some reactivation surfaces. Compare with Allen's Classes II, III and IV. (Fig. 10).

Figure 17. Photo taken from same location as Figure 16, but looking further to the right. Note scouring of one sandwave trough into another, and many reactivation surfaces.

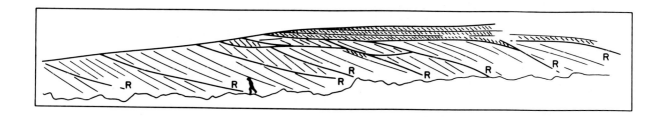

Figure 18. Sketch of the internal structure of the sandwave in Figure 16. R indicates reactivation surfaces.

Figure 19. Detail of main sandwave in Figure 16. Note angle of repose cross bedding, reactivation surfaces, and truncation of sandwave crest by smaller cross-bed sets. Some of these appear to curve downward to the right, indicating small megaripples moving over the sandwave crest and beginning to migrate down the lee side. Compare with Allen's Classes II and III (Fig. 10).

Figure 20. Detail of cross stratification in Figure 19. Note pebbles on some foresets, and abundant burrowing (especially below scale). Compare with Allen's Class III (Fig. 10), with bioturbated foresets.

Figure 21. Upper Woburn Sands, Chamberlain's Farm Pit. Regional bedding is horizontal. Note the roughly 5° dip to the left on set bounding surfaces, and the downstream inclined sets of cross stratification. Both the ridges (Houbolt, 1968; Fig. 8) and sand waves (McCave, 1971; Fig. 5) of the North Sea have "steep" faces inclined at about 5°, and this photograph may well be illustrative of the internal structure of the North Sea bedforms. Bob Dalrymple, one-time Chairman of the SEPM Research Group on Bedforms and Bedding Structures, is an appropriate scale.

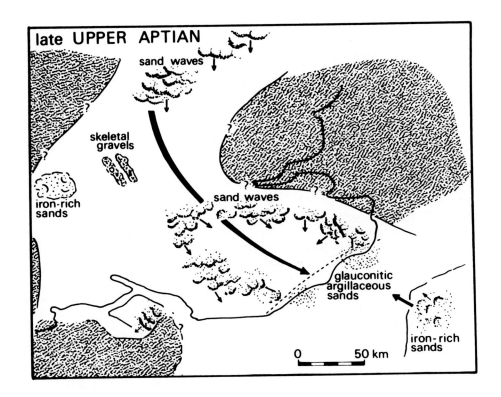

Figure 22. Paleogeographic reconstruction of depositional features in southeastern England during the late Upper Aptian. The Woburn Sands are shown as "sand waves", top left center of diagram. From Bridges, 1982.

---

The sandwave complex of the Upper Woburn Sands appears to have an essentially unidirectional paleoflow pattern (Fig. 11). South or southeastward flows also dominate in the horseshoe-shaped outcrop between Folkstone and Eastbourne. A paleogeographic interpretation is shown in Figure 22. The first person to suggest a tidal interpretation was (naturally!) Sorby in 1858. Narayan (1971) compared the Woburn Sands with the tidal deposits around the British coast (Stride, 1963). Bridges (1982, p. 184) simply commented that "a tidal interpretation is indeed attractive". The tidal interpretation can best be supported by reference to Allen's models. First, the mud-draped cross bedding of Figure 14 suggests strong tidal time-velocity asymmetry (Allen, 1981); second, the large sand waves (Fig. 16) with bioturbated bounding surfaces (Fig. 20) compare with Allen's (1980) Class III and IV models (Fig. 10), also suggesting considerable tidal velocity asymmetry.

Finally, the overall transgressive nature of the Upper Woburn Sands is emphasized by the overlying Gault Clay, which is an extensive marine clay full of ammonites. It presumably represents deepening and burial of the sands by marine muds. It is clear that both in the Weald and in the Woburn area, the Lower Greensand needs a detailed paleocurrent and facies study in the light of Allen's models, and in light of the fact that it appears to be one of the best candidates for a transgressive sand wave complex in the geological record.

The Upper Marine Molasse (Miocene) of western Switzerland has been described by Allen and Homewood (1984). The thickness of the unit is not stated, but it occurs above the Lower Freshwater Molasse, and is presumably transgressive. Nio (1976) mentions the Upper Marine Molasse briefly, also suggesting a transgressive situation, and showing two major sand wave complexes each about 5 m thick.

Allen and Homewood (1984; Fig. 23) illustrate "tabular" cross beds 1-2 m thick (the main beds in Fig. 23), overlain by smaller "trough" cross beds. The very curved and asymptotic foresets in the tabular unit suggest deposition in large scoured troughs, not in the lee of straight crested sand waves. The foresets alternate between groups of spring tide bundles (mostly sandy) and groups of neap tide bundles (with layers of darker, finer grained sediment). Measurement of 140 adjacent bundles, 4.4 to 14.8 m long, revealed a clear cyclicity of bundle thickness and drape spacing (Fig. 24), with the spacing showing a variation from zero to about 20 cm over a periodicity of about 27 drapes. This strongly suggests a semi-diurual tidal regime.

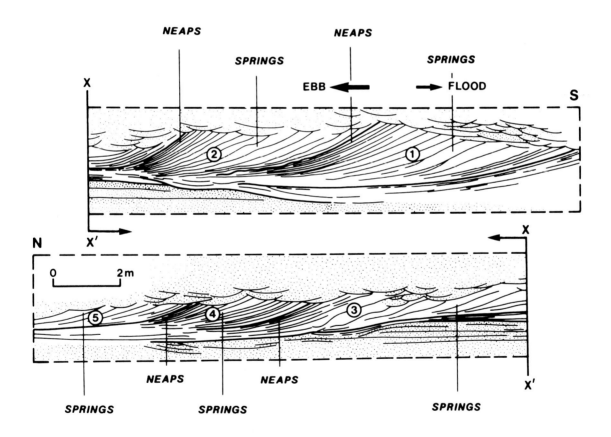

Figure 23. Upper Marine Molasse, Bois du Devin, Switzerland. The large cross beds (1-5) were described by Allen and Homewood (1984) as tabular, overlain by smaller trough cross beds. Note groups of bundles of spring-tide sandy laminae and neaptide laminae with darker, finer-grained drapes.

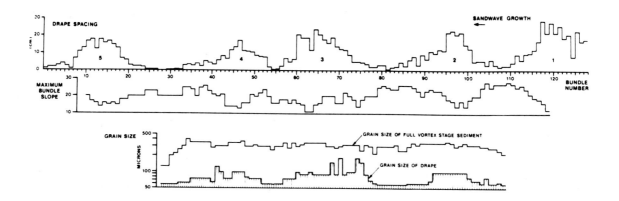

Figure 24. Cyclicity of drape spacing, from zero to about 20 cm with a periodicity of about 27 drapes. Note that maximum inclination of the bundles roughly coincides with low drape spacing. Upper Marine Molasse, Switzerland; from Allen and Homewood (1984).

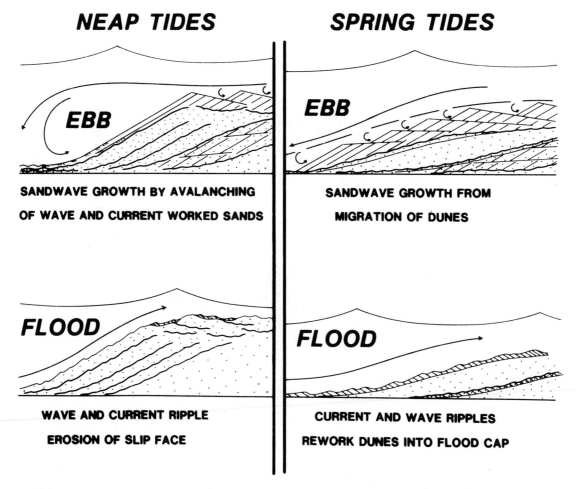

Figure 25. Model for the development of stratification in the Upper Marine Molasse, Switzerland. The four stages are explained in text. From Allen and Homewood (1984).

The overall dune height must have exceeded 2 m, with foreset lengths of about 6 m. Wavelengths were calculated to be about 30 m. In their model (Fig. 25), Allen and Homewood (1984, p. 69-70) suggest

1) at neap tides, ebb currents transported sand to the slip face. Avalanching took place, with some backflow ripples. This compares with Allen's Class IV (Fig. 10).

2) still during neap tides, flood currents scoured the slip face.

3) at spring tides, strong ebb currents transported sand in small dunes. The 10-20° dip of the bounding surfaces suggests incipient or weak flow separation (Allen's Class V, Fig. 10).

4) flood currents during spring tides eroded the lee slopes and produced trains of flood-oriented ripples and flood caps on top of the dominantly ebb-oriented sand waves.

The Roda Sandwave Complex (Paleocene) of the Pyrenees has been described by Nio (1976) and Nio and Siegenthaler (1978), and was compared to North Sea sand waves by Nio (1976, p. 36). Recently, however, Nio and Hommes (1982) have reinterpreted the Roda as an estuarine sand body. Although it remains an excellent example of the construction and subsequent degradational modification of a large sandwave complex, with individual angle-of-repose cross beds in sets up to 20 m thick, if we accept Nio's latest interpretation, it can no longer be used as an example of an open shelf, shallow marine sandwave complex.

The Viking Formation (Albian) of southern Alberta is not a classic example of a transgressive sandwave complex, but I include one example here because it is the only case I am aware of in which master bedding surfaces have been recognized in core (Fig. 26). The core is from Caroline Field, Location 13-8-35-6W5, and has been described by Reinson et al., 1983. The photograph in Fig. 26 is from a coarse sand body about 3.5 to 4 feet (1.10 to 1.22 m) thick, encased in marine shales. If indeed it does represent a sandwave complex with master bedding surfaces, it might be a situation in which the sandwave migrated down-dispersal rather rapidly into a muddy area (as could be envisaged in Fig. 1), later to be buried by more marine muds.

Figure 26. Possible example in core of dipping master bedding surfaces, with angle of repose cross bedding "marching" down the master bedding surfaces. Albian Viking Formation of southern Alberta, well 13-8-35-6W5, from Caroline Field, depth 8922 ft (2719.4 m).

## Transgressive Sandwave Complexes -- Summary

With so few examples, both recent and ancient, one can only suggest some broad generalities. The base of a transgressive sandwave complex may be sharp, erosive, and marked by a lag gravel, as in the recent sandwaves of the southern North Sea. Alternatively, there may be preservation of intertidal or very shallow subtidal deposits below the sandwave complex, as in the Woburn Sands. The nature of the base may thus depend on rate of transgression and rate of sand supply. Within the sandwave complex, one might predict the larger sets of cross bedding toward the base, with smaller sets truncating larger sets toward the top. As transgression continues, the supply of sand is likely to be sharply reduced, and the top of the sandwave complex might be marked by sets of cross bedding made by tidal currents reworking sand in situ. When this reworking stops, due to continued deepening, the sandwave complex will be abruptly buried by mud.

Sandwave complexes formed by transgression of relatively-easily reworked sediments (such as the transgressed Pleistocene tills in the North Sea and Georges Bank) may be very extensive but relatively thin. Maximum known dimensions for sands shaped by tidal currents in the seas of northwest Europe are about 400 km long, 50 km wide, and 12 m thick (Stride et al., 1982, p. 101). Dimensions of Georges Bank (Twichell, 1983, p. 696) suggest a sand sheet about 160 km long and 100 km wide; the sand ridges have a relief of 10-35 m, and seismic-reflection profiles (Twichell, 1983, p. 701) suggest an average sand thickness of 10-15 m above the Holocene marine unconformity (Table 1).

## 2. Thick Late-Precambrian/Cambrian Quartzites

Stratigraphic units of this type are known in many parts of the world, although few have been well described sedimentologically. Thicknesses are commonly in the 1000-3000 m range, and they can be traced laterally for hundreds of km. Some consist of quartzite throughout, and are dominated by small to medium-scale cross bedding (sets up to about 1 m); this suggests a wide, shallow sea with sedimentation keeping pace with subsidence. Others are more variable, with quartzites alternating with slates.

The Jura Quartzite (Anderton, 1976) is a late Precambrian unit which crops out on the islands of Jura and Islay, western Scotland, and covers an area of about 20 X 80 km (Fig. 27). Its maximum thickness is about 5300 m. Anderton recognized both coarse (quartzitic) and fine (interbedded sand and mud) facies. The coarse facies is dominantly cross bedded (Fig. 28), with climbing dune sets 3-25 cm thick, tabular sets 15-200 cm thick, thick tabular sets (200-450 cm) with backflow ripples (perhaps similar to Allen's (1980) Class I) and trough cosets 3-50 cm thick. There are various very broad, gentle erosion surfaces (mostly not channels), and some cosets of parallel lamination 10-400 cm thick interpreted as beach. The various cross bedded facies are interpreted as the deposits of a shallow tidal sea (Anderton, 1976, p. 439). The fine facies consist of sharp-based or erosively-based laminated and/or rippled sands 5-45 cm thick, alternating with mudstones. The parallel-to-rippled sequence of structures "is similar to that found in some turbidites... even when only a partial sequence is present the structures always occur in the same order. As in turbidites, the sequence is interpreted as due to deposition from a decelerating current" (Anderton, 1976, p. 441; Fig. 29, this paper). Anderton's interpretation, accepting "a shallow marine origin" is that deposition was "from tidally-dispersed, storm generated suspension clouds.

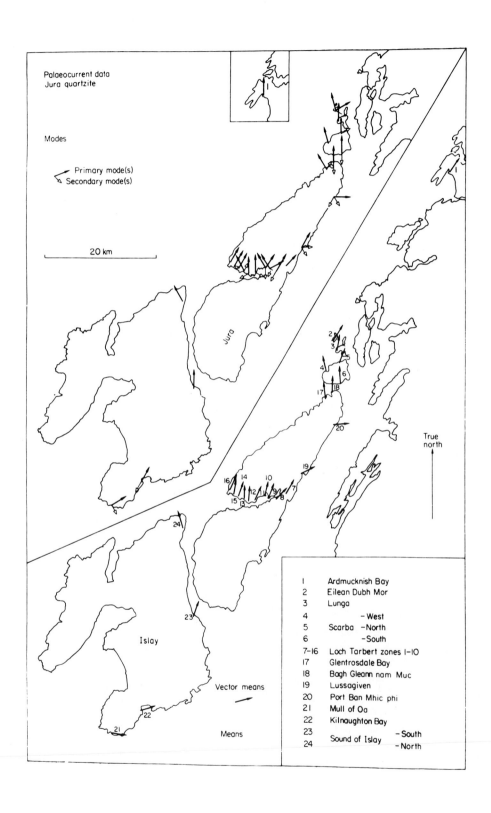

Figure 27. Paleocurrent patterns for the late Precambrian Jura Quartzite of western Scotland. From Anderton, 1976.

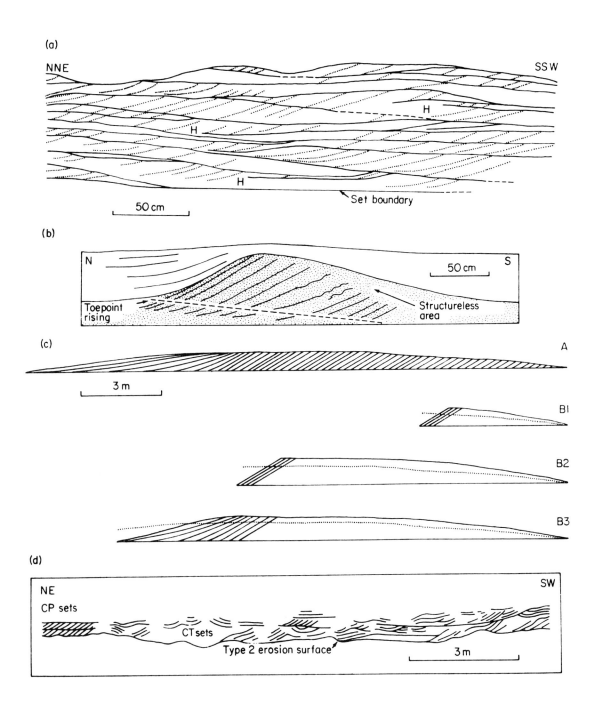

Figure 28. Styles of cross stratification in the late Precambrian Jura Quartzite of western Scotland. In (a), note climbing dunes. In (b), the rising toepoint also suggests climbing dunes. In (c), the lens-shaped tabular set may have evolved through stages B1, B2 and B3. In (d), there is a shallow irregular channel filled by trough sets and flanked by planar-tabular sets.

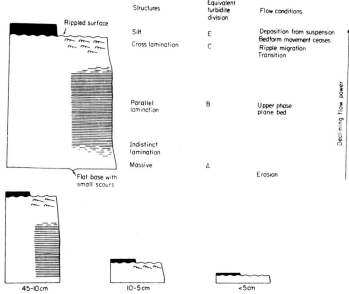

Figure 29. Various laminated facies from the late Precambrian Jura Quartzite of western Scotland. The full ABCE Bouma type sequence is present, as well as BC and C sequences. The beds clearly invite comparison with the turbidite sequence of Bouma (1962). From Anderton, 1976.

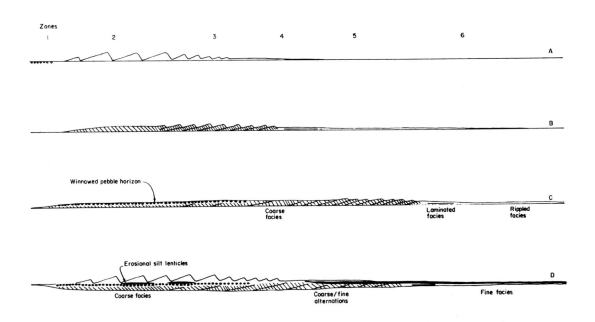

Figure 30. In (a), the fairweather zones 1-6 are shown: zone 1 = winnowed gravel lags, zones 2 and 3 = large and small sand waves, zone 4 = continuous flat bedded and rippled sand, zone 5 = sand patches in mud, and zone 6 = continuous mud. In (b), during moderate storms the bedforms of zone 2 are partly eroded, sand is deposited on migrating small dunes in zone 3 and as a thick storm-sand layer in zones 4 and 5. In (c), during severe storm surge conditions the dune zones are washed out and winnowed while climbing dunes form downcurrent of the fairweather dune field. A downcurrent thinning and fining storm layer is deposited. In (d), there is a return to fairweather conditions, and the storm-mud layer is partly eroded; the dune field becomes re-established. From Anderton, 1976.

The laminated and rippled facies may represent proximal and distal storm sand deposition, respectively" (Anderton, 1976, p. 445). Hydrodynamically, these beds are turbidites (i.e., the deposits of turbidity currents); the problem is how the density flows were generated and how far they have travelled to the final depositional site. These problems are reviewed in my later papers in this volume ("Geological evidence for storm transportation and deposition on Ancient Shelves", and "Comparison of shelf environments and deep-basin turbidite systems").

The various facies "alternate in vertical successions in an apparently random manner" (Anderton, 1976, p. 447). Predominant paleoflow is north-northeast (Fig. 27), but "many localities show a secondary mode or modes, and several localities have bipolar-bimodal paleocurrent patterns with the dominant mode always towards the northeast quadrant" (Anderton, 1976, p. 449). From all of these data, Anderton has derived the interpretation shown in Fig. 30. In the terminology of Stride's (1963) work around the British coast (Fig. 1), Zone 1 is the gravel lag or rock pavement, Zones 2 and 3 represent large and small sand waves, respectively, Zone 4 contains flat bedded and rippled sand, Zone 5 contains sand patches in mud, and Zone 6 is exclusively muddy. Thus, in Figure 30, A represents the Jura Quartzite during fair weather tidal conditions. B represents moderate storms, with thin storm sand layers in Zones 4 and 5. C shows severe storms, with winnowing and pebble lag development in the dune area (Zones 2 and 3), and storm sands in Zones 5 and 6. D represents a return to fair weather tidal conditions.

The Lower Sandfjord Formation (Levell, 1980) is another Late Precambrian sandstone, about 1500 m thick, from North Norway. It consists of crossbedded mature sandstones (about 98% of the formation) in single and compound sets (Fig. 31). The compound sets, "in which each small-scale set is separated from its neighbors by a convex upward surface [dips 5-15°] could represent either a large bedform with megaripples superimposed on its lee face [see Allen's model, Class V, Fig. 10, this paper] or a large bedform with extremely closely spaced periodic reactivation surfaces formed by reworking of a single angle-of-repose lee face" (see Allen's model, Class II, Fig. 10; Levell, 1980, p. 545). In addition to the previously noted types of cross bedding, there are minor amounts of plane-laminated sandstones, and pebble lags. The facies do not appear to occur in preferred sequences. Paleoflow is dominantly unimodal toward the east, "although a small number of sets indicate definite current reversals" (Levell, 1980, p. 549).

Levell (1980, p. 549) discusses but rejects a fluvial origin for the Lower Sandfjord, noting, for example, that "it seems unlikely that any river system could have deposited 1.5 km of sandstone with no channel fill deposits (for instance fining-upward deposits or clay plugs), virtually no fine-grained sediment, and with exclusively tabular bedding and planar erosion surfaces". Levell continues by discussing problems of sand supply, unidirectional paleoflows, and the significance of compound cross bedding in large sandwaves. This interpretation (Fig. 32) shows transfer of sand via a tidally influenced delta onto a tidal shelf. Gravel and sand ribbons form in starved areas. "In areas offshore from transgressed delta lobes....the coastal plain sands are transported offshore by tidal currents and reworked into large sandwaves" (Levell, 1980, p. 554). Levell concludes that "the enormous thickness of homogeneous sandstone, the unidirectional paleocurrent patterns, and the abundance of cross bedding are superficially more characteristic of fluvial than shallow marine deposits. This study, in conjunction with the work of Anderton (1976) on the Jura Quartzite, suggests that such deposits were formed on tidal shelves during Late Precambrian time" (Levell, 1980, p. 554).

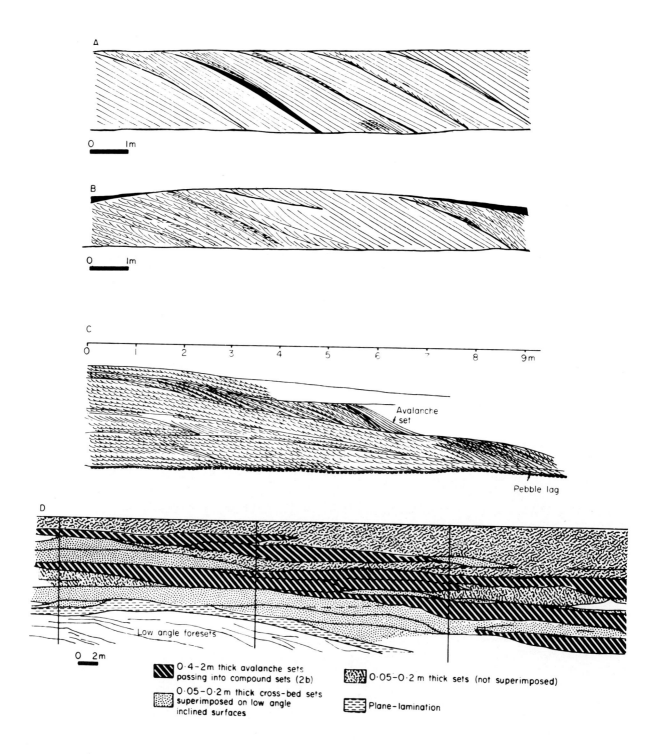

Figure 31. Various types of stratification in the late Precambrian Lower Sandfjord Formation, North Norway. In A, note the repeated reactivation surfaces, comparable to Allen's Class II (Fig. 10). In B, note compound cross-bedding with convex-up set boundaries, comparable to Allen's Class II or IV (Fig. 10). In C, the medium scale avalanche set has an inclined base, indicating superimposition on a still-larger bedform, perhaps most comparable to Allen's Class V (Fig. 10). In D, only the unusually irregular bedding and a general upward coarsening suggests that this unit represents an individual bedform complex. From Levell, 1980.

Figure 32. Model for the development of the Lower Sandfjord Formation, North Norway. Details in Levell, 1980. In areas offshore from transgressed delta lobes, the coastal plain sands are transported offshore by tidal currents and reworked into large sandwaves. From Levell, 1980.

3. Paleozoic Quartzites

The Eriboll Sandstone (Swett et al., 1971) is a Lower Cambrian unit which crops out in a long (160 km), narrow (16 km) strip in northwestern Scotland, and is about 220 m thick (Fig. 33). It is transgressive over older rocks (Torridonian Formation, Lewisian gneisses), and contains abundant Skolithos and Monocraterion trace fossils indicating a marine origin. No mudstones were reported, nor were preferred facies sequences noted. Cross stratification is the dominant structure, mostly in planar tabular sets averaging 50 cm thick (range 2-110 cm). Reactivation surfaces are present (Fig. 34), and the presence of small ripples on the lee faces of larger cross beds, with ripple crests oriented perpendicular to the strike of the cross beds, suggests emergence run-off during ebbing tides. Paleoflow measurements suggest "crudely bimodal" patterns (Swett et al., 1971, p. 406-408; Fig. 33), and herringbone cross bedding is present.

Swett et al. (1971, p. 410) conclude that the Eriboll Sandstone "represents a series of coalescing tidal sandbodies organized into a large regional sheet across a shelf, and is analogous to those reported (Jordan, 1962) from Georges Bank."

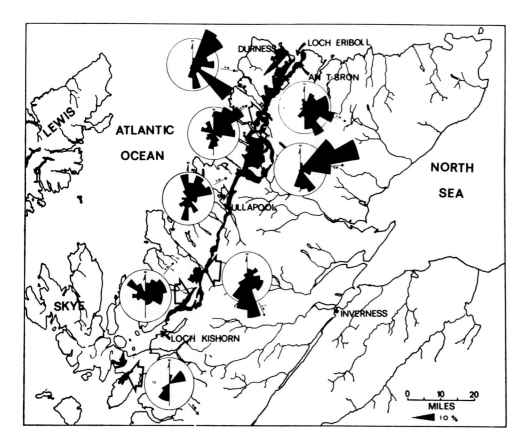

Figure 33. Outcrop pattern (shaded) and paleocurrents of the Lower Cambrian Eriboll Sandstone, northwest Scotland. From Swett et al., 1971.

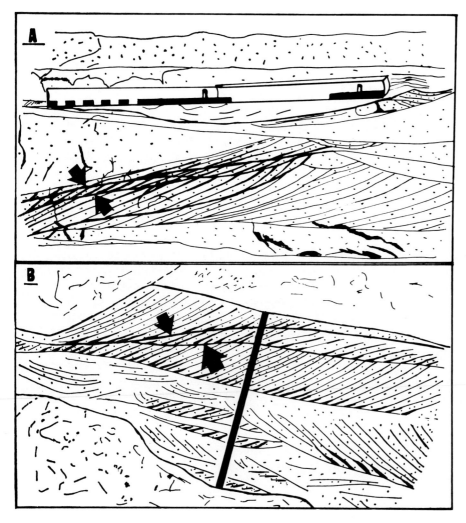

Figure 34. Cross stratification in the Lower Cambrian Eriboll Sandstone, northwest Scotland. In A, arrows show reactivation surfaces. In B, similar reactivation surfaces are shown from modern sandwaves in the Bay of Fundy, Canada, Scale is 1 m long. From Swett et al., 1971.

The Doulbasgaissa Formation (Banks, 1973) is about 550 m thick, and crops out in the Digermul Peninsula of North Norway. It covers a minimum area of about 10 X 25 km, and is Lower Cambrian in age. Banks describes four main facies:

1. Sandstones, 1-10 cm thick, with interbedded siltstones and mudstones. The sandstones have sole marks, slight grading, cross lamination and some "low angle cross bedding", parallel lamination, and symmetrical ripples.

2. Sandstones, 10-150 cm thick with interbedded siltstones and mudstones. The bases of the beds are loaded, the beds have parallel lamination and "low angle cross bedding", but small scale cross lamination is rare.

3. Sandstones, 1-100 cm thick, with subordinate siltstone and mudstone. The 1-10 cm beds have parallel lamination and cross lamination -- there is both high and low angle stratification in the thicker beds.

4. 10-400 cm thick sandstones, with trough cross beds mostly 20-50 cm thick, but some sets up to 4 m. A few fine-grained drapes and discontinuity surfaces occur, with some herringbone cross bedding. Paleocurrents are strongly bimodal and bipolar (Banks, 1973, Fig. 6i).

Banks interprets facies 4 as tidal in origin, and postulates a storm surge origin for facies 1-3. Future work on this unit should be directed to 1) determining if the "low angle cross bedding" is hummocky cross stratification, and 2) noting the sequence of sedimentary structures in facies 1-3, for possible comparison with the Bouma sequence for turbidites. The Duolbasgaissa Formation may well represent the same overall combination of tides and storms as the Jura Quartzite (Anderton, 1976).

The St. Peter Sandstone (Dott and Roshardt, 1972) covers a large area of the north-central U.S., of which about 1800 $km^2$ was studied by Dott and Roshardt southwest of Madison, Wisconsin. The St. Peter Sandstone is Middle Ordovician in age, and ranges from about 15-50 m in thickness in this area. It overlies older rocks unconformably, and is thus a transgressive sandstone. It contains large cross beds up to 10 m thick, with convex-up truncation surfaces that are nearly horizontal at the top of the set, steepening to 36° at the bottom of the set (Pryor and Amaral, 1971, p. 239, 241; Figs. 35, 36). Between these surfaces, which would now be termed reactivation surfaces (Allen, 1980, Class II, Fig. 10), there is a "secondary internal cross bedding, which is the tangential-trough type" (Pryor and Amaral, 1971, p. 240; Fig. 35). Paleocurrent patterns show a "bewildering array of directions" (Dott and Roshardt, 1971 -- see their Figs. 1 and 2; Fig. 37) unless care is taken to separate the various scales and types of cross bedding. The larger cross beds appear to show an east to west flow in the Madison area, and the cross beds were "deposited in complex submarine sand waves, dunes and ridges, as postulated by Pryor and Amaral (1971). The size and form of these must have been extremely variable, ranging up to heights in excess of 30 ft, thus rivalling modern sand waves on Georges Bank, in the North Sea, and in the English Channel" (Dott and Roshardt, 1972, p. 2593).

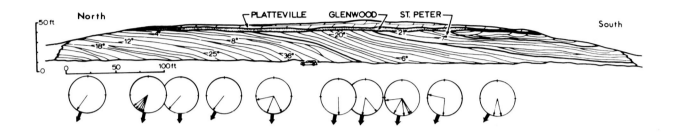

Figure 35. Sketch of roadcut outcrop of St. Peter Sandstone on Wisconsin Highway 69 near Monticello. Note dips of foresets, and paleoflow orientations. Car (lower center) for scale. Probably closest to Allen's Class IV (Fig. 10). From Pryor and Amaral, 1971.

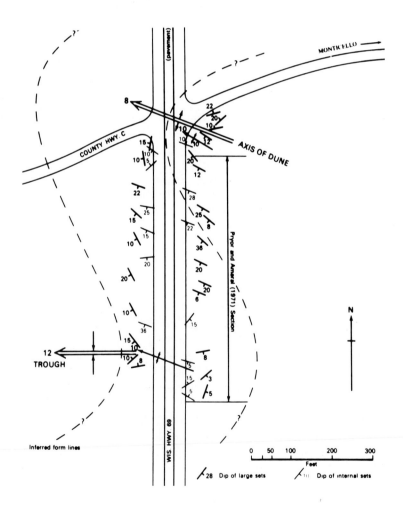

Figure 36. Location map for Figure 35, showing dips and strikes of the cross beds. Note dune axis, trough axis, and sinuous dotted lines showing the form of the major foresets. From Dott and Roshardt, 1972.

334

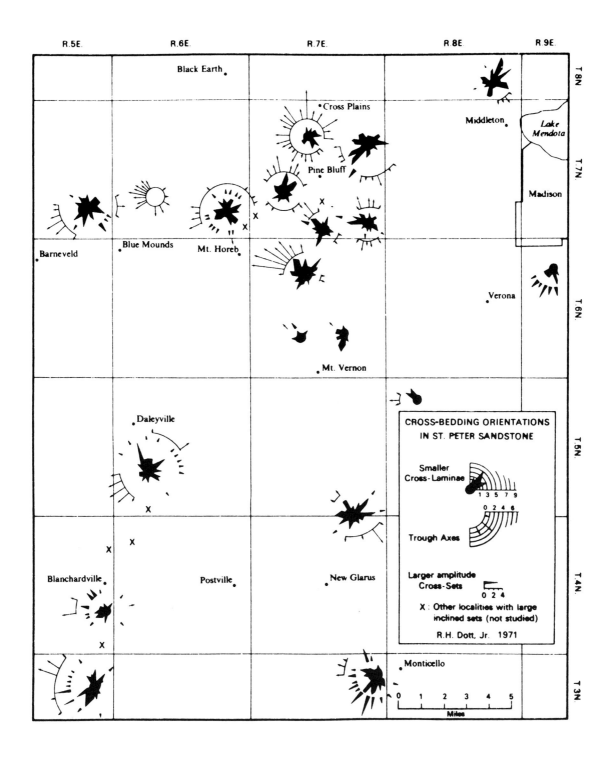

Figure 37. Cross bedding orientations in the St. Peter Sandstone southwest of Madison, Wisconsin. Note that small and large cross sets, as well as trough axis orientation, have all been plotted separately. From Dott and Roshardt, 1972.

The Peninsula Formation (Hobday and Tankard, 1978) is a lower Paleozoic unit which crops out near Cape Town, South Africa, and covers an area of about 12 X 45 km in the Cape Town area (Fig. 38). It is about 750 m thick, and is composed of five main sandstone facies. Facies 1-4 consist of various barrier-related sandstones -- washouts, shallow tidal channels and inlets, and tidal flats and deltas. Facies 5, interpreted as shallow marine sandbar complexes, is made up of complexly cross bedded sandstones in lenticular units 2-20 m thick. Large scale "master bedding surfaces" are "nonerosional -- deposition was probably related to migration of smaller scale bedforms on an inclined surface" (Hobday and Tankard, 1978, p. 1738; Fig. 39). Herringbone structures are occasionally observed. Unfortunately, descriptions are brief, and as details are given of how the sand was transported out onto the shelf. Hobday and Tankard's interpretation is shown in Figure 40.

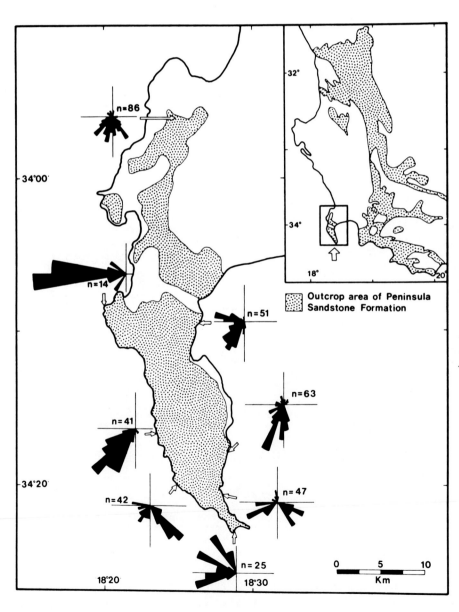

Figure 38. Outcrop area and paleocurrent pattern of the Lower Paleozoic Peninsula Formation near Capetown, South Africa. From Hobday and Tankard, 1978.

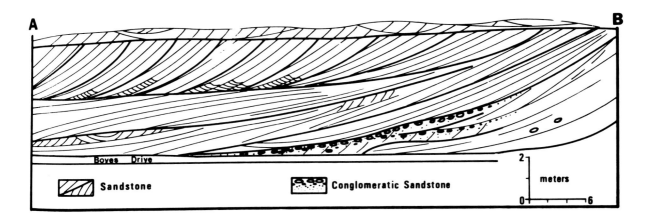

Figure 39. Compound cross stratification in the Lower Paleozoic Peninsula Formation, South Africa. Note reactivation surfaces, set truncation, and gravelly lag. Compare with Allen's Class IV (Fig. 10). From Hobday and Tankard, 1978.

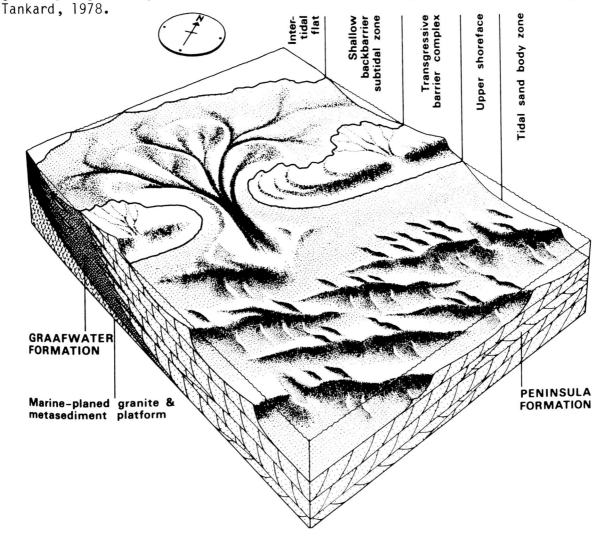

Figure 40. Interpretive block diagram of Peninsula Formation seaward facing sand waves. Tidal currents are locally unidirectional and parallel to the coast - tidal sandbars migrated mainly southwestward in a longshore direction. From Hobday and Tankard, 1978.

## SUMMARY

The sandstone portions of all of the formations described above share a complexity of cross bedding which is characterized by reactivation surfaces. Smaller sets have these reactivation surfaces as their bounding surfaces. The reactivation surfaces define compound sets up to about 10 m thick. Paleoflows tend to be dominantly unidirectional, although locally there may be bimodal-bipolar patterns. Preferred vertical facies sequences appear not to have been recognized.

In those units with fine-grained muddy portions, there may be thin, sharp-based sandstones with distinct descriptive similarities to turbidites (Anderton, 1976; Banks, 1973). These may represent sands transported into deeper, muddier areas by storm-generated flows. In both the Anderton and Banks studies, these beds need reexamination to document more carefully their possible turbidite nature, and the possible presence of hummocky cross stratification.

It appears that in these studies, sand is being widely dispersed by tidal currents. The areas of sand waves and ridges in the North Sea (McCave, 1971; Houbolt, 1968) and in Georges Bank (Twichell, 1983) are about 15,000 $km^2$, 5000 $km^2$ and 20,000 $km^2$, respectively. The preserved minimum outcrop areas for the Jura and Eriboll examples are about 1600 and 2500 $km^2$, respectively, so there is little problem with a comparison of ancient and recent in terms of scale. It is less clear

1. how the sand is being transported from the shoreline onto the shelf,

2. how subsidence and supply can balance each other so perfectly, such that shelf-to-shoreline vertical sequences are rare or absent. Deposition appears to be restricted to open shelf environments despite the great thicknesses of some of the units.

## Reactivation Surfaces and Master Bedding Surfaces

It is clear that these types of surfaces are characteristic of tidal sand bodies; however, one must be considered erosional and the other depositional. Reactivation surfaces form when the top of a large, angle-of-repose bedform is modified, commonly during falling tides. In deeper subtidal situations, the modification may be due to wave erosion of the bedform crest during neap periods of lower sediment transport rates, upstream of the bedform crest. In either case, the original topography of the bedform is rebuilt (that is, the bedform is reactivated) when sand transport toward the crest resumes (see particularly Allen's Class II).

The master bedding surfaces are probably depositional rather than erosional, that is, they do not originate as reactivation surfaces. In Allen's Classes V and VI, the sandwave complexes have been built under conditions of roughly equal ebb and flood flows, and the constantly alternating flows have not allowed the formation of a distinct crest, nor an avalanche slope, nor a zone of flow separation. Time-velocity asymmetries in flow may favour the preservation of cross stratification that marches up, or down, the master bedding surfaces. Alternatively, long term changes in the pattern of tidal currents may influence the preservation pattern of cross stratification oriented up, and/or down the master bedding surfaces.

CONCLUSIONS

The two tentatively-suggested groups of ancient tidal deposits differ mostly in their thickness and stratigraphic setting. The transgressive sand-wave complexes, exemplified by the Lower Greensand, are only a few tens of meters thick. The sand appears to have been reworked from older material during transgression, with little addition of new sand during the growth of the sandwave complexes. By contrast, the thick and extensive tidally-influenced quartzites typical of the Late Precambrian/Cambrian average 1400 m thick (this is the average of the examples given in this paper), and clearly required constant addition of sand. It is not clear how this sand was transported from the shoreline to the depositional site, although several authors have suggested storm modification of these extensive tidally-dominated bodies.

REFERENCES

Allen, J.R.L., 1970, Physical processes of sedimentation: Amsterdam, Elsevier, 248p.

Allen, J.R.L., 1980, Sand waves: a model of origin and internal structure: Sedimentary Geology, v. 26, p. 281-328.

_____, 1981, Lower Cretaceous tides revealed by cross-bedding with mud drapes: Nature, v. 289, p. 579-581.

_____ and J. Narayan, 1964, Cross-stratified units, some with silt bands, in the Folkstone Beds (Lower Greensand) of southeast England: Geologie en Mijnbouw, v. 43, p. 451-461.

Allen, P. A. and P. Homewood, 1984, Evolution and mechanics of a Miocene tidal sandwave: Sedimentology, v. 31, p. 63-81.

Anderton, R., 1976, Tidal shelf sedimentation: an example from the Scottish Dalradian: Sedimentology, v. 23, p. 429-458.

Banks, N. L., 1973, Tide-dominated offshore sedimentation, Lower Cambrian, North Norway: Sedimentology, v. 20, p. 213-228.

Bouma, A. H., 1962, Sedimentology of some flysch deposits: Amsterdam, Elsevier, 168 p.

Bridges, P., 1982, Ancient offshore tidal deposits, in Stride, A. H. (ed.), Offshore tidal sands: London, Chapman and Hall, p. 172-192.

Dalrymple, R. W., R. J. Knight and J. J. Lambiase, 1978, Bedforms and their hydraulic stability relationships in a tidal environment, Bay of Fundy, Canada: Nature, v. 275, p. 100-104.

Dike, E. F., 1972, Sedimentology of the Lower Greensand of the Isle of Wight: Ph.D. Thesis, Oxford University.

Dott, R. H., Jr. and M. A. Roshardt, 1972, Analysis of cross-stratification orientation in the St. Peter Sandstone in southwestern Wisconsin: Geological Society of America, Bulletin, v. 83, p. 2589-2596.

Hine, A. C., 1977, Lily Bank, Bahamas: history of an active oolite sand shoal: Journal of Sedimentary Petrology, v. 47, p. 1554-1581.

Hobday, D. K. and A. J. Tankard, 1978, Transgressive-barrier and shallow-shelf interpretation of the Lower Paleozoic Peninsula Formation, South Africa: Geological Society of America, Bulletin, v. 89, p. 1733-1744.

Houbolt, J.J.H.C., 1968, Recent sediments in the southern bight of the North Sea: Geologie en Mijnbouw, v. 47, p. 245-273.

Jordan, G. F., 1962, Large submarine sand waves: Science, v. 136, p. 839-848.

Klein, G. deV., 1970, Depositional and dispersal dynamics of intertidal sand bars: Journal of Sedimentary Petrology, v. 40, p. 1095-1127.

Levell, B. K., 1980, A late Precambrian tidal shelf deposit, the Lower Sandfjord Formation, Finnmark, North Norway: Sedimentology, v. 27, p. 539-557.

McCave, I. N. and D. N. Langhorne, 1982, Sand waves and sediment transport around the end of a tidal sand bank: Sedimentology, v.29, p. 95-110.

Reinson, G. E., A. E. Foscolos, and T. G. Powell, 1983, Comparison of Viking sandstone sequences, Joffre and Caroline Fields, in McLean, J.R. and G.E. Reinson (eds.), Sedimentology of selected Mesozoic clastic sequences: Canadian Society of Petroleum Geologists, Corexpo '83, p.101-117.

McCave, I. N., 1971, Sand waves in the North Sea off the coast of Holland: Marine Geology, v. 10, p. 199-225.

Narayan, J., 1971, Sedimentary structures in the Lower Greensand of the Weald, England, and Bas-Boulonnais, France: Sedimentary Geology, v. 6, p. 73-109.

Nio, S. D., 1976, Marine transgressions as a factor in the formation of marine sandwave complexes: Geologie en Mijnbouw, v. 55, p. 18-40.

_____ and J. J. Hommes, 1982, Geometry and sequential upbuilding of some large subtidal sandy complexes and their hydrodynamic interpretation: International Congress on Sedimentology, Hamilton, Canada, Abstracts, p. 163.

Nio, S. D. and J. C. Siegenthaler, 1978, A lower Eocene estuarine-shelf complex in the Isabena Valley: field guide to transgressive siliciclastic complexes in the southern Pyrenean basin, Spain: Sedimentology Group, State University of Utrecht, Report no. 18, 44 p.

Pryor, W. A. and E. J. Amaral, 1971, Large-scale cross-stratification in the St. Peter Sandstone: Geological Society of America, Bulletin, v. 82, p. 239-244.

Reineck, H. E., 1963, Sedimentgefuge in Bereich der Sudlichen Nordsee: Abh. Senckenberg. Naturforsch. Ges., no. 505, 64 p.

Sorby, H. C., 1858, On the ancient physical geography of the south-east of England: Edinburgh New Philosophical Journal, New Series, p. 1-13.

Stride, A. H., 1963, Current-swept sea floors near the southern half of Great Britain: Quarterly Journal, Geological Society of London, v. 119, p. 175-199.

_____, (ed.), 1982, Offshore tidal sands - processes and deposits: London, Chapman and Hall, 222 p.

_____, R. H. Belderson, N. H. Kenyon and M. A. Johnson, 1982, Offshore tidal deposits: sand sheet and sand bank facies, in Stride, A. H. (ed.), Offshore tidal sands: London, Chapman and Hall, p. 95-125.

Swett, K., G. deV. Klein and Smit, D. E., 1971, A Cambrian tidal sand body - the Eriboll Sandstone of northwest Scotland: an ancient-recent analog: Journal of Geology, v. 79, p. 400-415.

Swift, D. J. P., 1975, Tidal sand ridges and shoal retreat massifs: Marine Geology, v. 18, p. 105-134.

Twichell, D. C., 1983, Bedform distribution and inferred sand transport on Georges Bank, United States Atlantic Continental Shelf: Sedimentology, v. 30, p. 695-710.

# THE SHANNON SHELF-RIDGE SANDSTONE COMPLEX, SALT CREEK ANTICLINE AREA, POWDER RIVER BASIN, WYOMING

R. W. Tillman
Energy Resources Group
Exploration and Production Research Laboratory
Tulsa, Oklahoma

R. S. Martinsen
Consultant
Laramie, Wyoming

Two vertically stacked shelf-ridge (bar) complexes in the Shannon Sandstone member of the Cody Shale (designated upper and lower sandstones) crop out in the Salt Creek anticline of the Powder River Basin, Wyoming. The shelf-ridge complexes are composed primarily of moderately to highly glauconitic, fine- to medium-grained lithic sandstone and attain thicknesses of over 70 feet. The shelf-ridge complexes were deposited at least 70 miles from shore at middle to inner shelf depths by south to southwest-flowing shore-parallel currents intensified periodically and frequently by storms. Ridges in each sequence trend north-south, slightly oblique to current flow. A possible source of sediments for the shelf ridges was the Eagle Sandstone shoreline and deltaic deposits of southern Montana 200 miles to the northwest.

Eleven facies were defined in outcrop on the basis of physical and biologic sedimentary structures and lithology. Vertical and lateral changes in facies are relatively abrupt where observed in closely spaced outcrop sections, and, in general, facies are stacked in coarsening-upward sequences with Central Bar Facies commonly immediately overlying Interbar Sandstone Facies. Porous and permeable potential reservoir facies include: Central Bar Facies, a clean, cross-bedded sandstone; Bar Margin Facies (Type 1), a highly glauconitic, cross-bedded sandstone containing abundant shale and limonite (after siderite) rip-up clasts and lenses; and Bar Margin Facies (Type 2), a cross-bedded to rippled sandstone. These facies were formed by sediment transported and deposited in the form of medium- to large-scale troughs and sand waves on and across the tops of ridges by moderate to high energy shelf currents. Storm flow deposited Central Bar (Planar laminated) Facies are rare.

Finer-grained, non- to marginal-reservoir quality facies include Interbar Sandstone Facies (rippled to ripple-form bedded sandstone), Bioturbated Shelf Sandstone Facies, Bioturbated Shelf Siltstone Facies, Interbar Facies (interlaminated rippled sandstone and shale), Shelf Sandstone and Shelf Siltstone Facies (sub-horizontally laminated sandstone and siltstone). Interbar Sandstone Facies were most commonly deposited lateral to the higher energy portions of the ridges as well as near the base of the shelf ridges during their initial development. The two bioturbated facies most commonly occur near the base of the ridge complex and between the two vertically stacked ridge complex sequences, and probably represent periods of slow deposition. The Shelf Silty Shale Facies is actually a facies of the Cody Shale of which the Shannon Sandstone is a member.

The most common vertical sequence of sandstone facies is one in which a coarsening-upward sequence is formed where Central Bar Facies overlie Interbar Sandstone Facies; this contrasts with the sequence observed at Hartzog

Draw Field, 25 miles to the northeast, where the most common coarsening-upward sequence from bottom to top is Interbar Facies, Bar Margin Facies (Type 1) and Central Bar Facies. Relatively abrupt lateral changes in facies are observed in surface cross sections spaced from one-fourth to one mile apart. Thickness changes, but not facies changes, are readily observable on subsurface cross sections constructed using SP-resistivity logs.

The association of two vertically stacked shelf-ridge complexes at Salt Creek is atypical compared to other Shannon sequences in the Powder River Basin in several respects. The lower sandstone is correlative with productive Shannon sandstones in many of the Powder River Basin fields (e.g., Hartzog Draw); the upper sandstone sequence is only locally developed in the area of the Salt Creek anticline. Also, the spacing between ridges and the length to width ratios of the ridges at Salt Creek are much smaller than those in other areas. These differences are particularly apparent in the upper sequence wherein the sandstone bodies appear to have oblate geometries very unlike the strongly linear geometries typical of most shelf sandstone ridges. These differences are attributed to the presence during Shannon time of an actively growing paleo-high in the area of the present day Salt Creek anticline which localized sand deposition and ridge formation. Similar early structural growth and its influence on shelf sedimentation has been well-documented for the Lost Soldier anticline area by Reynolds (1976).

Baculites zones are commonly used for surface correlations in the study area and in Upper Cretaceous units throughout Wyoming. Subsurface cross sections paralleling surface sections at distances from one half to three miles away corroborate the surface sandstone correlations in the Salt Creek area. Bentonites above and below the Shannon form excellent subsurface correlation datums.

Foraminiferal data indicate that the shelf-ridge complexes were deposited in water depths ranging from the middle shelf to the outer part of the inner shelf. A wide diversity in size, orientation, and type of burrow-fill material suggests a relatively hospitable environment for burrowers in portions of the shelf-ridge complexes. Rare Teichichnus, Thallasinoides, Chondrites, and plural curving tubes were identified. Common Cretaceous shoreline traces such as Ophiomorpha, Asterosoma and Rhizocorallium were not observed. The Bioturbated Shelf Sandstone Facies and the Bioturbated Shelf Siltstone Facies range from 75 to 95% burrowed. Burrowing in the other facies averages from 5 to 27%. Glauconite is present throughout and is most abundant in association with shale rip-up clasts and limonite (after siderite) lenses and rip-up clasts in Bar Bargin Facies (Type 1).

Transport directions, determined by abundant high angle cross beds (mostly troughs), indicate a south-southwest transport direction (188°) for current deposition of the high energy facies. The range of variation in transport direction at individual outcrops and overall is relatively small (60°). Most current ripples in the Interbar Sandstone Facies also indicate a southerly transport direction. Only very locally, in the top foot or two of some Central Bar and Bar Margin Facies, trough orientations indicate transport directions strongly oblique (northeast) to the general south-southwest flow direction.

In both outcrop and in Hartzog Draw Field, ridge complexes trend nearly north-south, slightly oblique to current flow. Detailed subsurface correlations of the Shannon sand ridges throughout the Powder River Basin, using well-developed bentonite markers, show the reservoir facies to "rise and fall" parallel to their elongation, indicating that the ridges were not deposited in layer-cake fashion. In the Salt Creek area, the reservoir facies generally gradually rise in section to the south parallel to the direction of current flow. In the Hartzog Draw area, the ridges rise to the north, opposite to current flow; they also rise in paired-fashion laterally east and west. These stratigraphic patterns of development of the higher energy shelf-ridge facies are interpreted to reflect sea-floor topography during their deposition.

# BIBLIOGRAPHY

Asquith, D. O., 1970, Depositional topography and major marine environments, Late Cretaceous, Wyoming: AAPG Bull., v. 54, p. 1184-1224.

Asquith, D. O., 1974, Sedimentary models, cycles, and deltas Upper Cretaceous Wyoming: AAPG Bull., v. 58, p. 2274-2283.

Baker, F. E., 1957, History of Salt Creek Oil Field: Wyoming Oil and Gas fields, Wyoming Geol. Assn., p. 388-90.

Barlow, J. A. and J. D. Haun, 1966, Regional Stratigraphy of Frontier Formation and Relation to Salt Creek Field, Wyoming: AAPG Bull., v. 50, p. 2185-96.

Barratt, J. C. and A. J. Scott, 1982, Phayles Sandstone (Upper Cretaceous) Deltaic and Shelf Bar Complex, Central Wyoming: (abs.), AAPG Bull., v. 66, p. 545.

Berg, R. R., 1975, Depositional environments of the Upper Cretaceous Sussex Sandstone, House Creek Field, Wyoming: AAPG Bull., v. 59, p. 2099-2110.

Brenner, R. L., 1979, A sedimentologic analysis of the Sussex Sandstone, Powder River Basin, Wyoming: Wyo. Geol. Assn. Earth Sci. Bull., v. 12, no. 2, p. 37-47.

Costello, W. R. and J. B. Southard, 1981, Flume Experiments on Lower-Flow Regime Bed Forms in Coarse Sand: Jour. Sed. Pet., v. 51, p. 849-864.

Crews, G. C., J. A. Barlow, Jr., and J. D. Haun, 1976, Upper Cretaceous Gammon, Shannon and Sussex Sandstones, Central Powder River Basin, Wyoming: Wyoming Geol. Assoc. 28th Annual Field Conf., p. 9-20.

Davis, M. J. T., 1976, An Environmental interpretation of the Upper Cretaceous Shannon Sandstone, Heldt Draw Field, Wyoming: Wyom. Geol. Assoc., 28th Annual Field Conf., p. 125-138.

Eldridge, G. H., 1889, Some suggestions upon the methods of grouping the formations of the middle Cretaceous and the employment of an additional term in the nomenclature: Am. Jour. Sci., v. 38, p. 313-321.

Gill, J. R., and W. A. Cobban, 1966, The Red Bird section of the Upper Cretaceous Pierre Shale in Wyoming: U.S. Geol. Survey; Prof. Paper 393-A, 73 p.

Gill, J. R., and W. A. Cobban, 1973, Stratigraphy and geologic history of the Montana Group and equivalent rocks, Montana, Wyoming, and North Dakota: U. S. Geol. Survey; Prof. Paper 776, 37 p.

Martinsen, R. S. and R. W. Tillman, 1978, Hartzog Draw, New Giant Oil Field: (abs.), AAPG Bull., v. 62, p. 540.

Martinsen, R. S. and R. W. Tillman, 1979, Facies and Reservoir Chracteristics of Shelf Sandstones, Hartzog Draw Field, Powder River Basin, Wyoming: (abs.), AAPG Bull., v. 63, p. 491.

Meijer Drees, N. C., and D. W. Mhyr, 1981, The Upper Cretaceous Milk River and Lea Park Formations in Southeastern Alberta: Bull. Can. Petroleum Geology, v. 29, p. 42-74.

Parker, J. M., 1958, Stratigraphy of the Shannon Member of the Eagle Formation and its relationship to other units in the Montana Group of the Powder River Basin, Wyoming and Montana: Wyom. Geol. Assoc., 13th Annual Field Conf., p. 90-102.

Porter, K. W., 1976, Marine shelf model, Hygiene Member of the Pierre Shale, Upper Cretaceous Denver Basin, Colorado, in Studies of Colorado Field Geology, R. Epis and R. Weimer (eds.), Professional Contributions of Colorado School of Mines (Annual Meeting GSA, Denver), p. 251-263.

Reynolds, M. W., 1976, Influence of Recurrent Laramide Structural Growth on Sedimentation and Petroluem Accumulation, Lost Soldier Area, Wyoming: AAPG Bull., v. 60, p. 12-33.

Rice, D. D., and W. A. Cobban, 1977, Cretaceous stratigraphy of the Glacier National Park area, northwestern Montana: Bull. Can. Petroleum Geology, v. 25, p. 828-841.

Seeling, A., 1978, The Shannon Sandstone, a further look at the environment of deposition at Heldt Draw Field, Wyoming: The Mountain Geologist, v. 15, no. 4, p. 133-144.

Shelton, J. W., 1965, Trend and genesis of lowermost sandstone unit of Eagle Sandstone at Billings, Montana: AAPG Bull., v. 49, p. 1385-1397.

Shurr, F. W., (this volume), Geometry of shelf sandstone bodies in the Shannon Sandstone of Southeastern Montana: in R. W. Tillman and C. T. Siemers (eds.), SEPM Special Publication.

Spearing, D. R., 1975, Shannon Sandstone, Wyoming: in SEPM Short Course No. 2, Depositional environments as interpreted from primary sedimentary structures and stratification sequences, Dallas, p. 104-114.

Spearing, D. R., 1976, Upper Cretaceous Shannon Sandstone: an offshore, shallow marine sand body: Wyom. Geol. Assoc., 28th Annual Field Conf., p. 65-72.

Stubblefield, W. L., D. W. McGrail, and D. G. Kersey, (this volume), Recognition of Transgressive and Post Transgressive Sand Ridges on the New Jersey Continental Shelf: in C. T. Siemers and R. W. Tillman, eds., SEPM Special Publication.

Swift, D. J. P. and D. D. Rice, (this volume), Sand Bodies on Muddy Shelves. A Model for Sedimentation in the North American Cretaceous Seaway: in R. W. Tillman and C. T. Siemers (eds.), SEPM Special Publication.

Tillman, R. W. and R. S. Martinsen, 1979, Hartzog Draw Field, Powder River Basin, Wyoming: in R. W. Flory (ed.), Rocky Mountain High, Wyoming Geol. Assn. 28th Annual Meeting, Core Seminar Core Book, p. 1-38.

Tillman, R. W. and R. S. Martinsen, (in press), Hartzog Draw Field, Wyoming, in R. W. Tillman and K. Weber (eds.), Reservoir Sedimentology and Synergy, SEPM Special Publication.

Weimer, R. J. (1983), Relation of Unconformities, Tectonics, and Sea Level Changes, Cretaceous of the Denver Basin and Adjacent Areas, in Reynolds, M. W. and E. D. Dolly (eds.), Mesozoic Paleogeography of the West-Central United States, Rocky Mountain Paleogeography Symposium 2, Rocky Mountain Section SEPM, Denver, Colorado, p. 359-76.

Wilmarth, M. G., 1938, Lexicon of geologic names of the United States (including Alaska): U.S.G.S Bull. 896, 2396 p.

R. S. MARTINSEN, Cities Service Oil Co., 900 Colorado State Bank Building, Denver, Colorado, 80202, and R. W. TILLMAN, Cities Service Research and Exploration, Box 50408, Tulsa, Oklahoma, 74150.

HARTZOG DRAW, A NEW GIANT OIL FIELD

Hartzog Draw Field, located in the Powder River Basin, Wyoming, was discovered in August, 1975. It is one of the largest oil fields discovered in the Rocky Mountain province in recent years, with initial estimates of ultimate recovery exceeding 100,000,000 barrels of oil. Field development through the fall of 1977 extended more than 20 miles lengthwise in a northwest-southeast direction and up to three miles in width, encompassing in excess of 22,000 productive areas. Development drilling on 160 acre spacing has had a better than 95% success ratio and initial production rates commonly exceed 1,000 barrels of oil per day, with several wells having potentialed in excess of 3,000 barrels per day.

Production at Hartzog Draw is from the Upper Cretaceous Shannon Sandstone Member of the Cody Shale, at a depth of 9,000 to 9,600 feet. Oil accumulation is stratigraphically controlled, structure having almost no influence on entrapment. The reservoir sandstones (mostly MARINE CENTRAL BAR and BAR MARGIN FACIES) are quartzose and glauconitic, fine to medium grained, moderately well sorted, highly trough cross-bedded, and occur in stacked sequences up to 60 feet in thickness. Sideritic clasts and shale rip-up clasts occur locally in the high angle trough cross-bedded units.

In the reservoir facies, effective porosities average around 13% and permeabilities 12 md. There is no apparent water table, and net pay thickness closely parallels net sand thickness. The reservoir sandstones are associated with a 30 to 80 foot thick package of rippled interbedded very fine-grained sandstone and shale (INTERBAR and SHELF SILTSTONE FACIES) which in turn is completely enveloped in gray shale (SHELF SILTY SHALE FACIES). Shannon Sandstone deposition apparently occurred below effective normal wave base near the middle of the western shelf of the Pierre (Early Campanian) seaway, more than 100 miles from shore in water depths exceeding 60 feet.

Reference: AAPG Bulletin, Vol. 62, No. 3, March, 1978, p. 540.

R. S. MARTINSEN, Cities Service Oil Co., 900 Colorado State Bank Building, Denver, Colorado 80202, and R. W. TILLMAN, Cities Service Research and Exploration, Box 50408, Tulsa, Oklahoma 74150.

## FACIES AND RESERVOIR CHARACTERISTICS OF A SHELF SANDSTONE: HARTZOG DRAW FIELD, POWDER RIVER BASIN, WYOMING

Hartzog Draw Field is a stratigraphically controlled oil reservoir which produces from the Upper Cretaceous Shannon Sandstone at depths from 9000 to 9600 ft. The producing interval consists of a large mid-shelf sand bar complex deposited below effective normal wave base more than 100 miles from shore. The productive interval in the bar complex has a maximum thickness of 65 ft, is over 21 miles long and is up to 3 1/2 miles wide. Over 170 wells have been completed on 160 acre spacing since its discovery in 1975, and ultimate oil recovery may exceed 100,000,000 barrels.

The reservoir is completely enveloped in shale, has a solution gas drive, no water table and no produced formation water. Even zones that calculate water saturations of over 65% from logs do not produce water. Net pay is primarily a product of porosity, permeability and thickness of the sandstone, and is directly related to sedimentary facies. Of six facies observed in cores, only one, the *central bar facies*, a high angle trough cross-bedded glauconitic quartz sandstone, is a consistently high quality reservoir. Two others, the *bar margin facies*, a ripple to trough cross-bedded sandstone with abundant shale and siderite clasts, and the *interbar facies*, a rippled interbedded sandstone and shale, generally are marginal quality reservoirs.

Data from three cores indicates the *central bar facies* to have a significantly better average porosity and permeability (12.7%, 6.4 md) than either the *bar margin facies* (8.1%, 3.7 md) or *interbar facies* (6.2%, 2.1 md). In addition, wells with a thick central bar facies appear to maintain higher reservoir pressures. Recognition of the facies, as well as understanding their distribution and inter-relationships, are prerequisites to developing a program which will maximize oil recovery from the field.

Reference: AAPG Bulletin, Vol. 63, No. 3, March 1979; p.491

CARDIUM FORMATION 4. REVIEW OF FACIES AND DEPOSITIONAL
PROCESSES IN THE SOUTHERN FOOTHILLS AND PLAINS, ALBERTA, CANADA

Roger G. Walker
Dept. of Geology, McMaster University,
Hamilton, Ontario L8S 4M1, Canada

## REGIONAL SETTING

The Upper Cretaceous (Turonian) Cardium Formation (Figs. 1, 2) is a dominantly sandstone unit surrounded by marine shales of the Alberta (= Colorado) Group. It was deposited in the Western Interior Seaway in Alberta. In the subsurface, the Cardium is a major oil and gas reservoir (Table 1, on next page). It crops out abundantly in several thrust slices of the Rocky Mountain Foothills, between the U.S. Border in southern Alberta and Dawson Creek, B.C., a distance of about 800 km. The Alberta Group spans the Cenomanian to early Campanian, a period of about 15 million years (Fig. 2), and the Cardium belongs to the upper part of the Turonian, implying that deposition of the Cardium may have occurred in as little as one million years (Palmer, 1983).

In the Canadian Cordillera, there are two main clastic wedges, the late Jurassic-early Cretaceous Kootenay - Blairmore Assemblage, and the late Cretaceous - Paleocene Belly River - Paskapoo Assemblage. The Alberta Group presumably represents a period of relative tectonic quiescence between these two assemblages, with dominantly mudstones accumulating in the Alberta Basin.

## STRATIGRAPHY

In Alberta, the Alberta (= Colorado) Group is divided into three formations (Fig. 2);

top 1) Wapiabi Formation, about 600 m, mostly shales;
2) Cardium Formation, about 100 m, interbedded sandstones and shales;
base 3) Blackstone Formation, about 400 m, mostly shales.

The Cardium Formation is an approximate time equivalent of the Turonian Frontier, Ferron and Gallup Sandstones in the U.S. portion of the Western Interior Seaway. In terms of Kauffman's (1977, 1979) transgressive-regressive cycles, the Cardium is roughly equivalent to $R_6$, terminating the Greenhorn cycle.

## SUBDIVISION OF THE CARDIUM FORMATION

In outcrop, the Cardium has been subdivided into six members by Stott (1963). Five of these are exclusively marine (Fig. 3), but the Moosehound Member is non-marine, being characterized by thin coals and mudstones with root traces. Stott's member boundaries cross his cycle boundaries, a problem discussed by Walker (1983b). The cycles, or coarsening-upward sequences, are ubiquitous in the Cardium, and can be traced laterally more reliably than the members (Duke, 1985a). The number of sequences varies -- at the reference location (Seebe, Alberta) described below, there are 5 sequences.

The non-marine Moosehound Member can only be traced southeastward as far as the Clearwater River area (125 km N.W. of Calgary; Fig. 1) -- southeast of here the Cardium is exclusively marine.

TABLE 1

Major Cardium Reservoirs, With Initial
Recoverable Reserves and Year of Discovery

| FIELD | RESERVES (millions of barrels) | YEAR OF DISCOVERY |
|---|---|---|
| Pembina | 1500 | 1953 |
| Willesden Green | 231 | 1954 |
| Ferrier | 73.8 | 1965 |
| Ricinus | 18.1 | 1969 |
| Crossfield | 17.5 | 1956 |
| Garrington | 17.1 | 1954 |
| Cyn-Pem | 13.7 | 1962 |

Total Cardium, about 1.9 billion barrels. Data from Oil Pools of Western Canada: Geological Survey of Canada, Map 1559A, 1981.

## HISTORY OF CARDIUM INTERPRETATIONS

The discovery of the Cardium oil field at Pembina, Alberta, in 1953 ushered in the "modern" era of Cardium interpretations (reviewed by Walker, 1983a). Pembina is presently Canada's largest single oil field, with over 4000 producing wells, nearly 2000 of which have been cored -- these cores are all stored at the Core Research Laboratory of the Energy Resources Conservation Board in Calgary, and are available for public inspection. Widespread pebble horizons within the Cardium marine mudstones had been known before 1953, and had defied explanation. Floyd Beach (1955) applied the newly introduced turbidite concept (Kuenen and Migliorini, 1950) to explain the pebble horizons, but after a few years of controversy, interpretations of Cardium depositional environments settled into some version of delta/shoreline/offshore bar/barrier bar.

As the long, narrow trends of Cardium fields in the subsurface became better defined (Fig. 1) the "offshore" or "barrier" bar interpretation was strengthened (for example, Berven, 1966). Storm interpretations were first suggested by Michaelis and Dixon (1969), with more specific interpretations of Cardium gravels as bedload storm deposits by Swagor et al. (1976). A storm-generated turbidity current interpretation was first suggested by Wright and Walker (1981), based on the Seebe outcrop described below, and I now suggest that many of the sharp-based Cardium sandstones were swept into the basin by turbidity currents (Walker, 1983a, and this paper). However, my initial estimates of transport distances may have been over-estimated (Walker, 1983b, c), and sea-level fluctuations are now recognized as being more important than originally believed (A. G. Plint, personal communication, 1985).

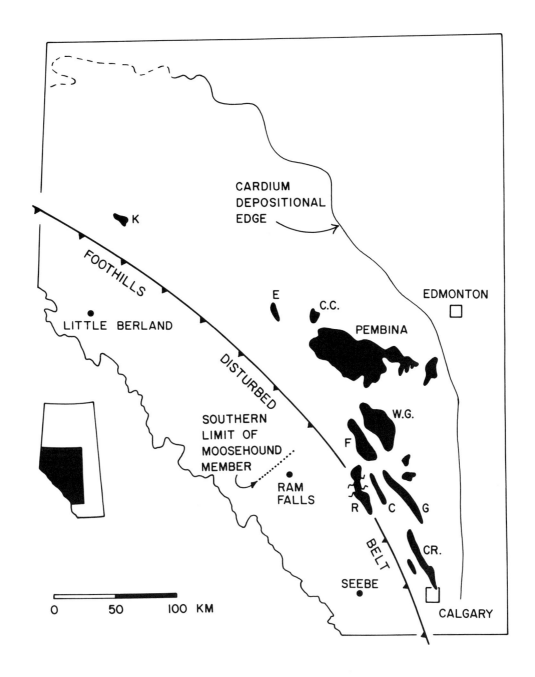

Figure 1. Map of major Cardium fields in the subsurface, with data from "Oil Pools of Western Canada, Geological Survey of Canada, Map 1559A, 1981". K = Kakwa; E = Edson; C.C. = Carrot Creek; W.G. = Willesden Green; F = Ferrier; R = Ricinus; C = Caroline; G = Garrington; CR = Crossfield. Note also the southern limit of the non-marine Moosehound Member. Between the Alberta-B.C. border and the eastern limit of the deformed belt, there are abundant Cardium outcrops. From Walker, 1983a.

## BASIC SEDIMENTOLOGY OF THE CARDIUM FORMATION

The marine portions of the Cardium can be subdivided into several coarsening-upward sequences (Fig. 3). These begin with dark shales or mudstones with a gradual upward increase in the proportion of thin (3-30 cm), fine sandstone beds. Toward the sequence tops, the proportion of sandstone increases, interbedded mudstones become much thinner (or are absent), and individual sandstone beds may be a meter or more in thickness. Conglomerate is commonly present at the top, and it is abruptly overlain by shales of the next sequence.

The shales or mudstones are normally thoroughly bioturbated. The trace fauna has been studied by Pemberton and Frey (1983, 1984) in outcrop and cores. Most traces display a grazing behavior and indicate deposition below fairweather wave base; see earlier paper in this volume, "Geological evidence for storm transportation and deposition on ancient shelves," for a discussion of fairweather wave base.

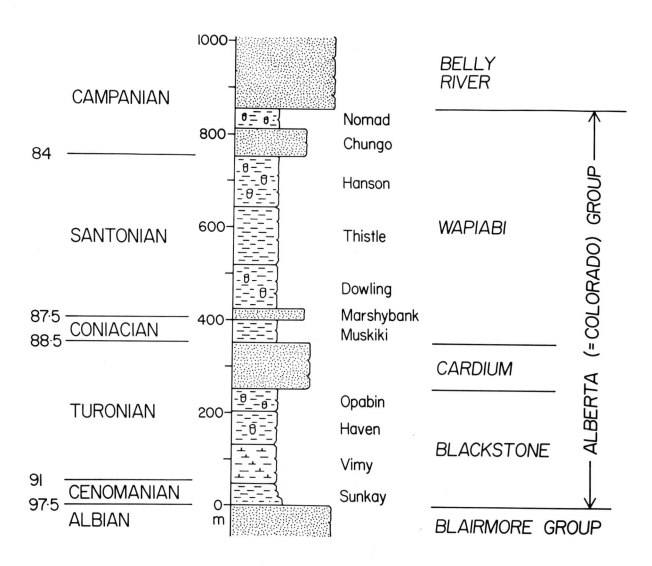

Figure 2. Stratigraphy of the Alberta (= Colorado) Group in the Alberta Foothills. Absolute ages (Palmer, 1983) are given at left.

The underlined sandstones are mostly fine to very fine grained, and occur in individual beds a few cm to over a meter in thickness. Beds have sharp (in places scoured) bases, but bed tops are commonly gradational due to bioturbation. The two ubiquitous sedimentary structures are hummocky cross stratification (HCS) and symmetrical ripples with wavelengths of a few cm. The lower portions of a few beds may be massive, or parallel laminated. There is no angle-of-repose cross bedding in sets thicker than a few cm, and hence no indication of any currents within the Cardium basin that could rework the HCS once it had been made by storm waves (see earlier paper in this volume on "Geological Evidence for Storm Transportation..."). Most, but not all, Cardium sandstones are fine to very fine grained, and hence it could be argued that sand wave and dune bedforms would not be expected. In finer sizes, the dune field is replaced by current ripples and upper flat bed (Southard, in Harms et al., 1982, p. 2-14). Parallel lamination (upper flat bed) is present, but uncommon in the Cardium, and current ripple cross-lamination is rare. The scarcity or absence of the features enhances the suggestion that no reworking takes place after formation of the HCS. I will present evidence in this paper that the Cardium sandstones were probably:

1) emplaced below fairweather wave base by storm-generated currents and turbidity currents; and

2) modified by storm waves to contain HCS rather than the normal turbidite Bouma (1962) sequence of sedimentary structures.

Thus, the sedimentology suggests that the bulk of Cardium deposition took place between storm and fairweather wave base (Duke, 1985a, b). The trace fossils also indicate deposition below fairweather wave base, and the only published foraminiferal paleoecology (Wright and Walker, 1981) suggests depths of several tens of meters for mudstones in two sequences at Seebe.

Many of the Cardium sequences in outcrop have conglomerate at the top, and less commonly within the sequence also. Conglomerates can be clast-supported in beds up to a few tens of cm thick (maximum conglomerate thickness is about 3 m in the Kananaskis River section), or mud-supported, as pebbly mudstones (Crowell, 1957) in beds up to 2-3 m thick (as at Seebe). The clasts are dominantly chert, with pebbles mostly in the 0.5 to 5 cm range. The only sedimentary structures in the conglomerates are straight-crested symmetrical gravel ripples (wavelengths about 1 m), as at Seebe (Duke and Leckie, 1984; Leckie and Duke, 1984).

Siderite concretions are common within Cardium sequences, and layers of siderite following bedding are commonly associated with gritty sandstones and conglomerates at sequence tops.

## THE SEEBE, ALBERTA, REFERENCE SECTION

The type section of the Cardium was designated by Stott (1963) as the Wapiabi Creek section, which is unfortunately very inaccessible. Stott (1963, p. 53) rejected the Seebe section, noting that "of the better exposed sections on the Bow River none is suitable for a type section... the Cardium section at Kananaskis [Seebe] is faulted". Careful mapping (Wright and Walker, 1981; Fig. 4) has shown that all faulting can be restored at Seebe, and a complete section of the first four sequences can be compiled (Fig. 3). The fifth sequence is well exposed at Horseshoe Dam, about 3 km downstream from Seebe. Because Seebe is only one kilometer north of the Trans Canada Highway, only 65 km west of Calgary, and because of its superb three-dimensional outcrop, I suggest it be used as a Cardium reference section. Detailed guidebooks to the outcrop have been prepared by Walker and Wright (1982; sedimentology) and Pemberton and Frey (1983, 1984; trace fauna).

The five sequences are shown in Figure 3, along with Stott's (1963) member terminology. The following comments only highlight the sequences - see Wright and Walker (1981) and Walker and Wright (1982) for details.

Sequence 1 is gradational from the Blackstone Formation. The first Cardium sandstones are sharp-based HCS beds (Fig. 5), and the proportion of sandstones increases upward. The top of the sequence consists of amalgamated HCS beds, but there is no capping of gravel.

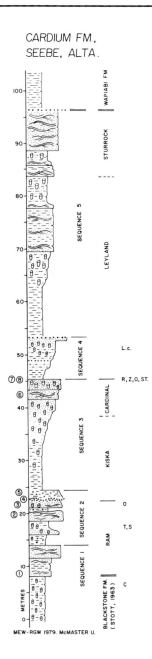

Figure 3. Cardium reference section at Seebe (sequences 1-4; location in Fig. 1) and Horseshoe Dam (section 5). All faulting has been restored, and circled location numbers refer to map in Figure 4. Stott's (1963) Member names are shown as well as the sequences. C = Chondrites; S = Skolithos; O = Ophiomorpha; T = Teichichnus; R = Rhizocorallium; Z = Zoophycos. The decapod Linuparus canadensis is designated L.c., and its possible burrows are designated as "subway tunnels", ST. Dashes = shales; dots = sandstones; dots and dashes = siltstones; large dots = conglomerates; undulating lines = hummocky cross stratification; vertical tubes = bioturbation From Wright and Walker, 1981.

358

Sequence 2 begins with a little over a meter of dark mudstones, with rare scattered chert pebbles up to 3-4 cm in diameter. Most of the coarsening-upward sequence is thoroughly bioturbated, but there are excellent plan views of HCS on both sides of the river (Fig. 6). On the south side, there is a beautiful plan view of three swales coming together at a point (Fig. 7). This outcrop is in the dam spillway channel WHICH MUST NEVER BE ENTERED BEFORE CHECKING AT THE CONTROL CENTER THAT DISCHARGE IS NOT IMMINENT.

The trace fauna is also well displayed in the spillway channel. Pemberton and Frey (1983, 1984) note that this sequence (2) does not contain Rhizocorallium or Zoophycos, and hence may represent shallower water than the other sequences (Fig. 8). However, HCS in the absence of medium scale cross bedding, asymmetrical ripple cross-lamination and abundant parallel lamination still indicates deposition below fairweather wave base. The top of the sequence shows a 3 to 30 cm gravel bed sitting directly on HCS sandstones. The gravel is moulded into straight-crested waves, height about 10 cm, wavelength about 1 m (Fig. 9). It is suggested that these gravel waves were formed below fairweather wave base by large storm waves (Leckie and Duke, 1984; Duke and Leckie, 1984) -- similar modern gravel waves have been described by Gillie (1979) and Yorath et al. (1979); see Wright and Walker (1981) for further discussion. The gravel in the waves is clast-supported, and is overlain by 2-3 m

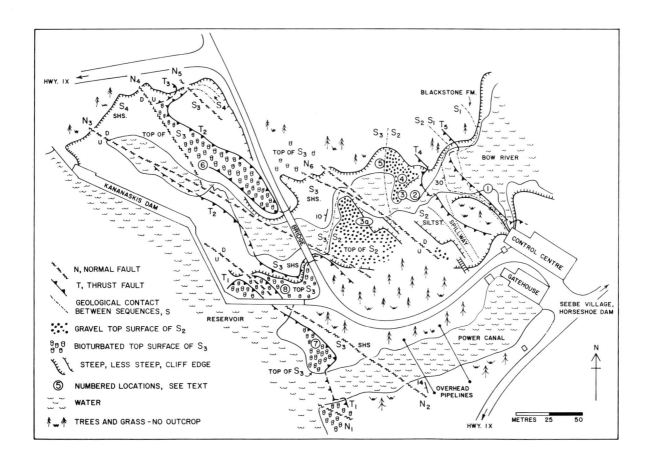

Figure 4. Map of Cardium outcrop below Kananaskis Dam at Seebe. Circled numbers refer to points on the measured section (Fig. 3). Sequences are designated $S_1$-$S_5$. From Wright and Walker, 1981.

of pebbly mudstone (Crowell, 1957). The pebbly mudstone could possibly be a debris flow, but I prefer the possibility that gravel was emplaced on top of uncompacted wet mud, with mixing and possibly slight flowage on the basin floor. This is the original mechanism proposed for pebbly mudstones by Crowell (1957), and in the Seebe situation, it suggests that the gravel waves were buried by mud (i.e., sequence 3 had begun) before the gravel that contributed to the pebbly mudstone was emplaced into the basin. If this interpretation is correct, the pebbly mudstone is not the coarse culmination of sequence 2, but represents a gravel flow that came into the basin and mixed with the lowest muds of sequence 3.

The pebbles gradually die out upward, and the pebbly mudstone grades into the main, non-pebbly mudstones of the next sequence.

Sequence 3 (Fig. 10) begins with mudstones, but the proportion of sandstone steadily increases upward. Except for one or two beds, the sequence is totally bioturbated, and those one or two beds invariably show HCS (Fig. 11). The top of the sequence is spectacularly burrowed, with excellent examples of Rhizocorallium, Zoophycos, and "subway tunnels" probably made by the decapod Linuparus canadensis (Pemberton et al., 1984). There is a one-pebble-thick veneer of gravel, occurring in irregular patches, on top of the sequence.

Sequence 4 is the thinnest (7 m), and consists mostly of bioturbated mudstones. The top is an unbedded (? bioturbated) gritty sandstone with scattered pebbles and irregular sideritic concretions. There is one important bed in the sequence -- a 2 cm thick sharp based gravel bed which occurs within mudstones about 4 m above the base of the sequence. The gravel occurs 1) as a coherent clast supported bed; in places the bed disappears and the gravel occurs 2) as scattered pebbles within the mudstones (? scattered by organisms such as Linuparus). The gravel also occurs 3) in discrete pockets a few cm below the base of the main bed -- these pockets appear to be detached load balls. If so, the loading of gravel into mud can be considered as the first stage of total mixing of gravel into mud, the process suggested by Crowell (1957) for pebbly mudstones (see comments about top of sequence 2).

Sequence 5 begins with black shales, which become increasingly silty and bioturbated upward. The upper half of the sequence contains HCS sandstones, with an 8 m thick amalgamated HCS bed at the top. There is a veneer, one-pebble-thick, of scattered chert pebbles on top of the HCS, and the veneer is abruptly overlain by black shales of the Wapiabi Formation.

## CARDIUM OUTCROP - CONCLUSIONS AND PROBLEMS

1. The depositional environment for marine members of the Cardium was consistently below fairweather wave base, as suggested by trace fossils, foraminiferal paleoecology, and sedimentology. The sedimentological features are the ubiquitous presence of HCS, complete absence of angle-of-repose cross bedding (in sets thicker than a few cm) and scarcity of asymmetrical ripple cross-lamination and parallel lamination.

2. This poses the problem of how the sand was initially transported to depths below fairweather wave base. Some of the possible mechanisms have been reviewed in "Geological evidence for storm transportation and deposition on ancient shelves" earlier in this volume. Evidence from outcrop and subsurface (see below) suggests incremental movement by storm flows, and/or turbidity currents as the two most likely mechanisms.

3. In either case, the flows might be expected to spread widely on the smooth basin floor and deposit individual beds with a sheet-like geometry. The turbidity current hypothesis does not, at first sight, appear to be compatible with the long, narrow subsurface Cardium fields such as Ricinus, Caroline, Garrington and Crossfield (Fig. 1). This problem is addressed below.

Figure 5. First well-preserved sharp-based sandstones in sequence 1, Seebe. Note excellent HCS in beds below and above notebook. Photo at about 8.0 m level, Figure 3.

Figure 6. Plan view of HCS in sequence 2, 20.0 m (Fig. 3). Note broad gentle curvature of stratification.

Figure 7. Plan view of intersecting swales in sequence 2 (20 m, Fig. 3, photo taken in spillway channel, see figure 4). Same stratigraphic horizon as Figure 6.

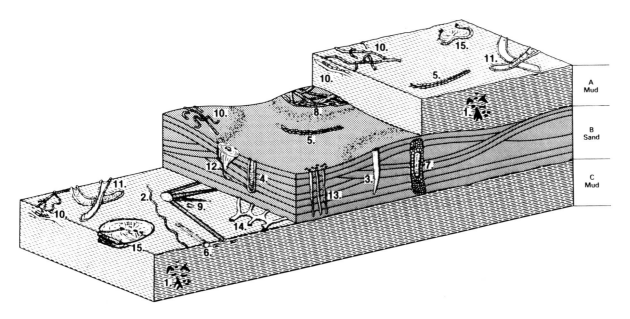

Figure 8. Schematic representation showing distribution of the common trace fossils in the Cardium Formation at Seebe. The Unit B sandstone is a hummocky cross stratified, storm-emplaced sand. 1, Chondrites; 2, Cochlichnus; 3, Cylindrichnus; 4, Diplocraterion; 5, Gyrochorte; 6, Muensteria; 7, Ophiomorpha; 8, Paleophycus; 9, Phoebichnus; 10, Planolites; 11, Rhizocorallium; 12, Rosselia; 13, Skolithos; 14, Thalassinoides; 15, Zoophycos. From Pemberton and Frey, 1983, p. 114.

Figure 9. Gravel waves on top of sequence 2, 22.5 m (Fig. 3). Waves have height of about 10 cm and a wavelength of about 1 m.

Figure 10. View of sequence 3, showing overall coarsening-upward and irregular bedding indicative of extensive bioturbation.

Figure 11. Three-dimensional view of the HCS in upper part of sequence 3 (43.0 m, Fig. 3). Arrows show two three-dimensional hummocks.

# SUBSURFACE CARDIUM

## Subsurface Cardium - Introduction

The Cardium field at Pembina, Canada's largest single oil field, was discovered in 1953. Major Cardium fields with their estimated reserves are shown in Table 1 (on second page of this paper) and Figure 1. Note that all fields south of Pembina are long and narrow, and are elongated parallel to present tectonic strike. North of Pembina, the fields appear to be "patchy" (Kakwa, Edson, Carrot Creek). Pembina itself is sheet-like, but within the field, individual pools and isopach maps of both sands and gravels all show a NW-SE trend.

It is noticeable (Fig. 1) that south of Pembina individual fields appear to become longer and narrower the farther south they occur. Also, there is a progressive offset of fields to the southwest, with Ferrier and Willesden Green offset from Pembina, and Ricinus-Caroline-Garrington in turn offset from Ferrier and Willesden Green. I know of no fully satisfactory explanation for the offset, nor the progressively longer and narrower shapes of the fields.

## Subsurface Cardium Stratigraphy

There is little published work on the correlation between individual Cardium fields. Berven (1966) attempted to correlate between Garrington and Crossfield, and my recent correlations in the Ricinus-Caroline-Garrington (RCG) area (Walker, 1983b; Fig. 12) are discussed below.

There is also no well established correlation between Stott's (1963) members in outcrop and any of the subsurface sandstones. Some of the suggested correlations have been reviewed by Walker (1983b).

In most Cardium fields, the main sandstone bodies are given letters (Cardium A, Cardium B, etc.). The top of the Cardium is taken at a normally prominent log marker known as the Cardium "Zone", which can be seen in the sonic and induction logs, but not the gamma ray or S.P. logs. The stratigraphic position of the "zone" varies somewhat with respect to the top of the A sand, but presently, there is no better marker that can be used over the whole Cardium basin. There are local gamma ray markers that are better than the "zone" (Walker, 1983b; Fig. 13). The base of the Cardium is drawn at the base of the lowest sand body (Fig. 16). In other areas, the Cardium "zone" appears to be a better marker, and can be used as a datum for a local stratigraphy (Krause and Nelson, 1984).

## Subsurface Cardium Structure

Most of the major Cardium fields (Fig. 1) are structurally undeformed, and presently have a gentle regional dip of about 0.5° to the southwest. Ricinus is the major exception -- it straddles the "triangle zone", which is the eastern limit of Foothills thrusting. Dips vary from a few degrees to vertical; in most cores, the dips are less than 10 degrees. Deformation is much more intense in south Ricinus than north Ricinus, with abundant fracturing of the sandstones evident in cores. The two transcurrent (wrench) faults that I show offsetting Ricinus into 3 parts (Fig. 12) probably are the result of relatively longer distances of thrusting in mid and north Ricinus compared with south Ricinus. In mid and south Ricinus, the main sandstone body can be repeated by thrusting 4 or 5 times in core, between depths of about 1900 m to 2900 m.

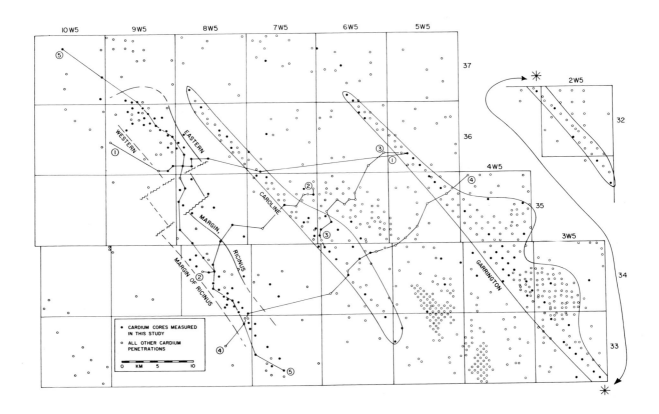

Figure 12. Map of Ricinus-Caroline-Garrington area, showing cores used in the study (black dots). Open circles indicate other Cardium penetrations. Cross sections 1-5 are located (see Figs. 13, 14, 15, and 18). Margins of fields drawn on sand distribution, and may not exactly coincide with area of oil production. Note fault offset of Ricinus, dividing field into north, mid, and south segments. From Walker, 1983b.

---

## RICINUS-CAROLINE-GARRINGTON AREA

### Sand Body Geometry and Stratigraphy, Ricinus-Caroline-Garrington Area

Locations of cores and well logs are shown in Figure 12, along with cross-section 1-4 (Figs. 13, 14, 15, 18). I measured and described 169 cores (shown as solid black circles, Fig. 12) and used about 300 gamma ray logs in this study. Log markers include a distinctive pair of gamma ray kicks termed K and L, and a less prominent marker in the Blackstone Formation, termed N. The correlations have been discussed in detail by Walker (1983b). Major conclusions are:

1. Sections 2, 3, 4 (Figs. 13, 14, 15) show that the Garrington A sandstone is correlative with the Caroline A sandstone; gamma ray logs between these fields indicate that the A sandstone is more or less continuous and has a sheet-like form. The long narrow fields are simply long narrow trends of production within the sheet sandstone. The reason for their form and orientation is unknown. The A sandstone has now been formally named the Raven River Member of the Cardium by Walker (1983c). The type section is shown in Figure 16.

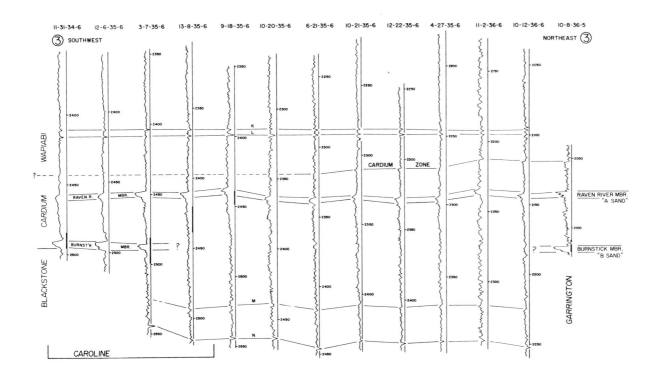

Figure 13. Cross section 3 (see Fig. 12), using the K and L markers as datum. All well logs are gamma ray, depths in meters. Note continuity of the Raven River (A sand) Member from Garrington to Caroline as a "ragged blanket". Note also the absence of the Burnstick (B sand) Member between Garrington and Caroline Fields. Dark bars indicate cored intervals. From Walker, 1983c.

---

2. Sections 3 and 4 (Figs. 13, 15) also show that the B sandstone at Caroline is at exactly the same stratigraphic position as the B sand at Garrington. The B sandstone is sheet-like beneath the Ricinus sand (Section 1, Fig. 18), but is absent between Caroline and Garrington. It has been formally defined as the Burnstick Member of the Cardium by Walker (1983c; Fig. 16).

3. The Ricinus sandstone scours into the Raven River Member, occupying a channel about 55 km long, 8 km wide (before palinspastic reconstruction) and 20-40 m deep. The form of the channel is shown by log correlations in Section 1 (Fig. 18). I have used a core marker as datum -- the facies involved are discussed below. However, the channel shape could be shown equally well by hanging the sections on the K and L markers (the anomalous positions of the markers in 11-5-36-8 is probably due to tectonic thickening of the section).

This is the first demonstration of the channelized nature of the Ricinus sandstone, and this interpretation differs from that of Almon (1979), who termed Ricinus a "bar-like buildup". The Ricinus Member has been formally defined by Walker (1985).

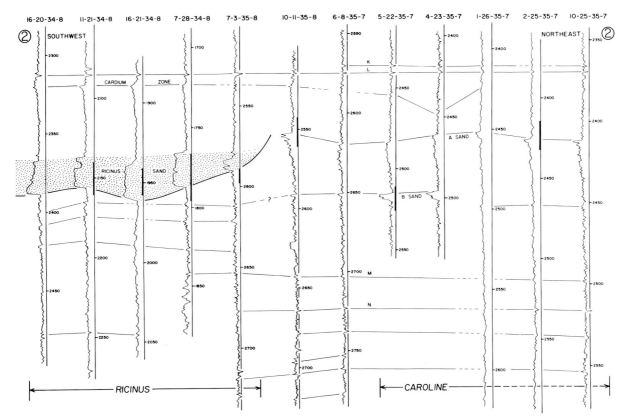

Figure 14. Cross section 2 (Fig. 12), using K and L markers as datum. Note the very different gamma ray response of the Ricinus sand, and its channelized situation with respect to the A sand. Depths in meters. Dark bars indicate cored intervals. From Walker, 1983b.

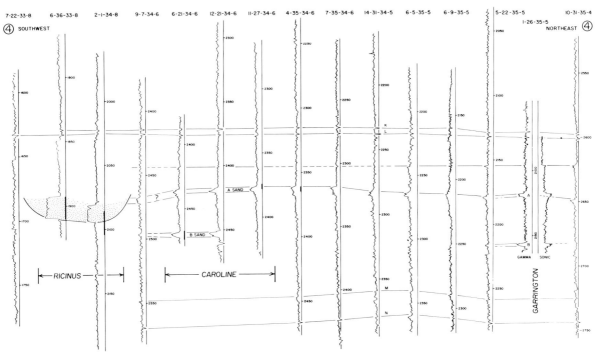

Figure 15. Cross section 4 (Fig. 12), using K and L markers as datum. All well logs are gamma ray, except in the Garrington well where the sonic log is also shown. The Cardium "Zone" shows up prominantly on the sonic log at about 2095 m. Note continuity of the A sandstone as a ragged blanket, and the channel contact of the Ricinus sandstone. Dark bars indicate cored intervals. Depths in meters. From Walker, 1983b.

Figure 16. Type section from Garrington (11-32-34-4W5), and reference section from Caroline (11-31-34-6W5), with gamma ray logs and cored intervals (black bars). The cored intervals show true facies distributions and thicknesses, but for the sake of completeness, the average facies thicknesses for Garrington and Caroline (see Fig. 17) are shown between cored intervals. Datum is the K and L marker, and the Cardium "Zone" marker is shown in its average position. It is hard to pick the true position of the "Zone" in these two wells. Numbers designate the facies described in this paper. The position of the Blackstone-Cardium faunal change is shown (data from Fig. 17), and the top of the Cardium is shown at the position of the Cardium "Zone". Note that the Raven River Member is defined by inflection of the gamma ray log, which does not always coincide with the base of the bioturbated sandstone facies (5). Core heights are adjusted to correspond with depths on the gamma ray log -- note shift in gamma ray trace at 6600 feet in the Garrington well. Scale is shown in feet, to facilitate reference to existing log and core measurements. From Walker, 1983c.

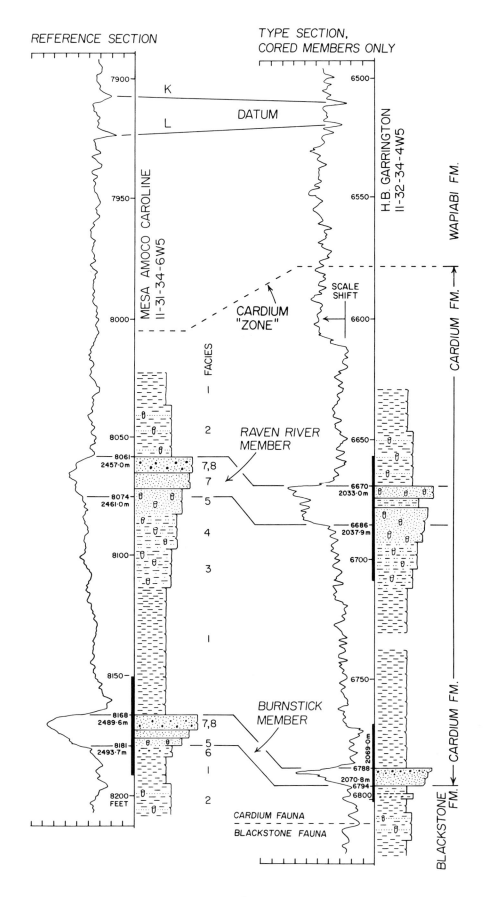

Figure 17. Composite stratigraphic sections for Caroline and Garrington Fields. The facies sequences are those observed in almost all cores, and thicknesses for each facies have been averaged from measurements in 48 Garrington cores and 30 Caroline cores. None of the Garrington cores examined totally penetrated the facies 1 mudstone blanket, but I assume that the gap at about 40 m is massive dark mudstone. Facies are designated by symbols and numbers. Vertical tubes indicate bioturbation, and hummocky cross stratification is shown in facies 7 -- all other symbols are self-explanatory. Member and facies correlations (datum -- top of Burnstick Member) show almost no changes (except thickness) between fields, but note the facies 2 to facies 1 change in the Blackstone Formation. The Cardium "Zone" is shown at its average position above the top of the Raven River Member. The base of the Cardium Formation is taken at the base of the Burnstick Member. From Walker, 1983c.

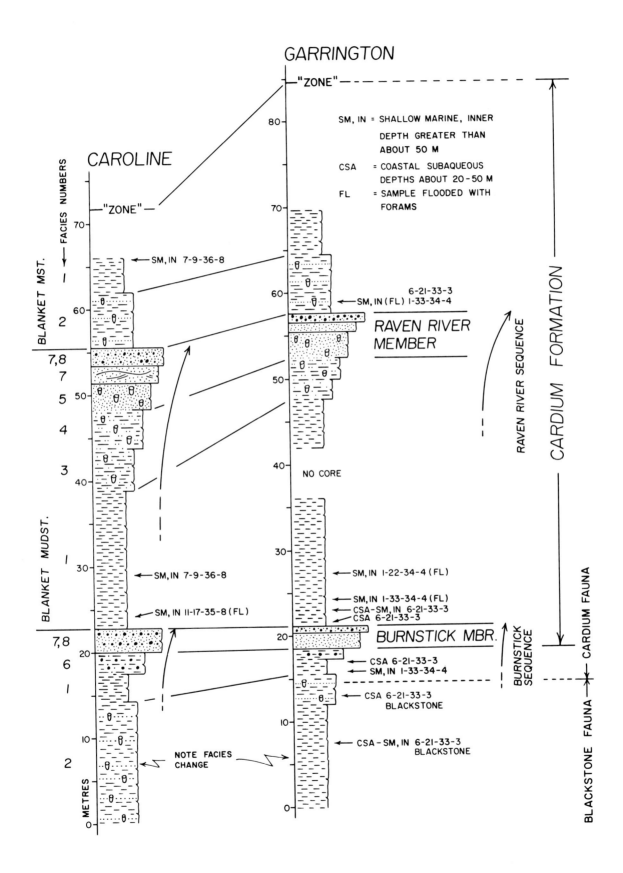

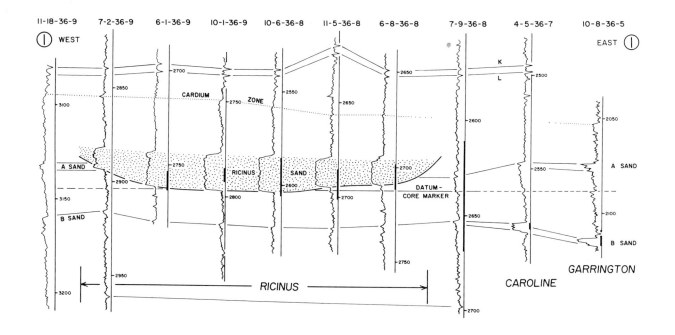

Figure 18. Cross section 1 (Fig. 12); prominent core marker used as datum - this is the change from facies 1 massive dark mudstones (Fig. 19) to facies 3 dark bioturbated muddy siltstones (Fig. 21). Note that the K and L markers parallel the core markers, except in 11-5-36-8 where the section is probably thickened tectonically. Channel nature of Ricinus sandstone shows prominently, cutting out the A sandstone. Note continuity of the B sandstone beneath Ricinus as a very ragged blanket. Dark bars indicate cored intervals, depths in meters. From Walker, 1983b.

---

Facies Descriptions and Interpretations, Caroline-Garrington Area

Both the Raven River (A) and Burnstick (B) sandstones occur at the top of progressively coarsening-upward sequences (Figs. 16, 17). These sequences can be subdivided into eight facies, defined by their lithology, primary sedimentary structures, and trace fauna. A full description and interpretation has been given by Walker (1983c).

Facies 1, massive dark mudstones (Fig. 19), shows no fissility and contains no distinct silty laminations or burrow forms. It blankets the Burnstick Member (Fig. 17), and also occurs above facies 2, above the Raven River Member.

Facies 2, laminated dark mudstones (Fig. 20), consists of dark mudstones with thin (1-4 mm) silty laminations. Some of the laminations have distinct sharp bases and fine internal parallel laminations. Small burrows and bioturbation are associated with the silty layers. This facies typically blankets the Raven River Member (Fig. 17), and grades upward into facies 1.

Facies 3, dark bioturbated muddy siltstones (Fig. 21), consists of a slightly silty dark mudstone with abundant mottling and bioturbation, but almost no distinct burrow forms. Chondrites occurs, but is rare. Facies 3 gradationally overlies facies 1 in the Raven River sequence (Fig. 17), and it then grades upward into facies 4.

Figure 19. Massive dark mudstones, Facies 1. Note absence of lamination, and very faint background bioturbation. Compare especially with Figures 20 and 21. Caroline, 10-33-34-6W5, 7851 feet (2293.0 m). Scale in cm.

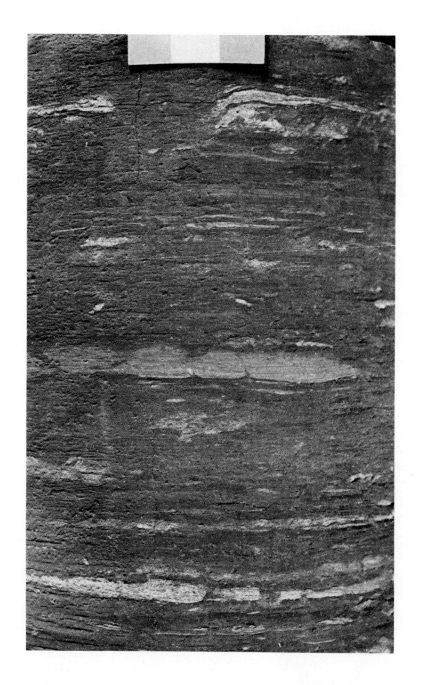

Figure 20. Laminated dark mudstones, Facies 2. Note presence of sharp-based, delicately laminated silty layers, partly to completely disrupted by organisms. Compare with Figures 19 and 21. Between Caroline and Garrington fields, 8-25-34-5W5, 2098.4 m. Scale in cm.

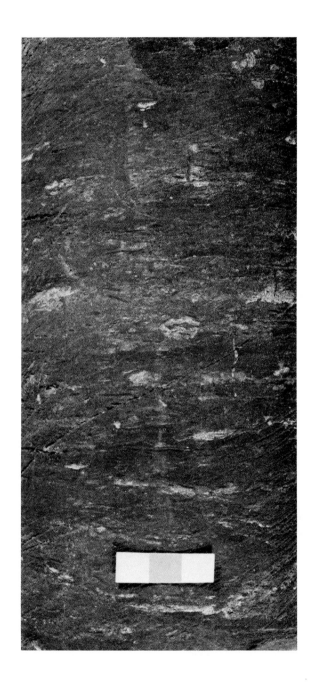

Figure 21. Dark bioturbated muddy siltstones, Facies 3. This facies overlies Facies 1. It contains more silt and very fine sand, but distinct lamination is rarely present -- the silt and mud are bioturbated ("stirred") together. Compare with Figures 19 and 20. Between Caroline and Ricinus, 10-17-34-7W5, 8489 feet (2587.4 m). Scale in cm.

Facies 4, pervasively-bioturbated muddy sandstones (Fig. 22), consists of abundant silt and very fine sand "stirred" into a muddy background sediment. It is coarser than facies 3, and contains distinct burrow forms including Zoophycos (Fig. 23), Teichichnus zigzagus and (rarely) Chondrites. Also, there are remnants of sharp-based, slightly graded 1-4 cm thick sandstone beds (Fig. 24).

Facies 5, bioturbated sandstones (Figs. 25, 26), is similar to facies 4 but contains a higher proportion of "stirred in" sand grains and more partially preserved sharp-based sandstones. Recognizable trace fossils include Zoophycos (Fig. 26), abundant Chondrites (Fig. 26), and Conichnus conicus (Fig. 26). The gradational junction between facies 4 and 5 is characterized by the proliferation of Chondrites, which seems to favor the sandier facies. Discrete sharpbased sandstones are commonly preserved within this facies (Fig. 27). They are mostly thicker than 4 cm (thinner ones tend to get destroyed by bioturbation), and contain parallel to gently inclined laminae.

Facies 6, speckled gritty mudstone (Fig. 28), only occurs in the Burnstick sequence (Fig. 17). It is similar to the pervasively bioturbated muddy sandstones of facies 4, but it contains abundant scattered quartz and chert grains (coarse to very coarse sand size), with rare granules and pebbles to about 1 cm.

Facies 7, non-bioturbated sandstones (Figs. 29, 30), consists of fine- to very fine-grained sandstones with thin bioturbated mudstone partings. Sedimentary structures are well preserved. Bases of sandstones are invariably sharp, overlain by low angle (<10°) sub-parallel lamination (Figs. 29, 30). Dips of laminae change slightly within the bed, and the structure is interpreted as HCS. These sandstones occur both in the Burnstick and Raven River Members (Fig. 17).

Facies 8, conglomerates (Figs. 31, 32, 33, 34), consists of three different types of conglomerate. The clast-supported types are unstratified, but have sharp bases with ripped-up mudstone or siderite clasts in places. Some beds display grading (Fig. 32). Most of the conglomerates, however, are mudsupported, and have no stratification, no preferred fabric, and no grading (Fig. 33). In some examples, the texture appears to have been disrupted by organisms. The thin graded layers occur within facies 2 laminated mudstones above the Raven River Member (Fig. 17), and consist of 1-2 cm beds commonly only one-pebble-diameter in thickness (Fig. 34).

Figure 22. Pervasively-bioturbated muddy sandstone, Facies 4. Totally bioturbated, and sandier than Facies 3 -- compared with Figure 21. Note long horizontal burrow above scale across full width of core -- probably Rhizocorallium. Between Caroline and Ricinus, 10-1734-7W5. 8390 feet (2557.3 m). Scale in cm.

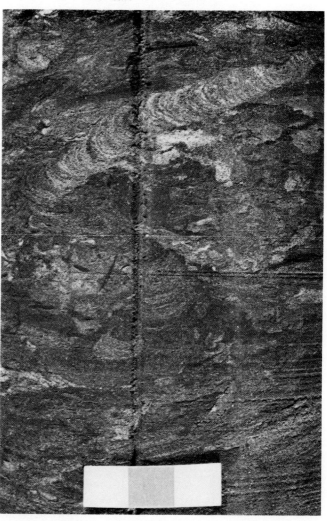

Figure 23. Pervasively-bioturbated muddy sandstones, Facies 4. Excellent example of <u>Zoophycos</u>, above scale, moving diagonally upward from center, to left, and then diagonally to the right. Between Caroline and Ricinus, 7-9-36-8W5, 8634 feet (2631.6 m). Scale in cm.

Figure 24. Pervasively-bioturbated muddy sandstones, Facies 4. Partially preserved sharp-based fine sandstone bed, top bioturbated. Note laminae with low dips compared to base of bed, and very low angle intersections between sets of laminae. This is interpreted as a very incomplete view view of hummocky cross stratification. Between Caroline and Ricinus, 6-2434-8W5, 8455 feet (2577.1 m). Core 7.5 cm wide.

Figure 25. Bioturbated sandstones, Facies 5. This is a sandier facies than the pervasively-bioturbated muddy sandstone (Facies 4, compare with Figures 22, 23, and 24). One thin sandstone is partially preserved, but mostly the facies is totally bioturbated. Between Caroline and Garrington, 10-20-37-7W5, 2293.3 m. Scale in cm.

Figure 26. Bioturbated sandstones, Facies 5. Excellent sample of diagonally penetrating, Z-shaped Zoophycos burrow. Note also small vertical tube at top of photo (Conichnus conicus), with later burrowing by Chondrites (white subcircular dots) (Pemberton and Frey, 1983). Between Caroline and Garrington, 10-20-37-7W5, 2294.1 m. Scale in cm.

Figure 27. Bioturbated sandstones, Facies 5. Note partial preservation of sandstone layers, the lower one having a very sharp flat base overlying bioturbated mudstones. Preserved lamination is apparently parallel (upper plane bed?, or possibly a partial view of rather flat lamination beneath a hummock). Between Caroline and Garrington, 8-25-34-5W5, 2106.8 m. Scale in cm.

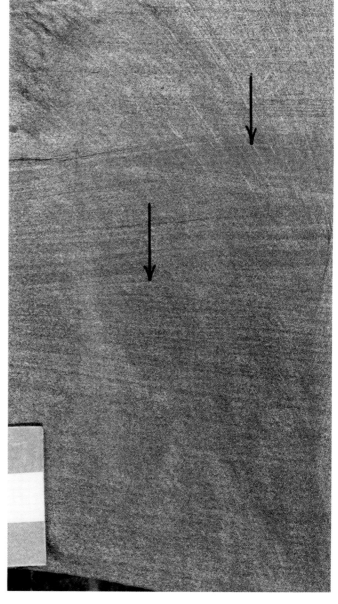

Figure 28. Speckled gritty mudstone, Facies 6. This facies is characterized by a bioturbated mixture of mudstone with floating coarse and very coarse sand grains, and chert pebbles up to about 1 cm in diameter. It occurs exclusively immediately below the Burnstick Member. Between Caroline and Ricinus, 7-9-36-8W5, 8701 feet (2652.1 m). Scale in cm.

Figure 29. Non-bioturbated sandstones, Facies 7. In this vertical core through horizontal beds, note low angle inclination of laminae, with subtle curvature and low angles of lamina intersections. Arrows show laminae with distinct convex-upward curvature. This is interpreted as hummocky cross stratification. Between Caroline and Ricinus, 10-11-35-8W5, 8395 feet (2558.8 m). Scale in cm.

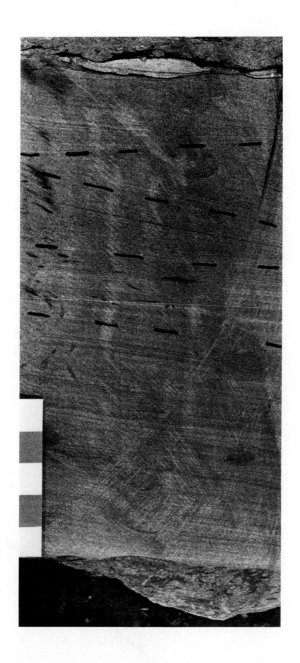

Figure 30. Well preserved sandstones, Facies 7. Note sharp base of bed on bioturbated mudstones, and low angle dips and lamina intersections within the sandstone. There are some flattened mud clasts lying on some laminae (especially between the first and second dashed lines upward), suggesting erosion of the substrate by the flow just before deposition. The structure is interpreted as hummocky cross stratification. Between Caroline and Ricinus, 4-22-35-8W5, 8799 feet (2681.9 m). Scale in cm.

Figure 31. Conglomerate, Facies 8. In this clast-supported conglomerate, note sharp sandy base to a layer at top of scale, and possible inverse grading toward top of bed. There also appears to be a preferred imbrication, with most elongate clasts dipping to the right. Pore spaces between chert clasts are sand filled, except near top where siderite occurs between the clasts. Caroline, conglomerate in Burnstick Member, 3-23-36-8W5, 8534 feet (2601.2 m). Scale in cm.

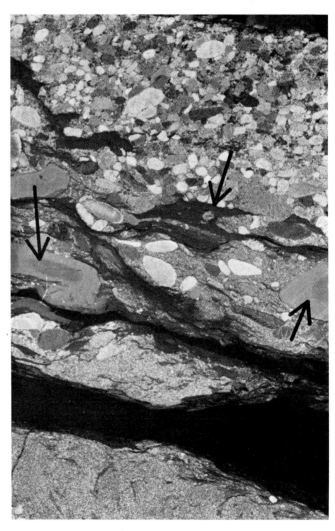

Figure 32. Conglomerate, Facies 8. In this clast-supported conglomerate, note the graded bedding, and the rimmed siderite and mudstone clasts in the base of the upper bed (arrowed). Between conglomerates is a partially bioturbated sandstone. The sharp base and grading indicate a flow which arrived suddenly, and gradually waned. The mud clasts indicate erosion by the emplacing flow, believed to be a turbidity current. Between Caroline and Ricinus, conglomerate in Raven River Member, 6-24-34-8W5, 8445 feet (2574.0 m). Core 7.5 cm wide.

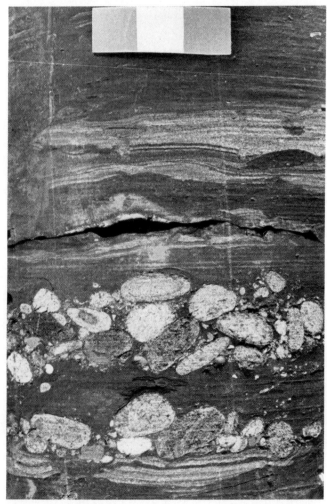

Figure 33. Conglomerate, Facies 8. In this mud-supported conglomerate, note disruption of bedding and random orientation of chert clasts. Much of the mudstone may have originally been in the form of clasts, suggesting erosive emplacement of gravel and sand. The substrate may have been poorly compacted, allowing sinking and local flowage of pebbles and mud, destroying any possible bedding. Between Caroline and Ricinus, conglomerate in Burnstick Member, 10-17-34-7W5, 8465 feet (2580.1 m). Scale in cm.

Figure 34. Conglomerate, Facies 8. These thin gravel layers are only one-to-two clasts in thickness, and mostly lack a sandy matrix. Layers such as these commonly occur immediately above the Raven River Member within laminated dark mudstones (Facies 2), as in this photo. Between Caroline and Ricinus, conglomerate immediately above Raven River Member, 10-11-35-8W5, 8375 feet (2552.7 m). Scale in cm.

## Facies Descriptions and Interpretations - Ricinus Area

Walker (1985) has recently extended the Caroline-Garrington facies scheme to include a suite of rather different facies at Ricinus. The Ricinus sand, as a whole, has a blocky gamma ray profile, and averages 16-19 m in thickness in North Ricinus, 17-23 m in South Ricinus (Fig. 35). This log response is quite different from that of the Burnstick and Raven River Members to the east. The different log responses also suggest that the ends of the Ricinus sand body are faulted, as suggested in Figure 35.

Facies 9, single event sandstones (Figs. 36, 37, 38, 39) consist of distinct sharp-based sandstones alternating with mudstones. The bases of the beds commonly contain mudstone rip-up clasts (Fig. 37) and a few beds also contain a layer of coarser grains or granules. Many of the single event sandstones are structureless, but others contain parallel lamination and current ripple cross-lamination; these would be described as BC beds in the Bouma (1962) terminology for turbidites (Figs. 36, 38). Rarely, the complete sequence from a massive base, through parallel lamination into ripple cross-lamination is seen, an ABC bed in Bouma's terminology (Fig. 39). Individual beds range in thickness from about 10 to 135 cm (average 49 cm, sample size 153 beds). Mean grain size is 2.6 phi, within the lower part of the fine sand range.

Facies 10, massive sandstones (Fig. 40), consist of thick sandstones that lack any sedimentary structures. The facies overlaps a little with single event sandstones, but beds tend to be considerably thicker, and have no parallel or cross-lamination. In North Ricinus, individual beds range from about 50 to 462 cm (average 144 cm), whereas in South Ricinus, the range is about 40 to 900 cm (average 248 cm). Even though the core is commonly broken, or cut for testing, beds are interpreted as single events if there are no mudstone partings, no grain size changes, and no rows of mudstone rip-up clasts to indicate more than one depositional event. Average grain sizes are a little coarser in North (2.4 phi) than South Ricinus (2.7 phi).

Facies 11, parallel laminated sandstones (Fig. 41) contain crude to well developed parallel lamination, with no changes in laminae dips upward through the bed. About 34% of the beds in South Ricinus belong to this facies, with thicknesses ranging from 50 to 424 cm (average 206 cm). Only 7% of the beds in North Ricinus belong to this facies, with thicknesses ranging from 50 to 336 cm (average 105 cm).

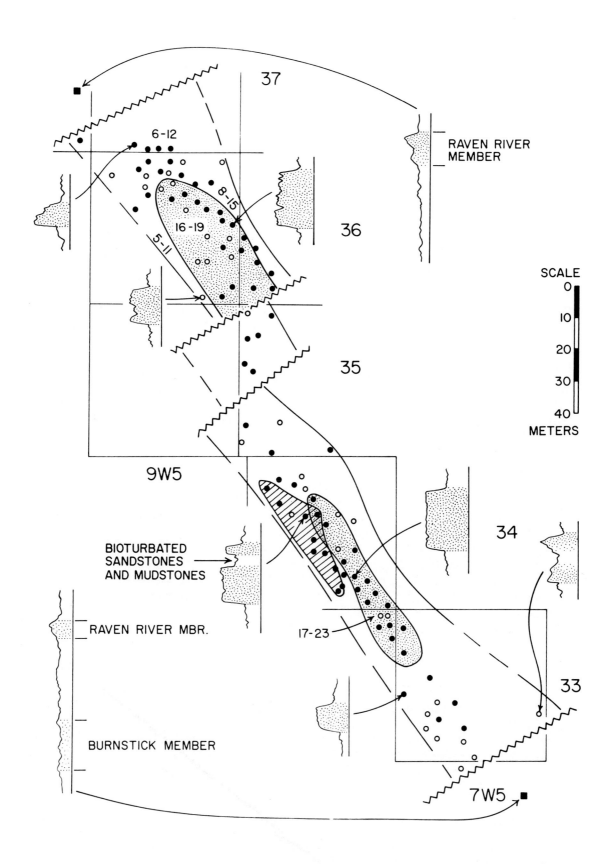

Figure 35. Ricinus Field, showing areas of maximum sand development (stippled), with thicknesses in meters. Diagonally ruled area contains most of the development of Facies 14, Bioturbated Sandstones and Mudstones. Gamma ray well logs show typical Ricinus sand responses, and contrast with Burnstick and Raven River log responses beyond the ends of the field.

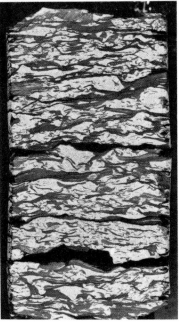

▲ Figure 36. Ricinus Facies B sandstone showing sharp base with broken mudstone beneath. Sandstone begins with parallel lamination grading up into small scale ripple cross lamination. Scale in cm. South Ricinus, 3-5-35-8W5, 8845 feet (2696 m).

Figure 37. Ricinus Facies B sandstone showing thick zone of ripped-up mud clasts, grading into crude parallel laminations. This in turn is overlain by cross lamination, with two mud clasts resting on the ripple foresets at the top of the core. Scale in cm. North Ricinus, 2-24-36-9W5, 2767 m. ▶

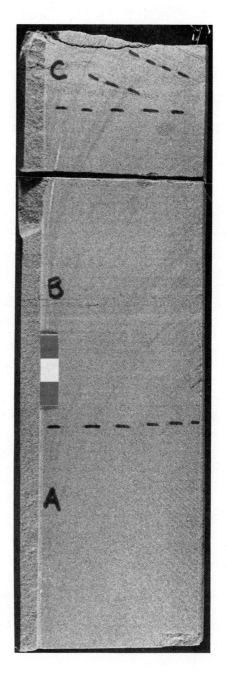

Figure 38. Ricinus Facies B sandstone showing parallel lamination (crude toward base but becoming better defined upward) grading into trough cross lamination at top of core. The trough cross lamination is made by migrating <u>current</u> ripples, not wave ripples. South Ricinus, 11-18-33-7W5, 8541 feet (2603 m). Core 7.5 cm wide.

Figure 39. Ricinus Facies B sandstone showing complete Bouma ABC sequence -- massive (A) to parallel laminated (B) to ripple cross laminated (C). South Ricinus, 10-3-34-8W5, 8685 feet (2647 m). Scale in cm.

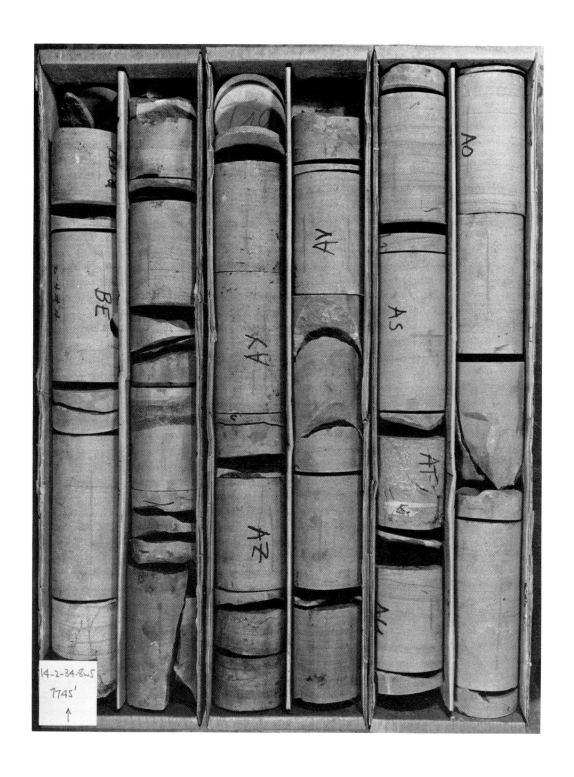

Figure 40. Ricinus Facies A structureless sandstones. Cores, although broken and cut, do not indicate any mudstone partings. Sandstones have no sedimentary structures except for faint parallel lamination in Core AO. Top of core is upward and to the right. South Ricinus, 11-2-34-8W5, base of core at 7745 feet (2361 m).

Figure 41. Parallel laminated sandstones, with inclined layers due to tectonic dip in well. From 6-36-33-8W5, 8223 ft. (2506.4 m).

Figure 42. Layers of clast supported conglomerate (two cores on left) grading up into massive sandstones (two cores on right). Note mudstone rip-up clasts. Top of core toward upper right, total core length 3 m. From 7-28-34-8W5, 5855 ft. (1784.6 m).

Facies 12, hummocky cross-stratified sandstones closely resembles the Facies 7 non-bioturbated sandstones of the Caroline-Garrington area. It is uncommon at Ricinus, and includes only about 6% of all beds.

Facies 13, conglomerates (Fig. 42) also resembles the conglomerate Facies 8 of the Caroline-Garrington area. They can occur at the base of the main Ricinus sand, or at the top of the sand, or at any point within the sand. Most commonly, the conglomerates occur as individual beds with sharp bases and tops, or as sharp based beds that grade up into massive sandstones (Fig. 42). In South Ricinus, beds range from 8 to 235 cm (average 40 cm); in North Ricinus, the range is 5 to 100 cm (average 23 cm). The conglomerates rarely display any primary stratification or preferred grain orientation. Maximum sizes were recorded for all beds; the average in North Ricinus (2.0 cm) is a little greater than in South Ricinus (1.5 cm).

Facies 14, bioturbated sandstones and mudstones (Fig. 43, 44) consists of thin alternations of very fine grained sandstones and mudstones, churned to varying degrees by organisms. It differs from the various bioturbated facies of the Caroline-Garrington area by containing abundant non-lined, sand-filled horizontal burrows about 1 cm in diameter (Fig. 43). The recognizable trace fauna includes Zoophycos, Skolithos, Ophiomorpha, Teichichnus, and one magnificent example of Rosselia (Fig. 44). This facies occurs mostly in South Ricinus (Fig. 35, diagonally ruled patern), and can be recognized in gamma ray well logs by the typical "bite" out of the log (Fig. 35 - see well in diagonally ruled area and well in southeast corner of field).

None of the Ricinus facies suggests reworking by fairweather currents, and the preservation of massive sandstones and Bouma sequences (rather than HCS) suggests deposition below storm wave base. Environments may have been deeper than those inferred for the Burnstick and Raven River Members.

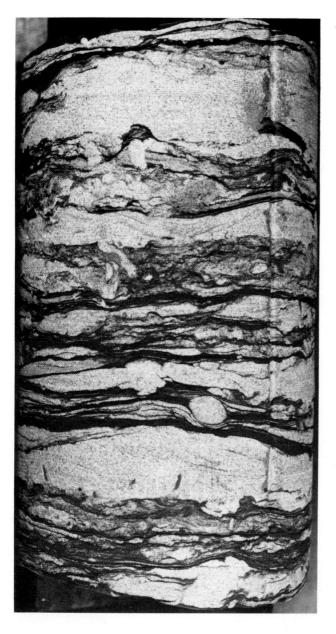

Figure 43. Typical view of Facies 14 Bioturbated Sandstones and Mudstones. Core width about 8 cm; from well 12-19-35-8W5, 9014 ft. (2747.5 m).

Figure 44. Example of trace fossil Rosselia from the bioturbated sandstone and mudstone facies, well 12-29-35-8W5, 9016 ft. (2748.1 m).

Ricinus Sand - Summary

Cross sections, such as that in Figure 18, demonstrate that the Ricinus sandstone rests in a channel at least 55 km long, 8 km wide (before palinspastic reconstruction) and 20-40 m deep. The major problems concern the mechanisms of channel cutting and filling.

Beds of facies 9, with Bouma ABC and BC sedimentary structures (Figs. 36, 37, 38, 39), strongly suggest deposition from turbidity currents. The structureless sandstones of facies 10 were probably deposited directly from suspension without storm reworking, and hence can be compared with the "massive sandstone" facies of Walker (1978); these massive sandstones are closely associated with classical turbidites in many basins worldwide. I tentatively suggest that the Ricinus facies A sandstones are also turbidity current deposits below storm wave base.

If we accept the turbidite fill mechanism, then the most likely <u>erosional</u> mechanism for cutting the Ricinus channel is by turbidity currents.

## REGIONAL CARDIUM PALEOGEOGRAPHY

The very simplified preliminary paleogeography suggested by Walker (1983) must now be considerably modified. This is largely due to the increasing recognition of widespread transgressions and regressions within the Cardium (A. G. Plint, pers. comm., 1985), making it difficult to draw on a map any single Cardium shoreline.

Dispersal directions (Fig. 45) are suggested by several different indicators. The orientation of individual pools and sandstone isopachs within Pembina field (Patterson and Arneson, 1957) is shown by stippled arrow 1. Arrow 2 indicates the trend of the Ricinus channel (Walker, 1985), and arrows 3 and 4 show summaries of paleocurrent indicators measured in outcrop (Duke, 1981).

In outcrop, non-marine Cardium facies prograde to a southeasterly limit in the Clearwater River area, just north of Ram Falls (Fig. 45). In the subsurface, upper shoreface and beach facies (indicated by core and by "blocky" well log responses) prograded northeastward in the Kakwa area to the limit shown in Figure 45. Four separate transgressions displaced this shoreface and beach back toward the southwest. The first transgression reached the general position shown by the dotted line just west of Kakwa (Fig. 40), and subsequent transgressions displaced the shoreline even farther southwestward.

The palinspastically restored locations of Seebe and Ram Falls (solid circles) are shown close to the Alberta-B.C. border. Both of these sections are fully marine, and at Seebe, there is no thick sand body that contains an upper shoreface facies. However, at Ram Falls there is an almalgamated hummocky and/or swaley cross stratified sandstone 14 m thick that closely resembles the upper shoreface facies at Kakwa. Thus the trend of the maximum progradation of the upper shoreface/beach (plotted by A. G. Plint, pers. comm. 1985, and shown by long dashes between Kakwa and Edson) must swing southwestward, passing through the palinspastically reconstructed position of Ram Falls. From there, it probably swings southeastward again to parallel the Alberta-B.C. border. The exact location of the maximum advance of this shoreline is a current research topic.

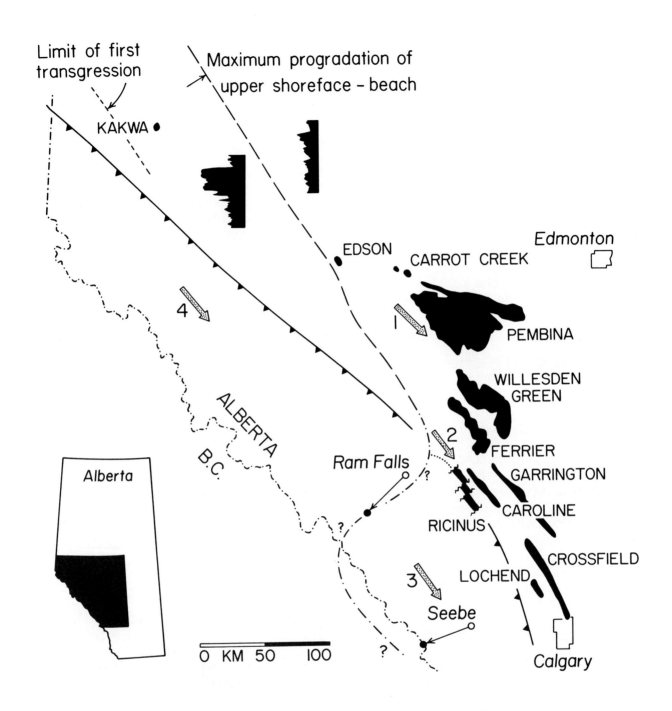

Figure 45. Paleogeography and dispersal of the Raven River Member ("A sand") of the Cardium Formation. Maximum advance of the upper shoreface/beach facies has been plotted by A. G. Plint (pers. comm., 1985), based on blocky versus serrated log profiles. Two typical gamma ray profiles are shown. The shoreline appears to swing southwestward through the palinspastically restored position of Ram Falls (solid circle). In outcrop south of Ram Falls, there are no thick upper shoreface sandstones, suggesting a shoreline somewhere west of the Alberta-B.C. border. Dispersal directions are shown by stippled arrows, explained in the text.

The large Cardium fields, especially Pembina, Ferrier and Willesden Green thus appear to have been deposited offshore and downdrift from this postulated shoreline. Sand may have moved several tens of kilometers from its point of origin at the shoreline. For the long, narrow fields of Caroline, Garrington, Crossfield and Lochend, the sand may have moved for over 100 km, both offshore and "downdrift"; that is, sub-parallel to the regional shoreline. Transport processes have been reviewed by several writers in this volume, but for the Cardium fields named above, incremental geostrophic dispersal by storm generated currents seems likely.

This is not the case for Ricinus, where the evidence still strongly supports the incision and filling of the channel by turbidity currents (Walker, 1985). It is possible that during a low stand of sea level, a major Cardium river became incised in one position. During subsequent transgression, turbidity currents generated in the estuary of the river (as in the modern Congo estuary, Heezen et al., 1964) may have flowed seaward, extending and down-cutting to form the Ricinus channel. The north end of Ricinus appears to be faulted (Fig. 35), but if the channel could be traced northwestward, it might swing a little more westerly, to meet up with the southward-trending segment of the Cardium shoreline. This hypothetical trend is shown by a short row of dots in Figure 45.

## REVIEW OF CARDIUM TRANSPORT PROCESSES

Marine members of the Cardium in outcrop and in the subsurface are dominated by HCS; medium scale angle-of-repose cross bedding is absent. The interbedded mudstones are extensively bioturbated by grazing organisms. Both the physical sedimentology and bioturbation strongly indicate deposition below fairweather wave base in a storm-dominated shallow marine setting a few tens of meters deep.

The main problem is the transport of sand into this environment. Dispersal is dominantly southeastward, parallel to the present tectonic strike of the Rockies. The regional Cardium shoreline was probably parallel to the present tectonic strike (Fig. 45). The southern tip of Garrington lies at least 100 km (63 mi) seaward of the nearest possible shoreline, and the southern tip of Crossfield may have been as much as 150 km from the shoreline.

Various transport mechanisms have been reviewed earlier in this volume in papers by Swift and by Walker. The most likely ones for the Cardium seem to be

1) wind-forced geostrophic currents, and
2) turbidity currents.

Set-up of water by storm winds against the Cardium shoreline could generate southeastward-flowing geostrophic currents. In the Middle Atlantic Bight of North America, such geostrophic flows have velocities of several tens of cm/sec, with most sand transport taking place during winter storms (Swift et al., 1981). Combined geostrophic flows and storm waves may well form HCS (Swift et al., 1983); the real problem is the initial emplacement of the sand.

A.  Incremental Sand Transport by Geostrophic Flows

Studies of recent sediments suggest that fine and very fine sand can be moved incrementally by geostrophic flows, and dispersed parallel to shoreline. Gradual across-isobath movement may also disperse the sand seaward. Most transport occurs during storms. The dominant sedimentary structures are probably current ripple cross lamination; although there is little published core information, the resulting cross laminated sandy beds are probably fairly thin (cms). Although there are beds of this type in Cardium outcrops, they tend to occur toward the bottom of coarsening upward sequences. The bulk of Cardium sand beds are sharp-based, 10-100 cm thick, and dominated by HCS and symmetrical wave ripples. In an "incremental" interpretation, this would imply times when the bottom is extensively reworked, combining the sand from many thin rippled sand layers, winnowing away the fines, and redepositing the sand in thicker HCS beds.

The dispersal directions probably constitute the most powerful argument in favor of incremental geostrophic sand movement for the long, narrow fields such as Caroline, Garrington and Crossfield. However, geostrophic flows are not known to cut long, deep channels in the modern shelf, so an alternative mechanism should be examined, especially for the Ricinus channel with its fill of massive sandstones, and Bouma BC and ABC beds.

B.  Sand Transport by Turbidity Currents

There are several positive lines of evidence in favor of powerful unidirectional flows - these can be listed as:

1. sharp-based, graded sandstones abruptly resting on bioturbated mudstones

2. sharp-based, graded gravels within mudstone intervals (sequence 4 at Seebe, for example; Wright and Walker, 1981)

3. the 20-40 m deep Ricinus channel

4. the presence of massive sandstones, and Bouma BC and ABC sequences within the Ricinus channel

5. the abundance of ripped-up mud clasts within the lower parts of Ricinus sharp based sandstones and graded conglomerates

6. the apparent long distance from shoreline of the sandstones and conglomerates at Ricinus-Caroline-Garrington.

In most cases, the deposits of these flows have been reworked by storm waves to form HCS. If we are to contrast these powerful, erosive, unidirectional flows with the geostrophic flows discussed above, it is clear that whatever the Cardium flows were, they must have been capable of transporting sand and gravel for long distances. At the depositional site, the flows must be capable of ripping up the substrate to form mud clasts, and capable of forming parallel lamination (upper flat bed). Long distance sand transport in quiet mud basins is commonly associated with turbidity currents, and the Bouma ABC and BC sequences at Ricinus are the best positive evidence for Cardium turbidity currents.

We end up with a final dilemma. Does the <u>geological</u> evidence for powerful unidirectional channelized flows (probably turbidity currents) overwhelm the evidence that sand is dispersed on modern shelves by storm-generated geostrophic currents which flow relatively slowly parallel to the isobaths? Is the geological evidence sufficiently convincing that we must continue to search for a turbidity current generating mechanism, or does the evidence from modern shelves force us to reassess our interpretation of the geological evidence?

At the moment, there are no answers; perhaps posing the above questions will help to focus our research efforts.

## ACKNOWLEDGEMENTS

Research on the Cardium has been funded by operating and strategic grants from the Natural Sciences and Engineering Research Council of Canada. Many of the ideas have been developed with students, particularly Bill Duke, Brent Ainsworth, and Marsha Wright. Dr. A. G. Plint, postdoctoral fellow at McMaster, has emphasized the importance of sea level changes, and has developed the paleogeographic ideas discussed in the paper. My subsurface work was done while I was a Visiting Scientist at Amoco Canada. Many individuals and companies (particularly Home Oil) have freely helped with ideas and logistics. This paper reflects my ideas, but would not exist without the help of all those mentioned above. The manuscript has been improved by the comments of R. W. Tillman and D. J. P. Swift; they do not necessarily agree with all my interpretations.

## REFERENCES

Almon, W. R., 1979, Petrophysical evidence of cementation differences in the Cardium Formation: Canadian Well Logging Society, 7th Formation Evaluation Symposium, Calgary, Alberta, p. K1-K13.

Beach, F. K., 1955, Cardium a turbidity current deposit: Journal of the Alberta Society of Petroleum Geologists, v. 3, p. 123-125.

Berven, R. J., 1966, Cardium sandstone bodies, Crossfield-Garrington area, Alberta: Bulletin of Canadian Petroleum Geology, v. 14, p. 208-240.

Bouma, A. H., 1962, Sedimentology of some flysch deposits: Amsterdam, Elsevier, 168 p.

Crowell, J. C., 1957, Origin of pebbly mudstones: Bulletin of the Geological Society of America, v. 68, p. 993-1009.

Duke, W. L., 1981, Internal stratigraphy and paleogeography of the Upper Cretaceous Cardium Formation in the Alberta Foothills: Geological Association of Canada, Abstracts (Calgary, 1981), v. 6, p. A-16.

_____, 1985a, Sedimentology of the Upper Cretaceous (Turonian) Cardium Formation in outcrop in southern Alberta: Ph.D. Thesis, McMaster University, Hamilton, Canada, 720 p.

_____, 1985b, Hummocky cross stratification, tropical hurricanes, and intense winter storms: Sedimentology, v. 32, p. 167-194.

_____ and D. A. Leckie, 1984, Origin of hummocky cross-stratification; Part 2. Paleohydraulic analysis indicates formation by orbital ripples within the wave-formed flat-bed field: Sedimentology of shelf sands and sandstones, University of Calgary, Abstracts, p. 32.

Gillie, R. D., 1979, Sand and gravel deposits of the coast and inner shelf, East Coast, Northland Peninsula, New Zealand: Ph.D. Dissertation, Univ. of Canterbury, Christchurch, New Zealand.

Harms, J. C., J. B. Southard and R. G. Walker, 1982, Structures and sequences in clastic rocks: Society of Economic Paleontologists and Mineralogists, Short Course 9.

Kauffman, E. G., 1977, Geological and biological overview, Western Interior Cretaceous basin: Mountain Geologist, v. 14, p. 75-99.

_____, 1979, Cretaceous, in Treatise on Invertebrate Paleontology: Boulder, Colorado, Geological Society of American and University of Kansas Press, Pt. A, Introduction, p. A418-A487.

Krause, F. F. and D. A. Nelson, 1984, Storm event sedimentation; lithofacies association in the Cardium Formation, Pembina area, west-central Alberta, Canada, in Scott, D. F. and D. J. Glass (eds.), The Mesozoic of Middle North America: Canadian Society of Petroleum Geologists, Memoir 9, p. 485-511.

Kuenen, P. H. and C. I. Migliorini, 1950, Turbidity currents as a cause of graded bedding: Journal of Geology, v. 58, p. 91-127.

Leckie, D. A. and W. L. Duke, 1984, Origin of hummocky cross-stratification; Part 1. Straight crested symmetrical gravel dunes: Sedimentology of shelf sands and sandstones, University of Calgary, Abstracts, p. 51.

Michaelis, E. R. and G. Dixon, 1969, Interpretation of depositional processes from sedimentary structures in the Cardium sand: Bulletin of Canadian Petroleum Geology, v. 17, p. 410-443.

Palmer, A. R., 1983, The decade of North American geology. 1983 geological time scale: Geology, v. 11, p. 503-504.

Patterson, A. M. and A. A. Arneson, 1957, Geology of Pembina field, Alberta: Amer. Assoc. Petroleum Geologists Bull., v. 41, p. 937-949.

Pemberton, S. G. and R. W. Frey, 1983, Biogenic structures in Upper Cretaceous outcrops and cores: Canadian Society of Petroleum Geologists, Conference on the Mesozoic of Middle North America, Field Trip Guidebook 8, 161 p.

_____, 1984, Ichnology of storm-influenced shallow marine sequence: Cardium Formation (Upper Cretaceous) at Seebe, Alberta, in Stott, D. F. and D. J. Glass (eds.), The Mesozoic of Middle North America: Canadian Society of Petroleum Geologists, Memoir 9, p. 281-304.

_____, R. W. Frey and R. G. Walker, 1984, Probable lobster burrows in the Cardium Formation (Upper Cretaceous) of Southern Alberta, Canada, and comments on modern burrowing lobsters: Journal of Paleontology, v. 54, p. 1422-1435.

Stott, D. F., 1963, The Cretaceous Alberta Group and equivalent rocks, Rocky Mountain Foothills, Alberta: Geological Survey of Canada, Memoir 317, 306 p.

Swagor, N. S., T. A. Oliver and B. A. Johnson, 1976, Carrot Creek field, central Alberta, in The sedimentology of selected clastic oil and gas reservoirs in Alberta: M. M. Lerand (ed.), Canadian Society of Petroleum Geologists, Calgary, p. 78-95.

Swift, D.J.P., R. A. Young, T. L. Clark, C. E. Vincent, A. Niedoroda and B. Lesht, 1981, Sediment transport in the Middle Atlantic Bight of North America: synopsis of recent observations, in Nio, S. D., R.T.E. Schuttenheim and T.C.E. Van Weering (eds.), Holocene marine sedimentation in the North Sea Basin: Blackwells, Oxford, Int. Assoc. of Sedimentologists, Special Publication 5, p. 361-383.

_____, A. G. Figueiredo, G. L. Freeland and G. F. Oertel, 1983, Hummocky cross stratification and megaripples: a geological double standard?: Journal of Sedimentary Petrology, v. 53, p. 1295-1318.

Walker, R. G., 1978, Deep water sandstone facies and ancient submarine fans: models for exploration for stratigraphic traps: Amer. Assoc. Petroleum Geologists Bull., v. 62, p. 932-966.

_____, 1983a, Cardium Formation 1. "Cardium a turbidity current deposit" (Beach, 1955): a brief history of ideas: Bulletin of Canadian Petroleum Geology, v. 31, p. 205-212.

_____, 1983b, Cardium Formation 2. Sand body geometry and stratigraphy in the Garrington-Caroline-Ricinus area, Alberta - the "ragged blanket" model: Bulletin of Canadian Petroleum Geology, v. 31, p. 14-26.

_____, 1983c, Cardium Formation 3. Sedimentology and stratigraphy in the Garrington-Caroline area: Bulletin of Canadian Petroleum Geology, v. 31, p. 213-230.

_____, 1985, Cardium Formation 5. Channel cut and filled by turbidity currents in the Cretaceous Western Interior Seaway, Ricinus Field, Alberta: American Association of Petroleum Geologists Bulletin.

_____ and M. E. Wright, 1982, Cardium Formation at Seebe, Alberta: Walker, R. G. (ed.), Clastic units of the Front Ranges, Foothills and Plains in the area between Field B.C. and Drumheller, Alberta: International Association of Sedimentologists, 11th International Congress on Sedimentology (Hamilton, Canada, 1982), Guidebook for Excursion 21A, p. 72-87.

Wright, M. E. and R. G. Walker, 1981, Cardium Formation (U. Cretaceous) at Seebe, Alberta - storm transported sandstones and conglomerates in shallow marine depositional environments below fairweather wave base: Canadian Journal of Earth Sciences, v. 18, p. 795-809.

Yorath, C. J., B. D. Bornhold and R. E. Thompson, 1979, Oscillation ripples on the northeast Pacific continental shelf: Marine Geology, v. 31, p. 45-58.

# THE TOCITO AND GALLUP SANDSTONES, NEW MEXICO, A COMPARISON

R. W. Tillman
Consulting Sedimentologist
4555 S. Harvard
Tulsa, Oklahoma 74135

## ABSTRACT

Five depositional models have been utilized during the last 30 years to explain the deposition of the Upper Cretaceous Gallup and Tocito Sandstone and Tocito Sandstone Lentil in the San Juan Basin in New Mexico. It has generally been recognized that most of the true Gallup sandstones were deposited as strand plain and beach deposits. There is a continuing controversy as to the relationship of the sand ridge (offshore bar) deposits to the Gallup shoreline sandstones. The offshore deposits have been designated as Gallup by some writers and as the Tocito Sandstone Lentil of the Mancos Shale by others. Scenarios such as described in this paper for the Gallup Sandstones may be more common for shelf sandstones than is presently recognized.

The Gallup Sandstone may be divided into two major depositional units. Most of the Gallup Sandstone is a strand plain deposit with a typical transition zone, shoreface, foreshore vertical sequence. Hummocky cross stratification marks the base of the Gallup sequence in some areas. Most of the fluvial portions of the Gallup are designated as the Torrivio Sandstone Member.

Some of the earliest correlations (Model I) suggested that the Gallup Sandstone was younger than the Tocito. Other correlations (Model II) indicated that the Gallup consisted of a series of synchronous shoreline and offshore deposits.

The third, fourth and fifth models include a major unconformity which separates the shoreline Gallup sandstones from offshore-bar sandstones designated as the Tocito Sandstone Lentil of the Mancos Shale. The third model stresses the importance of the generally transgressive nature of the Tocito, the fourth infers that Tocito offshore sand-ridges were localized during the transgression in northwest-southeast trending topographic lows.

A fifth model includes one major unconformity separating the Tocito and Gallup Sandstones as well as multiple local unconformities on which a series of synchronous Gallup shoreline and offshore sandstones were deposited. The unconformities within the Gallup are interpreted to be the result of local submarine erosion. The major post-Gallup unconformity in this model is interpreted to be subaerial to the southwest and submarine to the northeast. A series of Tocito sand ridges were deposited on unconformity terraces during pauses or stillstands during the transgression.

Pre-Tocito unconformity surfaces can be recognized in outcrop and cores. Subsurface correlations indicate that time markers have been truncated by a pre-Tocito unconformity and a time gap is indicated below the Tocito by both microfauna and macrofauna.

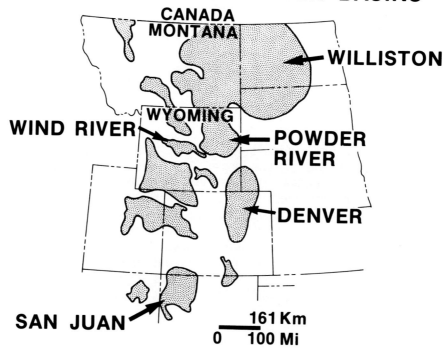

Figure 1. Major basins of the Rocky Mountains. The San Juan Basin is located in northwestern New Mexico and southwestern Colorado.

Facies names used in Tocito outcrop and core descriptions are derived from similar facies which comprise the offshore sand ridges in the Shannon Sandstone in Wyoming. High-angle cross-bedded sandstones and burrowed to bioturbated shaly sandstones comprise most of the Tocito sandstones.

Production within the "Gallup" is primarily from stratigraphic trap sandstones, almost all of which may be correlated with the Tocito. The true Gallup Sandstone may produce only from closed structures.

INTRODUCTION

The Upper Cretaceous Tocito and Gallup Sandstones in the San Juan Basin in northwestern New Mexico (Fig. 1) were deposited during the Turonian and Coniacian and Coniacian (only) respectively. The Gallup Sandstone was named in 1925 by Sears for outcrops near the town of Gallup, New Mexico. The name Tocito was first used for certain Upper Cretaceous sandstones in 1924 by Reeside.

The purpose of this paper is to contrast the origins, depositional environments and stratigraphic relationships of the Gallup and Tocito sandstones. The "Gallup" Sandstone crops out around the west, south and southeast side of the San Juan Basin and it occurs in the subsurface over extensive portions of the Basin. The emphasis in this paper will be on outcrops along the west and southwest flank of the San Juan Basin (Fig. 2) and will include some subsurface control. The Gallup Sandstone is equivalent to the Carlile and Niobrara sections in southern Montana and parts of Colorado and Wyoming. As will be discussed in a subsequent section of the paper, the stratigraphic section shown in Figure 3 is incomplete. The Gallup Sandstone, as indicated in Figure 3, is primarily a regressive deposit; the unit designated by most workers as the Tocito Sandstone Lentil of the Mancos Shale (Fassett and Jentgen, 1978) is not included on this 1979 section by Campbell.

The term Gallup Sandstone is the legal terminology in New Mexico for all productive sandstones between the top of the Greenhorn Limestone and the base of the Mesaverde Formation (Fig. 3; Fassett and Jentgen, 1978). In this paper this interval is designated as the "Gallup" and the term Gallup is restricted to Turonian and Coniacian sandstones deposited as strandline, fluvial and brackish water deposits during the R-2 regression of Molenaar (1983a). The Tocito Sandstone Lentil of the Mancos Shale, as used in this paper, is entirely Coniacian in age, restricted to marine offshore sand-ridge deposits and is younger than the true Gallup Sandstone.

Ship Rock (Fig. 4) is one of the more prominent topographic features in the northwestern part of the San Juan Basin. Many of the Tocito and Gallup outcrops are located along the west side of the basin within sight of Ship Rock (Fig. 5). The true Gallup Sandstone is primarily a series of regressive shoreline sandstones trending northwest-southeast. The Gallup Sandstone consists of several vertically stacked sequences of shoreface, foreshore and some nonmarine facies deposited during the early (R2) regression designated by Molenaar (1983a) (Fig. 6). The Gallup Sandstone is late Turonian to early Coniacian in age and is about 89 to 90 million years old. Most workers recognize that the Tocito Sandstone Lentil of the Mancos Shale is younger (88.2 million years) than the Gallup.

The Cretaceous sediments in the San Juan Basin are exposed along the edges of the basin and in the deeper part of the basin are 6500 feet thick (Fig. 7A). Early drilling in the San Juan Basin was primarily for gas in some of the shallower formations; however, a number of "Gallup" sandstone oil discoveries were made in the late 50's. Two of these fields, Horseshoe and Bisti, which have produced 37 and 35 million barrels of oil, respectively (Fassett, 1983), will be discussed in this paper.

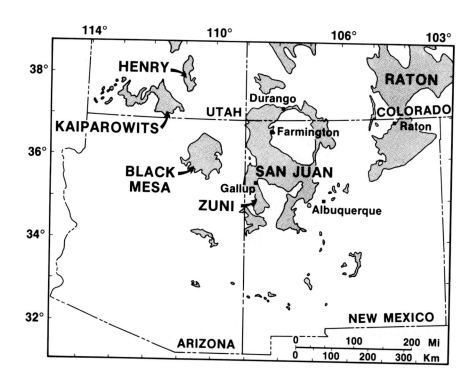

Figure 2. Major Tertiary basins in a part of the southwestern United States. Upper Cretaceous outcrops are shown by shaded pattern. Central portion of San Juan Basin is covered by Tertiary deposits (from Molenaar, 1983).

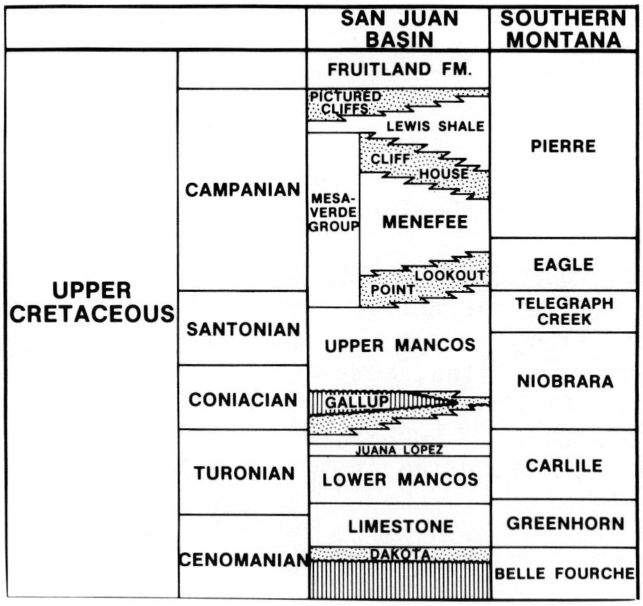

Figure 3. San Juan Basin stratigraphic column. Gallup Sandstone, as shown, includes primarily regressive sandstones. Minor transgressive deposits above the unconformity (vertical lines) portion of the diagram are also included in the Gallup Sandstone by some workers. (Modified from Campbell, 1979).

Figure 4. Ship Rock, New Mexico. Many outcrops of Tocito Sandstone Lentil are within sight of this landmark.

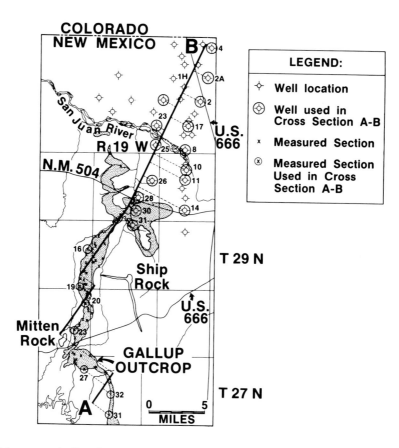

Figure 5. Gallup and Tocito outcrops along the west side of the San Juan Basin, New Mexico (from Campbell, 1973).

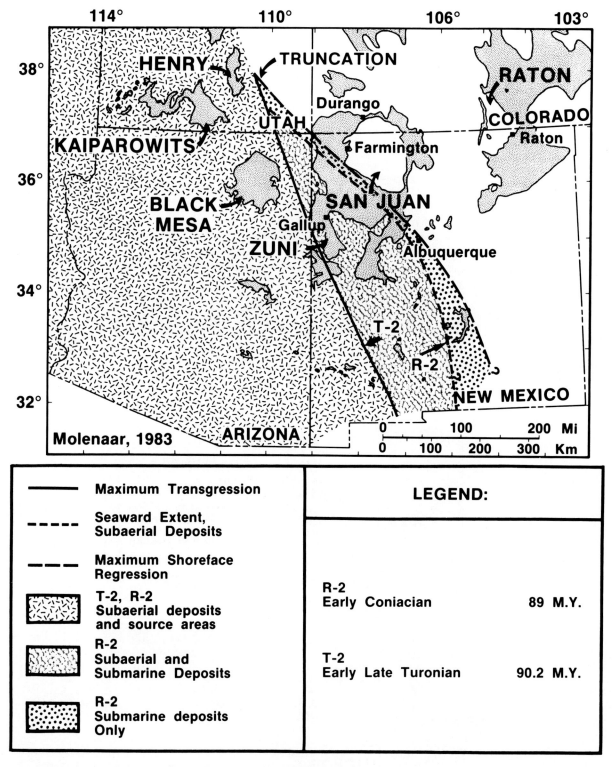

Figure 6. Map showing restored positions of the shorelines at maximum transgression and regression of the T2-R2 (Gallup Sandstone) cycle. The solid line is landward extent of shoreline deposits of the T2 transgression. The short dashed line is the seaward extent of subaerial-plain deposits of the R2 regression, and the long dashed line is the seaward extent of shoreface sandstones of the regression. (Modified from Molenaar, 1983).

## EARLY GALLUP-TOCITO DEPOSITIONAL MODEL (MODEL I)

Early stratigraphic sections such as Figure 7B (Bozanic, 1955), which included both outcrop and subsurface sections, suggested that the Gallup was significantly younger than the Tocito. Stoney Butte Field (Fig. 7A), which produces from the Gallup Sandstone, was one of the early small "Gallup" discoveries. Doswell field, which was indicated to produce from an "older" sandstone, was informally recognized as having its main production from the Tocito sandstones. Since the mid-1950's, sandstones designated as Tocito have been recognized by most workers as marine or offshore-bar (sand-ridge) sandstones.

Bozanic and other early workers used the Greenhorn Limestone as a horizontal datum in subsurface cross sections. The Juana Lopez Member of the Mancos Shale (Fig. 7B) is also an excellent time-stratigraphic marker throughout most of the San Juan Basin. subsurface cross sections. The Tocito was indicated to lie only a short distance above the Juana Lopez. The Gallup was designated to be "younger" and separated by a shale sequence from the somewhat "older" Tocito (Fig. 7C). This interpretation incorrectly indicated, as will be discussed later, that the Tocito was only slightly younger than the Juana Lopez and that the Gallup is younger than the Tocito.

## THE GALLUP SANDSTONE

The Gallup Sandstone, where it outcrops at the principal reference section near the town of Gallup, New Mexico, contains quite different facies from many of the outcrops farther to the northeast. Wide variations in facies are expected in a regressive and northeast prograding sequence such as the Gallup. The principal reference section of the Gallup, designated by Molenaar (1983) east of Gallup New Mexico may be divided into a marine and a non-marine interval (Fig. 8). The lower sandstones and interbedded shales are shallow marine and marine related. Some of the upper sandstones are thick fluvial channel sandstones and have been designated by Molenaar (1973, 1983b) as the Torrivio Member. The upper fluvial portion grades into brackish and shoreline deposits to the northeast (Molenaar, 1983b).

Production from the Tocito or "Gallup" Sandstones occurs in a belt which is downdip from the outcrops in the northwest and central part of the basin (Fig. 9). Most of the producing fields, including Horseshoe and Bisti fields, trend northwest-southeast. A structure map on top of the "Gallup" indicates a basinal slope from southwest to northeast to the seaward pinch out of the "Gallup" (Fig. 10).

Several different origins for the Tocito and "Gallup" offshore sand ridges (bars) in the San Juan Basin have been suggested. The true Gallup Sandstone is known to have produced 13 million barrels of oil as Hospah Field (King and Wengerd, 1957); however, it was not yielded significant additional production (Molenaar, 1983b; Fassett, 1983), and for that reason alone it is important to assess the relationship of the Gallup Sandstone to the Tocito. The depositional setting and stratigraphic relationships of the true Gallup Sandstone will be discussed first.

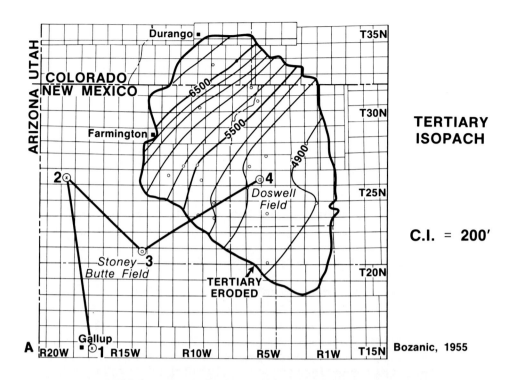
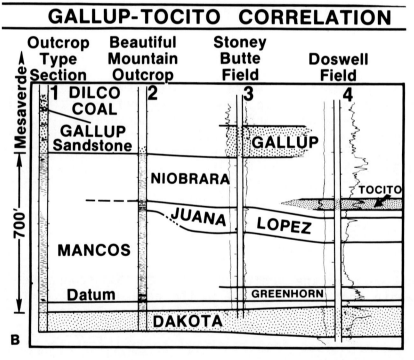

Figure 7. (A) Map showing total Cretaceous isopach thickness in the San Juan Basin. Also shown are locations of outcrop (1 and 2) to subsurface (3 and 4) cross section (Fig. 8). Doswell field is among the first fields to be designated as Tocito production (Lily, 1952). Stoney Butte field produced some oil from the Gallup Sandstone. (Modified from Bozanic, 1955). (B) Gallup-Tocito correlation by Bozanic (1955) showing stratigraphic position of Tocito and Gallup Sandstones relative to the Greenhorn and Juana Lopez Members. Note that in this cross section the Tocito appears to be older than the Gallup.

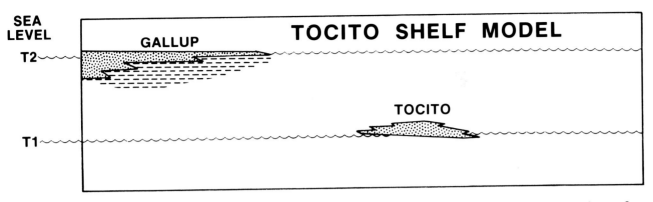

Figure 7C. Diagrammatic model (Model I) of Bozanic's (1955) interpretation of the depositional sequence which includes the Tocito and Gallup Sandstones. According to this interpretation, the Tocito is older than the Gallup Sandstone. Subsequent work indicates that this model is incorrect.

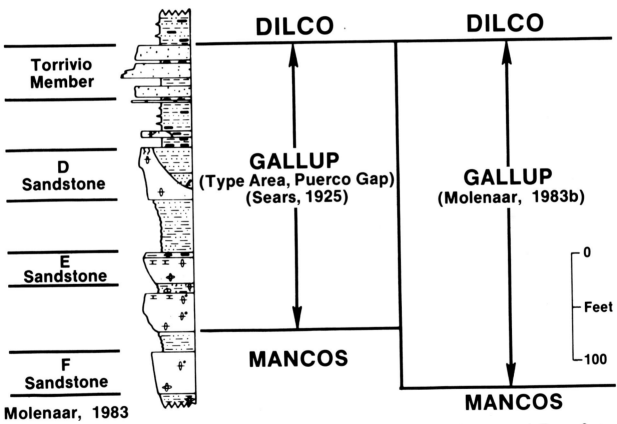

Figure 8. Principal reference section of the Gallup Sandstone and Type Section of Torrivio Member of the Gallup. Location: SE NE Sec. 13, T15N, R18W, McKinley County, New Mexico. Gallup sandstone tongues designated by letters (D, E, F) after Molenaar (1973). In the area where the Gallup was first named by Sears in 1925 in Puerco Gap, the lowest Gallup tongue (F) is poorly exposed and was not included in his definition of the Gallup Sandstone interval. Sandstones with "bisected circles" pattern are burrowed and are interpreted as marine. Roots occur at top of D sandstone. Torrivio Member sandstones and a portion of the D sandstone are fluvial. Modified from Molenaar (1983).

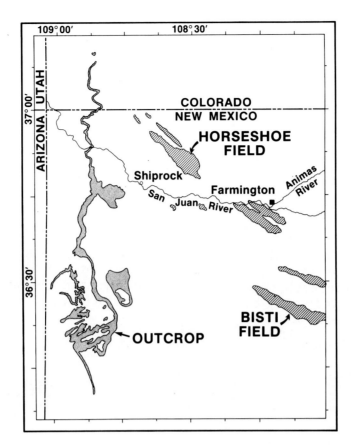

Figure 9. Oil fields producing from "Gallup Sandstone" (diagonal pattern) and outcrop of "Gallup Sandstone" along west side in San Juan Basin, New Mexico. Note elongation of oil fields is northwest-southeast. Modified from McCubbin, 1969.

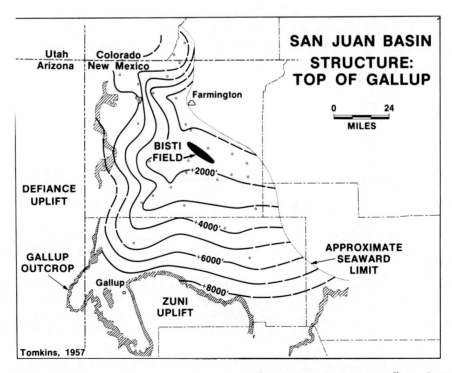

Figure 10. Structural contours on top of "Gallup Sandstone". Data points shown by small circles. Approximate seaward limit of Gallup Sandstone as interpreted by Tompkins is indicated. "Gallup" outcrops (diagonal pattern) around south and west edges of basin. (modified from Tomkins, 1957).

The model proposed by Sabins (1963), Model II (Fig. 11), involves syncronous deposition of Gallup shoreline and offshore sandstones. Sabins (1963) applied this depositional setting in explaining the reservoir at Bisti field which has produced about 37 million barrels of oil from the "Gallup" Sandstone. In the 1950's and early 1960's, Gallup beach and shoreface sandstones were considered to be time equivalent of productive offshore bar sandstones (Fig. 11).

The coastal strand plains, possible barrier sequences and offshore bars of the "Gallup Sandstone" have been studied by a number of authors (Fig. 12), especially McCubbin (1969, 1983), Campbell (1973) and Molenaar (1973, 1983b).

The Gallup vertical sequence exhibited in outcrop is most commonly one attributable to a strand-plain shoreline (Fig. 13). From bottom to top the sequence is: (1) inner shelf; (2) transition zone, thin and commonly containing hummocky cross stratification (Fig. 14); (3) lower shoreface, burrowed to bioturbated poorly laminated sandstones; (4) upper shoreface, predominantly trough cross-bedded sandstones (Fig. 15); and (5) foreshore, primarily sets of nearly horizontal low-angle laminations truncated by subsequently deposited laminated cosets (Fig. 16). Some of the non-marine time equivalents of the Gallup, which lie southwest of the Gallup shorelines, are included in the Gallup Sandstone; others are included in the Dilco Coal Member of the Crevasse Canyon Formation (Molenaar, 1973). McCubbin (1983) discusses the Gallup shoreline deposits in his comprehensive discussion of shoreline sandstones.

## Bisti Field

There is no structural closure at Bisti field (Fig. 17); the trap is entirely stratigraphic and is elongate parallel to the structural strike. Bisti field produces from the "Gallup" Sandstone and was considered by Sabins (1963) to be productive from sand ridges (offshore bars) which were deposited seaward of and contemporaneous with strand plain (beach) Gallup sandstones (Fig. 18). Sabins (1963) recognized a northwest-southeast trend in which beach sandstones were deposited landward of "nearshore bars" (Fig. 19). The "nearshore bars" were bounded on the shoreward side by what he termed "backbar facies" and on the marine side by "forebar facies". Sabins also described the variations in petrographic aspects of these different facies. The "offshore sand facies", below the producing sandstones at Bisti (Fig. 18), are probably similar to the transition zone facies in the outcrop (Fig. 13) and consist of thinly interbedded sandstone and shale.

The reservoir at Bisti was subdivided by Sabins into three offshore bars, the Marye bar, the Huerfano bar, and the Carson bar (Fig. 20). Details of the production worked out by Sabins indicate that the youngest offshore bar (Marye) is much more excessive and extends the length of the field (Fig. 21). The Huerfano and Carson bars both underlie the Marye bar. The major features of the reservoir observed by Sabins are listed in Table 1. Sabins' model (Figs. 19 and 22) suggests that landward of the bars a time-equivalent beach extended northwest-southeast and that the offshore bar formed as a part of a regressive sequence.

# TOCITO SHELF MODEL

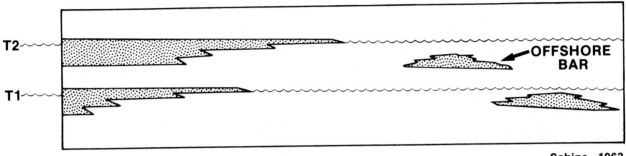

Figure 11. Synchronous Gallup shoreline and offshore bar sandstones model (Model II). Shoreline sandstones deposited contemporaneously with offshore bar sands at times T1 and following a minor "transgression" at T2. Model is based on Sabins (1963) interpretations.

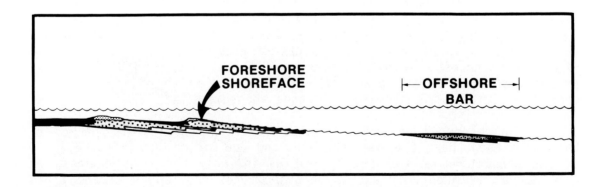

Figure 12. Synchronous transgressive "Gallup" offshore bar and shoreline sandstones. Modified from Campbell (1979).

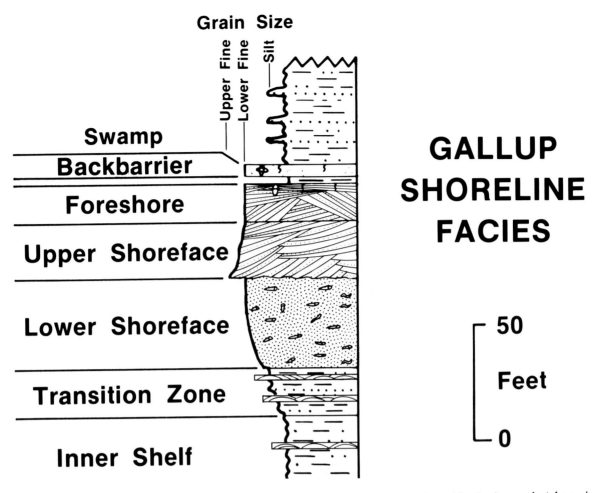

Figure 13. Gallup shoreline-facies sequence. From detailed description by Tillman of Campbell's (1979) Stop 4. (modified from Molenaar, 1973).

Figure 14. Hummocky Cross Stratification (HCS) in transition zone at base of Gallup Sandstone. Campbell (1979) Stop 4.

Figure 15. Cross-bedded upper-shoreface deposits in Gallup Sandstone. Scale in feet. At Campbell's (1979) Stop 4.

Figure 16. Foreshore (beach) lamination. Low angle sets truncated at low angles. Gallup Sandstone, Campbell (1979) Stop 4.

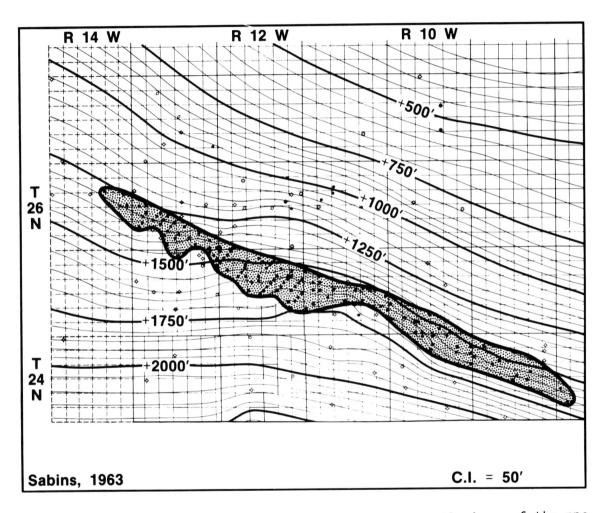

Figure 17. Structure map on surface which is at or near the base of the producing sandstone interval at Bisti field. Mapped horizon is base of "Low SP Facies" in Figure 18. Bisti production is independent of structure. (Modified from Sabins, 1963).

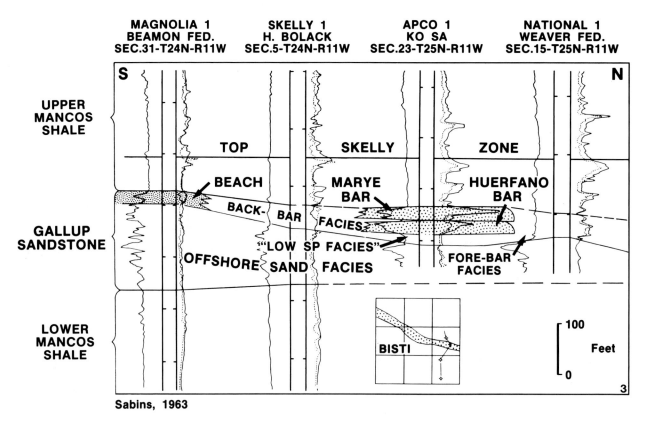

Figure 18. Facies nomenclature in area of Bisti field, New Mexico. Note stacked bar facies and inferred time equivalent beach sandstone. See inset map for orientation. (From Sabins, 1963).

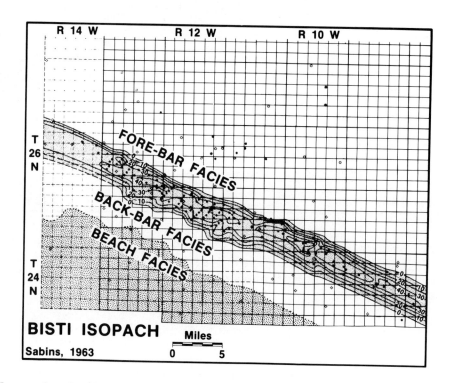

Figure 19. Isopach of all offshore-bar sandstones at Bisti field. Shown are relationships inferred by Sabins (1963) of bar sandstones to fore- and back-bar facies and to beach facies. Refer to Figure 21 for outline of individual bars.

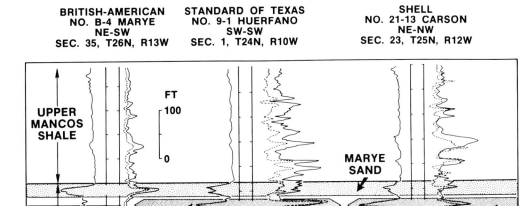

Figure 20. Type logs for three offshore bar sandstones which produce at Bisti field. Marye bar, the upper bar, extends the whole length of field as shown in Figure 18. Section is oriented northwest-southeast and follows trend of the field.

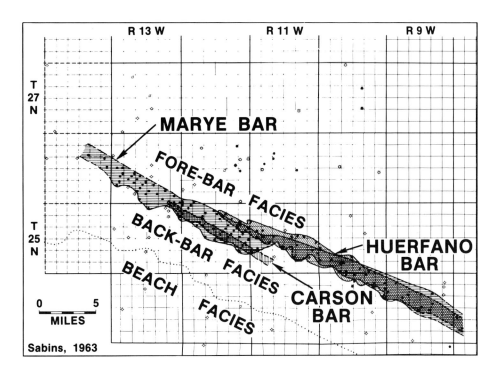

Figure 21. Outlines of three offshore bar sandstones at Bisiti field. Note inferred beach facies to lower left. (Modified from Sabins, 1963, and G.M. Nevers).

# BISTI FIELD

| | |
|---|---|
| BASAL CONTACT | COARSENS UPWARD FROM SANDY SHALE |
| UPWARD SIZE GRADATION | COARSENS |
| LATERAL FACIES | FORE-BAR & BACK BAR |
| DEPOSITIONAL ENVIRONMENT | NEARSHORE MARINE BAR |

Table 1. Summary of significant features of Bisti field, Gallup Sandstone, New Mexico. (From Sabins 1972).

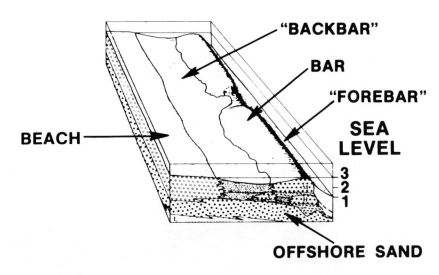

Sabins, 1963

Figure 22. Model for deposition of three offshore bars (H = Huerfano bar, C = Carson bar, M = Marye bar). In cross section beach sand (small circles), "Low SP facies" (brushy pattern) and "offshore (shoreface?) sand" are also shown. (From Sabins, 1963).

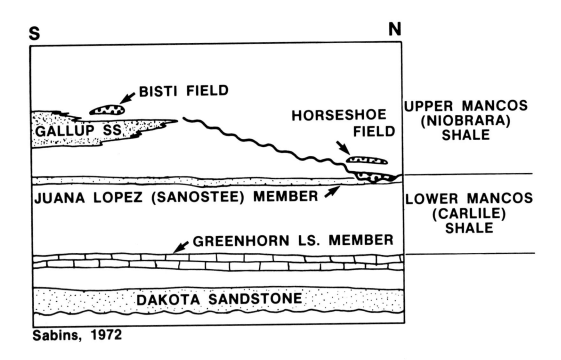

Figure 23. Regional stratigraphic cross section oriented northeast-southwest through the San Juan Basin inferring time equivalency of Bisti sandstone offshore bars and Gallup "beach" sandstones. (Modified from Sabins, 1972).

---

Several years later Sabins (1972) published another paper in which he reaffirmed his belief that Bisti field was a nearshore bar deposited concurrently with the Gallup Sandstone (Fig. 23). He indicated that the lower contact of the sandstones in the field "showed a coarsening upward from sandy shale below." Compare Figure 23 with the correlation by Bozanic (1955; Fig 7B) whose interpretation suggested that the Tocito, which occurs just above the Juana Lopez, was somewhat older than the Gallup.

## POST-GALLUP UNCONFORMITY MODELS

Lamb (1968), who studied the microfauna of the Gallup Sandstone and the Tocito Sandstone Lentil of the Mancos Shale, concluded that the Gallup and Tocito were separated by a major unconformity (Fig. 24A); a conclusion also reached by Molenaar (1973, 1983a,b). The Gallup Sandstone, which lies below the unconformity, is younger to the northeast as a result of northeastward progradation. If the unconformity recognized by Lamb and Molenaar is ignored, one might come to the false conclusion that the Tocito is actually older than the Gallup. The pre-Tocito unconformity cuts down into the lower Mancos Shale and locally into the Juana Lopez Member. Lamb (1968) indicated that the Juana Lopez is younger to the northeast (Fig. 24A) while Molenaar (1985, personal communication and Fig. 24B) believes the Juana Lopez is basically time synchronous. In the southwest part of the San Juan Basin the northwest-southeast trending Gallup shoreline prograded northeastward about 89 to 90 million years ago. (Molenaar, 1983) (Fig. 6). Shoreface and nearshore sandstones prograded slightly farther east than the foreshore deposits. The deposits of the Dilco Coal Member were in part synchronous with the Gallup and accumulated southwest of the Gallup shoreline.

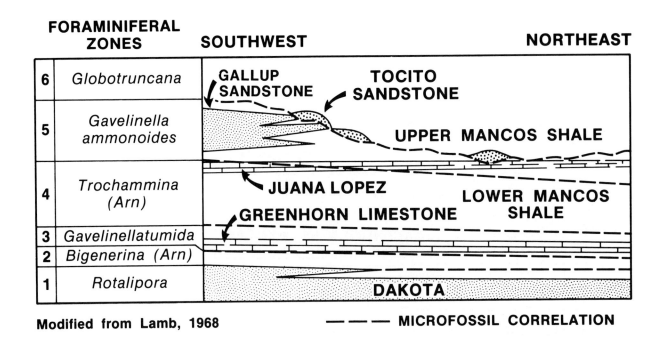

Figure 24A. Microfossil zonation and correlation of stratigraphic interval from Gallup through Tocito in San Juan Basin. Dashed lines are foraminifera zone boundries. From this the writer infers that the Juana Lopez Member crosses time lines; see text for discussion. (modified from Lamb, 1968).

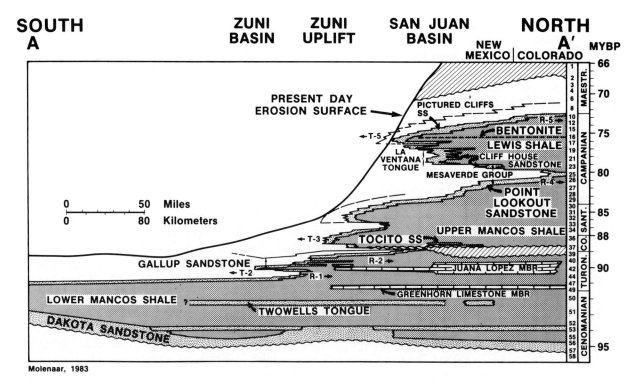

Figure 24B. Time-stratigraphic cross section from southwestern New Mexico to the northern part of the San Juan Basin. See Molenaar's Figure 4 for exact location. Note Gallup Sandstone is Turonian and Coniacian. Tocito Sandstone Lentil is indicated to be above an unconformity. (modified from Molenaar, 1983a).

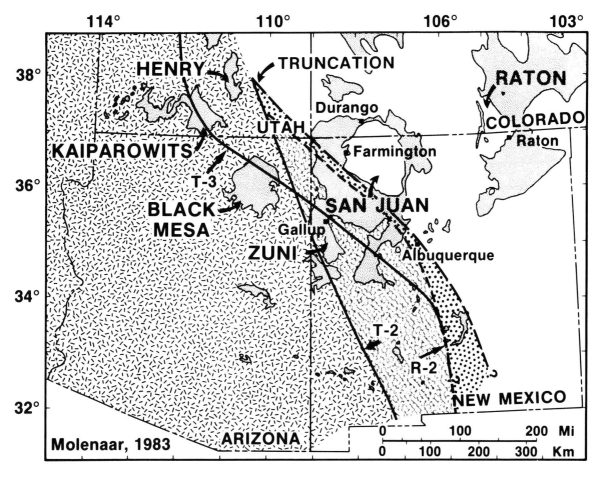

Figure 25. Map showing restored positions of the shorelines at maximum transgression and regression of the T2-R2 (Gallup Sandstone) cycle and maximum transgression T3 (Tocito) superimposed. One solid line is landward extent of shoreline deposits of the T2 transgression. Short dashed line is seaward extent of subarial-plain deposits of the R2 regression, and long dashed line is seaward extent of shoreface sandstones of the regression. Solid line T3 marks the most landward position of shoreline deposits of the T3 transgression. (Modified from Molenaar, 1983).

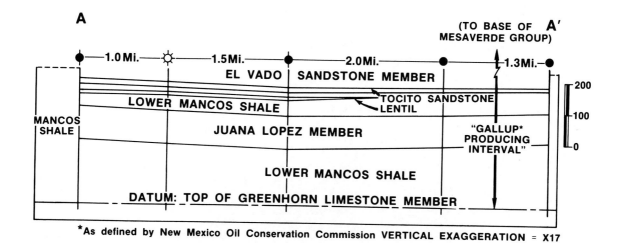

Figure 26A. Legal "Gallup Sandstone" producing interval (arrows) from top of Greenhorn to base of Mesaverde. Includes true Gallup, El Vado Sandstone Member and Tocito Sandstone Lentil of Mancos Shale. Section is through Blanco Tocito South oil field T26N R6W, New Mexico. (from Fassett and Jentgen, 1978).

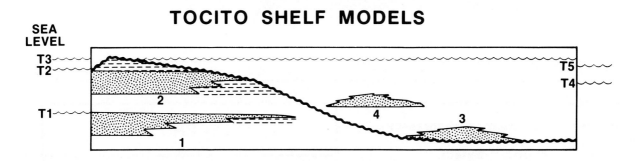

Figure 26B. Tocito Sandstone Lentil depositional model (Model IV). Gallup shoreline sandstones numbered 1 and 2 were deposited at times T1 and T2. The Dilco Coal was deposited locally above the shoreline facies prior to erosion at time T3. Following erosion offshore sand bars were deposited at times T4 and T5. This in general follows the post-unconformity transgressive model developed by Molenaar (1973, 1983b) and Penttila (1964).

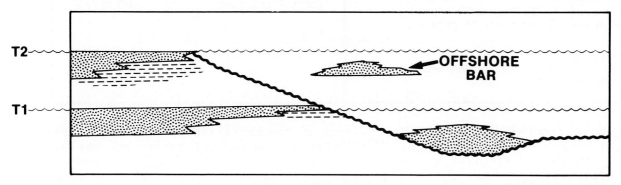

Figure 26C. Shelf sandstone model III showing lower Tocito deposited in topographic low on unconformity surface and upper Tocito (offshore bar) sandstone surrounded by shale. Based on models developed by McCubbin (1969) and Penttila (1964).

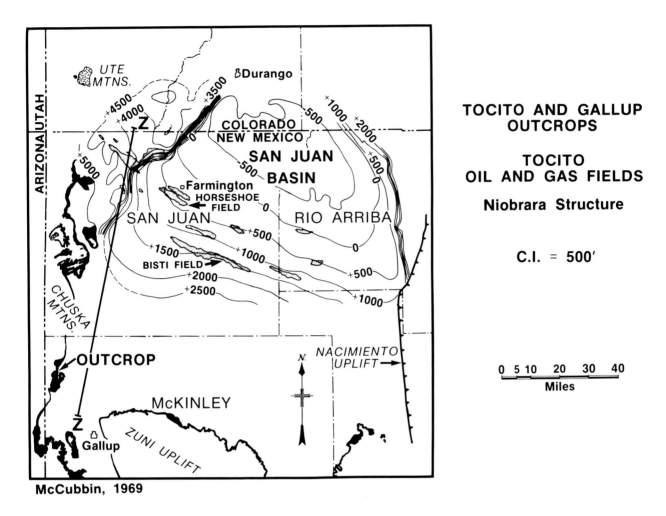

Figure 27. Map of San Juan Basin. Structure map on top of lower Tocito sandstone. Gallup and Tocito outcrops are shown in black on south and west side of Basin. (modified from McCubbin, 1969).

---

Molenaar also mapped the southwestern extent of the transgressive marine deposits that overlie the Gallup Sandstone. In the San Juan Basin the Tocito sandstones extend farther southwest than the northeastern limit of the Gallup shoreline (Fig. 23). The transgression in which the Tocito was deposited took place about 88 million years ago in Coniacian time (Fig. 24B). Detailed work by Molenaar (1973, 1983a) demonstrated that the regressive Gallup sequence is located between two transgressions which he termed transgressions T2 and T3. He recognized a major unconformity between the R-2 regression and the T-3 transgression as depicted in Model III (Fig. 26B).

The Tocito sandstones were deposited during the T3 transgression (sea level rise) and occur either within the Mancos Shale or overlie the Gallup Sandstone and the Dilco Coal Member.

Why is it important to recognize the difference between the Tocito and the Gallup? The "Gallup" produced, through 1983, 128 million barrels of oil, 0.5 million barrels on condensate and 67 billion cubic feet of gas (Fassett, 1983). Most of the oil production in the San Juan Basin comes from Tocito sandstones which are included in the "Gallup" as defined by the New Mexico Oil Conservation Commission (Fassett and Jentgen, 1978; Fig. 25). If the Tocito can be separated from the Gallup, the probabilities are increased for discovery of significant oil production.

A variation on Molenaar's model is one developed by McCubbin (1969) and Penttila (1964) (Fig. 26C). Both authors studied Horseshoe field (Figs. 9 and 27) and outcrops along the west side of the San Juan Basin. Horseshoe field trends northwest-southeast and is adjacent to Many Rocks and Verde fields (Figs. 28 and 29). The thicker portions of the lower of the two sandstones at Horseshoe field is limited to a very narrow northwest-trending area. Verde field produces from fractured sandstones and shales from the upper part of the Tocito, probably an off-bar facies (Molenaar, personnel communication, 1985), and shouldn't be included with the Tocito sandstones for purposes of this discussion. Cross sections normal to the strike of the field (Fig. 28) were constructed by Penttila (1964) (Fig. 30) and McCubbin (1969) and all of the cross sections indicate topographic depressions in which the thicker parts of the lower Tocito sandstones accumulated. At Horseshoe field the unconformity on which the lower Tocito sandstone rests cuts deeply into the lower Mancos Shale and locally into the Juana Lopez Member of the Mancos Shale. The lower Tocito sandstone lies for the most part directly on the surface eroded into pre-unconformity deposits and extends in a northwest-southeast direction parallel to the strike of the underlying beds (Fig. 29).

A series of maps of Horseshoe field were prepared by McCubbin (1969). One map (Fig. 31) shows the geology of the surface underlying the unconformity. The northwest-southeast trends observed in the field are also recognizable on this pre-unconformity map. Lithology on the pre-unconformity surface varies from shale to sandy shale to sandstone and locally to limy sandstone. The variations in resistance to erosion allow deeper erosion of the softer sediments. This erosion probably occurred in a submarine environment; however, some workers, including Sabins (1963), believe that the unconformity was subaerially exposed. It is possible that the unconformity was in part submarine and in part subaerial, becoming more subarial to the southwest. Two sandstones, designated as lower and upper sandstones produce in the field.

A number of bentonite markers are observed above and below the Tocito. One of these markers, the M2, has been referred to by a number of workers (especially McCubbin, 1969) (Fig. 32). An isopach map of the interval from the M2 marker to the unconformity below the Tocito sandstone indicates areas of thins and thicks (Fig. 33). As would be expected, the productive sandstones are mostly in the thicker areas. This type of map may be useful, where adequate control is available, in locating production in Tocito reservoirs such as Horseshoe field. The topography on the surface resulting from

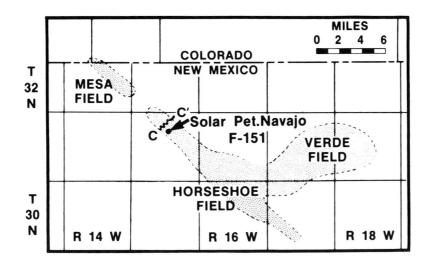

Figure 28. Outline map of Horseshoe, Verde and Mesa oil fields, San Juan Basin, New Mexico. Horseshoe and Mesa fields produce from the Tocito. (Penttila, 1964).

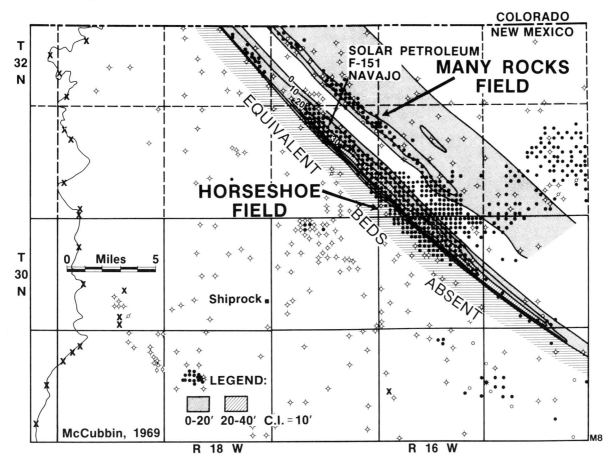

Figure 29. Isopach map of lower Tocito sandstone at Horseshoe and Many Rocks fields. Beds truncated on southwest by onlap. Beds thin to east by facies change. (modified from McCubbin, 1969).

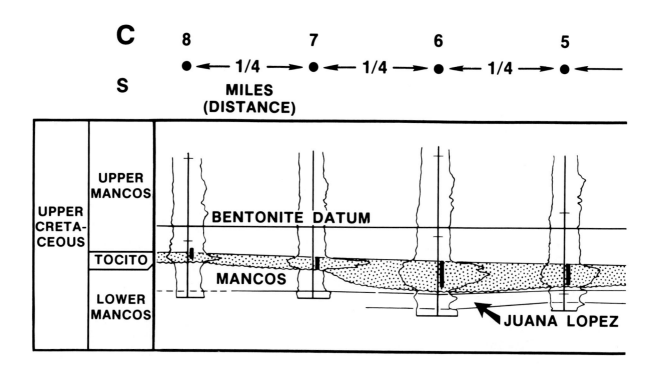

Figure 30. North-south stratigraphic section through Horseshoe field showing upper and lower Tocito sandstones. The lower sandstone rests on pre-Tocito unconformity and is thickest in the area of topographic lows. (modified from Penttila, 1964).

differential erosion resembles a strike-valley fill as defined by Busch (1959) (Fig. 34). McCubbin's interpretation infers that when the strike valley was transgressed an offshore bar was deposited in the topographic lows parallel to the strike valley (Fig. 35).

At Horseshoe field the upper Tocito sandstone is not strongly influenced by the pre-Tocito unconformity. An east-west cross section (Fig. 36), based on cores and logs and constructed through Horseshoe and Verde fields (Fig. 37) by McCubbin (1969), shows the difference in degree of control the pre-Tocito surface exerted on the lower and upper Tocito sandstones. The upper sandstone may have locally been limited laterally on the shoreward side by topography. The upper sandstone has only a moderately strong northwest-southeast lineation and produces also from a large circular area northeast of the northwest-southeast topographic low which is filled by the lower Tocito Sandstone (Fig. 38).

Sabins (1972) compared the major features of the lower and upper Tocito at Horseshoe field (Table 2). The major features of Tocito at Horseshoe field may be summarized. The basal contact of the lower sandstone is sharp and locally unconformable on the pre-Tocito unconformity surface. The lower sandstone below and on the landward side rests on an unconformity; on the seaward side the sandstone grades into a silty shale. The underlying contact for the upper Tocito is also relatively sharp and, where it does not abut previous topography on the landward side, it grades laterally into shale both landward and seaward. Sabins (1972) indicated that the lower sandstone shows no upward coarsening; the upper Tocito sandstone he believes coarsens upward (Table 2).

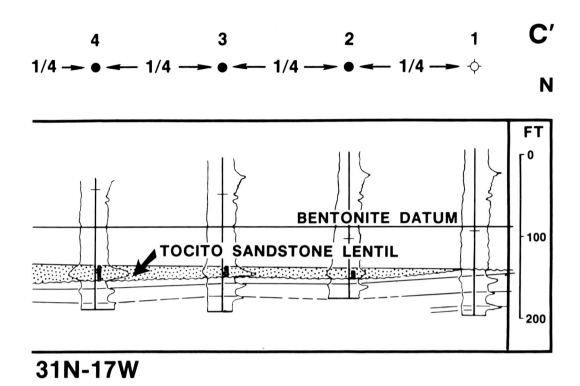

31N-17W

## GALLUP AND TOCITO OUTCROPS AND CORES

A south to north cross section (Figs. 39 and 40) showing sections measured by a variety of workers prior to 1957 includes correlations which cut across the pre-Tocito unconformity which was not recognized until somewhat later. These sections are also correlated without regard for the facies changes within the "Gallup" interval.

Studies by Penttila (1964) and Molenaar (1973) indicated that a significant break in deposition occurs within the same set of north-south trending sections (Fig. 41). The Gallup Sandstone and Dilco Coal Member were recognized as being deposited below an unconformity; above the unconformity in the northern sections is the Tocito Sandstone Lentil of the Mancos Shale. The seaward edge of the Gallup, on a cross section which runs north-south parallel to the New Mexico-Arizona state line, is observed in outcrop at Rattlesnake anticline (Fig. 41). The transgressive model (Model III) preferred by Molenaar (1973) which involves a single transgressive unconformity (Fig. 26B) is given preference in this study.

Outcrops of the Tocito sandstone exhibit an interesting pattern from the air (Fig. 42). Individual sand bodies are observed to be curvilinear and each of these curvilinear features can be traced for tens of feet (Fig. 43). Dip on each of these 0.5 to 2-foot-thick planar-tangential curvilinear ridges is consistently to the southeast (130-140°). A number of workers (Swift and

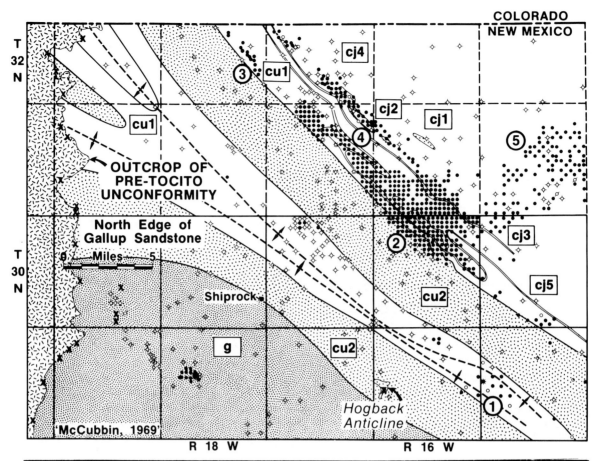

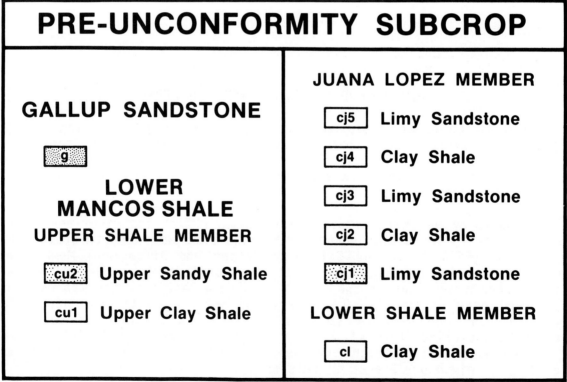

Figure 31. Pre-Tocito subcrop map showing distribution of units underlying pre-Tocito unconformity. Axis of anticlines trends northwest-southeast parallel to lithologic units. Units exposed in map area are all Upper Cretaceous and range from lower Mancos Shale to Gallup Sandstone (see Fig. 3). Note that the of north edge of Gallup Sandstone is southwest of Horseshoe field (northeast of "2"). (modified from McCubbin, 1969).

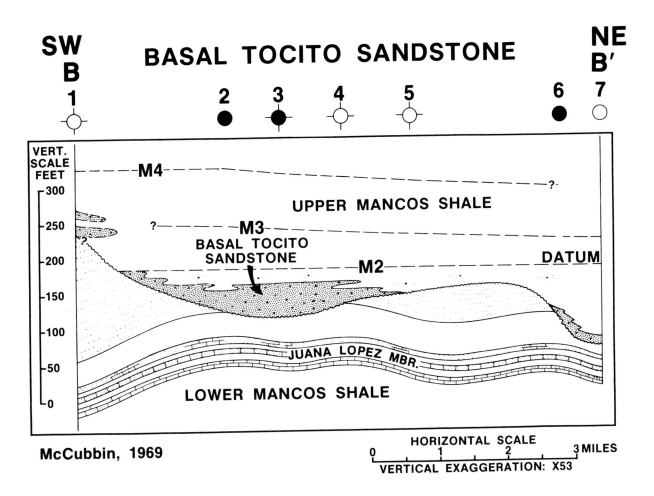

Figure 32. Stratigraphic cross section showing relationship of basal Tocito sandstone to underlying pre-unconformity units in area of Horseshoe field. Section is hung on M2 bentonite marker which is readily recognized over a large area of the San Juan Basin. Cross section is oriented southwest-northeast about half way between field numbers "1" and "2" in Figures 31 and 33 (modified from McCubbin, 1969).

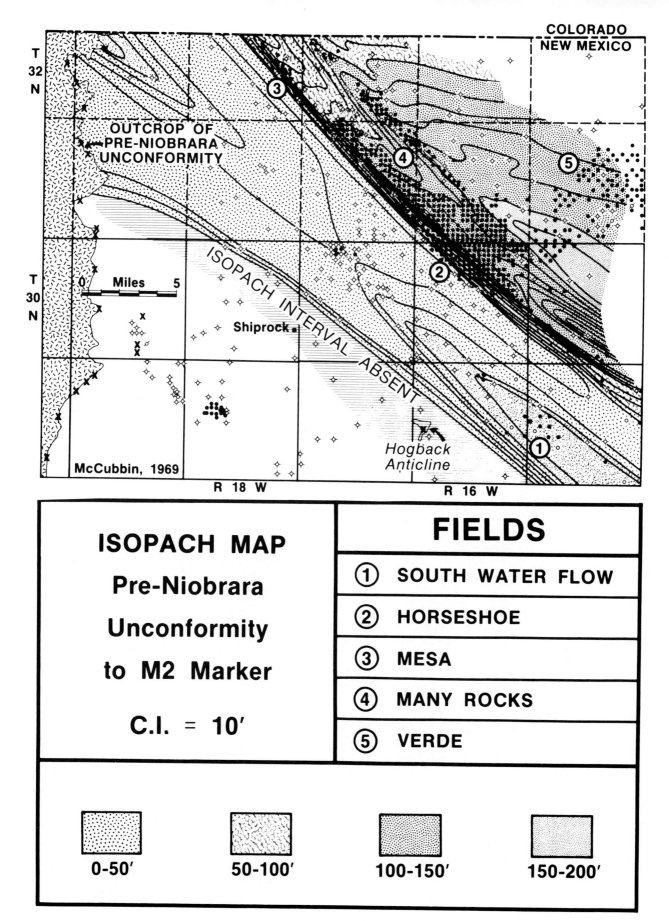

Figure 33. Isopach map of interval from pre-Tocito unconformity to M2 bentonite marker (see Fig. 32). This type of map is useful to interpret paleo-topography on unconformity surface. Areas which are thick are interpreted to have been topographic lows. (modified from McCubbin, 1969).

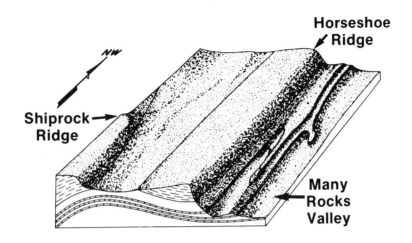

Figure 34. Model of strike valley topography on pre-Tocito unconformity surfaces in Horseshoe field area. Erosion on unconformity surface is interpreted by McCubbin to be subaerial. Evidence for subaerial, instead of submarine erosion is, however, not conclusive. (McCubbin, 1969).

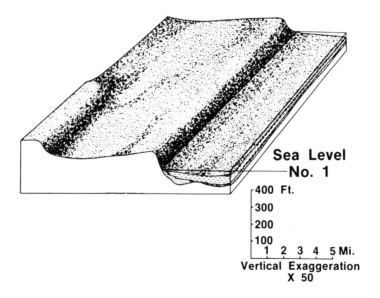

Figure 35. Model showing deposition of lower Tocito sandstone in topographic low on unconformity. Also shown is shale which overlies basal Tocito sandstone; the M1 bentonite (see Fig. 36) is at the top of this shale. Interval above shale is water. (McCubbin, 1969).

433

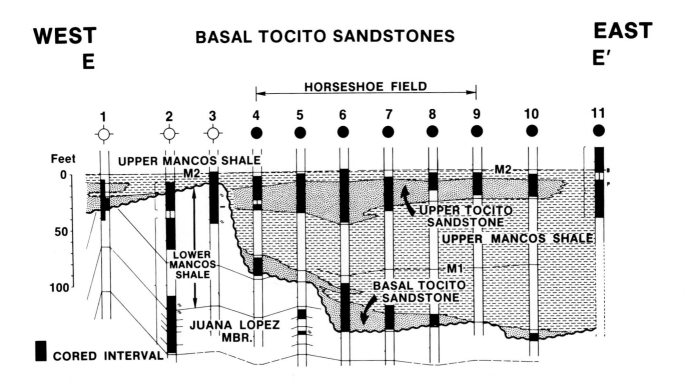

Figure 36. Detailed lithologic cross section through Horseshoe and Verde fields (Fig. 37). Correlation based on cores (black) and M1 and M2 bentonites. (modified from McCubbin, 1969).

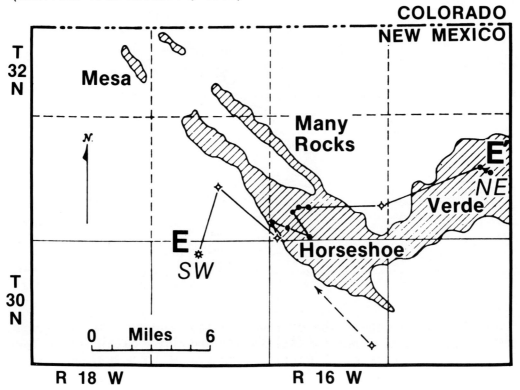

Figure 37. Location of cross section and control points for section E-E' (Fig. 36) (modified from McCubbin, 1969).

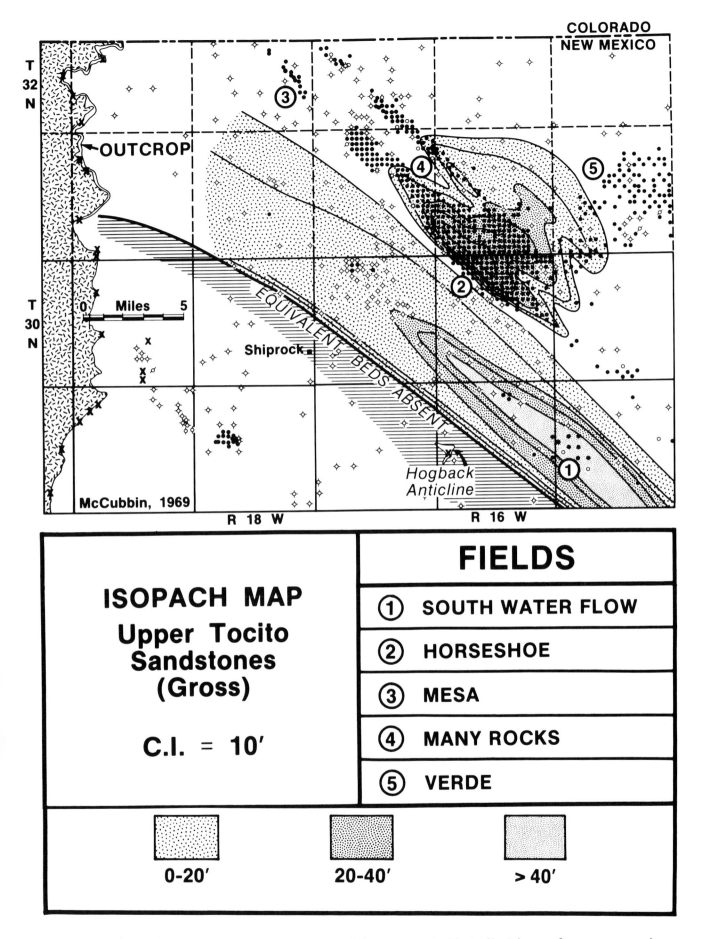

Figure 38. Isopach map of gross thickness and distribution of upper sandstone beds of Tocito. Beds thin by onlap to southwest. Location of fields keyed to numbers. (modifed from McCubbin, 1969).

# HORSESHOE FIELD

|  | LOWER TOCITO | UPPER TOCITO |
|---|---|---|
| BASAL CONTACT | SHARP, UNCONFORMABLE ON PRE-NIOBRARA | SHARP TO GRADATIONAL |
| PHOSPHORITE & COLLOPHANE | ABUNDANT AT BASE | ABSENT |
| LATERAL FACIES | UNCONFORMITY | GRADES TO SHALE |
| UPWARD SIZE GRADATION | NONE | COARSENS |
| DEPOSITIONAL ENVIRONMENT | POST-UNCONFORMITY SAND | NEARSHORE BAR |

Table 2. Comparison of significant features of lower and upper Tocito sandstones in Horseshoe field. (Sabins, 1972).

---

Nummedal, personal communication) argue that these ridges were the result of tidal currents operating in an inner shelf setting. Others lean toward bottom currents enhanced by storms as the major depositional processes.

The outcrops of the Tocito sandstones on the west side of the San Juan Basin (Fig. 44) contain several recognizable facies. Bedding is primarily horizontal. Within the beds, cross-laminated rippled and burrowed sandstone facies are recognized. The burrowed beds commonly are greater than 75% burrowed and are designated as bioturbated (Fig. 45). Cross-laminated beds occur as stacked beds and as interbeds between bioturbated beds. Locally, individual beds may be up to 8 feet in thickness and the steeply dipping foreset laminations may be in excess of 10 feet in length (Fig. 46). Thick sets usually occur by themselves or amalgamated with relatively thinner cross-laminated beds. The sand to shale ratio observed in some outcrops is similar to that recognized in the Tocito sandstones in Horseshoe field. The Tocito is locally very coarse grained (Fig. 47) and may contain small pebbles.

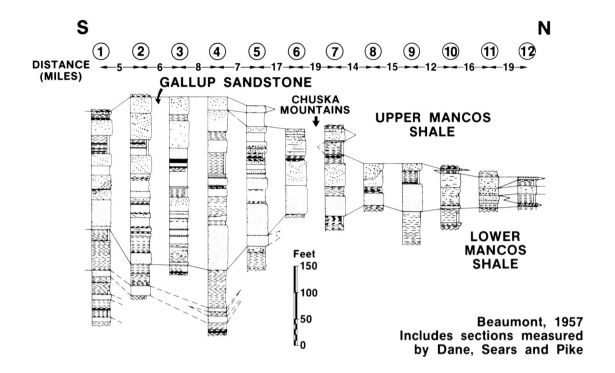

Figure 39. Stratigraphic sections of the "Gallup Sandstone" (and Tocito Sandstone Lentil of the Mancos Shale) and associated rocks between Interstate 40 (U.S. 66) and the San Juan River. "Gallup Sandstone" was interpreted to include all units between solid correlation lines. Compare with Figure 41. Location of measured sections shown in Figure 40. (from Beaumont, 1957).

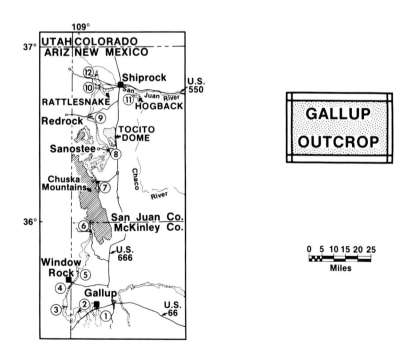

Figure 40. Map of Gallup and Tocito outcrop distribution and locations of measured sections along west flank of San Juan Basin. Sections measured by Beaumont, Dane, Sears and Pike and included in Figure 39. (modified from Beaumont, 1957).

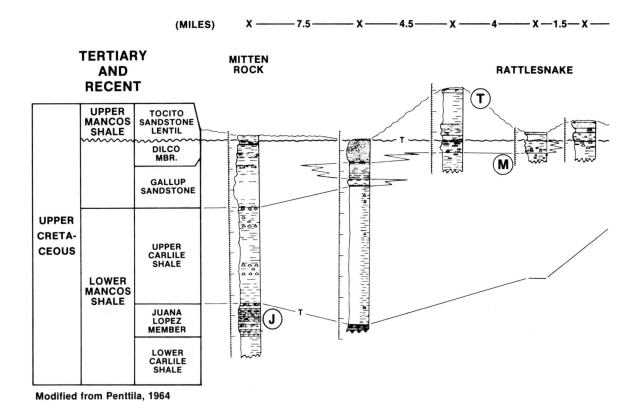

Figure 41. Stratigraphic section along northwest side of San Juan Basin showing relationship of Tocito Sandstone Lentil to Gallup Sandstone as interpreted by Penttila (1964). A major unconformity separates the two sandstones.

The transport directions measured on steeply cross-laminated Tocito sandstones are quite consistent over the whole area. McCubbin (1969), on the basis of 265 outcrop measurements, concluded that the mean transport direction for Tocito sandstones is to the southeast. The range of transport directions indicated for numerous outcrops is less than 40° (Fig. 48). Note that the transport direction is roughly parallel to the trend of the Gallup shoreline and to the trend of the oil fields producing from the "Gallup" and Tocito sandstones.

A series of facies names, developed to describe the Shannon Sandstone in the Powder River Basin of Wyoming (Tillman and Martinsen, 1984), has been adapted to the Tocito Sandstone Lentil of the Mancos Shale (Figs. 49 and 50). Most high-angle cross-bedded and cross-laminated sandstones (Fig. 46) are designated as Central Bar Facies. Highly burrowed beds (Fig. 45) are designated as Bioturbated Siltstones, Bioturbated Sandstones and Bioturbated Shales. The other common type of sedimentary structure recognized in the Tocito is ripples. In the Shannon Sandstone, the rippled Interbar Facies occurs between sand ridges as well as below them. Whether the rippled facies occupies a similar niche in the Tocito has yet to be determined.

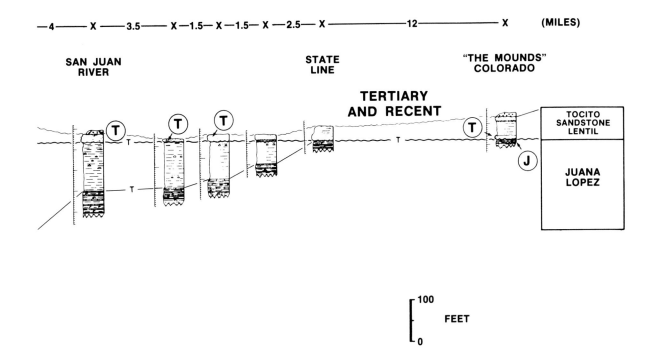

Subsurface cores which penetrate the unconformity contain some fairly major changes in microfauna (Lamb, 1968) (Fig. 24A) and macrofauna (Penttila, 1964). Variations in the type of Inoceramus are recognized across the unconformity (Penttila, 1964) (Fig. 51). In parts of the basin the unconformity may span over a million years and as expected the fauna above and below the unconformity contain different suites of fossils. The Solar Petroleum Navajo F-151 core from Horseshoe field (location in Fig. 28) illustrates the type of contact between the Tocito and the underlying lower Mancos Shale (Fig. 52; Tillman and Martinsen, Tocito Sandstone Core..., this volume). The lower Mancos is observed in outcrop and in core to be a very silty to sandy shale consisting of ripple-formed horizontal beds which are internally rippled to horizontally laminated (Fig. 53).

The Tocito-Mancos contact is commonly abrupt, suggesting erosion at the contact. Clasts including Inoceramus fragments are common just above the

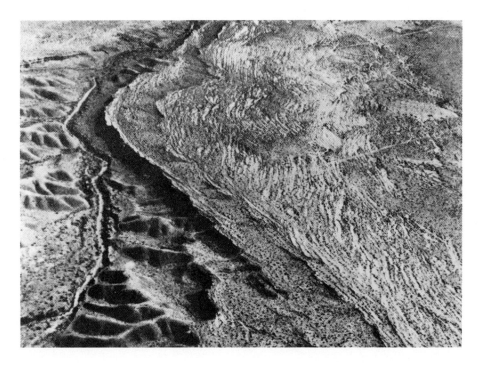

Figure 42. Airview looking northeast at Tocito offshore-bar sandstones at Campbell's (1979) Stop 1. Dip of cross laminations is to right. According to Campbell (1979) most of the apparent curvature of the sets is due to "differential" erosion.

Figure 43. Curvilinear offshore bar sandstones in Tocito Lentil of the Mancos Shale at Campbell's (1979) Stop 1. Flow direction at this location is toward 130° (to right).

Figure 44. Typical 50-foot-thick Tocito sandstone outcrop showing alternating resistant, <u>Central Bar Facies</u> and non-resistant facies, <u>Bioturbated Sandstone Facies</u>. Photo is from near Campbell's (1979) Stop 3. See also Measured Section NM3 (Fig. 49).

Figure 45. Typical <u>Bioturbated Shelf Sandstone Facies</u> at base of Tocito at Measured Section NM-3 (Fig. 49). Note that ripple-form bedding surfaces are horizontal. Burrowing in this sandstone facies is more than 75% by volume.

Figure 46. Seven-foot-thick Tocito planar-tangential cross-laminae in single set which immediately overlies a silty shale. This set is continuous in a down flow direction for several hundred yards. Flow direction is southerly. Tocito sandstone offshore-bar. Location is 2 miles northwest of town of Shiprock, New Mexico.

Figuer 47. Pebbly coarse-grained sandstone in 7-foot-thick planar tangential beds of Tocito offshore bar (Fig. 46).

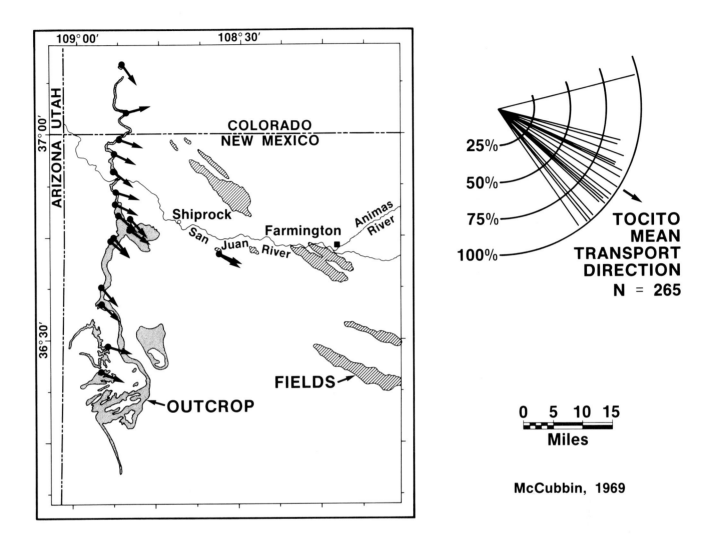

Figure 48. Flow directions for Tocito offshore-bar sandstones calculated by McCubbin (1969) from 265 cross-bed measurements. Mean flow directions for individual outcrops are indicated in outcrop area (to left). All mean flow directions are plotted (to right). All but one mean direction fall within a 40° angle. The mean of all flow directions is 127°.

MEASURED SECTION NM3-81(84)
S 1/2 SW 1/4 Sec. 27 T30N R19W
Tocito Sandstone outcrops west of Shiprock Wash in Lechii Wash

Measured and described by: R. W. Tillman
T. Moslow
Location on NW side of wash where stream intersects west canyon wall.
Rattlesnake, N. Mex. Quadrangle Map (1.62500, 1934) and Rattlesnake, N. Mex. 15" Quadrangle (1983)
Reference: Campbell, 1979

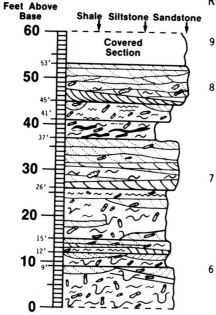

9  60'  60'+ Covered; post-Paleozoic debris (volcanic?). Across-canyon view indictes that stratigraphic interval may extend higher across canyon.

8  15'  45.0-60.0' Sandstone (350-450 μ), trace to 5% glauconite. 90% trough to planar-tangential cross-bedded (transport direction 135°). 5% ripples on troughs. 5% burrowed, trace of Ophiomorpha and Condrites. Average bed thickness 1.0' (maximum 2.5'); gradational base. Moderately well exposed. CENTRAL BAR FACIES (95%).

7  4.0'  41.0'-45.0' Sandstone (300 μ); trace of glauconite. Unit is mostly interbedded trough cross-bedded and burrowed sands; 30% wavy bedded, 20% trough cross-bedded, 10% rippled. 40% burrowed, 20% bioturbated (10% distinct burrows), trace of Ophiomorpha. Average bed thickness 1.0' (maximum 2.0'); sharp base. LOW ENERGY BAR-MARGIN FACIES (75%).

6  4.0'  37.0'-41.0' Sandstone (200 μ), trace of glauconite, 25% siltstone, (interbedded lenses of silty shale). 50% wavy bedded, 30% rippled, 20% burrowed. Recessive beds; sharp base. INTERBAR FACIES (80%).

5  11.0'  26.0'-37.0' Sandstone (300 μ), 2-20% glauconite, 10% in sigmoids, tr-5% above; 75% cross-bedded, 10% planar-tangential (transport direction 135°), 10% rippled; 5% burrowed, trace of Ophiomorpha. Beds resistive, average thickness 0.5'-1.0' (maximum 2.0'); abrupt base. Prominent sigmoid bed at base (tidal). CENTRAL BAR FACIES (90%).
(*Section moved 50 yds. to NNW)

4*  11.0'  15.0'-26.0' Sandstone (150 μ), 10-20% glauconite. 60% burrowed (50% bioturbated, 10% distinct); 20% trough cross-bedded, 10% rippled, 10% wavy bedded. Recessive and resistive beds; sharp lower contact. BIOTURBATED SHELF SANDSTONE (95%).

3  3.0'  12.0-15.0' Sandstone (200-300 μ), 5-10% glauconite. 30% cross-bedded, 30% planar-tangential. Maximum bed thickness 4". Transport directions approximately 140°. 20% rippled (5% symmetrical). 20% burrowed (includes some oblique to vertical burrows. 1/4" diameter). CENTRAL BAR FACIES (90%).

2  3.0'  9.0'-12.0' Sandstone (175 μ) 80%. 10% glauconite, 20% gray silty shale; 60% burrowed (55% bioturbated, 5% distinct). 25% trough cross-bedded, 10% rippled, 5% wavy bedded. Beds less than 5' thick. Recessive, poorly exposed; sharp base. BURROWED SHELF-SANDSTONE (75%).

1  9.0'+  0.0'-9.0' Sandstone (350 μ), 10% glauconite; 80% burrowed (75% bioturbated, 5% distinct). 10% rippled, 5% low-angle trough cross-bedded, 5% sub-horizontal bedding. Beds alternately recessive to resistive, average bed thickness 0.5' (maximum 1.0'). BIOTURBATED SHELF-SANDSTONE (95%).

Figure 49. Tocito sandstone, Measured Section NM3-81(84). Facies terminology modified from usage of Tillman and Martinsen (1984).

Figure 50. Lower part of Measured Section NM 3-81(84) (Fig. 49). The lower 9 feet (below arrow) is a Bioturbated Shelf Sandstone Facies, Unit 1 (Fig. 49). The resistive beds below the middle arrow and above the upper arrow are designated as Central Bar Facies, Units 3 and 5. An 11-foot thick Bioturbated Shelf Sandstone crops out between two upper arrows. See Figure 44 for general view of this outcrop.

Figure 51. Stratigraphic succession of Inoceramus perplexus Whitfield in lower Mancos Shale (Carlile) overlain disconformably by Inoceramus thick shelled, which according to Penttila (1964) is typical of the Tocito (Niobrara). Fossils were collected in the west offset to the Atlantic Ute No. 19, Sec. 26 T31N 16W, New Mexico.

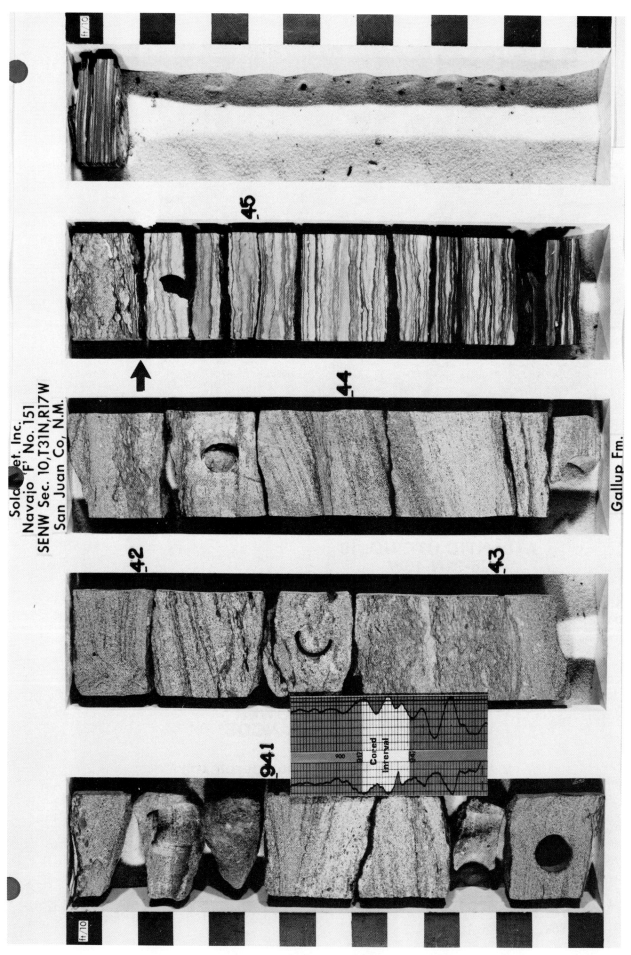

Figure 52. Tocito-lower Mancos Shale contact (arrow) in Solar Petroleum, Navajo F No. 151 core from Horseshoe field. See discussion in paper on Tocito core (in this volume). Location of well is shown in Figure 28. Typical lithologies: (1) lower Mancos Shale 944.7 feet to base of core, (2) planar-tangential cross bedding 943.7 to 944.5 feet, (3) Bioturbated Sandstone Facies 942.6 to 943.0 feet, (4) Pebbly cross-bedded sandstone 941.0 to 941.5 feet.

unconformity (Fig. 53). There also may be clasts from the Mancos Shale incorporated in the lower portion of the Tocito indicating at least a minor local reworking during deposition. Steeply cross-laminated sandstones are assumed to form the best reservoir facies (Fig. 54). The Tocito sandstone, where it produces at Horeshoe field, commonly consists of burrowed and bioturbated sandstones interbedded with high-angle cross-bedded sandstones as illustrated in the discussion by Tillman of the Solar Petroleum Navajo F-151 core in this volume. The burrowed portions of the sandstones are generally more shaly and have lower quality reservoir properties.

As discussed earlier (Model III, Fig. 26B), Molenaar (1973, 1983a,b) interpreted the Gallup-Tocito relationship in terms of a single unconformity model (Fig. 55). He interprets the Gallup to be a regressive sandstone and the Tocito sandstone to be a post unconformity sandstone deposited contemporaneously with the Upper Mancos Shale. The Borrego Pass Lentil or post-unconformity "stray sands" are indicated to be shoreline or nearshore marine equivalents of the Tocito Sandstone Lentil (Fig. 55). The Gallup is considered to be a regressive sequence of shoreline sandstones; these marginal marine sandstones are in part time equivalents of the Torrivo Member sandstones which were deposited in northeast flowing fluvial channels (Fig. 56). Field evidence and limited subsurface work by the writer suggests that Molenaar's model is the preferred one.

## TOCITO AND GALLUP OFFSHORE BAR (SAND RIDGE) MODEL V

Still another model (Model V) for deposition of the Tocito sandstones was discussed by Campbell (1979) (Fig. 57). Campbell studied the Gallup-Tocito relationship along the west flank of the San Juan Basin (Fig. 58) utilizing both outcrop and core data. His model is well illustrated in a southwest to northeast cross section which utilized both outcrop and subsurface data (Fig. 59). According to his model, the Tocito Sandstones were deposited above a major unconformity which separates them from the Gallup Sandstone. Individual Tocito sand ridges were deposited during short period stillstands which were superimposed on a longer term transgression. The Tocito sand-ridge (offshore bar) facies overlying the major unconformity are the (1) Bisti-Gallegos, (2) Ship Rock and (3) Rock Ridge sand ridges (Fig. 59). The presence of aeolian and associated sediments below the offshore bar and above the unconformity (i.e. core hole 23) suggested to Campbell (1979) that at least the landward portion of the major unconformity was cut by subaerial (rather than submarine)
erosion.

A second category of sand ridges (offshore bars) is recognized by Campbell in the Gallup deposits below the major unconformity. Campbell interprets these Gallup shoreline and offshore bar sandstones to have been deposited above multiple local unconformities. The Gallup shoreface sandstones below the Bisti-Gallegos offshore bar sandstones are, according to Campbell (1979), time equivalent to preserved offshore bar sandstones in core holes 10 and 11 (Fig. 59). He projects from core hole 31 the bentonite markers closest above and closest below the Bisti-Gallegos offshore bars to

Figure 53. Base of coarse grained (locally burrowed) Tocito sandstone resting unconformably (lower arrow) on lower Mancos Shale. Note Inoceramus fragment (upper arrow) just above contact. 944.5 feet, Solar Petroleum F-151, Horseshoe field, San Juan County, New Mexico.

Figure 54. Planar-tangential cross-bedding at 943.6 feet, Solar Petroleum. Navajo F-151 core in Horseshoe field.

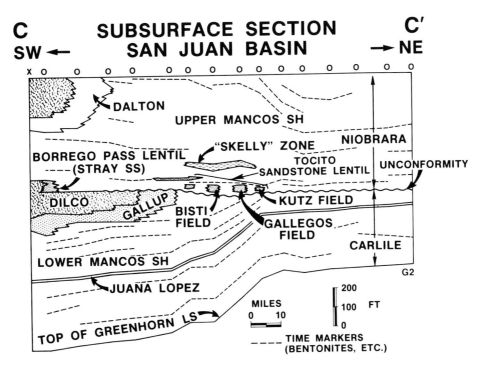

Figure 55. Subsurface stratigraphic cross section oriented southwest-northeast across central San Juan Basin. Dashed lines are inferred to be time marker bentonites or calcareous silty zones. Note truncation of some markers below the unconformity. Shown are well locations (circles) and outcrop location (x) used as control points. (modified from Molenaar, 1973). For exact wells used see Molenaar (1973). Location of cross section is shown in Figure 60.

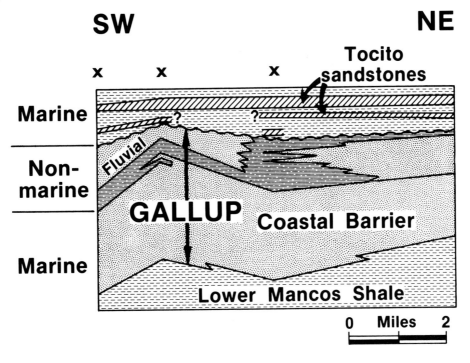

Figure 56. Facies relationships across T26N R19W on western side of San Juan Basin, New Mexico. On the southwest side of the section a single tongue of the Gallup Sandstone includes shoreline and possible coastal barrier sandstones, nonmarine shale (broken horizontal lines) and fluvial sandstones (Torrivio Member). In the northeast part of the section a younger tongue of the Gallup extends westward into non-marine carbonaceous shales. The Gallup is overlain unconformably by upper Mancos Shale and offshore bars of the Tocito Sandstone Lentil of the Mancos Shale (modified from Molennar, 1973).

bracket the interval that contains the unconformity separating the transgressive bars and the regressive beach (Campbell, 1985, personal communication) south of core hole 31. Any unconformity underlying the offshore bar is subparallel to the bentonite markers and because of lack of disparity in dip direction difficult to recognize.

Figure 57 shows Campbell's interpretation of the sequences involving the Tocito and Gallup sandstones. He recognizes a major unconformity between the Tocito and Gallup Sandstones. Gallup Sandstone shoreface (1A) and offshore sand ridges (1B) are synchronous as are Gallup Sandstone shoreface (2A) and offshore sand ridge (2B). These Gallup sandstones are interpreted to be unconformable at the base in the offshore area. The Tocito sand ridges were deposited in an offshore to onshore sequence (3, 4 and 5) above the major unconformity.

Summary of Depositional History (Campbell, 1985 personal communication)

The history of the Juana Lopez, Gallup and Tocito stratigraphic section may be summarized as follows:

1. "Warping of the Juana Lopez was followed by local erosion over at least the area of the present San Juan Basin. Erosion probably occurred below sea level.

2. Northeastward progradation of the Gallup beach complex. This deposition was episodic resulting in repetitive units of beach facies which are bounded by unconformities of small areal extent, short duration and minor erosion. Seaward of some beach episodes bentonites accumulated and less frequently small offshore bars developed. The unconformities bounding these beach episodes represent mini-regressions or mini-transgressions and the small bars [sand ridges] probably formed on the mini-transgressions.

3. Periodical lowering of sea level occurred during the progradation. The result of the lowering was cutting of terraces, each successively lower than the previous one.

4. Transgession with deposition of offshore bars [sand ridges]. Sand accumulated on the wave cut terraces abutting the terrace edges. The bars formed first on the lowest terrace and subsequently on higher and higher terraces as the transgressing waters deepened."

# TOCITO SHELF MODEL

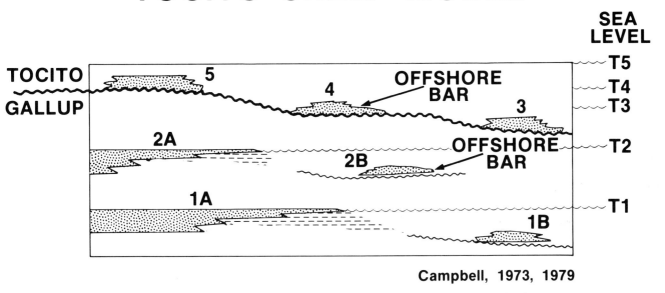

Figure 57. Tocito shelf model (Model V) based on Campbell's (1979) transgression and stillstand interpretation. A sequence of synchronous beach and offshore sandstones were interpreted to have been deposited following a succession of sea level rises and stillstands.

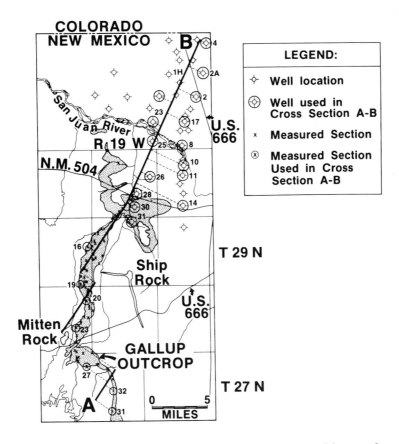

Figure 58. Gallup and Tocito sandstone outcrops and line of section (A-B) along west side of basin used by Campbell (1979) to construct Figure 59. Note several east to west offsets are indicated. Control points are circled. (modified from Campbell, 1973).

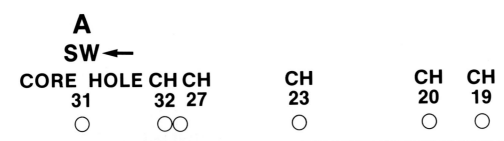
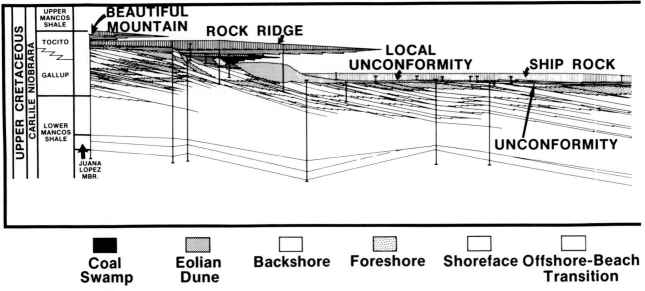

Figure 59A. Interpretive stratigraphic section for Upper Cretaceous Gallup beach and offshore sandstone deposits (modified from Campbell, 1979). Measured sections and core hole control points are indicated by vertical lines. Backshore deposits (white) are always to the left of foreshore sandstones; shoreface deposits (white) are always seaward of foreshore sandstones. Offshore siltstones and mudstones (white) are always seaward of foreshore-shoreface lateral sequence. Note aeolian sandstone above the unconformity at core hole 32 and extending from core hole 23 to core hole 16. Above the aeolian deposit are "offshore-bar transition" deposits which represent the initial deposits of a transgression-stillstand.

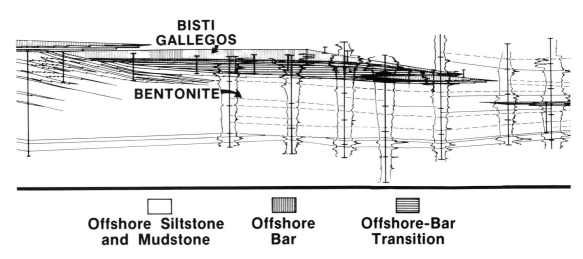

Middle section of three part panel. Dashed lines are inferred to be time lines (bentonites?). Note that top two bentonites in core hole 31 are interpreted by Campbell to bracket the shoreface deposits encountered in measured section midway between CH16 and CH31. This implies that the offshore bar sandstone in core holes 10 and 11 is synchronous with the shoreface deposits in the measured seciton. This model requires a series of four or more transgressions and accompanying stillstands. The youngest offshore bars are those to the southwest.

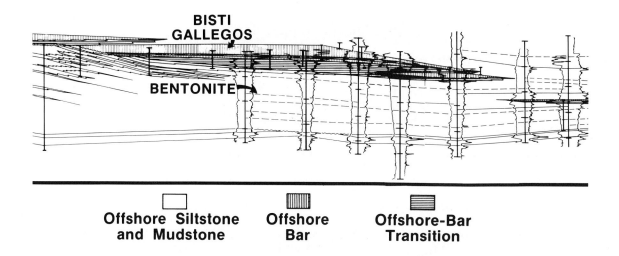

Figure 59B. Most seaward portion of interpretive section. Dashed lines are inferred time lines (bentonites?). Note McCubbin's M2 marker which can be correlated from near the outcrop to Horseshoe field.

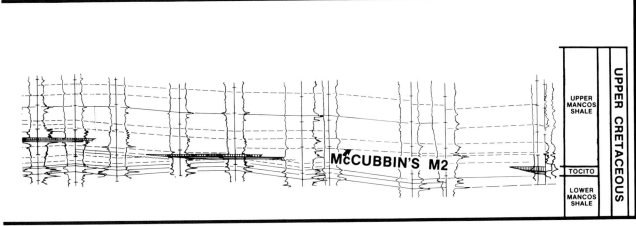

## CONCLUSIONS

Five different models showing the relationship of the Gallup Sandstone to the Tocito Sandstone Lentil of the Mancos Shale have been discussed (Fig. 62).

Early workers such as Sears (1925), Reeside (1924) and others designated all sandstones occurring within the Mancos Shale above the Juana Lopez Member below the coaly deposits now identified as the Dilco Coal Member as Gallup Sandstone. The area where the Gallup was first defined (Sears, 1925) included both shoreline and fluvial-channel sandstones. Offshore sandstones were assumed to be time equivalents of the shoreline Gallup sandstones and the whole vertical sequence was assumed to have formed during a series of regressions. An even broader usage of the term Gallup is that by the New Mexico "State Board" in legal matters pertaining to both the true Gallup Sandstone and the Tocito Sandstone Lentil (Fasset and Jentgen, 1978) (Fig. 26A). They include all production from the top of the Greenhorn to the base of the Mesaverde as "Gallup".

Bozanic (1955) suggested that the Tocito was older than the Gallup Sandstone because it occurs only a short distance (at Doswell field) above the Greenhorn Limestone Member which is assumed to be a time marker in the San Juan Basin. The Gallup was assumed to overlie the Tocito because it occurs 700 feet above the Greenhorn (Fig. 7B). Additional control has negated this early correlation.

McCubbin (1969) and Penttila (1964) did analyses of the Tocito in the northwest part of the San Juan Basin and emphasized the origin of the reservoir at Horseshoe field. Both authors indicated that a regional unconformity existed between the Tocito and Gallup Sandstones and that the lower Tocito sandstone reservoir at Horseshoe field was deposited in post-unconformity lows which trend parallel to the subcrop of the pre-unconformity lithologic units. Differential erosion of various lithologies was attributed by McCubbin as the cause of the very sharp changes in topography. Lamb (1968) detailed the time-stratigraphic relationships of the pre-Tocito unconformity, on the basis of microfauna.

Molenaar in 1973 outlined in detail the stratigraphy of the Gallup and associated strata in the San Juan Basin. On the basis of both surface and subsurface work he recognized a relatively major regional unconformity which cuts out portions of the Gallup Sandstone, Mancos Shale, and locally the Juana Lopez Member. He played down the importance of local topography as being important in localizing sandstone deposition, and emphasized the transgressive nature of the Upper Mancos sea in which the Tocito Sandstone Lentil was deposited. Why almost all the Tocito production is located northeast of the seaward pinchout of the Gallup (Fig. 60) is still unexplained. To the southwest, where the Tocito overlies the Gallup, fluid communication between the two resulted in the oil moving updip to the outcrop on the Zuni uplift and to Hospah field (Molenaar, 1977).

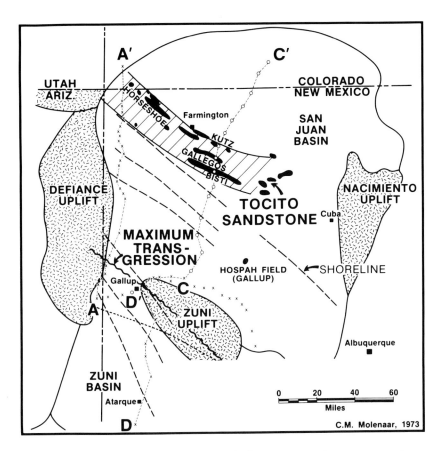

Figure 60. San Juan Basin, New Mexico. Fairway northwest-southeast elongated Tocito sandstone producing fields are indicated. "Shoreline" immediately southwest of Tocito production is most seaward extent of Gallup Sandstone shorelines. The "Maximum Transgression" line marks the approximate landward limit of T-3 transgression. See Molenaar (1983) for explanation of other dashed lines. Hospah field is one of very few fields that produces from the true Gallup Sanstone (Molenaar, 1983). C-C' is line of section of Figure 55.

## HOSPAH FIELD
### Gallup (Hospah) Sandstone

DISCOVERY:   March, 1927    1610'    8 BBLS/DAY
             October, 1927  3282'    161 BBLS/DAY

PRODUCED OIL:   13 MILLION BBLS.

TRAP:   ANTICLINE/FAULT  (2½ x 1½ Miles), 100' +Closure

RESERVOIR:   25-60' (35')   $\phi$ = 24-30%   Perm. 200-500 md.

WELLS:   40   10 ACRE SPACING

DRIVE:   Water (?)

Table 3. Hospah Field Production Data. Production in this field is from Hospah (Gallup equivalent) Sandstone. Compiled from King and Wangerd (1957).

Campbell (1973), using surface outcrop, shallow cores and logs, developed a relatively complicated model (Model V) which recognizes both Tocito and Gallup offshore bars (sand ridges). The Tocito sand ridges lie above a major conformity which he believes was for the most part subaerial. He based this on the occurrence of what he interpreted as aeolian sandstones immediately above the unconformity. A series of westward "younging" offshore bars (sand ridges) were deposited on terraces on the unconformity. Using bentonite markers which bracketed several of the subsurface offshore bars, he was able to "correlate" with Gallup shoreface deposits occurring in outcrop. Each Gallup offshore bar he assumed had at the time of deposition a shoreline sandstone equivalent.

A variety of processes have been considered for deposition of the Tocito Sandstone. There is strong evidence that the Tocito is an inner to middle shelf deposit (Fig. 61). the types of possible currents that could be expected on the middle to inner shelf area are numerous (Tillman, A Spectrum of Shelf Sandstones, this volume). Sedimentary structures suggest that tidal currents may have had at least local effects and that permanent and semi-permanent currents augmented by storm currents were important in other areas.

From an exploration and production point of view, separating the Tocito and Gallup sandstones is very important. All of the oil production in the San Juan Basin, to date, is in the northwest and north-central part of the basin in sandstones designated by Molenaar (1983) as Tocito sandstones (Fig. 60). One field, Hospah Field (Fig. 60 and Table 3), which produces from a structural anticline and was discovered in the 20's, is interpreted as being a true Gallup Sandstone producer (Molenaar, 1983 and personal communication). All the other oil production is considered by Molenaar (1983 and personal communication) and Fassett (1983) to be from Tocito sandstones.

This Gallup-Tocito problem will continue to be of interest for some time to come. There is still extensive gas production in some of the shallower units in the San Juan Basin. It is expected that additional wildcats will continue to be drilled in the Basin and many of these will be drilled deep enough to penetrate the Tocito. Oil production from the Tocito may be discovered in some very unexpected part of New Mexico during the next decade.

ACKNOWLEDGMENTS

This paper was reviewed by D. J. P. Swift, R. G. Walker, C. M. Molenaar and D. Nummedal. Discussion with D. Nummedal, C. M. Molenaar and C. V. Campbell and B. Kofeld have helped in understanding the Gallup-Tocito problem. Drafting was provided by Cities Service Oil and Gas Company Research drafting department, primarily Jeannie Sommers and Sherrie Franklin. Fred Mason did the darkroom work. The manuscript was typed by Janice Brewer.

## SHELF PROCESSES

|  | Tidal Currents | Wind Induced Alongshore Currents | Wave Modified (Combined flow) Currents | Storm Currents | Permanent or Semi-Permanent Currents | Turbidity Currents |
|---|---|---|---|---|---|---|
| Shoreface Attached | X | X | X | X | X | — |
| Inner Shelf | X | X | X | X | X | ? |
| Middle Shelf | ? | ? | X | X | X | X |
| Outer Shelf | — | — | — | X | X | X |

Figure 61. Shelf processes and their distributions on the shelf.

# TOCITO SHELF MODELS

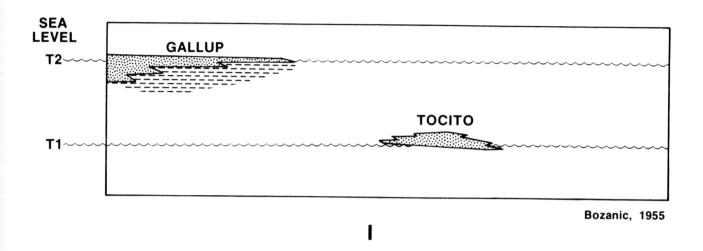

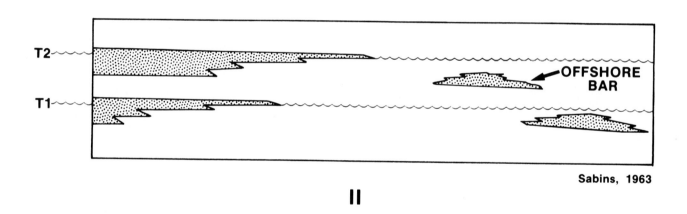

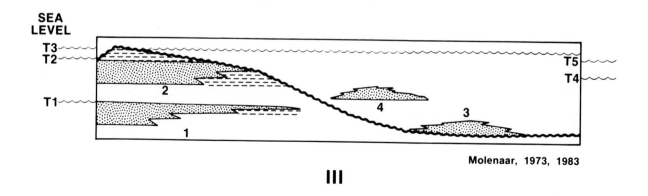

Figure 62. Five Gallup-Tocito summary depositional models. See Figures 7C, 11, 26B, 26C, 57 and accompanying text for explanation of each model.

## TOCITO SHELF MODELS

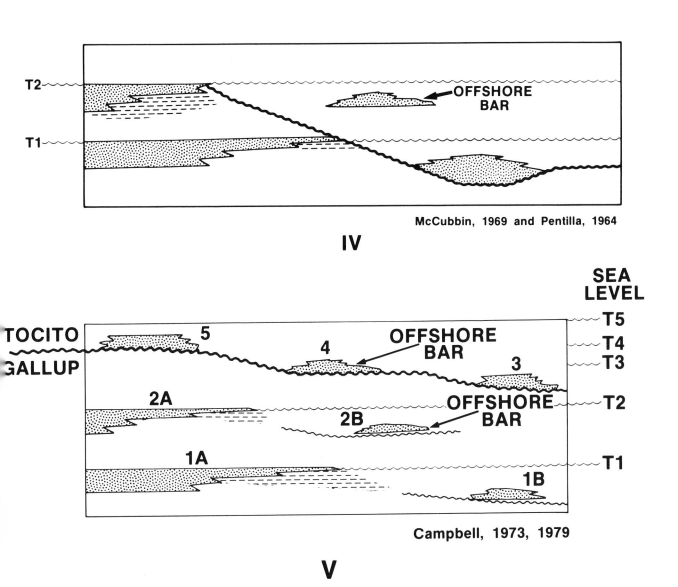

Figure 62 (cont.). Tocito shelf models.

# REFERENCES

Beaumont, E. C., 1957, The Gallup Sandstone as exposed in the western part of the San Juan Basin: in Second Annual Field Conference Guidebook: Four Corner Geological Society, p. 114-120.

Bozanic, D., 1955, A brief discussion of the subsurface Cretaceous rocks of the San Juan Basin, in 1st Annual Field Conference Guidebook: Four Corners Geological Society, p. 89-107.

Busch, 1959, Prospecting for stratigraphic traps: American Association of Petroleum Geologists Bulletin, v. 43, p. 2829-2843.

Campbell, C. V., 1979, Model for beach shoreline in Gallup Sandstone (Upper Cretaceous) of northwestern New Mexico: New Mexico Bureau of Mines and Mineral Resources Circular 164, 32 p.

Fassett, J. E. and R. W. Jentgen, 1978, Blanco Tocito oil field, in Fassett, J. E. (ed.), Oil and gas fields of the Four Corners area: Four Corners Geological Society, p. 233-240.

Fassett, J. E., 1983, Stratigraphy and oil and gas production of northwest New Mexico updated through 1983, in Oil and Gas Fields of the Four Corners Area: Four Corners Geological Society, p. 849-863.

King, V. L. and S. A. Wengerd, 1957, The Hospah oil field, McKinley County, New Mexico, in Second Annual Field Conference Guidebook: Four Corners Geological Society, p. 155-166.

Lamb, G. M., 1968, Stratigraphy of the lower Mancos Shale in the San Juan Basin: Geological Society of America Bulletin, v. 79, p. 827-854.

Lily, O. J., 1952, The Doswell oil field, Rio Arriba County, New Mexico: Geological Symposium of the Four Corners Region, p. 99-103.

McCubbin, D. G. 1969, Cretaceous strike valley sandstone reservoirs, northwestern New Mexico: American Association of Petroleum Geologists Bulletin, v. 53, p. 2114-2140.

McCubbin, 1983, Barrier island and strand plain facies, in Sandstone Depositional Environments: American Association of Petroleum Geologists, Tulsa, Oklahoma, p. 247-280.

Molenaar, C. M., 1973, Sedimentary facies and correlation of the Gallup Sandstone and associated formations, northwestern New Mexico, in Fassett, J. E. (ed.), Cretaceous and Tertiary rocks of the southern Colorado Plateau: Four Corners Geological Society Memoir, p. 85-110.

_____, 1977, The Pinedale Oil Seep - an exumed stratigraphic trap in the southwestern San Juan Basin, in New Mexico Geologic Society Guidebook, 28th Field Conference, San Juan Basin III: p. 243-246.

_____ 1983a, Major depositional cycles and regional correlations of Upper Cretaceous rocks, southern Colorado Plateau, in M. W. Reynolds and E. D. Dolly (eds.), Mesozoic Paleogeography of West Central United States: Rocky Mountain Section of Society of Economic Paleontologists and Mineralogists Symposium No. 2, p. 201-224.

_____ 1983b, Principal reference section and correlation of Gallup Sandstone, northwestern New Mexico, in Contributions to mid-Cretaceous paleontology and stratigraphy of New Mexico, part II: New Mexico Bureau of Mines and Mineral Resources Circular 185, p. 29-40.

Penttila, W. C., 1964, Evidence for the Pre-Niobrara unconformity in the northwestern part of the San Juan Basin: Mountain Geologist, v. 1, p. 3-14.

Reeside, J. B. Jr., 1924, Upper Cretaceous and Tertiary formation of the western part of the San Juan Basin, Colorado and New Mexico: U. S. G. S. Professional Paper 134, 70 p.

Sabins, F. F. Jr., 1963, Anatomy of a stratagraphic trap, Bisti field, New Mexico: American Association of Petroleum Geologists Bulletin, v. 47, p. 193-228.

Sabins, F. F., 1972, Comparison of Bisti and Horseshoe Canyon stratigraphic traps, San Juan Basin, New Mexico, in R. E. King (ed.), Stratigraphic oil and gas fields: American Association of Petroleum Geologists Memoir 16, p. 610-622.

Sears, J. D. 1925, Geology and coal resources of the Gallup-Zuni basin, New Mexico: U. S. G. S. Bulletin 767, 53 p.

Tillman, R. W. and R. S. Martinsen, 1984, The Shannon shelf-ridge sandstone complex, Salt Creeek anticline area, Powder River Basin, Wyoming, in R. W. Tillman and C. T. Siemers (eds.), Siliciclastic Shelf Sedimentation: Society of Economic Paleontologists and Mineralogists Special Publication 34, p. 85-142.

Tillman, R. W., 1985, (this volume) Tocito sandstone core, Horseshoe field, San Juan County, New Mexico, in R. W. Tillman, D. W. J. Swift and R. G. Walker (eds), Shelf sands and sandstone reservoirs: Society of Economic Paleontologists and Mineralogists Short Course, 20 p.

Tomkins, J. Q., 1957, Bisti oil field, San Juan County, New Mexico: American Association of Petroleum Geologists Bulletin, v. 41, p. 906-922.

COMPARISON OF SHELF ENVIRONMENTS
AND DEEP-BASIN TURBIDITE SYSTEMS

Roger G. Walker

Department of Geology, McMaster University,
Hamilton, Ontario L8S 4M1, Canada

## INTRODUCTION

Comparisons commonly help sharpen our observations and interpretations in depositional environments. It is appropriate here to compare shallow marine/shelf environments with the next major sandstone depositional environments to be found in a seaward direction -- classical deep water turbidite systems. The need for such a comparison is apparent from the problems raised in my paper on "Geological Evidence for Storm Transportation and Deposition on Ancient Shelves," this volume.

We may decide to compare individual beds, groups of beds traced laterally, or groups of beds in vertical sequences. However, it has been shown earlier in this volume that the turbidity current process can operate on the shelf, and that preservable turbidites can be deposited in shallow seas. Deciding what to compare is not so simple as it might first appear.

## BASIC SUBDIVISION OF SHELF SANDSTONE FACIES

Existing classifications of shallow marine/shelf systems emphasize the comparison between tidally dominated and storm dominated shelves. This classification is broad, but useful. However, from a geological point of view, I believe that current research is converging on the system shown below (Fig. 1). This system emphasizes three main environmental areas: the shoreface, the area between fairweather and storm wave bases, and the area below storm wave base. The three components of the system can be readily identified in the geological record, and can be related to major shelf processes:

1) Cross-Bedded Sandstone Facies. These are dominated by various scales of planar-tabular and trough cross bedding, but lack abundant mudstone interbeds. Depositional environments are mostly, but not exclusively above fairweather wave base and above tidal current base (Fig. 1).

2) Interbedded Hummocky Cross Stratified Sandstone and Mudstone Facies. The sandstones, commonly 5-100 cm thick, normally have sharp and/or scoured bases, and are interbedded with bioturbated mudstones. Depositional environments are below fairweather wave base, but above storm wave base (Fig. 1): see "Geological Evidence for Storm Transportation", this volume.

3) Interbedded Turbidite-Sandstone and Mudstone Facies. These sandstones, also commonly 5-100 cm thick, are characterized by the Bouma turbidite sequence -- sharp base, massive division A, parallel laminated division B, and cross-laminated division C. The sandstones are monotonously interbedded with mudstones, Bouma's divisions D and E. Deposition is consistently below storm wave base (Fig. 1). In Figure 1, the three subdivisions of shallow marine/shelf systems are shown in lateral relationship to each other. I emphasize that the

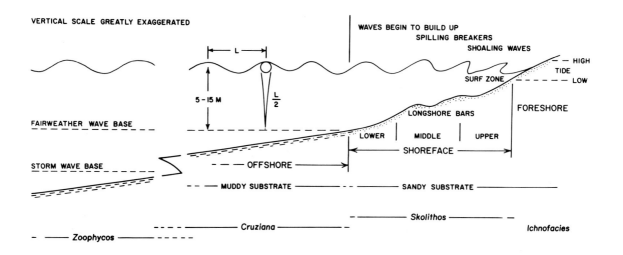

Figure 1. Conceptual diagram showing three main areas of accumulation in shallow marine/shelf settings - above fairweather wave base with abundantly cross bedded sandstones, between fairweather and storm wave base with mudstones and interbedded storm-emplaced hummocky cross stratified sandstones, and below storm wave base with mudstones and some classical turbidites. Note that the bottom of the shoreface is here taken at fair weather wave base.

environments can all exist independently from one another. For example, large areas of the North Sea (Stride, 1982) contain tidally-dominated sands with no immediately adjacent mudstones containing storm sand layers. Conversely, large areas of the Cretaceous Alberta Basin are dominated by storm deposits with no cross-bedded sandstones even close to the shoreline (Walker, "Cardium Formation 4", this volume).

The major differences between these three broad facies are related to shelf processes; the numbers refer to the numbered facies above.

Differences between 1 and 2. Facies 1 is dominated by tidal and fairweather currents (such as relatively gentle wind-controlled alongshore currents). Hence, facies 1 is dominantly sandy, and grades into facies 2 at fairweather wave base or tidal current base. At fairweather wave base, the currents that move sand on a day-by-day basis no longer influence the substrate. Cross bedding, therefore, becomes proportionally much less abundant, and mudstones become the main day-by-day deposits. When sandstones are emplaced (probably by wind-forced currents or turbidity currents - see "Geological Evidence for Storm Transportation", this volume), the dominant sedimentary structure may be storm-formed hummocky cross stratification. This is seldom reworked into angle of repose cross bedding by fairweather currents. Alternatively, incremental sand movement by wind-forced currents may result in ripple cross lamination or medium scale cross bedding.

Differences between 2 and 3. In these deeper shelf environments, below fair-weather wave base, sands are mostly emplaced by the processes listed above. Turbidity currents and storm-generated geostrophic currents are particularly important, and turbidites can be preserved in shelf environments below storm wave base. These types of deposits, associated with hummocky cross stratification, have only recently been recognized in the geological record (Hamblin and Walker, 1979; Walker, 1983a and the "Cardium Formation 4" in this volume). The distinction between the two sandstone-mudstone facies (HCS vs. Bouma-sequence-dominated) can be drawn at storm wave base - above, the waves make HCS, but below, the turbidity currents make Bouma sequences (Fig. 1).

## BASIC SUBDIVISION OF CLASSICAL, DEEP WATER SANDSTONE FACIES

The group of rocks included here are the "classical turbidites" and associated facies -- massive and pebbly sandstones, and conglomerates (a basic classification, with references to other classifications, is given by Walker, 1978).

Depositional environments are consistently below storm wave base, commonly in several hundred meters of water or deeper. The depositional area is commonly separated from adjacent shallow water areas by slope facies. The slope may be a major delta foreset, or more commonly, a slope akin to modern Continental Slopes or the slopes that define basins in the California Borderland.

Turbidity currents are normally funnelled past the slope environments in channels, which may range from minor feeder channels (a few tens of meters deep) to major canyons hundreds of meters deep. At the bottom of the channel, at the junction of slope and basin floor, the turbidity currents begin to deposit their sediment, building up a submarine fan (Normark, 1978; Walker, 1978). Fans are normally divisible into three parts (Fig. 2):

1) An inner or upper fan characterized by a single main channel with prominent levees,

2) A mid fan built up of switching depositional lobes (or suprafans). The upper part of the lobes may be channelized (several shallow channels, low levees or none at all), but the lower part is topographically smooth, and

3) An outer or lower fan which is topographically smooth, and is transitional from the suprafan lobes to the very flat and extensive basin plain (in small basins, the fan can fill the entire basin and there may be no basin plain).

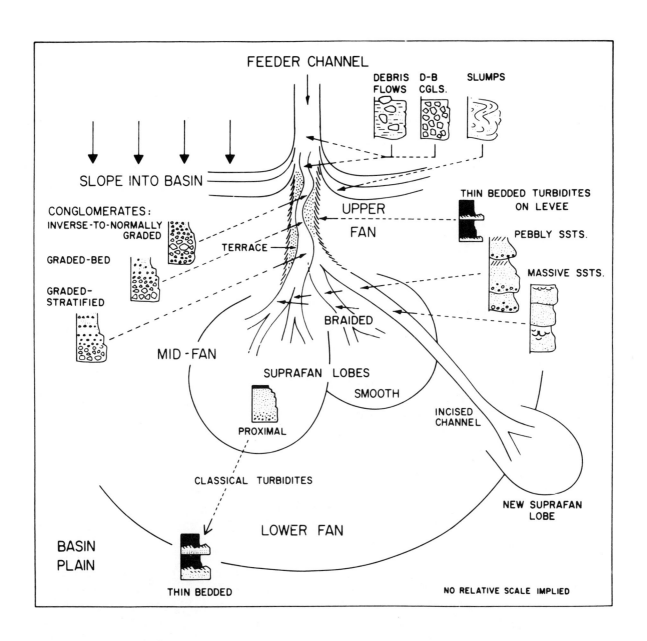

Figure 2. Model of submarine fan deposition, relating facies (D.B. = disorganized bed conglomerates), fan morphology and depositional environment. Details discussed in text. From Walker, 1978.

The deposits of the fan environment can be broadly classified into five main facies (Walker, 1978);

1) <u>Classical turbidites</u> - monotonous interbeds of sandstone and shale, with little or no substrate erosion (greater than a few tens of cm), and abundant incomplete or complete Bouma sequences (Fig. 3). Although the Bouma sequence contains <u>current</u> ripple cross lamination, symmetrical <u>wave</u> ripples are absent from the classical turbidite facies.

2) <u>Massive sandstones</u> - very thick (0.5 to many meters) sandstone beds with few (or no) thin shale partings (Fig. 4). Bedding lenticularity and channelling on a scale of a few meters is common. Bouma sequences are rare -- most beds lack sedimentary structures and are described as massive or structureless.

3) <u>Pebbly sandstones</u> - thick, commonly graded beds which begin with pebbles up to a few cm diameter, and grade up into sandstone before being sharply overlain by the next bed. Mudstone partings are rare or absent, and Bouma sequences are extremely uncommon.

4) <u>Conglomerates</u> - Individual beds vary from a few cm to several meters in thickness. Beds can be normally graded, inversely graded, stratified, or completely structureless -- the various types have been described by Walker (1975).

5) <u>"Exotic" facies</u> - slumps, slides, debris flows, pebbly mudstones and olistostromes. These chaotic facies are normally associated with base-of-slope environments.

Figure 3. Complete Bouma turbidite showing graded division A, parallel-laminated division B, ripple cross laminated division C and thin drape mudstone (D and/or E) on top. Levis Formation (Cambrian, Quebec). Bed is 5 cm thick.

Figure 4. Massive sandstone facies, Annot Sandstone (Eocene-Oligocene, southern France). Individual beds appear to be several meters thick, and interbedded mudstones are rare. Total thickness of outcrop shown in photo about 140 m.

## COMPARISONS - WHAT SHOULD BE COMPARED?

At first it might seem easy to compare and contrast beds of the shelf and deep water (fan) environments. However, it has been emphasized that turbidites occur in both of these environments. Individual turbidite beds, or even groups of beds in fan and shelf environments could not necessarily be differentiated in outcrop, let alone in core. With well log signatures only, some fan and shelf situations cannot be distinguished. With the aim of highlighting important descriptive and interpretive differences, I therefore propose to discuss the following comparisons:

1) Interbedded sharp-based hummocky cross stratified sandstones and mudstones (abbreviated herein to HCS sst/mst facies) compared with interbedded sharp-based Bouma-sandstones and mudstones abbreviated to Bouma sst/mst facies, i.e., classical turbidites), on the scale of individual beds.

2) Stratigraphic sequences containing hummocky cross stratified sandstones, compared with stratigraphic sequences containing classical turbidites from acknowledged submarine fan environments.

3) Prograding turbidite lobes compared with prograding shelf-to-shoreline sequences.

4) Broader aspects of sand body geometry in shallow seas, compared with sand body geometry in submarine fans.

5) Comparison of generating mechanisms for shelf compared with deep marine turbidity currents.

### 1. COMPARISON: HCS-Sst/Mst FACIES AND BOUMA-Sst/Mst FACIES.

These two facies are superficially rather similar, and can both be characterized by four main features.

1) Both contain sharp-based sandstones commonly in the 5-100 cm thickness range (Geological Evidence for Storm Transportation, this volume, Fig. 3).

2) Sandstones commonly have gradational tops in both facies.

3) Sandstones may be interbedded with thoroughly bioturbated mudstones in both facies.

4) On the scale of an outcrop, beds of both facies tend to be parallel and continuous (Figure 3 in Geological Evidence for Storm Transportation, this volume).

Both facies can occur in shallow marine/shelf settings, hence this particular comparison is basically one of hummocky cross stratified shelf sandstones compared with shelf classical turbidites.

## Base of beds

In both facies, beds are normally very sharp based, with minor scouring (scale of cms) on some beds. Directional sole marks (tool and scour) occur equally commonly in both facies, with scour marks commonly forming due to flow separation in small burrows or depressions in the underlying mudstones. Organic markings also occur in both facies, and consist of infilled trails and burrows from the underlying substrate, or burrows made at the sand/mud interface after deposition of the sand.

## Internal sedimentary structures

The classical turbidite facies needs little description. Complete (ABC, Fig. 3) and incomplete (BC, C) Bouma sequences occur, and grading is present in the coarser-grained beds (medium sand or coarser). Grading is very difficult to detect in fine- and very fine-grained sandstones. The rippled division C is characterized by asymmetrical current ripples, and asymmetrical wave ripples are absent from the classical-turbidite facies.

The HCS sst/mst facies is characterized by the occurrence of HCS, which has been described by many authors (see Walker, 1982; Dott and Bourgeois, 1982; Walker et al., 1983; Dott and Bourgeois, 1983; and see Figure 2 in Geological Evidence for Storm Transportation, this volume). Recently an idealized sequence of internal structures was suggested by Dott and Bourgeois (1982; Fig. 5 of this paper), and modified by Walker et al. (1983; Fig. 6 of this paper). In Figure 6, our modification shows a massive, graded basal division (which is very uncommon), and a division of parallel lamination. This may be deposited from rapid unidirectional flows (upper plane bed, Figure 18 in Geological Evidence for Storm Transportation, this volume), or from oscillating flows of high orbital velocity (Figure 16 in Geological Evidence for Storm Transportation, this volume). In many beds, the parallel lamination appears to grow progressively upward into hummocky bedforms, although Dott and Bourgeois (1982) have challenged this observation and believe that hummocky and swaley lamination drapes pre-existing scours rather than growing from a planar bedform. The hummocky and swaley topography commonly flattens upward (Division F, Figs. 5 and 6), and may be overlain by small scale symmetrical oscillation ripples. Alternatively, the oscillation ripples may directly overlie the hummocky division (H, Fig. 6). It is emphasized that asymmetrical current ripples are extremely uncommon in the HCS sst/mst facies.

In a few beds, parallel lamination is the dominant structure, but the top of the bed is erosively sculpted into a hummocky and swaley form without further sand deposition. This is indicated by "wave scour" in Figure 6.

The HFX part of the sequence represents decreasing orbital velocities and orbital diameters on the bed, and when the storm waves no longer feel bottom, day-by-day quiet conditions return and mud is deposited.

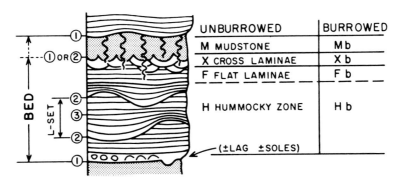

Figure 5. Idealized "hummocky sequence" proposed by Dott and Bourgeois (1982). Note distinction of hummocky beds from lamina sets, and heirarchical order of first order boundaries (1), second order truncations within hummocky units (2), and third-order lamina surfaces (3).

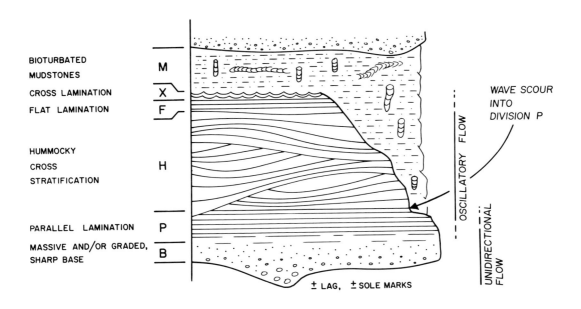

Figure 6. Development of the idealized sequence shown in Figure 5, as proposed by Walker, Duke and Leckie (1983). Note that our added parallel laminated division may originate from unidirectional or bi-directional flows, and that in this sequence, it is suggested that hummocky cross stratification can grow upward from a flat bed. This contrasts with the Dott and Bourgeois (1982) model of Figure 5, where the hummocky and swaley laminae are shown as invariably draping a previously-scoured bed.

## Bioturbation

In both the HCS sst/mst and Bouma sst/mst facies of the shelf, the interbedded mudstones are normally extensively bioturbated. The faunal behavior is typically grazing and mining. In the Cretaceous rocks of the Western Interior Seaway, deposit feeders (Pemberton and Frey, 1983, p. 85) include Planolites, Rosselia, Zoophycos, Muensteria, Phoebichnus, Cylindrichnus, Chondrites, Thalassinoides, Rhizocorallium, ?Gyrochorte and ?Cochlichnus. Single vertical tubes are sparsely developed in the sandstones, and concave up spreite suggest these are "escape burrows" formed by organisms that were rapidly buried by sand, and which survived the experience and tried to escape to the new sediment-water interface.

## Conclusions

These two facies are not normally difficult to separate -- the wave-formed structures (HCS, small scale oscillation ripples) characterize the HCS sst/mst and unidirectional flow structures (upper plane bed, current ripples) characterize the Bouma sst/mst facies. In many cases, the mechanism of sand emplacement may be the same, namely turbidity currents (see Geological Evidence for Storm Transportation, this volume, for the evidence). It must also be emphasized that storm-generated geostrophic flows may deposit sediment from suspension in a manner indistinguishable from turbidity currents. In fact, such flows would be behaving _locally_ just like turbidity currents, except that they would be driven by storm generated forces, not solely by gravity. My use of the term Bouma sst/mst facies therefore describes a deposit, but does not necessarily imply that it is solely the deposit of a gravity-driven turbidity current.

## 2. COMPARISON: SHELF AND SUBMARINE FAN VERTICAL SEQUENCES

This comparison involves sequences of beds. In submarine fan environments, sequences involve various turbidite facies but the sequence as a whole represents environments consistently below storm wave base. In ancient shallow marine situations, the sequences are mostly progradational, and involve changes (seaward to landward) from turbidite through hummocky cross stratified beds into fairweather-dominated sands, and thence commonly into shoreline and sub-aerial facies.

## Submarine Fan Facies Sequences

Fan sequences were first emphasized by Mutti and Ghibaudo (1972). They defined two different types, now normally termed 1) thickening-upward and 2) thinning-upward. These terms refer to trends of sand bed thickness; commonly, as individual sand beds become thinner, the sand/shale ratio also decreases, and vice versa.

## 1. Thickening-upward sequences

Thickening-upward sequences (Fig. 7) are normally interpreted as the result of lobe progradation and/or aggradation. Two lobes, called suprafan lobes, are shown in Figure 2. Local progradation of the lobe fringe may develop a thickening-upward sequence only a few meters thick, and composed mostly of relatively thin classical turbidites. Progradation of an entire lobe may result in a sequence a few tens of meters thick, which begins with classical turbidites, but may pass upward into massive or pebbly sandstones.

Figure 7. A thickening-upward turbidite sequence from the Ordovician Cloridorme Formation, Grande Vallee, Quebec. The relatively thin beds involved in this sequence, and the rather thin sequence itself suggest that this is a small lobe-fringe prograding sequence. Beds are overturned, and stratigraphic top is to left.

## 2. Thinning-upward sequences

Thinning-upward sequences (Fig. 8) are normally interpreted as channel fills. They begin with thick massive sandstones and pass upward into thin-bedded classical turbidites, in sequences up to several tens of meters in thickness. However, there are some thinning upward sequences which involve only classical turbidites, with spectacular parallel bedding and no indication of topography (channelling) on the sea floor (Fig. 9). Rather than suggesting channels, these sequences may indicate overbank deposition adjacent to a major channel, with progressive lateral shift of the channel away from the depositional area. Alternatively, the thinning-upward may be due to the gradual shifting of a depositional lobe away from a specific point in the basin, the sequence thereby representing a change from lobe center to lobe fringe.

There is now a large literature on turbidite sequences (Mutti and Ghibaudo, 1972; Ricci Lucchi, 1975; Walker, 1978; Ghibaudo, 1980; Hiscott, 1981); for our purposes here, the important point is that rarely, if ever, do submarine fan facies sequences shallow to the point of approaching storm wave base. The sequences are dominated by classical turbidites, massive sandstones and pebbly sandtones (Walker, 1978), with few indications of other current activity.

Figure 8. A thinning-upward turbidite sequence from the Cretaceous rocks of the Simi Hills, California. Sequence begins with the massive sandstone facies, and passes upward through classical turbidites into mudstones.

Figure 9. Thinning-upward sequence of turbidites at Shelter Cove, Point San Pedro, California (Paleocene). Note the lateral continuity of the beds and the fact that they are all relatively thin. In this photograph, there is no overall suggestion of channelling, nor are the beds at the base of the sequence suggestive of channel filling (compare with Fig. 8). It is possible that, in contrast to channelling, the sequence could imply overbank deposition adjacent to a major channel, with gradual lateral shift of the channel away from the depositional area. Alternatively, the sequence could imply the the gradual lateral shift of a depositional lobe. The thicker beds at the base would then represent a central position on the lobe, whereas the thinner, higher beds would represent the lobe margin.

## Shallow marine/shelf facies sequences

Shelf sequences are most commonly developed during shoreline progradation, although there are many examples of shelf coarsening-upward sequences which were deposited tens of kilometers from the closest shoreline; see for example, the two papers in this volume by Tillman and Martinsen on the Shannon Sandstone and Hartzog Draw Field. Some of the coarsening-upward sequences may have formed during regional sea floor aggradation rather than shoreline progradation.

I will consider first prograding, storm-influenced sequences, because these are the most closely comparable to thickening-upward turbidite sequences. I will later consider coarsening-upward sequences which do not appear to show any storm influence.

### Prograding storm-influenced sequences

Prograding storm-influenced sequences (Fig. 1) can be broadly subdivided into four parts.

Top   4)   Shoreline

         3)   Shoreface, above fairweather wave base

         2)   Storm wave base to fairweather wave base section (with HCS sst/mst facies)

Base  1)   Below storm wave base (with Bouma sst/mst facies)

The basal part, the shelf turbidites below storm wave base, may not always be present in these sequences and parts 2, 3, and 4 can be taken as defining the sequence. I suggest that there are two end-member sequences with a gradation between end members, depending on the nature of the shoreface facies.

    1)   In fairweather-shoreface systems, the tidal and alongshore fairweather currents dominate, and any storm disruption of the shoreface is repaired by the day-by-day processes. Such a shoreface is dominated by angle of repose, small to medium scale cross bedding. The overall sequence might therefore be:

        Top   4)   Beach, penetrated by roots

                3)   Cross-bedded shoreface sandstones

                2)   Interbedded hummocky cross stratified sandstones and mudstones

        Base 1)   Turbidites and/or offshore mudstones.

This sequence is typified by the Gallup Sandstone sequence described by Harms et al. (1975) and shown here in Figure 10.

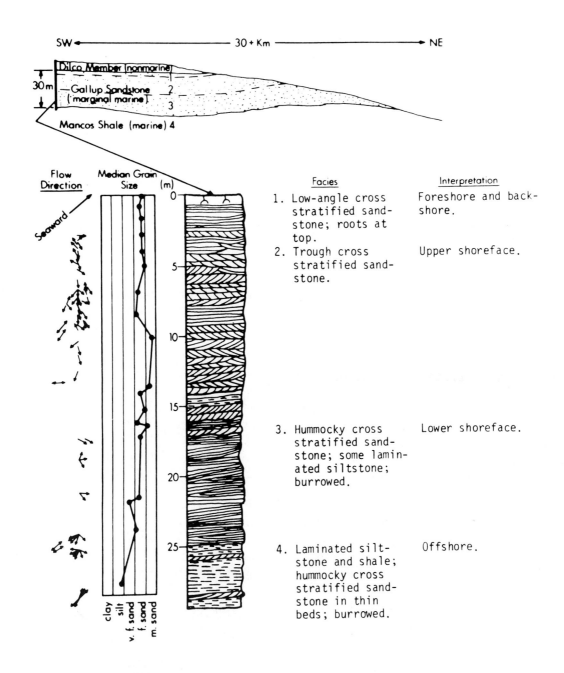

Figure 10. Coarsening-upward prograding shoreline sequence from the Gallup Sandstone, New Mexico, from Harms et al. (1975). The sequence begins with alternations of hummocky cross stratified sandstones and bioturbated siltstones and shales, grades up into cross bedded shoreface sandstones, and terminates in low angle cross stratified (beach) foreshore sandstones penetrated by root traces.

2) In storm-shoreface systems, the shoreface is so dominated by storms that there is little preservation of the fairweather record. Day-by-day tidally formed cross bedding is thoroughly and regularly reworked by storm waves, and instead of cross bedding, amalgamated hummocky cross stratification and swaley cross stratification (Leckie and Walker, 1982; Walker, 1982) is preserved. If these storm-formed structures are reworked by the fairweather currents, they are thoroughly reworked again by the next storm. The overall sequence (Fig. 11) might therefore be:

Top 4) Beach, penetrated by roots

3) Amalgamated HCS and swaley cross stratification, of the shoreface

2) Interbedded HCS-sandstones and mudstones

Base 1) Turbidites and/or offshore mudstones.

There will be all gradations between these two end members, depending on the relative proportions of fairweather and storm sedimentary structures preserved on the shoreface (Fig. 12).

Two facies terms, amalgamated hummocky cross stratification and swaley cross stratification, have been introduced here without definition. Amalgamated HCS results from the stacking of individual HCS sands with little or no preservation of interbedded mudstones. Sand bodies several meters thick can be developed (Figure 5 in Geological Evidence for Storm Transportation, this volume). It has been our experience in the Cretaceous Seaway of Alberta that as amalgamated HCS sand bodies become thicker, the sedimentary structures gradually change. The convex-up hummocky surfaces tend to disapper, leaving only the concave-up swales -- hence the term swaley cross stratification (SCS) (Figs. 13, 14). Within SCS sandstones, which can also be up to about 20 meters thick, there may be a few convex-up surfaces suggesting a descriptive link, and probably a genetic link with hummocky cross stratification. There may also be some angle of repose, medium scale cross beds indicating some preservation of the fairweather record. A fuller comparison of the geometries of HCS, SCS and angle of repose cross bedding is given by Walker (1982).

Several examples of the turbidite-HCS-SCS-shoreline sequence (or THSS sequence) have been identified in Alberta, particularly:

1) U. Cretaceous Wapiabi-Belly River transition (Fig. 11) at Lundbreck Falls (Walker and Hunter, 1982, p. 70), Trap Creek and the Highwood River (Fig. 12; Walker and Hunter, 1982, p. 61-71).

2) U. Jurassic Fernie-Kootenay transition (Passage Beds), Hamblin and Walker, 1979. Note that the lower part of facies D (Fig. 15), originally identified as beach, is now believed to be SCS -- it is well exposed in 1983 new road cuts at Banff, Alberta. This example is discussed in detail in "Geological Evidence for Storm Transportation and Deposition on Ancient Shelves", this volume.

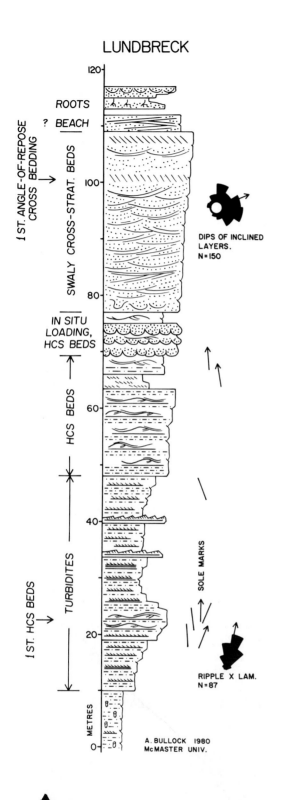

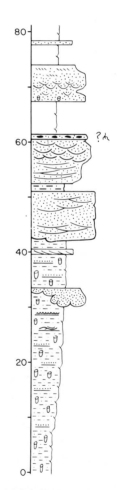

Figure 12. Coarsening-upward prograding shoreline sequence at the Highwood River, Alberta. Section shows the Upper Cretaceous Hanson-Chungo transition, Wapiabi Formation. Note the mudstone (0-8 m), turbidite (8-23 m), HCS (23-42 m), SCS (42-57 m) sequence, with a thin trough cross bedded sequence representing tidal or alongshore currents (57-61 m). Above 61 m, the section is fluvial. Note channeling and loading at about 33 m, which is taken as the base of the Chungo. From Walker and Hunter, 1982, p. 70.

Figure 11. Coarsening-upward prograding shoreline sequence at Lundbreck Falls, southern Alberta. The section is equivalent to the Chungo Member of the Wapiabi Formation (U. Cretaceous). Note the overall progression from Wapiabi black shales into turbidites, hummocky cross stratified sandstones, thick swaley cross stratified sandstones (see Fig. 14) (with rare angle-of-repose cross bedding near the top) and finally non-marine rocks. From Bullock, 1981.

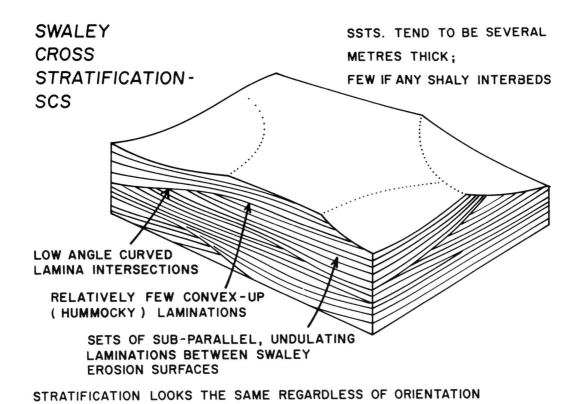

Figure 13. Block diagram of swaley cross stratification, showing broad swales, very few hummocks, and sets of sub-parallel to undulating laminations between the swaley scours. From Walker, 1982.

Figure 14. Swaley cross stratification (swales arrowed, portion of hummock to left of scale) from Permian rocks of New South Wales, Australia.

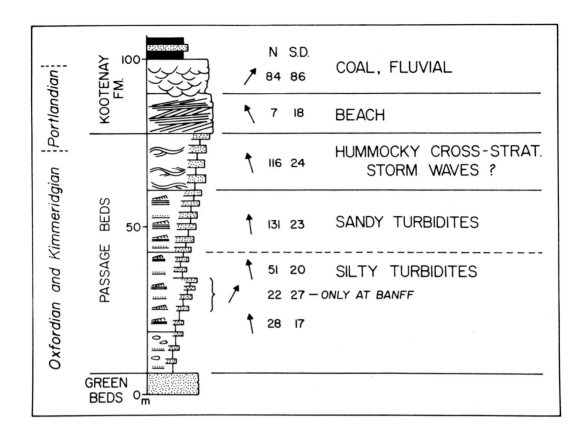

Figure 15. Stratigraphic section of the Upper Jurassic "Passage Beds" (transitional from Fernie black shales to Kootenay coal bearing rocks) at Banff, Alberta. Note sequence from turbidites into HCS sandstones, with identical paleoflow directions for both of these facies. New (1983) road cuts show that facies D, originally interpreted as beach, is probably swaley cross stratified and contains an in situ coal a meter or two thick within the sand body. From Hamblin and Walker (1979).

---

3) Upper Cretaceous Cardium Formation in outcrop north of Grande Cache, Alberta (unpublished).

4) Lower Cretaceous Moosebar-Gates transition, northeastern B.C. (Leckie and Walker, 1982).

On a recent field trip, I observed a very similar sequence in the Kenilworth Member of the Blackhawk Formation (Mesaverde Group) near Woodside, Utah. The lower beds resembled turbidites and contained Bouma sequences, but some beds also contained HCS and symmetrical small scale ripples. These beds were overlain by the HCS sst/mst facies, which in turn was overlain by a swaley cross stratified lower shoreface sequence. The upper shoreface contained some bi-directional angle-of-repose cross bedding below a possible beach. The ?beach was abruptly and transgressively overlain by bioturbated marine sandstones. Prominent developments of swaley cross stratification in lower shoreface environments above hummocky cross stratification have also been noted in the Chungo Member (Wapiabi Formation) at Mt. Yamnuska (Lerand, 1982) and in the Milk River Sandstone at Writing-on-Stone Provincial Park, Alberta (McCrory, 1984).

The turbidite-hummocky-swaley-shoreline (THSS) sequence suggests a genetic link between facies -- that is, the storm domination of the shoreface would facilitate the generation of turbidity currents that could then emplace the HCS sst/mst and Bouma sst/mst facies. This is basically what the model shown in Figure 16 suggests. If so, the sequence has important predictive possibilities, namely that the recognition in core or outcrop of several meters of swaley cross stratification directly overlain by beach facies penetrated by roots, predicts HCS sst/mst and Bouma sst/mst facies in the seaward direction, or stratigraphically down section. This may open up many exploration possibilities, especially as turbidity-current-emplaced hummocky cross stratified sandstones are known to form important reservoirs tens of km offshore (Cardium Formation 4, Walker, this volume). Thus, a storm-dominated, swaley cross stratified shoreface could predict sand bodies in an otherwise unexplored marine mudstone basin, way offshore.

---

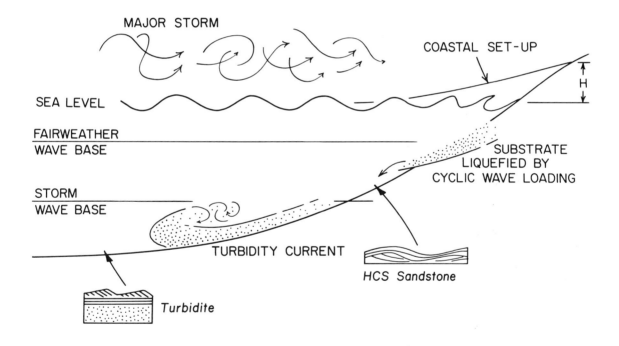

Figure 16. Model for the generation of a turbidity current by a major storm or hurricane. Storm waves have caused a coastal set-up (the 7 m figure is from Hurricane Carla), and have liquified the substrate. Sediment has flowed, accelerated, and developed into a turbidity current. This modification of an earlier model is discussed in more detail by Walker (Geological Evidence for Storm Transportation, this volume). In stratigraphic sequence, storm-dominated SCS sands of the shoreface may predict HCS sands and turbidites farther out in the basin, or lower in the stratigraphic sequence.

## 3. IDENTIFICATION OF PROGRADING TURBIDITE LOBE, AND SHELF TO SHORELINE SEQUENCES

The thickening-upward turbidite sequences and shelf-to-shoreline sequences are all commonly a few tens of meters thick. They all begin with interbedded sandstones and mudstones, and become progressively sandier upward. It is unlikely that they could be distinguished using well logs alone.

In outcrop, the shoreface and beach facies would be easily distinguished from the massive or pebbly sandstones at the top of the turbidite sequence, as would the abundance of wave formed sedimentary structures in the shelf-to-shoreline sequence. In cores, the distinction might not be so easy. If the shoreface were fairweather-dominated, the abundance of medium scale cross bedding in the cores would identify a shelf-to-shoreline sequence. At the same point in a submarine fan prograding lobe sequence, the sandstones would be dominantly massive. In the THSS sequences, with storm-dominated upper shoreface, the abundance of low angle (<10°) stratification in the cores should separate HCS and SCS from the structureless turbidite massive sandstones.

### Thinning- and Fining-Upward Sequences

These are well known from submarine fans, and have commonly been associated with channel filling by analogy with delta plain channels (Mutti and Ghibaudo, 1972). These sequences are up to a few tens of meters in thickness, and may begin with a channel lag -- either a conglomerate, or a pebbly or massive sandstone. The turbidite origin is normally obvious, because the beds are either massive sandstones or contain Bouma sequences, and wave-formed structures are absent. In the thinning-upward sequences associated with lobe switching, the same comments regarding recognition of turbidite beds apply.

In shallow marine/shelf situations, thick transgressive (or fining-upward) sequences are uncommon. This is because normally, transgression results in a base-level rise for the rivers, hence aggradation of the alluvial plain, and cut off of clastic supply at the shoreline. Transgression tends to be associated with the development of relatively thin (few meters at the most) sequences, and with erosion and winnowing of the transgressed surfaces.

The one outstanding example of a thinning- and fining-upward storm influenced shelf sequence is the Upper Cretaceous Cape Sebastian Sandstone of Oregon (Bourgeois, 1980; Hunter and Clifton, 1982). The sequence is about 200 m thick (Fig. 17), and can be divided into four parts:

Top 4) <u>Parallel laminated and burrowed facies</u>, characterized by "alternating parallel laminated very fine sandstone and burrowed, organic-rich sandy siltstone" (Bourgeois, 1980, p. 691). Symmetrical ripples are absent (p. 698).

3) <u>Upper hummocky bedded and burrowed facies</u>, characterized by "hummocky bedded fine-grained sandstone [grading] up into very fine-grained, finely laminated sandstone, succeeded by burrowed silty sandstone" (Bourgeois, 1980, p. 689).

2) <u>Lower hummocky bedded facies</u>, characterized by "hummocky bedded, medium- to fine-grained sandstone with no burrowed zones" (p. 685) (i.e., amalgamated HCS).

Base 1) <u>Conglomerate facies</u>, characterized by a "basal conglomerate overlain by trough cross bedded, plane bedded and pebbly coarse sandstone" (p. 682).

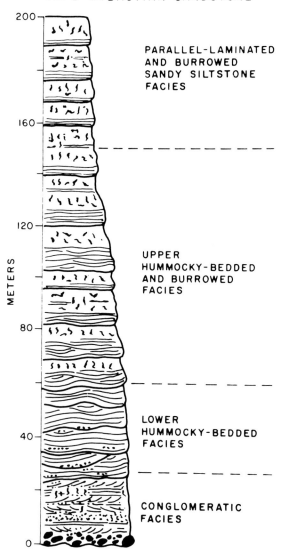

Figure 17. Composite section of the Cape Sebastian Sandstone, southwestern Oregon. From Bourgeois, 1980.

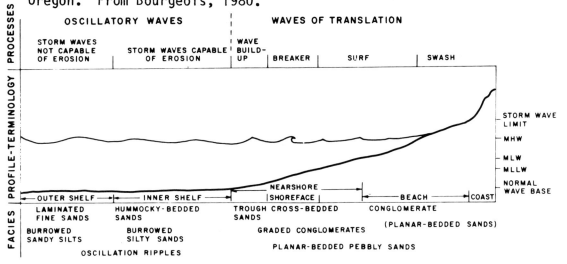

Figure 18. Cape Sebastian Sandstone facies related to shelf, nearshore, beach and coastal depositional environments. From Bourgeois, 1980.

Bourgeois (1980, p. 693; Fig. 18) interprets the conglomeratic facies as beach to shoreface deposition, the HCS facies as a "shelf-storm feature" (p. 695), with the "hummocky bedded to burrowed sequence" (p. 693 - presumably the same as the "upper hummocky bedded and burrowed facies") representing "sedimentation in the shelf zone where the bottom is scoured only by major storms". The parallel laminated and burrowed facies "is believed to have been deposited on the outer shelf, seaward of the erosional influence of nearly all storm waves" (p. 697-8). Bourgeois concludes (1980, p. 699) that the "laminated fine sands of the Cape Sebastian Sandstone probably accumulated relatively rapidly during or following storms, by combined wave, wind-drift and post-storm-surge density-driven mechanisms".

The parallel laminated and burrowed facies might perhaps be emplaced by turbidity currents, and the whole 200 m thick sequence could be compared to an _inverted_ version of a THSS sequence:

Top 4) parallel laminated and burrowed facies - turbidites

3) upper hummocky bedded and burrowed facies - HCS

2) lower hummocky bedded facies - amalgamated HCS and SCS

Base 1) conglomerate - shoreline gravels and breaker-surf zone cross bedded sands.

Bourgeois (1980, p. 699) convincingly argues for the transgressive nature of the sequence and, realizing how uncommon it is, suggests that "on an active continental margin, high rates of sediment supply and tectonic activity may easily produce deposition during a transgression".

A sequence such as the Cape Sebastian Sandstone could easily be distinguished from a thinning-upward turbidite sequence in outcrop or core, using the criteria discussed above. However, from well logs alone, a sequence with diminishing amounts of sandstone progressively upward in a suspected marine environment is _much_ more likely to be a submarine fan channel fill or lobe switching sequence, than a rare analogy of the Cape Sebastian Sandstone.

## 4. COMPARISON: OVERALL GEOMETRY OF SUBMARINE FAN AND SHELF SANDSTONE FACIES

So far, I have compared submarine fan and shelf environments at the level of individual beds, and of bedding sequences. A brief comparison of broad geometrical aspects may also be useful.

### 1. Submarine Fan Facies

Figure 2 shows an idealized submarine fan. It is based on a large number of studies of ancient and modern fans. Although there is still considerable discussion of variations from this model, it serves to highlight the tripartite structure of submarine fans. The genesis of fan models has been discussed by Nilsen (1980) in a discussion of Walker (1978), with a commentary on fan model evolution by Walker (1980).

Fans are fed by feeder channels or submarine canyons. Deposition begins where turbidity currents slow down at the foot of the slope, and begin to flow on the flatter basin floor. The foot-of-slope depocenter (fan) has an <u>upper (or inner) portion</u> characterized by a <u>single channel with levees.</u> Here, channel depths range from about 50 to 500 m. The bottom of the main inner fan channel may be smoothed by sedimentation, with an incised meandering thalweg. In La Jolla Fan, California, the thalweg is about 20 m deep and 300 m wide. The main inner fan channel is only a few km wide, and each levee may also be a few km wide. Slopes on the back of the levees may be quite steep -- a few degrees, as on Laurentian Cone (Normark et al., 1983). Levees are fed by spillover of suspended turbidity current fines (mostly silt and clay), and are mostly fine grained. The inner fan channel is probably too deep to fill with a single, progressively fining-upward sequence; one of the thickest single fining-upward sequences in the geological record is only of the order of 50-60 m thick (Grosses Roches, Ordovician, Gaspe, Quebec; Hendry, 1978). Levees are characterized by relatively thin and fine-grained turbidites, commonly with evidence of local substrate erosion (rip-up clasts), climbing ripples, convolute lamination and local slumping (Walker, 1985; Advocate and Link, 1981).

The <u>mid-fan area</u> is characterized by suprafan lobes (Normark, 1978; Normark, Piper and Hess, 1979) a few km to a few tens of km in diameter. The preserved thickness of beds associated with these lobes is of the order of several tens of meters. The upper parts of the lobes are commonly channelized, with channel depths decreasing from a few tens of meters to zero by mid-lobe. Likewise, levee heights decrease to zero and the outer parts of the suprafan lobes are smooth. Broad, smooth fan areas are the depositional sites for regionally extensive classical turbidites. The mid-fan area is gradually built up by growth of a lobe, and then rapid lobe switching to a new area lower on the fan surface. The best example is Navy Fan (Normark et al., 1979), where six distinct suprafan lobes can be delineated in the mid-fan area.

The <u>lower (or outer) fan area</u> is gradational between the mid-fan and basin plain. Areas in front of active mid fan lobes build up whilst other areas may be inactive, receiving only fine suspended mud. Consequently, active lower fan areas probably develop thickening-upward turbidite sequences by a combination of aggradation and progradation.

Hypothetically, an entire fan may prograde, giving a sequence similar to that in Figure 19. This sequence shows outer fan deposits overlain in turn by suprafan lobe deposits and an inner fan channel fill. If progradation continued, the entire fan could be buried by slope deposits (Fig. 20, Walker, 1966), which in turn may be overlain by shallow marine or deltaic sands.

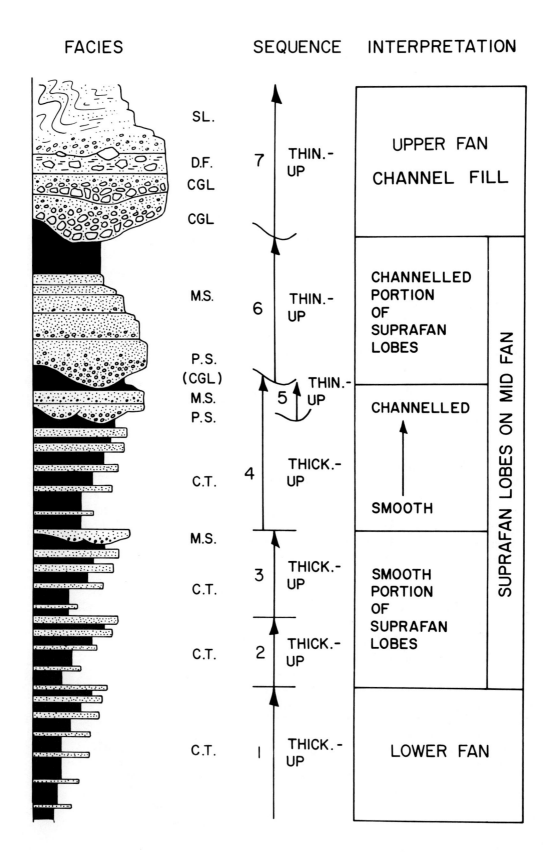

Figure 19. Hypothetical sequence produced by progradation of a submarine fan. F-U indicates fining (or thinning) upward sequences, C-U indicates coarsening (or thickening) upward sequences. Facies are classical turbidites (C.T.), massive sandstones (M.S.), pebbly sandstones (P.S.), conglomerates (CGL.), debris flows (D.F.) and slumps (SL.). Compare with fan model of Figure 2. From Walker, 1978.

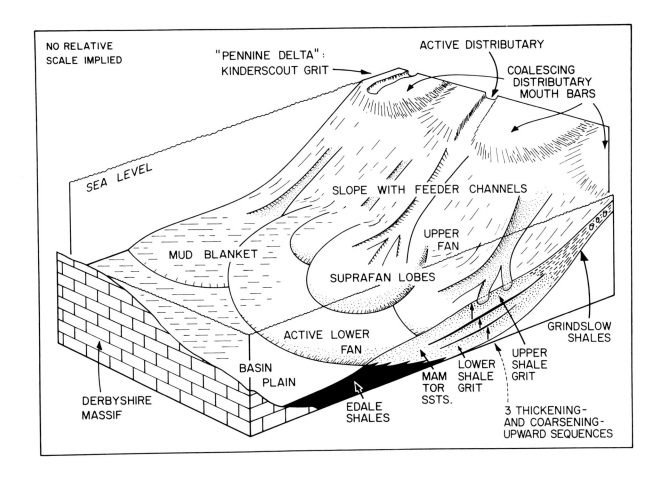

Figure 20. Interpretation of the Shale Grit (Pennsylvanian, northern England) as a submarine fan complex. The Grindslow Shales, which overlie the Shale Grit, are interpreted as prograding slope deposits about 100-120 m thick -- they are overlain in turn by the shallow water deltaic complex of the Kinderscout Grit (redrawn from Walker, 1966 in Walker, 1978). Note that in true turbidite systems, the main sand depocenters may be separated from their source sands (deltas, etc.) by a prograding dominantly muddy slope. This may not be the case for shelf sands (see Fig. 22).

Submarine Fan Associations - Conclusions

Prograding fan sequences can be recognized by superimposed thickening-upward sequences (fan fringe, suprafan lobe), cut by thinning-upward sequences (lobe and inner fan channels). Varying proportions of lobe and channel deposits may suggest variations on the simple fan model of Figure 2 (see Mutti, 1979; Link and Nilsen, 1980). This type of fan development can possibly be identified from electric logs, as in the Devonian example from Bradford Field, Pennsylvania (Fig. 21) -- the environmental interpretations are my suggestions.

In an unconfined turbidite basin, input from one point (the base of the feeder channel) will tend to product dip-oriented sand bodies, especially the channelized ones. This contrasts with the strike-oriented sand bodies which are more characteristic of the shelf.

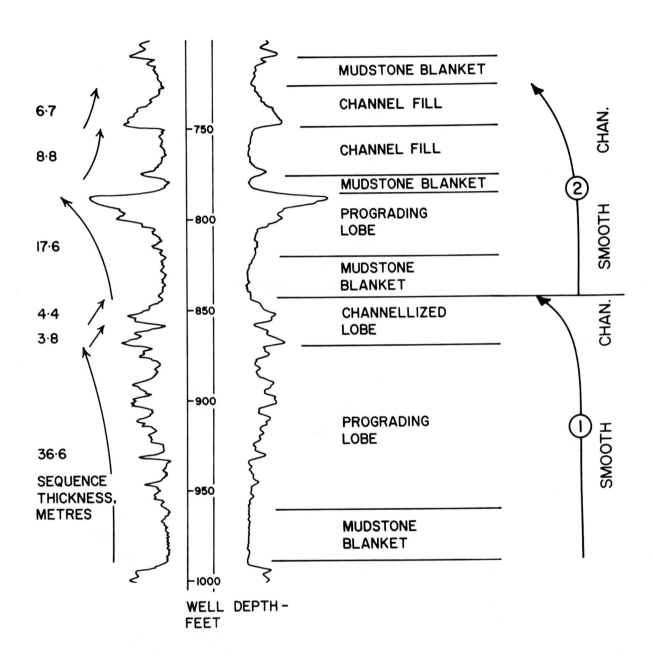

Figure 21. SP and resistivity logs of a Devonian sequence from Pennsylvania. I have suggested an overall prograding fan interpretation based on log shape and thickness of interpreted lobes and channels.

## 2. Shelf Sand Facies and Geometry

In shallow marine/shelf systems, channels and distinct switching lobes appear to be rare. Instead, many shallow marine sand bodies are described as "offshore bars" (Fig. 22), implying elongate bodies a few km wide, a few tens of km long, and perhaps 10-40 m thick (Shannon and Sussex sand bodies in Wyoming; Spearing, 1976; Berg, 1975; Brenner, 1978; Hobson et al., 1982; Tillman and Martinsen, this volume) and Viking and Cardium sand bodies in Alberta (Walker, 1983b, c; Koldijk, 1976). These sand bodies commonly lie many tens of km from the nearest possible shoreline (Shannon and Sussex, Spearing, 1976; Cardium, Walker, 1983c), but there is little concensus of opinion as to 1) how the sands were transported so far offshore, and 2) why they end up in long narrow sand bodies. Some answers are now possible in the case of the Cardium, although the answers in turn raise more problems.

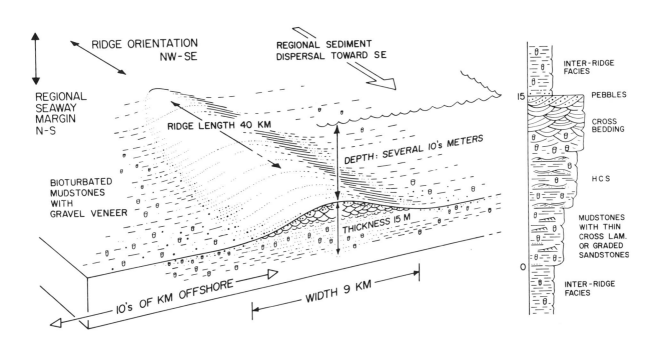

Figure 22. A summary of the general characteristics of shelf "bars" or "ridges" from the Cretaceous Western Interior Seaway. Data base for summary given by Walker, 1984.

Consider the following:

1) Cardium outcrops and cores suggest sand emplacement by turbidity currents, with storm reworking of the deposits above storm wave base to form hummocky cross stratification (Walker; in press; and this volume).

2) If this is correct, why should the turbidity currents form long narrow sand bodies? -- one would expect the flows to spread laterally on the shelf to form extensive sheet like deposits.

3) Detailed correlations between Cardium fields (Walker, 1983b) shows that some of the sands (especially the "A sand" or Raven River Member) are in fact sheet-like. It is only the zones of oil and gas production that are long and narrow within the sheet sands.

4) The sheet sands can be envisaged as a large floor covered in newspapers. Each newspaper is a single bed, but the whole accumulation of overlapping newspapers makes up the sheet sand.

5) The sheet-like, rather than long and narrow geometry, is much more compatible with a turbidity current origin for sands in the sheets. But it changes the fundamental question one asks -- instead of "what makes long narrow sand bodies?" (offshore bars), the question becomes "what controls reservoir quality within the sheet sands?". For the Cardium, this question remains unanswered.

It has been shown (Walker, 1983c and this volume) that the Cardium sheet sands may lie some 140 to 200 km downdrift. It has also been suggested (Walker, this volume and in press) that the Cardium field at Ricinus is a turbidity-current-filled channel, the channel most likely having been cut by turbidity currents. As well as raising the possibility of a new sheet-like Cardium field beyond the end of the Ricinus channel, it also raises the possibility of other as yet undiscovered Cardium channels.

Although the data base is very small, there are some important contrasts with respect to turbidite channels. Ricinus is very broad compared to its length -- it is about 55 km long and 5-8 km wide before palinspastic reconstruction; with reconstruction the width could double, or even triple. Its depth varies from 20-40 m, making it much wider and shallower than typical submarine fan channels. For example, the La Jolla inner fan channel is 1 to 2 km wide and 100-140 m deep (Buffington, 1964). There is no evidence of levees associated with Ricinus, and the fill shows no evidence of an overall thinning- and fining-upward sequence (see gamma ray logs in Walker, 1983b, and in press, and "Cardium Formation 4" paper, this volume, Fig. 18).

The most perplexing aspect of Ricinus is its setting within the Alberta Basin, especially with respect to the upstream end of the channel. In submarine fans, the channels can normally be traced back to existing or former positions of major feeder channels or canyons. Ricinus is not known to be related to a major feeder, nor to a "slope facies". It simply seems to appear on the basin floor, cutting into the Cardium "A sand" (or Raven River Member, Walker, 1983c). It may well be that much more drilling is needed at the northern end to help resolve the problem, although current work (Walker, in press) suggests that the upstream end of the channel is faulted off and should be present in the Foothills Disturbed Belt rather than in the subsurface of the Plains.

The geometry of cross-bedded shelf sands is difficult to generalize. Some appear to be sheet-like, but others (Shannon, for example, Spearing, 1976) are long and narrow. However, they pose less of a problem in a shelf/ fan comparison because the abundance of cross bedding immediately identifies the shelf situation, as compared with the massive or Bouma-sequence-sandstones of the submarine fan. The very generalized model of Figure 22 could be improved by the work reported in this volume on various subdivisions of Shannon sand bodies. Here, a central bar facies and two distinct bar margin facies have been defined, as well as various interbar facies (Tillman; Gaynor and Swift; this volume).

## Overall Geometry - Conclusions

Fans are essentially fed by one feeder channel or canyon. The upper (or inner fan) channel in turn feeds suprafan lobes, which build up and then switch position, rather like the switching positions of Mississippi Delta lobes. An abandoned lobe is draped with a blanket of suspended fines as a new lobe grows elsewhere. The overall geometry might, therefore, be succinctly summarized as

1) upper fan; a leveed channel, bounded by fine-grained levees, feeding

2) mid-fan; an area of switching suprafan lobes, each one being a few km to a few tens of km in diameter, and up to a few tens of m thick. As one lobe is abandoned, a new one grows and the old one is draped by fines.

3) lower fan; a smooth area seaward of the suprafan lobes where beds are expected to be individually extensive.

This type of geometry and channel/lobe relationship is unknown in shallow marine/shelf situations. Here, sands are not apparently supplied to the depocenter by stable, long lived feeder channels. Thus, there is not a consistent channel-feeding-lobe situation (although Ricinus might turn out to be an exception). In the storm-generated turbidity current model (Fig. 16), sands are derived from different parts of the shoreline, depending on where the maximum turbulence and storm surge tide is developed. Beds may, therefore, be expected to spread over various parts of the seafloor (sheets of newspaper on the floor) rather than being consistently funnelled into specific lobes.

## 5. COMPARISON OF GENERATING PROCESSES FOR TURBIDITY CURRENTS ON SHELVES AND IN DEEP SEA BASINS

One of the major problems concerning the possibility of turbidity currents in basins such as the Cretaceous Western Interior Seaway is the problem of flow generation. Gradients near the shoreline and slopes into the basin both seem inadequate for major soft sediment slumps that could evolve into turbidity currents.

In modern deep ocean turbidity current settings, flows are characteristically generated in at least three distinct ways. The reader is referred to the literature cited for details.

1. <u>Earthquakes</u>. There are several examples of earthquakes that have caused major slumps which evolved into turbidity currents. The best known one was the Grand Banks (Newfoundland) earthquake of 1929, which generated a current that broke a series of submarine telegraph cables (Heezen and Ewing, 1952; Heezen et al., 1954; Heezen and Drake, 1964; Piper and Normark, 1982). The Orleansville (Algeria) earthquake of 1954 also generated a cable-breaking turbidity current (Heezen and Ewing, 1955); other turbidity currents were generated by earthquakes in Fiji (Houtz, 1960), Norway (Terzahgi, 1957) and Japan (Shepard, 1933).

2. <u>Rivers in Flood</u>. The classical example is the Congo River in Africa. Heezen et al. (1964) have shown that turbidity currents are generated at times when the river is in flood, during those years when the river is establishing a new path through its estuarine sand bodies. Sand is swept into the canyon head, generating turbidity currents that continue down canyon, breaking submarine cables in depths up to at least 2800 m. Slopes in the canyon head, averaged over the first 10 km, are only 1.43 degrees, compared with a slope of about 2.3 degrees in the epicentral area of the Grand Banks earthquake (see "Geological Evidence for Storm Transportation", this volume).

3. <u>Spontaneous Sediment Failures</u>. Unstable, rapidly deposited piles of sediment can fail spontaneously, generating slumps which may accelerate into turbidity currents. The theory has been discussed by Moore (1961) and Morgenstern (1967), and one possible example is the slump off the mouth of the Magdalena River (Colombia) in 1935 (Heezen, 1956; Kolla et al., 1984). This slump is known to have turned into a cable-breaking turbidity current. Most of the spontaneous sediment failues occur during times of lowered sea level.

In all of the above examples, the turbidity currents had a major slope (Continental Slope, delta front slope) on which to accelerate after initial flow generation. It is this major slope that appears to be absent in basins such as the Cretaceous Western Interior Seaway. However, possible slump-generating mechanisms <u>do</u> exist, and I will examine these before speculating on whether the slumps might accelerate and turn into turbidity currents.

1. <u>Earthquakes</u>. The Western Interior Seaway lay to the east of a Cordillera actively undergoing uplift during most of the Upper Jurassic and Cretaceous. Earthquakes were almost certainly present, associated both with accretion of exotic terranes and basinward advance of thrust slices from the rising mountains. The earthquake slump-generating mechanism seems likely.

2. <u>Rivers in Flood</u>. The rising mountains were probably the source of steep, rapid streams bringing large amounts of sediment eastward into the Seaway. There would seem to be no reason why a Congo-like generating mechanism should not have been effective.

3. **Spontaneous Sediment Failure.** Most of the sandstones in the Seaway are fine to very fine grained. It seems reasonable to suggest that the rivers flowing from the rising mountains might have constructed fine grained deltas, with rapid sedimentation, large volumes of trapped pore fluids, and hence high susceptibility to spontaneous failure (see, for example, Coleman et al., 1983). Failure would, of course, be aided and abetted by earthquakes.

These mechanisms all operate independently of major storms. In many stratigraphic units, the turbidite-like beds under discussion are all characterized by the presence of hummocky cross stratification. It is therefore tempting to search for processes that might 1), generate turbidity currents to transport sand into the basin and 2) imprint hummocky cross stratification on the deposit. These processes would be essentially related to major storms.

4. **Storm Surge Relaxation.** This mechanism for turbidity current generation was originally proposed by Hayes (1967), and was applied to the interpretation of ancient rocks by Hamblin and Walker, 1979, and Walker, 1979. Problems have been discussed in "Geological Evidence for Storm Transportation" (this volume); essentially, Swift (personal communication, 1984) argues that when storms suspend sand at the shoreline, the densities would be insufficient to generate turbidity currents. Instead, the relaxation flow would evolve into a shore-parallel geostrophic flow rather than a down-slope-flowing turbidity current.

5. **Cyclic Wave Loading.** I suggest one modification of the storm process discussed above which might overcome some of the problems of storm generation of turbidity currents near the shoreline. In 1969, Hurricane Camille indirectly caused the overturning of Shell's Platform B in Block 70 of the South Pass area, Mississippi Delta (Sterling and Strohbeck, 1975; Fig. 23). Surveys of the sea flow after the hurricane suggested that large volumes of sediment had been liquefied and had flowed, undermining some of the legs of the platform. The contours in Figure 23 indicate that parts of the seafloor were lowered, and that the sediment that was removed piled up in ridges elsewhere. A minimum volume of about $6.7 \times 10^7$ $m^3$ of sediment was involved. This example is discussed in more detail in "Geological Evidence for Storm Transportation" (this volume), and a model is presented there (Fig. 8) for turbidity current generation by cyclic wave loading. In the case of Block 70, the gradient on which the sediment flow took place was only about 0.4 degrees. It is not known whether sediment movement was restricted to flowage at the surface, or whether any of the liquefied sediment moved completely out of the area as a turbidity current.

Conditions in the Western Interior Seaway were probably ideal for such a mechanism to be important. The Cordillera would have rapidly supplied large volumes of fine grained sediment. This may have accumulated in large deltas

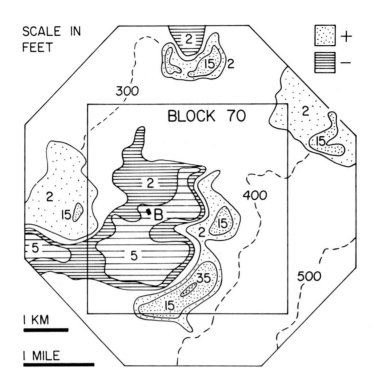

Figure 23. Lowering of sea floor in area of Platform B, South Pass Block 70, with sediment accumulation in nearby areas. Contours in feet, thicknesses of sediment removed or added in feet. Redrawn from Sterling and Strohbeck, 1975.

---

with high pore pressures, and appreciable delta front slopes. It seems reasonable to suggest that at times, storm-generated cyclic wave loading could have caused liquefaction, flowage, and possibly bulk slumping; these sediment movements may in turn have evolved into turbidity currents.

In order for any of the plausible slump-generating mechanisms to evolve into turbidity currents, the sediment mass must slump and accelerate to velocities capable of suspending sand before the excess pore pressure is dissipated (Morgenstern, 1967). Important controlling factors include the thickness of the slump, the pore pressure, the rate at which the slump is drained, and the slope. In the absence of sufficient data from recent sediments, we are forced to combine:

1) the geological evidence for turbidity currents in shallow seas with

2) facts regarding known slopes on which turbidity currents have been generated. For example, the slope in the epicentral area of the Grand Banks slump is 2.3 degrees, and the slope of the first 10 km of the Congo Canyon averages only 1.4 degrees. The slope reported by Coleman and Wright (1975, p. 141) for the area offshore of the

Sao Francisco Delta of Brazil is an amazing 11.2%, or about 6.4 degrees. It therefore does not seem so unreasonable to suggest that a slope of just 2 or 3 degrees might have existed at the margin of the Western Interior Seaway, particularly at times of active uplift (and hence sediment supply) in the Cordillera. Given one of the many generating mechanisms discussed above, it does appear possible that a turbidity current could be sustained, and even accelerate on a slope of 2 to 3 degrees. If it originated at the shoreline and flowed downslope for about 2 km, it would be in a depth of 70 to 105 m -- probably not outrageous depths for parts of the Western Interior Seaway.


## FINAL CONCLUSIONS

1) On the shelf, cross-bedded sands above fairweather wave base can probably be consistently distinguished from interbedded sandstone-mudstone facies.

2) Two different shelf sandstone-mudstone facies can be recognized: a) HCS sst/mst, and b) Bouma sst/mst. These can also be consistently distinguished by the wave-formed structures (HCS, small scale oscillation ripples) in the HCS sst/mst facies, and their absence in Bouma sst/mst facies.

3) Shelf and submarine fan sediments can be distinguished on the basis of their vertical facies sequences. In shelf sequences, a high proportion are prograding and at maximum development pass from turbidites into HCS sst/mst facies, then beds which indicate deposition above fairweather wave base, and finally shoreline deposits. This type of coarsening-upward sequence can be distinguished in core or outcrop from the turbidite coarsening-upward (or thickening-upward) sequence, which rarely or never approaches fairweather wave base let alone the shoreline.

4) Thinning-upward (or fining-upward) sequences develop in at least two situations on fans: a) channel fills, and b) prograding switching suprafan lobes. Thinning-upward sequences are extremely rare in shelf sequences.

5) Fans show a gross morphology due to their being fed from one feeder channel (or canyon), the fan building up by switching suprafan lobes. This contrasts with shelf sand morphologies, where feeder channels and canyons are almost unknown (the exact position of Ricinus channel with respect to shorelines and depositional lobes is unknown).

6) Generating processes include earthquakes, rivers in flood, spontaneous sediment failure, and storm-induced sediment failure. The latter may be particularly important as a generating mechanism for turbidity currents in shelf/shallow marine settings.

# REFERENCES

Advocate, D. M. and M. H. Link, 1981, Summit mobil home park field trip stop, in Link, M. H., R. L. Squires, and I. P. Colburn (eds.), Simi Hills Cretaceous turbidites, southern California: Pacific Section, Society of Economic Paleontologists and Mineralogists, Field Trip Guidebook (Oct. 1981), p. 125-128.

Berg, R. R., 1975, Depositional environment of Upper Cretaceous Sussex Sandstone, House Creek Field, Wyoming: American Association of Petroleum Geologists Bulletin, v. 59, p. 2099-2110.

Bourgeois, J., 1980, A transgressive shelf sequence exhibiting hummocky stratification: the Cape Sebastian Sandstone (Upper Cretaceous), southwestern Oregon: Journal of Sedimentary Petrology, v. 50, p. 681-702.

Brenner, R. L., 1978, Sussex sandstone of Wyoming - example of Cretaceous offshore sedimentation: American Association of Petroleum Geologists Bulletin, v. 62, p. 181-200.

Buffington, E. C., 1964, Structural control and precision bathymetry of La Jolla submarine canyon, California: Marine Geology, v. 1, p. 44-58.

Bullock, A., 1981, Sedimentation of the Wapiabi-Belly River transition (Upper Cretaceous) at Lundbreck Falls, Alberta: B.Sc. Thesis, McMaster University, Hamilton, Canada, 94 p.

Coleman, J. M., D. B. Prior, and J. F. Lindsay, 1983, Deltaic influences on shelfedge instability processes, in Stanley, D. J. and G. T. Moore, (eds.), The Shelfbreak: Critical Interface on Continental Margins: Society of Economic Paleontologists and Mineralogists, Special Publication 33, p. 121-137.

_____ and L. D. Wright, 1975, Modern river deltas: Variability of processes and sand bodies, in Broussard, M. L. (ed.), Deltas, Models for Exploration: Houston Geological Society, p. 98-149.

Dott, R. H., Jr. and J. Bourgeois, 1982, Hummocky stratification: significance of its variable bedding sequence: Bulletin of the Geological Society of America, v. 93, p. 663-680.

_____ and J. Bourgeois, 1983, Hummocky stratification: significance of its variable bedding sequence: reply: Bulletin of the Geological Society of America, v. 94, p. 1249-1251.

Ghibaudo, G., 1980, Deep sea fan deposits in the Macigno Formation (Middle-Upper Oligocene) of the Gordana Valley, northern Appennines, Italy: Journal of Sedimentary Petrology, v. 50, p. 723-742.

Hamblin, A. P. and R. G. Walker, 1979, Storm dominated shallow marine deposits: the Fernie-Kootenay (Jurassic) transition, southern Rocky Mountains: Canadian Journal of Earth Sciences, v. 16, p. 1673-1690.

Harms, J. C., D. R. Spearing, J. B. Southard, and R. G. Walker, 1975, Depositional environments as interpreted from primary sedimentary structures and stratification sequences: Tulsa, OK, Society of Economic Paleontologists and Mineralogists, Short Course 2, 161 p.

Hayes, M. O., 1967, Hurricanes as geological agents - case studies of Hurricanes Carla, 1961 and Cindy, 1963: Texas Bureau of Economic Geology, Report of Investigations 61, 56 p.

Heezen, B. C., 1956, Corrientes de turbidez del Rio Magdalena: Societa Geografica Colombia, Bull., v. 51/52, p. 135-143.

_____ and C. L. Drake, 1964, Grand Banks slump: American Association of Petroleum Geologists, Bulletin, v. 48, p. 221-225.

_____, D. B. Ericson, and M. Ewing, 1954, Further evidence for a turbidity current following the 1929 Grand Banks earthquake: Deep Sea Research, v. 1, p. 193-202.

_____ and M. Ewing, 1952, Turbidity currents and submarine slumps, and the 1929 Grand Banks earthquake: American Journal of Science, v. 250, p. 849-873.

_____ and M. Ewing, 1955, Orleansville earthquake and turbidity currents: American Association of Petroleum Geologists, Bulletin, v. 39, p. 2505-2514.

_____, R. J. Menzies, E. D. Schneider, W. M. Ewing, and N. C. L. Granelli, N.C.L., 1964, Congo submarine canyon: American Association of Petroleum Geologists, Bulletin, v. 48, p. 1126-1149.

Hendry, H. E., 1978, Cap des Rosiers Formation at Grosses Roches, Quebec - deposits of the mid-fan region on an Ordovician submarine fan: Canadian Journal of Earth Sciences, v. 15, p. 1472-1488.

Hiscott, R. N., 1981, Deep sea fan deposits in the Macigno Formation (Middle-Upper Oligocene) in the Gordana Valley, northern Appennines, Italy. Discussion: Journal of Sedimentary Petrology, v. 51, p. 1015-1021.

Hobson, J. P., Jr., M. L. Fowler, and E. A. Beaumont, 1982, Depositional and statistical exploration models, Upper Cretaceous offshore sandstone complex, Sussex Member, House Creek field, Wyoming: American Association of Petroleum Geologists Bulletin, v. 66, p. 689-707.

Houtz, R. E., 1960, The 1963 Suva earthquake and tsunami: Geol. Surv. Dept., Fiji Ref. Geol., Series 61, 13 p.

Hunter, R. E. and H. E. Clifton, 1982, Cyclic deposits and hummocky cross stratification of probable storm origin in Upper Cretaceous rocks of the Cape Sebastian area, southwestern Oregon: Journal of Sedimentary Petrology, v. 52, p. 127-143.

Koldijk, W. S., 1976, Gilby Viking B. A storm deposit, in Lerand, M. M. (ed.), The sedimentology of selected clastic oil and gas reservoirs in Alberta: Calgary, Alberta, Canadian Society of Petroleum Geologists, p. 62-77.

Kolla, V., Buffler, R. T. and Ladd, J. W., 1984, Seismic stratigraphy and sedimentation of Magdalena Fan, southern Colombian Basin, Caribbean Sea: American Association of Petroleum Geologists Bulletin, v. 68. p. 316-322.

Leckie, D. A. and R. G. Walker, 1982, Storm- and tide-dominated shorelines in Cretaceous Moosebar-Lower Gates interval -- outcrop equivalents of Deep Basin Gas Trap in western Canada: American Association of Petroleum Geologists Bulletin, v. 66, p. 138-157.

Lerand, M. M., 1982, Chungo (sandstone) Member, Wapiabi: Formation, at Mt. Yamnuska, Alberta, in Walker, R. G. (ed.), Clastic units of the Front Ranges, Foothills and Plains in the area between Field, B. C. and Drumheller, Alberta: Intl. Association of Sedimentologists, 11th International Congress on Sedimentology (Hamilton, Canada), Guidebook for Excursion 21A, p. 96-116.

Link, M. H. and T. H. Nilsen, 1980, The Rocks Sandstone, an Eocene sand-rich deep sea fan deposit, northern Santa Lucia Range, California: Journal of Sedimentary Petrology, v. 50, p. 583-601.

McCrory, V., 1984, Storm- and tide-dominated shoreface deposits, Milk River Formation (U. Cretaceous), southern Alberta: B.Sc. Thesis, McMaster University, Hamilton, Ontario, Canada, 101 p.

Moore, D. G., 1961, Submarine slumps: Journal of Sedimentary Petrology, v. 31, p. 343-357.

Morgenstern, N. R., 1967, Submarine slumping and the initiation of turbidity currents, in Richards, A. F. (ed.), Marine Geotechnique: Urbana, University of Illinois Press, p. 187-220.

Mutti, E., 1979, Turbidites et cones sous-marins profonds, in Homewood, P. (ed.). Sedimentation detritique (fluviatile, littorale et marine): Switzerland, Universite de Fribourg, Institute de Geologie, p. 359-419.

_____ and G. Ghibaudo, 1972, Un esempio di torbiditi di conoide sottomarina esterna: le Arenarie di San Salvatore (Formazione di Bobbio, Miocene) nell' Appennino di Piacenza: Memorie dell' Accademia delle Scienze di Torino. Classe di Scienze Fisiche, Matematiche e Naturale, Serie 4A, no. 16, 40 p.

Nilsen, T. H., 1980, Modern and ancient submarine fans: discussion of papers by R. G. Walker and W. R. Normark: Bulletin of the American Association of Petroleum Geologists, v. 64, p. 1094-1101.

Normark, W. R., 1978, Fan valleys, channels and depositional lobes on modern submarine fans: characters for recognition of sandy turbidite environments: American Association of Petroleum Geologists Bulletin, v. 62, p. 912-931.

_____, D.J.W. Piper, and G. R. Hess, 1979, Distributary channels, sand lobes, and mesotopography of Navy Submarine Fan, California Borderland, with applications to ancient fan sediments: Sedimentology, v. 26, p. 749-774.

_____, D.J.W. Piper, and D.A.V. Stow, 1983, Quaternary development of channels, levees and lobes on Middle Laurentian Fan: American Association of Petroleum Geologists Bulletin, v. 67, p. 1400-1409.

Pemberton, S. G. and R. W. Frey, 1983, Biogenic structures in Upper Cretaceous outcrops and cores: Calgary, Alberta, Canadian Society of Petroleum Geologists, Mesozoic of Middle North America, Field Trip Guidebook 8, 161 p.

Piper, D.J.W. and W. R. Normark, 1982, Effects of the 1929 Grand Banks earthquake on the Continental Slope off eastern Canada: Geological Survey of Canada, Paper 82-1B, p. 147-151.

Ricci Lucchi, F., 1975, Depositional cycles in two turbidite formations of northern Appennines: Journal of Sedimentary Petrology, v. 45, p. 3-43.

Shepard, F. P., 1933, Depth changes in Sagami Bay during the great Japanese earthquake: Journal of Geology, v. 41, p. 527-536.

Spearing, D. R., 1976, Upper Cretaceous Shannon Sandstone - an offshore shallow marine sand body: Wyoming Geological Association, 28th Annual Guidebook, p. 65-72.

Sterling, G. H. and G. E. Strohbeck, 1975, The failure of South Pass 70 Platform B in Hurricane Camille: Journal of Petroleum Technology, v. 27, p. 263-268.

Stride, A. H. (ed.), 1982, Offshore tidal sands: New York, Chapman and Hall, 222 p.

Terzaghi, K., 1957, Varieties of submarine sloape failure: Norwegian Geotech. Inst., Publication 25, p. 1-16.

Walker, R. G., 1966, Shale Grit and Grindslow Shales - transition from turbidite to shallow water sediments in the Upper Carboniferous of northern England: Journal of Sedimentary Petrology, v. 36, p. 90-114.

_____, 1975, Generalized facies models for resedimented conglomerates of turbidite association: Bulletin of the Geological Society of America, v. 86, p. 737-748.

_____, 1978, Deep water sandstone facies and ancient submarine fans: models for exploration for stratigraphic traps: American Association of Petroleum Geologists Bulletin, v. 62, p. 932-966.

_____, 1979, Facies models 7. Shallow marine sands, in Walker, R. G., (ed.) Facies Models: Geoscience Canada Reprint Series 1, p. 75-89.

_____, 1980, Modern and ancient submarine fans: reply: American Association of Petroleum Geologists Bulletin, v. 64, p. 1101-1108.

Walker, R. G., 1982, Hummocky and swaley cross stratification, in Walker, R. G. (ed.), Clastic units of the Front Ranges, Foothills and Plains in the area between Field B.C. and Drumheller, Alberta: International Association of Sedimentologists, 11th International Congress on Sedimentology (Hamilton, Canada, 1982). Guidebook for Excursion 21A, p. 22-30.

_____, 1983a, Cardium Formation 1. "Cardium a turbidity current deposit" (Beach, 1956) - a brief history of ideas: Bulletin of Canadian Petroleum Geology, v. 31, p. 205-212.

_____, 1983b, Cardium Formation 2. Sand body geometry and stratigraphy in the Garrington-Caroline-Ricinus area, Alberta - the "ragged blanket" model: Bulletin of Canadian Petroleum Geology, v. 31, p. 14-26.

_____, 1983c, Cardium Formation 3. Sedimentology and stratigraphy in the Garrington-Caroline area, Alberta: Bulletin of Canadian Petroleum Geology, v. 31, p. 213-230.

_____, 1984, Shelf and shallow marine sands, in Walker, R. G., (ed.), Facies Models, 2nd Edition: Geoscience Canada, Reprint Series 1, p. 141-170.

_____, 1985, Mudstones and thin bedded turbidites associated with the Upper Cretaceous Wheeler Gorge conglomerates, California: a possible channel-levee complex: Journal of Sedimentary Petrology, v. 55, p. 279-290.

_____, in press, Cardium Formation at Ricinus Field, Alberta: a channel cut and filled by turbidity currents in the Cretaceous Western Interior Seaway: American Association of Petroleum Geologists Bulletin.

_____, W. L. Duke, and D. A. Leckie, 1983, Hummocky stratification: significance of its variable bedding sequence: Discussion: Bulletin of the Geological Society of America, v. 94, p. 1245-1249.

_____ and D. F. Hunter, 1982, Transition, Wapiabi to Belly River Formation at Trap Creek, Alberta, in Walker, R. G. (ed.). Clastic units of the Front Ranges, Foothills and Plains in the area between Field, B.C. and Drumheller, Alberta: International Association of Sedimentologists, 11th International Congress on Sedimentology (Hamilton, Canada, 1982), Guidebook for Excursion 21A, p. 61-71.

SHELF SANDSTONES IN THE WOODBINE--EAGLE FORD INTERVAL,
EAST TEXAS: A REVIEW OF DEPOSITIONAL MODELS

Sandra Phillips and Donald J. P. Swift

ARCO Exploration and Technology Company, Exploration and
Production Research, Plano, TX

ABSTRACT

This paper reviews studies of Woodbine--Eagle Ford reservoir sandstones from the subsurface of East Texas and evaluates shelf sand depositional models in the light of recent studies of fluid and sediment dynamics on modern shelves. The application of fluid and sediment dynamical principles has reaffirmed some shelf depositional models, traditionally applied to the East Texas basin, but modifies or discredits others; in these cases, new models are proposed. Three distinct types of reservoir-quality shelf sandstones can be recognized in these studies; (1) sand ridge deposits, (2) tabular or sheet sandstones, and (3) lenticular (topographically controlled) sandstones. This preliminary classification is based on external sand body geometries, facies associations and facies distributions.

Sand ridge deposits occur at Kurten Field as stacked, en echelon, linear sandstone bodies deposited on the muddy shelf of the east side of the Cretaceous Interior Seaway. Sandstone bodies are asymmetric in cross-section with steeper eastern flanks and are elongated in a north-south direction. Sand ridge deposits at Kurten Field occur stratigraphically adjacent to deposits of the Harris Delta. Sand ridge deposition probably occurred in an inner to middle shelf environment during small scale transgressive episodes, possibly associated with the abandonment of delta lobes (autocyclic transgression). Intermittent, alongshelf, geostrophic flows appear to be the most likely mechanism of sand transport and deposition.

Tabular shelf sandstones occur in the lower Woodbine at Damascus Field as a complex of single to multistory thin beds within a dominantly shale section. Cores display stacked, massive to laminated, fining-upward sandstone sequences, with abundant soft-sediment deformation and primary structures indicating rapid sedimentation. Sandstones form a series of thin sheet-like deposits elongated across the strike of the paleoshoreline. Sand deposition took place in an inner to middle shelf environment, during a general period of shoreline regression. Deposition is suggested to have occurred in localized zones of alongshelf flow deceleration and expansion during storms. Bouma-like vertical sequences of primary structures in Damascus sandstones indicate that these beds are tempestites (i.e. suspension deposits produced by storm flows).

Lenticular shelf sandstones are present in the uppermost Eagle Ford (Sub-Clarksville) section in Grimes County, Texas. Fining-upward sandstone sequences consist of amalgamated, massive to cross-stratified beds with erosional bases, overlain by bioturbated shaley sandstones. Individual sandstone bodies have restricted areal extents and deposition appears to have been controlled by local, salt-related topographic lows. These Sub-Clarksville sands were apparently deposited during a regional transgression which succeeded a phase of sea level stillstand. Remobilization of the substrate by wind-forced storm currents during transgression appears to have formed broad erosional surfaces, accompanied by deposition of sands swept into zones of local flow deceleration. Transgressive sand ridges may have formed contemporaneously on other parts of the late Eagle Ford shelf.

## INTRODUCTION

Sands of the Woodbine--Eagle Ford groups of East Texas were deposited in a wide range of environments including terrigenous, fluvio-deltaic, strandplain, open marine shelf, and deep marine basin. These Upper Cretaceous sandstones have been prolific hydrocarbon producers in the East Texas area since the discovery of the giant East Texas Field in 1930. Hydrocarbon accumulations have been found in Woodbine--Eagle Ford sandstones from the entire range of depositional environments mentioned above. However the discovery of the estimated 100 million barrel Kurten Field in 1976 has stimulated a great deal of interest in Gulf Coast sandstones deposited in the open marine shelf sedimentary regime. This paper summarizes the results of a review of the literature concerned with the sedimentology and stratigraphy of shelf sandstones in the Woodbine--Eagle Ford section. Shelf sand depositional models proposed in the literature are subjected to a critical analysis in light of recent advances in fluid and sediment dynamics, and are reaffirmed or modified on this basis. This paper constitutes the preliminary results of an ongoing research project concerned with the sedimentology and seismic-stratigraphy of Upper Cretaceous clastics in East Texas.

Examples of shelf sandstone bodies in the Woodbine--Eagle Ford section have been described in the literature from subsurface data in Brazos, Grimes, Houston, Leon, Madison and Polk counties, Texas (Fig. 1). These studies have been largely concerned with the description of sandstone characteristics and reservoir properties from cores, and the determination of depositional environments (Bell, 1980; Turner and Conger, 1981, 1984; Barton, 1982a,b; Theiss, 1983). Review of these papers suggests a need to develop depositional models which are constrained not simply by the physical stratigraphy and sandstone characteristics as are those from the literature, but are also realistic in terms of the fluid and sediment dynamics of the shelf regime. Interpretation of the genesis of these deposits must be based on analysis of primary structures seen in subsurface cores, and also on actualistic analogy with depositional processes observed in modern shelf regimes.

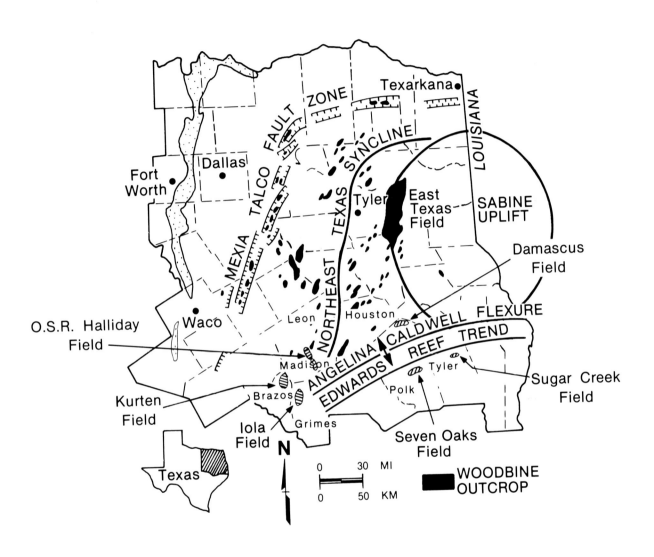

Fig. 1. Index map of northeast Texas showing major structural features and the Edwards reef trend. Woodbine--Eagle Ford shelf sandstone bodies discussed in this paper are from subsurface field areas indicated in Brazos, Grimes, Houston, Leon and Madison counties. Modified from Foss (1978).

The analysis of these ancient shelf deposits must be approached at two separate spatial and temporal scales. The analysis of individual sand bodies and facies relationships between sand bodies using closely-spaced well control in developed fields, undertaken in this paper, provides a basis for classification of different shelf sand body types. As a next step in our study, we will develop a regional time-stratigraphic and rock-stratigraphic framework. This will serve as the foundation for a comprehensive model for the depositional history and sequence stratigraphy of the Woodbine--Eagle Ford shelf sequence within the larger-scale prograding shelf-slope sedimentary wedge deposited on the subsiding continental margin.

This paper constitutes a first step toward such a model. It presents a preliminary classification of Woodbine--Eagle Ford shelf sandstones (Fig. 2), summarizes their characteristics and evaluates depositional mechanisms in light of our present understanding of fluid and sediment dynamics on modern continental shelves.

## STRATIGRAPHIC FRAMEWORK

Sands and shales of the Woodbine and Eagle Ford groups of East Texas (Fig. 3) were deposited in a regressive clastic wedge bounded above and below by the dominantly carbonate Cretaceous section. During Gulfian Cretaceous time a broad epeiric sea extended across the structurally low East Texas embayment, bounded to the west by the Central Texas platform, to the east by the Sabine uplift and separated from open ocean and deeper water to the south by the Angelina Caldwell flexure (Fig. 1). Planktonic-benthonic foraminifera ratios and seismic data (Stehli et al., 1972) show that the Angelina-Caldwell flexure and Edwards reef trend marked the seaward edge of the Lower Cretaceous continental shelf and the beginning of the continental slope. The exact configuration of the seaway during Cenomanian through Coniacian time, and the periods during which it was connected with the Western Interior Seaway are still problematical.

Throughout a large part of the East Texas basin the Woodbine and Eagle Ford clastic section is underlain unconformably by the Buda Limestone and overlain unconformably by the Austin Chalk. The regional extent of these unconformities, as well as many of the details of the physical and time-stratigraphy adjacent to these boundaries are not yet well defined. The clastic interval has been divided into numerous formal and informal units (Fig. 3) defined primarily on the basis of gross lithologic changes in updip areas (outcrop and landward margins) of the basin. The Woodbine and Eagle Ford sections have been clearly differentiated in updip areas by well log and biostratigraphic correlations. However, considerable uncertainty remains in recognizing the Eagle Ford--Woodbine boundary in downdip (basinward) regions (Barton, 1982a). Consequently, the term Woodbine--Eagle Ford will be used to refer to the interval from the base of the Austin Chalk to the

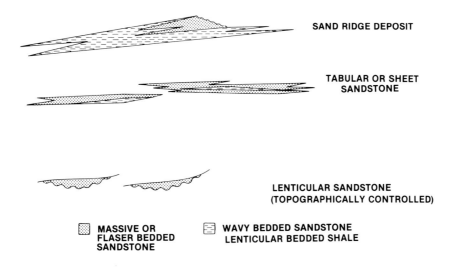

Fig. 2. Preliminary classification of shelf sandstone bodies recognized in the Woodbine--Eagle Ford section of East Texas. Sandstone types are delineated based on external geometries, facies associations and facies configurations.

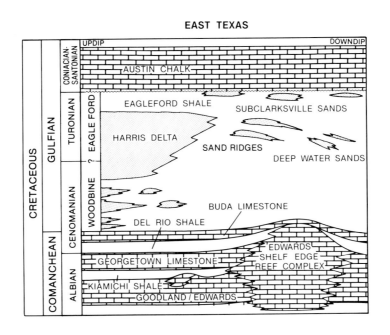

Fig. 3. Schematic stratigraphic nomenclature chart of upper Comanchean and lower Gulfian strata of East Texas. Stratigraphic relationships of Woodbine and Eagle Ford sandstones are shown diagramatically.

top of the Buda Limestone. This interval reaches a maximum thickness of 365 m (1,200 ft) of dominantly nonmarine sediments in the northern and central parts of the East Texas basin. The interval thins to less than 15 m (50 ft) in thickness above the Edwards reef trend in some areas (Siemers, 1978), and expands rapidly downdip of the Edwards shelf-slope break to over 457 m (1,500 ft) of dominantly shale in Polk and Tyler counties (Siemers, 1978; Foss, 1979).

Initial Woodbine deposition followed a period of major uplift in the East Texas area. Early Woodbine sediments were eroded from the Paleozoic rocks of the Ouachita foldbelt in southern Oklahoma and Arkansas (Nichols, 1964; Oliver, 1971), and deposited in the subsiding East Texas embayment and on the peneplain of Lower Cretaceous strata of the Sabine uplift (Halbouty and Halbouty, 1982).

Oliver (1971) concluded that the updip Woodbine consists of three principal depositional systems, the Dexter fluvial system, the Freestone wave-dominated delta system, and the Lewisville shelf-strandplain system (Fig. 4). Downdip Woodbine deposition (south of the Angelina-Caldwell flexure) was characterized by a mud-dominated clastic wedge associated with the prograding shelf margin. Slope sandstones within this wedge have been interpreted as coalescing submarine "fan lobes" in Tyler County (Siemers, 1978) and isolated channel and overbank turbidite deposits in Polk County (Foss, 1978).

During late Woodbine and early Eagle Ford time, the Sabine uplift was reactivated (Halbouty and Halbouty, 1982), resulting in erosion of previously deposited Woodbine strata from the crest and flanks of the uplift. Subsequent early Eagle Ford deposition was contemporaneous with the rise of the Sabine uplift, and the gradual shifting of its structural axis westward (Nichols, 1964). Eroded Woodbine sediments were transported to the southwest and have been interpreted alternatively as a prograding delta system (Oliver, 1971; Fig. 4 this paper), or as inner neritic and littoral deposits (Nichols, 1964). These sandstones, termed the "Harris Sand" or "Harris Delta" (Nichols, 1964; Oliver, 1971; Bell, 1980), have been the subject of considerable debate, and have been correlated with both the Woodbine and Eagle Ford formations. The sandstones form a fan-shaped clastic wedge which prograded southwestward from the west flank of the uplift (Granata, 1963) extending for more than 257 km (160 mi). The wedge varies in thickness from 0 to a maximum of 137 m (450 ft) in the Houston, Walker, and Madison counties area. To the southwest in the vicinity of Kurten Field, stratigraphically equivalent, pod-like sandstone bodies have been interpreted as prograding "offshore bars" (sand ridge deposits) which formed by current and wave action several miles west of the Harris Delta (Bell, 1980; Turner and Conger, 1981). To the northeast in Leon, Houston and Madison counties, Woodbine and Eagle Ford shelf sandstones are similar, but are suggested to have been deposited in deeper water (Theiss, 1983).

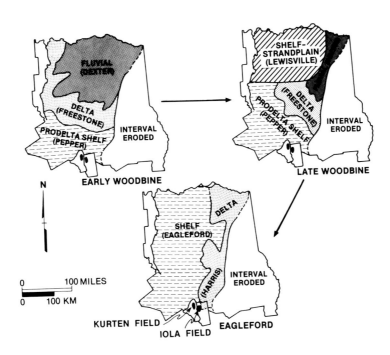

Fig. 4. Paleogeographic map of northeast Texas showing distribution of Woodbine--Eagle Ford depositional environments and facies, and area of erosion related to movements of the Sabine uplift. Outline of Brazos county with location of Kurten Field and Grimes county with location of Iola Field are shown at the southern limit of each map (after Oliver, 1971). See Figure 1 for details of East Texas geography.

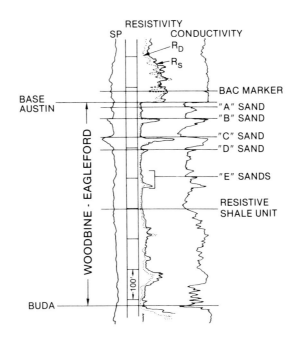

Fig. 5. Type log of the Woodbine--Eagle Ford section at Kurten Field illustrating electric log markers and sandstone nomenclature. Composited from Amalgamated Bonanza Lloyd No. 1 and Cayuga Cobb No. 2 wells. The "B", "C" and "D" units shown are related to sand ridge deposition at Kurten Field (modified from Turner and Conger, 1981).

Apparent onlap of younger Eagle Ford sandstones onto the Harris Delta in Madison county suggests a marine transgression succeeded Woodbine deposition. Clastic sediment influx ceased as transgression continued, transforming deltas into sediment-trapping estuaries. The uppermost sandstones preserved in the Eagle Ford section immediately underlying the Austin Chalk have been referred to as Sub-Clarksville (Forgotson, 1958; Nichols, 1964; Barton, 1982a). Barton studied cores of the Sub-Clarksville sands from Grimes County, and concluded that the sands were deposited rapidly from suspension by storm-generated bottom currents. Barton suggested that the distribution and morphology of the Sub-Clarksville sandstones were controlled by shelf topography caused by deep-seated salt movement. He depicts sands as accumulating in shelf topographic lows of restricted areal extent. Deposition of upper Eagle Ford sandstones in the Pleasant Ridge Field, Leon County, has also been attributed to storm-generated currents (Theiss, 1983). As in the case of the Sub-Clarksville, sandstone distribution was inferred to be controlled by shelf topographic lows. Renewed transgression eventually brought to a close the Eagle Ford clastic deposition, and caused widespread carbonate deposition of the Austin Chalk.

## CHARACTERISTICS OF WOODBINE--EAGLE FORD SHELF SAND RIDGE DEPOSITS

### Woodbine Deposits, Kurten Field

Recent subsurface studies of the Kurten Field sandstones (Bell, 1980; Turner and Conger, 1981) describe the characteristics and morphology of the first sand ridge deposits to be recognized in the Woodbine--Eagle Ford section of East Texas. Although these sandstones have been most commonly referred to as "subtidal bars" or "offshore bars" (Bell, 1980; Turner and Conger, 1981), Turner and Conger (1981, p. 228) did allude to them as sand ridges. The confusion in terminology appears to simply reflect the differences in perspective of students of ancient sedimentary deposits versus those of modern shelves. The term "bar" generally refers to a feature built by breaking waves on oceanic beaches. It is typically 1-2 m (3-6 ft) high and its crest lies about 100-500 m (300--1500 ft) seaward of the shoreline. Shelf sand bodies with 10 m (30 ft) of relief, and widths of 2-4 km (1-2 mi) will be referred to as "sand ridges" in this paper.

Five sandstone units occur in Kurten Field, and have been given different names or letter designations by several workers. In this paper, the terminology used will be the "A" through "E" designations, from top to bottom, as used by Turner and Conger (1981) (Fig. 5). Of these five units, the "B", "C" and "D" units have been interpreted in the literature as "sand ridge deposits" based on body geometry, composition, texture, sedimentary structures, degree of bioturbation and regional setting. The "C", "D" and "E" sandstones (Fig. 5) were correlated by Turner and Conger

(1981) as stratigraphic equivalents to the Harris Delta (Fig. 6). This correlation implies that the sand ridges formed from reworked deltaic sediments (Fig. 7). The sand ridges were inferred to have formed in front of and offshore from the delta, at a distance estimated at 3 km (2 mi) to 6 km (4 mi) by Bell (1980, p. 920) and 19 km (12 mi) by Turner and Conger (1981, p. 214). The "C" and "D" units have an elongate geometry trending north-south, are 4.5 mi (7.2 km) wide, over 16 km (10 mi) long, and 12 m (40 ft) thick (Figs. 8 and 9). The "B" and "E" units are described as somewhat thinner, with a northeast-southwest trend.

The sand ridge interpretation for formation of the Kurten sand bodies has been generally accepted by most workers familiar with the area. However, although several cross-sections across Kurten Field have been published (Bell, 1980; Turner and Conger, 1981) they typically show only the gross sand body geometry and do not attempt to show the details of the internal facies relationships for individual sand bodies (Fig. 10). Additional detailed mapping of both the external body geometry and internal facies relationships is required in order to confirm the sand ridge hypothesis and document more explicitly the three-dimensional geometry of discrete sand bodies.

## Kurten Field Facies

Results from the analysis of twelve conventional cores from the Kurten Field area were incorporated in Bell's (1980) study. He characterizes the shelf sandstones as highly quartzose, extensively burrowed, glauconitic, and interfingering both laterally and vertically with shale. Sedimentary structures and textures are reported to be virtually identical in each of the cored sandstone sequences examined by Bell (1980), leading him to infer a common depositional environment. Four different facies (Figs. 11 and 12) were recognized from the Kurten cores by both Bell (1980) and Turner and Conger (1981). Descriptive facies reported by Turner and Conger (1984) will be reviewed here, followed by remarks on depositional models which have been proposed in the literature. Turner and Conger's (1984) facies were based on core control from the "C" and "D" units in the Amalgamated Bonanza No. 1 Lloyd and No. 1 Smith Unit 4 wells in Kurten Field. The vertical sequence described is typical of the sand ridge deposits, observed in the Kurten Woodbine--Eagle Ford section.

*Laminated bioturbated sandstone.*--The lowermost unit in the sequence is a laminated, bioturbated sandstone, typically 1 to 5 m (3 to 16 ft) thick. This facies was observed in the "C" (Fig. 11, F-H) and "D" (Fig. 12, I-N) units in the Amalgamated Bonanza No. 1 Smith Unit 4 core, and the No. 1 Lloyd core (Fig. 11, K-N). The upper and lower contacts of the unit as a whole are gradational over a few feet, with a marine shale below and clayey, intensely bioturbated sandstone above. The sandstone is dark-gray, very

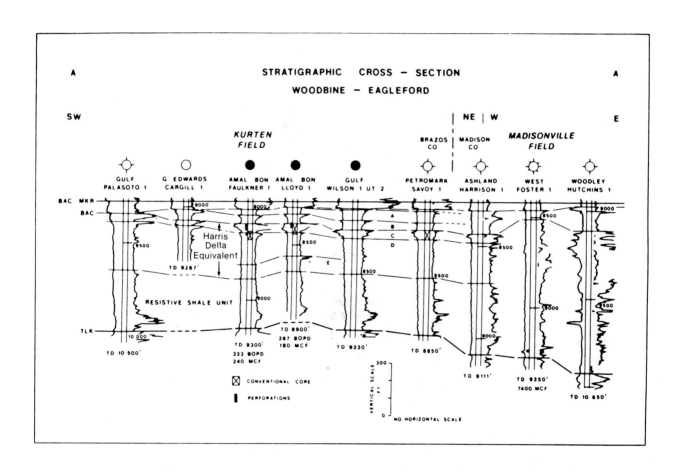

Fig. 6. Stratigraphic cross-section A-A' across Kurten Field to Madisonville Field showing interpreted relationship of Kurten sandstones to Harris Delta and two-dimensional sandstone geometry. Datum is BAC marker, BAC = Base of Austin Chalk, TLK = Top of Lower Cretaceous. Location of cross-section is shown in Figure 8 (From Turner and Conger, 1984).

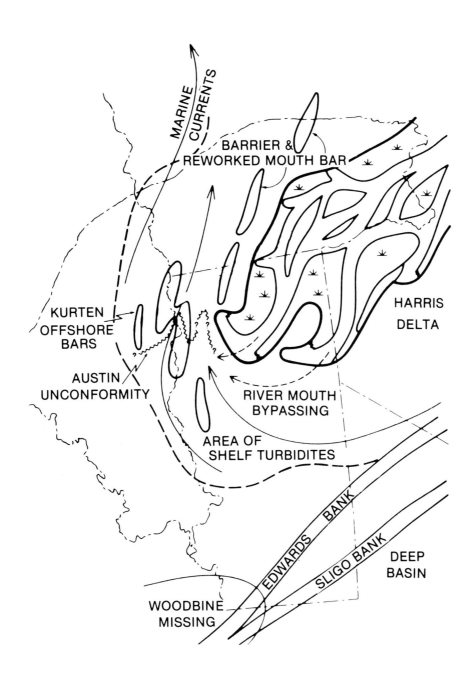

Fig. 7. Depositional model proposed by Turner and Conger (1981) for Kurten Field sandstones. Diagram shows southwest prograding Harris Delta, inferred processes of river mouth by-passing, shelf turbidity current deposition and shelf bars modified by storm and tide currents (from Turner and Conger, 1981).

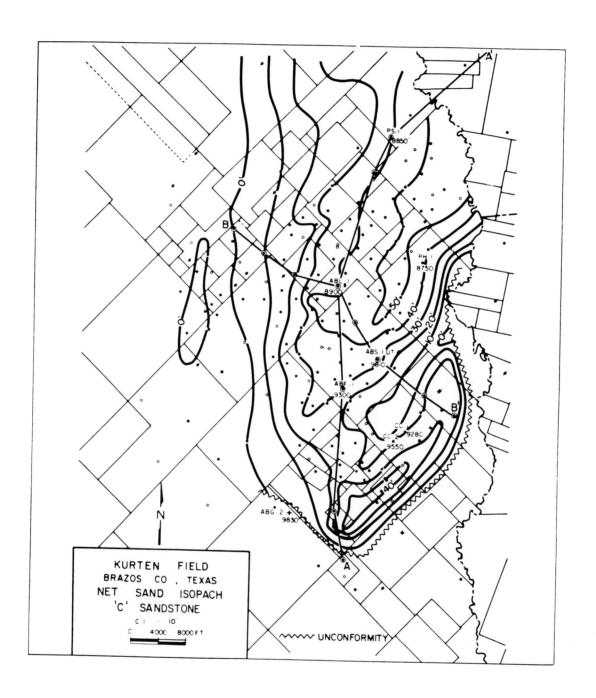

Fig. 8. Net sandstone isopach of Woodbine "C" sandstone unit Kurten Field, showing an en echelon pattern of "thicks." Location of cross-sections A-A' (Fig. 6) and B-B' (Fig. 10) are also shown (From Turner and Conger, 1984).

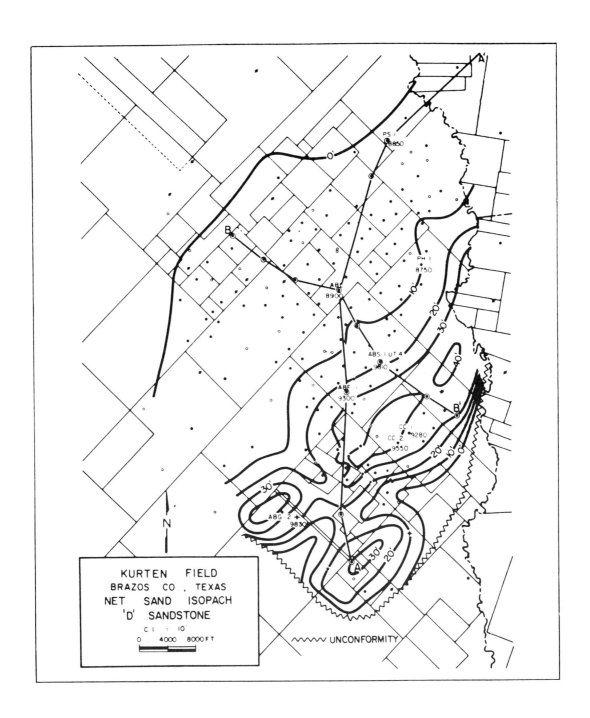

Fig. 9. Net sandstone isopach of Woodbine "D" sandstone unit, Kurten Field, showing elongate trend of sandstone bodies. Location of cross-sections A-A' (Fig. 6) and B-B' (Fig. 10) are also shown (From Turner and Conger, 1984).

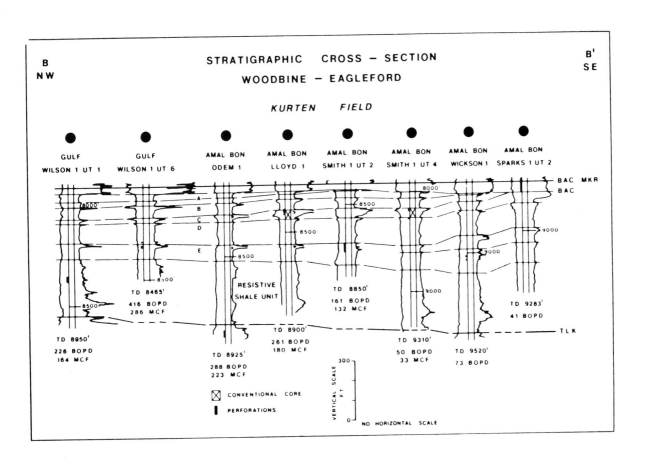

Fig. 10. Stratigraphic cross-section B-B' across Kurten Field from northwest to southeast, showing asymmetrical sandstone bodies with steep sides facing east. A, B and C sandstones are missing to the east. Datum is BAC marker, BAC = Base of Austin Chalk, TLK = Top of Lower Cretaceous. Location of cross-section is shown in Figure 8 (From Turner and Conger, 1984).

Fig. 11. Sedimentary structures in Woodbine "C" and "D" units, Amalgamated Bonanza No. 1 Lloyd, Kurten Field. A: "C" unit; burrowed (arrow) flaser cross-bedded sandstone; 2547 m (8356 ft). B: "C" unit; burrowed (b), flaser cross-bedded (f) sandstone with sharp basal contact (c) with overlying bedset; 2548 m (8359 ft). C: "C" unit; flaser cross-stratified unit (f) showing stacked lenses of sand, becoming bioturbated (b) upward; 2550 m (8364 ft). D: "C" unit; bedset of massive to indistinctly laminated sandstone (i), distinct contact (c), and flaser cross-bedded sandstone (f); 2551 m (8366 ft). E: "C" unit; bedset of massive to indistinctly laminated sandstone with shale classes (arrow) overlain by flaser cross-beds (f), followed by wavy irregular laminated unit (w) with large burrows (b); 2612 m (8567-68 ft). F: "C" unit; clean flaser cross-bedded sandstone with shale clasts; 2552 m (8369 ft). G: "C" unit; intensely bioturbated (b) clayey, silty sandstone, sedimentary structure largely destroyed; 2552 m (8373 ft). H: "C" unit; intensely bioturbated (b) clayey, silty sandstone; inclined shale lined burrow (arrow); 2553 m (8375 ft). I: "C" unit; intensely bioturbated (b) clayey sandstone, with sinuous burrows; 2554 m (8378 ft). J: "C" unit; laminated to bioturbated unit with siltstone-shale bedsets (s); 2559 m (8392 ft). K: "D" unit; wavy laminated, clayey, bioturbated sandstone; 2563 m (8408 ft). L: "D" unit; bioturbated, silty, clayey sandstone with irregular contact (c); 2564 m (8409 ft). M: "D" unit; clayey, bioturbated siltstone; 2564 m (8410 ft). N: Laminated sparsely burrowed marine shale underlying "D" unit; 2564 m (8411 ft) (from Turner and Conger, 1984).

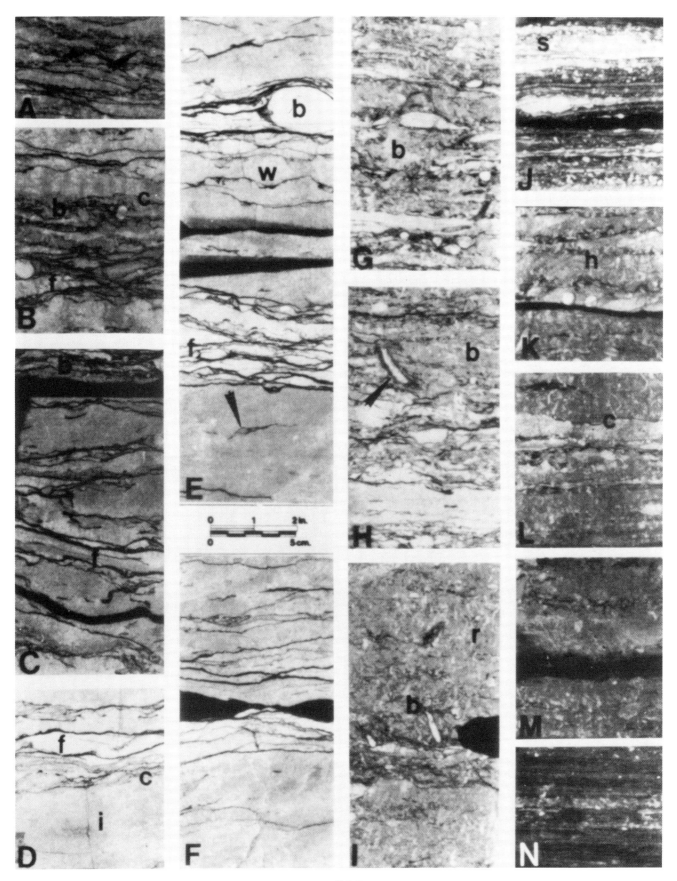

Fig. 12. Sedimentary structures in cores of Woodbine "C" and "D" units, Amalgamated Bonanza No. 1 Smith Unit 4, Kurten field. A: "C" unit; clean to burrowed sandstone (b) with wavy shale laminations; 2615 m (8579 feet). B: "C" units; clean burrowed sandstone with sharp basal contact (c) on overlying sand unit, wavy (w) shaley zones probably represent flaser cross-bedding; 2616 m (8580 ft). C: "C" unit; clean burrowed sandstone with shale clasts (arrow); 2616 m (8582 ft). D: "C" unit; intensely bioturbated clayey, silty sandstone with small sinuous Scalarituba-like burrows (r); 2617 m (8585 ft). E: "C" unit; intensely bioturbated clayey, silty sandstone with vertical and horizontal burrows; 2618 m (8586 ft). F: "C" unit; laminated bioturbated unit showing repeating siltstone-shale bedsets (x) and small ripple lenses (p); 2621 m (8597 ft). G: "C" unit; laminated bioturbated unit exhibiting soft sediment deformation (c); 2626 m (8614 ft). H: "C" unit, laminated to bioturbated sets; 2627 m (8615 ft). I: "D" unit; alternating intensely bioturbated and laminated sandstones with sharp, wavy upper contacts (c) and large round horizontal burrows (b); 2628 m (8620 ft). J: "D" unit; intensely bioturbated clayey sandstone with large and small burrows; 2630 m (8627 ft). K: "D" unit; intensely bioturbated to laminated clayey siltstone; 2631 m (8631 ft). L: "D" unit; clayey, bioturbated sandstone containing a large sandstone clast (t); 2632 m (8633 ft). M: "D" unit; bioturbated to laminated sandstone with pellet-lined Ophiomorpha-type (o) burrow; 2632 m (8634 ft). N: "D" unit; bioturbated to laminated siltstone with Rhizocorallium-type burrow (z) 2633 m (8635 ft) (from Turner and Conger, 1984).

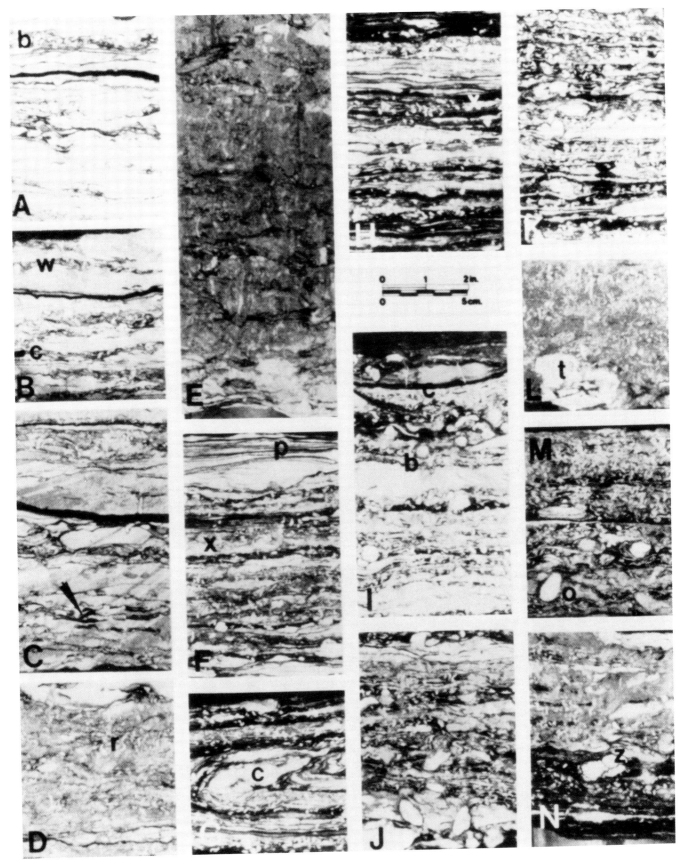

fine-grained, silty to clayey, with shaley laminae and interbedded shale stringers. Laminae are distinct, continuous and slightly wavy. Individual bedsets are thin (2.5 to 5 cm; 1 to 2 in), consisting of very fine-grained sand and silt with small ripple lenses, abruptly overlain by shale. Turner and Conger (1981) suggest these units resemble "cde" divisions of the Bouma turbidite sequence (Bouma, 1962). The bedsets have sharp basal contacts but typically lack evidence of erosion or soft-sediment deformation. Zones of small-scale bioturbation are present, alternating with the lamination. Turner and Conger (1984) infer that the pattern of alternating bioturbation and lamination implies episodic deposition, which they suggest is similar to depositional patterns found in turbidity flows.

Intensely bioturbated sandstone.--This second unit in the vertical sequence consists of 4 to 5 m (12 to 17 ft) of intensely bioturbated gray to dark-gray clayey, silty sandstone. It gradationally overlies the laminated sandstone in the "C" unit. This facies was observed in the No. 1 Lloyd core (Fig. 11, G-I) and the No. 1 Smith Unit 4 core (Fig. 12, D-E). The bioturbation is reported to be pervasive, comprised of larger-scale burrows than in the underlying unit, and virtually obliterating primary sedimentary structures. Trace fossils recognized by Turner and Conger include; Diplocraterion, Asterosoma, Chondrites (Fig. 12), Ophiomorpha (Fig. 12M), Rhizocorallium or Teichichnus (Fig. 12N), Scalarituba and Arenicolities (Fig. 12, D-G). The upper contact of the unit is also gradational.

Flaser-bedded to cross-bedded sandstone.--Overlying the intensely bioturbated sandstone is 3 to 5 m (11 to 16 ft) of clean, light-gray to white, flaser- to cross-bedded sandstone. This facies represents the highest energy of deposition, and was observed only in the "C" unit (Fig. 11, D-F). Turner and Conger (1981) report alternating flaser-bed sets and zones of indistinctly laminated sandstone. Laminae are inclined about 20° and contain occasional shale clasts. They describe the basal contacts of the indistinctly laminated beds as sharp, irregular and slightly scoured (Fig. 11E). The upper contacts are gradational (Fig. 11, D-E).

Bioturbated, flaser- to cross-bedded sandstone.--The uppermost unit in the sequence is a 3 m (9 ft) thick zone of light-gray, clayey, bioturbated flaser-bedded sandstone (Fig. 11, A-C; Fig. 12A), in gradational contact with the unit below. The primary differences from the underlying unit are (1) increased bioturbation, (2) increased frequency of flaser- or cross-bed sets, and (3) increased matrix (Turner and Conger, 1981). Individual bedsets are approximately 30 cm (1 ft) thick, formed by flaser-bedded and indistinctly laminated sandstones. Marine shale overlies the sandstone sequence in gradational contact.

Depositional Models, Kurten Field

## Fluid Dynamical Regime

Shelf storm currents.--Turner and Conger (1981) infer a shelf depositional setting for the Kurten Field sand bodies on the basis of their ridge-like geometry, proximity to the Harris Delta deposits, and trace fossils assemblage (Rhizocorallium, Asterosoma, and Diplocraterion). The processes of transport and deposition inferred by Turner and Conger (1981) for formation of the sand bodies are a combination of rivermouth by-passing, storm-surge turbidity flows, and longshore currents. This interpretation differs significantly in some aspects from observations of the fluid and sediment dynamic processes of sand ridge formation on modern continental shelves. Results of recent studies of the shelf dynamic regime on the Atlantic Continental Shelf (summary in Swift, in press a) indicate that the primary mechanism of shelf sediment transport and deposition on storm-dominated shelves is alongshelf geostrophic flow. These wind-driven currents dominate the shelf regime of approximately 80% of modern continental shelves, and very likely the Woodbine--Eagle Ford shelf as well. The remainder of modern shelves are either tide-dominated (characterized by macrotidal or mesotidal ranges) or dominated by intruding oceanic currents (Swift, in press a). Although a comprehensive review of the fluid and sediment dynamics of continental shelves is beyond the scope of this report, a few brief comments on the shelf hydrodynamic regime are warranted.

In view of the fact that deposition of the Kurten sandstones has been attributed to deposition by storm-surge turbidity flows, two aspects of the shelf hydrodynamic regime are particularly noteworthy: the relationship between geostrophic storm currents and storm surge, and the role of turbidity flows in shelf sedimentation. A more comprehensive discussion is presented in Swift et al., (in press a). Geostrophic currents are wind-driven currents which develop in response to the passage of atmospheric low pressure cells (storms) across the shelf. In mid-latitudes, these storm-induced currents have a frequency of about 3 to 10 days. In these areas, major storm flows, capable of transporting very large amounts of sediment, occur 3 to 5 times a year primarily in response to winter storms (Swift et al., 1984). The storm flows are driven by the pressure field associated with the set-up (or set-down) of the sea surface against the shore (Fig. 13). Because of the Coriolis effect, fluid transport trends along the contours of the sea surface slope, rather than normal to them. The resulting currents are therefore called geostrophic (earth-turning) currents. Consequently, although the storms themselves are circular wind systems, the induced currents flow parallel to both the contours of the shelf floor and to the shoreline.

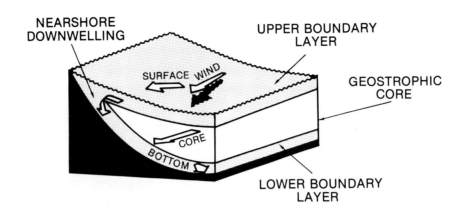

Fig. 13. Block diagram representation of geostrophic flow on the continental shelf. Flow components shown are landward moving upper boundary layer (stippled), alongshelf directed fluid interior (clear), and seaward veering bottom boundary layer (stippled) (from Swift and Rice, 1984).

## GRADED TEMPESTITE SEQUENCE

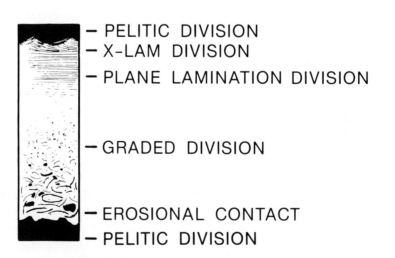

- PELITIC DIVISION
- X-LAM DIVISION
- PLANE LAMINATION DIVISION
- GRADED DIVISION
- EROSIONAL CONTACT
- PELITIC DIVISION

Fig. 14. Idealized tempestite sequence showing vertical sequence of stratigraphic divisions. Note similarity to a Bouma sequence. From Aigner, 1982.

The term "storm surge" refers simply to the vertical rise of water associated with a storm. Storm surge is commonly a consequence of wind set-up. During the initiation of set-up there is a landward component of flow, and there is a seaward component in the relaxation flow after the storm (Swift et al., in press a). However, in both cases, the generally shore-normal velocities are less than alongshore velocities so that flow is at most oriented obliquely offshore. When the winds shut off, or move inland, the bulge of water that they had been pressing against the coast does not rush back seaward, but instead flows alongshelf as a "topographically trapped shelf wave" (Swift et al., in press a). As the wave propagates alongshelf, it is damped by friction. Water is returned seaward as it loses amplitude, but large-scale offshore-directed "storm-surge ebb currents" are not generally observed in field studies of storm circulations on modern continental shelves (Swift et al., in press a). Furthermore, the initial or final discharge associated with the set-up is trivial when compared to the alongshelf discharge of a geostrophic flow of shelf width and depth, which may be prolonged for several days (Morton, 1981).

Storm currents vs. turbidity currents.--The inference of density current deposition in the Kurten sequence by Turner and Conger (1981) is based on the observation of primary sedimentary structures resembling Bouma sequence "cde" divisions in cores of the Kurten "C" unit and Bouma "bcde" bedsets in the "E" unit. It should be recognized, however, that the sequence of sedimentary structures and textures in the Bouma sequence, as well as associated scoured contacts and load features are not unique to density current deposits (Nelson, 1982). The Bouma sequence simply reflects deposition from waning flow, as do similar appearing divisions of tempestites (i.e. storm beds, Ager, 1974). The same vertical sequences then might be as reasonably described as divisions of an Aigner (1982) tempestite sequence (Fig. 14). Consequently, without additional information such as reliable paleocurrent indicators, the vertical sequence itself is not diagnostic of deposition by turbidity currents rather than geostrophic storm flows.

It is noteworthy that density underflows have never been observed on modern continental shelves. The Gulf Coast study by Hayes (1967) is often cited as evidence for the presumed role of density underflows in inner shelf sedimentation (Hamblin and Walker, 1979; Turner and Conger, 1981; Dott and Bourgeois, 1982; Balsley, 1982). Hayes (1967) described graded beds up to 9 cm thick at water depths up to 37 m (120 ft) on the Texas inner shelf in the aftermath of Hurricane Carla (1967). He interpreted the depositional mechanism to have been a density underflow triggered by a reflux of sediment-laden water back out over the barrier island from the surge-filled lagoon. However, Morton (1981) recently reevaluated the data from Hurricane Carla and disagreed with Hayes' density flow interpretation, concluding that deposition was a result of alongshore geostrophic flows. Walker (this volume) suggests that liquefaction of the sea floor during severe storms can result in slumping, which leads to generation of turbidity currents. Although liquefaction of the substrate may occur due to cyclic wave loading during storms, the types of resulting mass movement as observed off the Mississippi Delta are generally a variety of subaqueous landslides (Prior and Coleman, 1978a,b). These mass movements commonly occur as mudflows and are frequently characterized by disturbed debris and extensive soft-sediment deformation. Movement in such mud flows is described as "creep," and the velocities required for high density turbidity currents do not occur. The actual transformation of such subaqueous slumps into density underflows has not yet been observed in a continental shelf setting. Furthermore, the lack of significant soft-sediment deformation features in the Kurten sandstone sequences suggests that slump-related shelf turbidity currents are an unlikely depositional mechanism for these sandstones.

In a recent modification of their Kurten paper, Turner and Conger (1984) substitute the term storm-generated "sheet flow" for their previously suggested storm-surge turbidity current mechanism. They do not elaborate on their meaning for the term except to invoke it as a depositional mechanism for the Bouma-like sequences in the Kurten "E" unit. Turner and Conger (1984) may be referring to high-concentration traction transport, of the kind described by Bagnold (1963) as supported by dispersive pressure. The Glossary of Geology (Bates and Jackson, 1980) identifies a sheet flow as a phenomenon occurring on bare subaerial hillsides during a rain, or as a hydraulic term meaning laminar flow. Neither definition is applicable to density-driven transport. Shelf storm flows do develop high concentrations near the bed (Lavelle et. al., 1978; Clarke et al., 1982; Vincent et al., 1982), but flow parallel or sub-parallel to the shore. Such shore-parallel shelf current flow is also implied in Turner and Conger's (1981) depositional model (Fig. 7).

Pantin (1979) and Parker (1982) have recently considered the conditions required for turbidity current transport from a theoretical point of view. Sustained transport can only be accomplished if "ignition" occurs; that is if the velocity and density are sufficient for autosuspension to occur. The term autosuspension (Bagnold, 1962) refers to a feedback process whereby the turbulence of the flow is sufficient to suspend the particles, and the resulting suspension is dense enough to drive the fluid and create the turbulence. While there is some difference of opinion as to exactly how to determine the autosuspension criteria (Southard and Mackintosh, 1981), most computations (e.g., Pantin, 1979, Parker, 1982) indicate that exceedingly high density suspensions (on the order of $1gl^{-1}$), high stirring velocities ($>30$ cm sec$^{-1}$) and relatively steep slopes ($>1.5°$) are required. These requirements tend not to occur together on continental shelves; high density storm suspensions are characteristic mainly of the very flat mud-accumulating on shelves, whose gradients are measured in tenths or hundredths of a degree; while coarse-grained, high relief shelves would require prohibitively high stirring velocities. The turbidity current interpretation is a particularly awkward mechanism to apply to the Kurten sand bodies in view of their depositional geometry. Modern shelf sand ridges typically exhibit up to 10 m of relief and are nourished by storm or tidal currents which flow obliquely across them (Parker et al., 1982; Kenyon et al., 1981). Seaward-oriented turbidity currents would somehow have to flow up one side of a ridge and down the other.

In contrast, geostrophic storm flows have been repeatedly observed on modern continental shelves (Vincent et al., 1982, Clarke et al., 1982, Drake et al., 1980), with wave orbital and mean flow velocity components sufficient to carry dense sand suspensions, and graded sand beds have been repeatly shown to be the product of such flows, on shelves without sand ridges (Aigner and Reineck, 1982, Nelson, 1982, Morton, 1981) and on shelves with sand ridges (Swift et al., in press b). Consequently, we feel obliged to apply the scientific Principle of Parsimony to the interpretation of the depositional mechanism of the Kurten sand bodies. There is no need to call on a mechanism which has never been observed (shelf turbidity currents) in order to explain these deposits when a mechanism which has been abundantly observed (geostrophic storm flows) is capable of forming their characteristic features.

## Sand Body Geometry

The geometry, distribution and vertical sequence of sedimentary units in the Kurten Field sandstones bears numerous similarities to sand ridge deposits on the modern Atlantic Continental Shelf (Slatt, 1984) as well as shelf sandstone bodies described in the Cretaceous of Western North America (Summary in Swift and Rice, 1984). The Kurten sandstones are elongate, up to 15 m (50 ft) thick, 8.8 km (5.5 mi) wide and 20.1 km (12.5 mi) long (Turner and Conger, 1981; Figs. 8, 9 and 10, this report). In comparison, Holocene sand ridges on the Georges Bank-Nova Scotia-Newfoundland Labrador shelf system exhibit similar elongate geometries and dimensions, measuring up to 35 m (116 ft) in thickness, tens of kilometers in length, with spacings up to 15 km (9 mi) apart in parallel sets (Slatt, 1984). Although the Cretaceous shelf sand bodies are also similar in terms of geometry and dimension (Table 1), both they and the Kurten sandstones have been described as being encased in shale, in contrast to modern ridges which are rooted in the Holocene transgressive sand sheet. However, recent studies suggest that this distinction is more apparent than real. Cretaceous sandstones are now known to rise from thin sandy horizons within the shelf shales (Gaynor and Swift, in preparation) and the same may be true of Kurten sand ridge deposits. Both modern and ancient sand ridge deposits that have been subjected to close scrutiny appear to have formed in transgressive settings, and a transgressive model should be tested for the Kurten Field deposits.

TABLE 1. APPROXIMATE DIMENSIONS OF SOME CRETACEOUS SAND RIDGE DEPOSITS[1]

| Sandstone | | MAXIMUM THICKNESS | | Dimensions | Reference |
|---|---|---|---|---|---|
| | | Ridge Sandstones* | Ridge Sandstone Complexes** | | |
| Shannon | Salt Creek Area, Powder River Basin, Wyoming | | 15 m (49 ft) | 50x30 km (32x19 mi) | Spearing, 1976 |
| Shannon | Salt Creek Area, Powder River Basin, Wyoming | | 20 m (66 ft) | >30x10 km (>19x6.3 mi) | Tillman and Martinsen, 1983; 1984 |
| Shannon | Heldt Draw, Powder River Basin, Wyoming | | 18-20 m (59-66 ft) | 15x2.5 km (9.4x1.6 mi) | Seeling, 1978 |
| Shannon | Southeastern Montana | 10-20 m 33-66 ft) | 15-25 m (49-82 ft) | 50 km$^2$* (20 mi$^2$) 1500 km$^2$** 600 mi$^2$ | Shurr, 1984 |
| Sussex | House Creek Field, Powder River Basin, Wyoming | | | 45x1.5 km 28 x 0.9 mi) | Hobson et al., 1982 |
| Viking | Joffre-Joarcam Field Areas, Alberta | 6 m (20 ft) | | 110x80 km** (69x50 mi) | Beaumont, 1984 |
| Duffy Mountain | Northwestern Colorado | 20 m (66 ft) (pre-compaction= 27 m [89 ft]) | | >50x10-15 km** >32x6.3-9.0 mi) | Boyles and Scott, 1982 |
| First Frontier | Spearhead Ranch Field, Powder River Basin, Wyoming | 6 m (20 ft) | | | Tillman and Almon, 1979 |
| Mosby, Phillips & Second White Specks | Central Montana, Alberta | 6 m (20 ft) | | 5x3 km* (3x2 mi) 80x55 km** (50x35 mi) | Rice, 1984 |
| Teapot | | | | 25 m (82 ft) | Curry, 1976 |
| Cardium | Alberta | 4 m (13 ft) | | 115x3 km (72x2 mi) | Walker, 1979 |

[1] Modified from Slatt, 1984.

Depositional models proposed to explain the geometry of ancient shelf sand bodies have in many instances lacked an understanding of the fluid and sediment dynamic processes of shelf sedimentary regimes. Processes of sediment transport and deposition observed on the storm-dominated North American Atlantic Shelf in numerous studies (summary in Swift et al., in press a, b) have provided insights into a process-model approach for understanding the geometry of ancient shelf sand bodies. For example, Bell's (1980) model for the Kurten sand bodies fails to recognize that they are asymmetric in cross-section. Studies of modern ridges (Parker et al., 1981; Figueiredo et al., 1981) have demonstrated that sand ridges are inherently asymmetrical because they form at an oblique orientation to the prevailing storm flow direction. Turner and Conger's (1981) model does depict the asymmetry of the sand bodies, steep sides facing east (Fig. 10), and they infer from this an easterly net transport of sand by "marine currents" (Turner and Conger, 1981; p. 228). However, this conclusion does not seem to be consistent with other examples of modern and ancient sand ridges. The asymmetry of modern sand ridge profiles appear to be in response to the grain size gradient over the sand ridges rather than direct responses to flow dynamics. On the modern sandy Atlantic Shelf, the steep flank of the sand ridges is in fact the downcurrent easterly flank (Swift et al., in press b). On the Atlantic Shelf, the downcurrent flanks of the sand ridges consist of fine sand, winnowed from the eroding upcurrent sides. In contrast, the sand bodies of the Campanian section in Colorado (Boyles and Scott, 1982) were deposited in a muddy shelf environment, probably similar to the muddy Cretaceous Woodbine--Eagle Ford shelf, and they exhibit steeper upcurrent flanks. Silt and mud swept over the ridge crest by impinging storm currents travels much further before coming to rest than does the fine sand. Consequently, on muddy shelves, the slope of the downcurrent flank is so low that the asymmetry of the ridge is effectively reversed. This logic suggests that the steep east-facing flanks of the Kurten sand bodies are in fact the upcurrent flanks, indicating a net transport to the northwest not the east. Sedimentation was probably driven by westward storm flows, a pattern which is also characteristic of the modern Texas Shelf (Holmes, 1982).

Facies Associations

Two of the most important criteria for the recognition of shelf sand ridge deposits in the subsurface are the characteristic lateral and vertical facies relationships. Sandstone bodies of the Cretaceous shelf section of the Western Interior Seaway are well-studied in both the surface and subsurface and serve as a model for comparison of shelf sandstone and mudstone facies. The three predominant lithofacies which characterize the Western Interior shelf sequence are (1) sandstone lithofacies consisting of elongate sandstone lenses, (2) transitional or thin-bedded lithofacies gradational between sandstone lenses and surrounding shale, and (3) shale lithofacies (Spearing, 1976; Rice, 1980; and Rice and Shurr, 1980; Boyles and Scott 1982; Gaynor and Swift, in preparation). The sequences of sedimentary structures and the facies associations which characterize the Sussex and Shannon Western Interior shelf sandstones are generally recognized as indicative of deposition by geostrophic storm flows in middle to outer shelf environments (Gaynor and Swift, in preparation).

The Woodbine--Eagle Ford shelf facies in the Kurten Field area exhibit a gradational lithofacies association of the four sedimentary units previously described (Turner and Conger, 1981). These facies associations closely resemble those of the Western Interior Shelf sequences. A comparison of the facies classifications for the Woodbine--Eagle Ford to those of the Western Interior Seaway shelf sections illustrates their similarities (Fig. 15) and general correspondence with the spectrum of stratification types described by Reineck and Wunderlich (1968). This comparison suggests that Woodbine--Eagle Ford lateral facies relationships and depositional processes may also be similar to those of the Western Interior shelf sandstones.

Detailed correlations based on closely spaced Shannon outcrop sections in the Salt Creek Anticline area (Tillman and Martinsen, 1984; Gaynor and Swift, in preparation) revealed relatively abrupt vertical and lateral changes in facies. In the subsurface, however, Tillman and Martinsen were only able to observe thickness changes: SP-resistivity logs did not reveal facies changes on their subsurface cross-sections. This suggests that similar rapid lateral and vertical facies changes and interfingering might be possible in the Woodbine--Eagle Ford sand ridge deposits, although they are clearly not shown in Turner and Conger's model (1981, Fig. 16 this paper). The "layer-cake" facies configuration depicted in Turner and Conger's (1981) model (Fig. 16) probably is accounted for by the available log data being limited to only SP-resistivity logs in many instances in the Kurten area, and the difficulty in using only these logs to resolve facies changes on subsurface cross-sections. However, by comparing facies characteristics revealed in Kurten cores with their associated log responses, it is possible to improve resolution of the vertical and lateral facies relationships of the Kurten sandstones. Initial results applying this method to Kurten Field reveal a facies configuration similar to the asymmetric, time-transgressive facies model (Fig. 17) of Gaynor and Swift (in preparation).

The vertical distribution of the sand ridge deposits in Kurten Field is also comparable with those of the Cretaceous Interior Seaway. The most common vertical sequence of sandstone facies is a coarsening upward profile for the individual sand ridges, and either a vertical or en echelon stacking of the individual bodies into larger sand ridge complexes. The Kurten sand ridges appear to be stacked en echelon (Fig. 8) in a manner similar to the Shannon ridges in the Hartzog Draw area (Gaynor and Swift, in preparation). The stacking reflects a downcurrent shift in facies as the shelf surface aggraded.

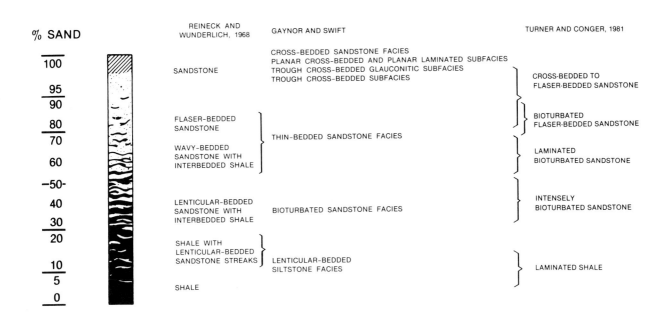

Fig. 15. Comparison of facies units for Kurten sand ridge deposits (Turner and Conger, 1981) and Shannon sand ridge deposits (Gaynor and Swift, in preparation) to the Reineck and Wunderlich (1968) descriptive classification related to sandstone - shale percentages.

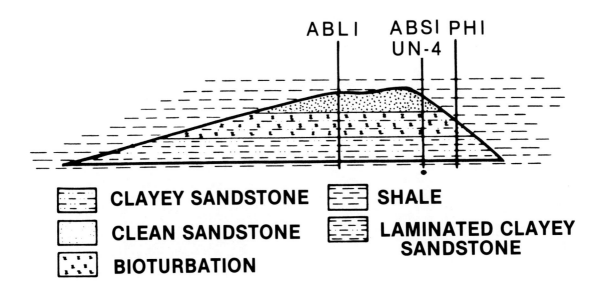

Fig. 16. Sand ridge facies model interpreted by Turner and Conger (1981) for Woodbine "C" sandstone at Kurten Field. Asymmetrical ridge geometry and inferred facies relationships are illustrated. Facies from cored wells Amalgamated Bonanza No. 1 Lloyd (ABL1) and Amalgamated Bonanza No. 1 Smith Unit 4 (ABS1 UN4) are shown in Figs. 7 and 8 (from Turner and Conger, 1981).

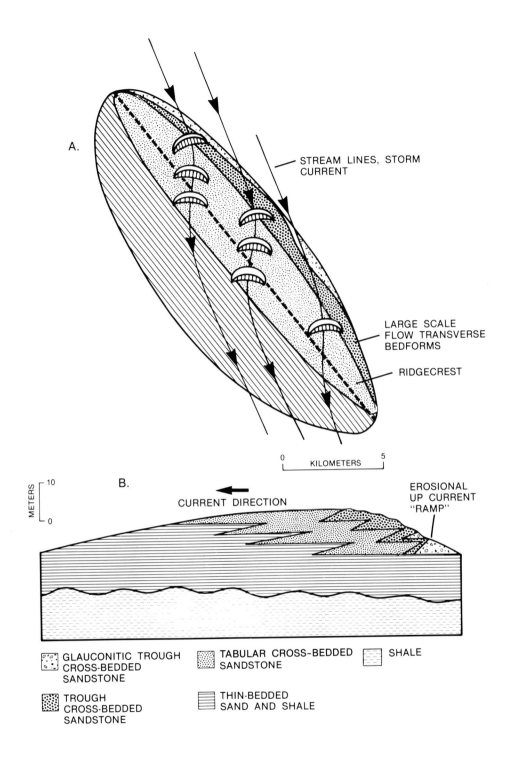

Fig. 17. Idealized sand ridge facies model based on sedimentary facies of Campanian Shannon Sandstone at Hartzog Draw Field and Salt Creek Anticline, Powder River Basin, Wyoming (from Gaynor and Swift, in preparation). A) Plan view showing relationship between facies distribution and current flow pattern. B) Cross-sectional view showing vertical and lateral facies relationships.

Depositional Model

Recent stratigraphic correlations suggest that the Kurten sand ridges are closely associated with the adjacent deposits of the Harris Delta (T. Gracinin, personal communication). It seems likely that the sand ridges may have formed in an inner shelf environment, by reworking during local transgressions resulting from abandonment and subsidence of lobes of the wave-dominated delta. The development of a marine sand by the trangression of a subsiding, abandoned delta lobe was first described by Scruton (1960) in his study of the stratigraphy of the Mississippi Delta. Sand ridge development was not part of the original model, but more recently Cuomo (1984) has described Ship Shoal off the Mississippi Delta as an example of a sand mass associated with delta lobe switching (Fig. 18). The shoal may have originated as an overstepped barrier, but now appears to be reorienting itself as a submarine sand ridge, in response to local easterly flows that characteristically develop during winter storms on the west side of the Mississippi Delta (D. Nummedal, personal communication). This process has been termed "autocyclicity," meaning that the transgression is the result of a shift in the balance of process variables (rate of relative sea level rise, here due to subsidence, versus rate of sediment supply). This shift results from events within the depositional system, as opposed to a shift in the balance imposed from without by global eustatic sea level fluctuations. A lobe switching model for Kurten Field is presented in Figure 19.

This depositional scenario is similar to Slatt's Model 3 (Slatt, 1984; p. 1115) for sand ridge development. His model suggests formation of a topographically high surface resulting from delta or shoreline progradation during periods of relatively shallow water or subaerial exposure, followed by transgression and sand ridge formation during reworking. This type of model adequately accounts for the association of deltaic deposits with the shelf deposits and implies local erosion of shoreline sediments as the sand source for ridge formation.

A second prodelta depositional model potentially applicable to the Kurten Field ridge deposits is the "plume model" proposed by Patterson, (1983) based on ideas originally suggested by Coleman et al. (1981) from their study of the plume-like deposit of sand on the shelf in front of the Damietta mouth of the Nile (Fig. 20). Sand is piled on the Damietta mouth bar when the river is in flood stage. The sand is spread both east and west along the deltaic shoreface, carried by wave-driven littoral currents, which also reverse on a seasonal (and sometimes on a daily) basis. During flood stage however, fresh water builds up along the Nile shoreface and the density contrast between this nearshore fresh water band and the saltier offshore water drives a powerful coastal current. Prior to the construction of the Aswan dam and subsequent reduction of flood discharge, such alongshore density currents had been measured at 150 cm sec$^{-1}$ (Sharaf Al Din, 1977).

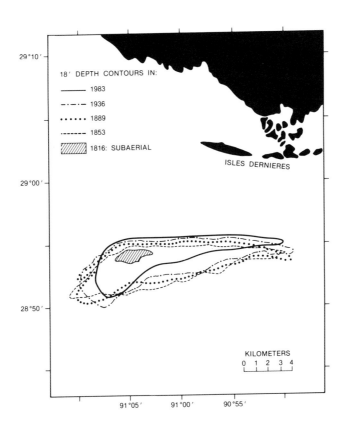

Fig. 18. Map tracing submergence and landward migration of Ship Island Shoal since 1816 (from Cuomo, 1984).

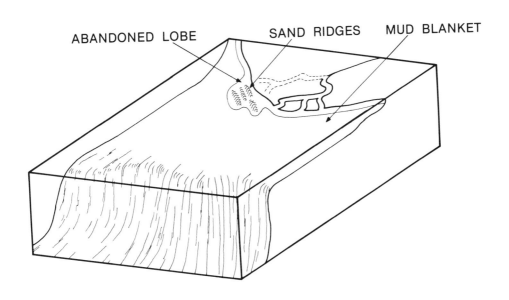

Fig. 19. Proposed model for deposition of Kurten sand ridge deposits. Model shows sandstone deposition resulting from winnowing and reworking of delta front sands following abandonment of lobes of the Harris Delta.

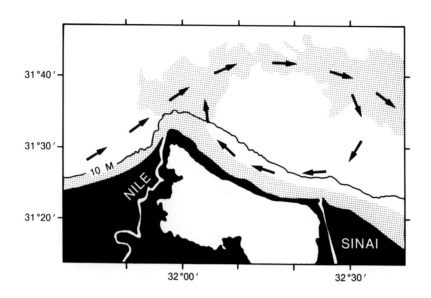

Fig. 20. Sand plume deposit associated with the Damietta Mouth of the Nile. From Coleman et al., 1981.

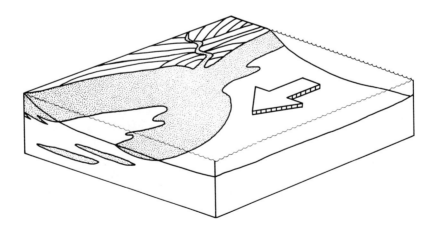

Fig. 21. Generalized depositional model for prodelta plume sandstones.

Coleman et al (1981) have shown that such alongshore flows do not hug the shoreline everywhere, but undergo boundary layer separation and the development of an eddy on the downcurrent side of such promontories as the Damietta mouth. As a result, during periods of strong flow, sand streams seaward and downcurrent from the Damietta mouth, and a plume-like bottom deposit is found on that part of the sea floor on a year around basis.

Features resembling sand ridges occur on the Damietta mouth plume, but it is not clear to what extent they date from the period of delta front progradation and the extent to which they are the result of reworking of the prodelta plume during the post Aswan Dam "transgression." A generalized model for Kurten Field as a prodelta plume deposit is presented in Figure 21.

### Woodbine in Leon, Houston and Madison Counties

A recent study of conventional cores from several Fields in Leon, Houston and Madison counties (Fig. 22), Texas has documented the occurrence of Woodbine--Eagle Ford shelf sandstones similar to the Kurten Field sand ridge deposits (Theiss, 1983). Theiss subdivided the clastic section into three units, the Woodbine, lower Eagle Ford and upper Eagle Ford. He inferred the presence of sand ridge deposits based on lithofacies associations and sandstone geometry in the Woodbine and lower Eagle Ford intervals, and suggested that deposition occurred in relatively deep water (middle neritic) on a marine shelf. The two major lithofacies which characterize these shelf deposits are a (1) heavily-bioturbated sandstone facies and (2) wavy-bedded sandstone facies.

The heavily-bioturbated sandstones are described as fine-grained quartz wackes exhibiting increases in both grain size and percent quartz upward in vertical sequence (Theiss, 1983). Trace fossils typical of these sandstones include Terebellina, Teichichnus, and small Ophiomorpha. This trace fossil assemblage (Chamberlin, 1978) and degree of bioturbation are interpreted by Theiss as indicative of an offshore shelf environment. After considering similarities in composition, texture and sedimentary structures to the Kurten sandstones, Theiss inferred an elongate broadly lenticular morphology, parallel to depositional strike for these sandstones. Due to a lack of well control, Theiss did not construct isopach maps or cross-sections to delineate the geometry of the bioturbated sandstones cored in Houston county. Consequently, additional mapping is needed to confirm his conclusions regarding morphology.

The wavy-bedded Woodbine sandstone lithofacies was described by Theiss from cores in southeast Halliday and Pleasant Ridge Fields in Madison and Leon counties respectively. These sandstones are also fine-grained quartz wackes but display no systematic changes in composition or texture through the section. Subsurface isopachs and cross-sections from the Pleasant Ridge

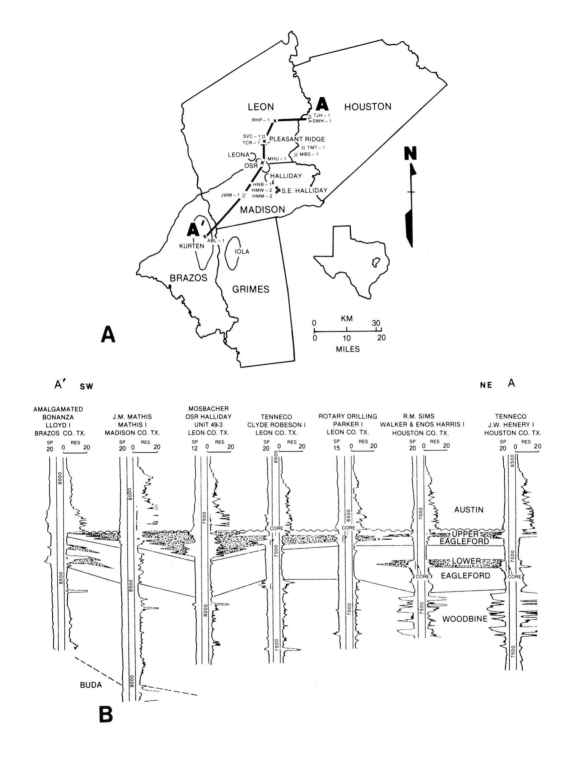

Fig. 22. A) Index map of Theiss' study area of Leon, Houston, Madison, Brazos and Grimes counties showing the location of cored wells and major oil fields in the area (from Theiss, 1983). B) Stratigraphic cross-section A-A' from Houston to Brazos counties showing the stratigraphic subdivisions and selected cored wells used in Theiss' study. Location of cross-section shown above in 22A (from Theiss, 1983).

show north-south elongate lenticular sandstone lenses, overlapping in a successively offset fashion vertically. This pattern resembles the en echelon or imbricate stacking of the Kurten sandstones (Fig. 10) and of the Shannon sandstones (Gaynor and Swift, in preparation). Theiss' cross-sections also illustrate rapid lateral and vertical facies changes for the sandstones which pinch out into surrounding shales. He interprets these wavy-bedded sandstones as "prograding offshore bars" (shelf sand ridges), deposited by low-flow regime currents in relatively deep water on the Woodbine--Eagle Ford shelf.

Comparison of the Kurten sandstones with those in Theiss' study area to the north raises several important questions. Some of the major problems include (1) what are the spatial and temporal relationships between the shelf deposits of these two adjacent areas? (2) why are the sand ridge deposits described by Theiss comprised of only one or two facies, in contrast to the Kurten sandstones which are composed of a vertical sequence of four distinct facies? (3) do the differences between the Kurten sand ridges and those to the north reflect differences in depositional processes, sediment supply and shelf configuration, or perhaps a process-response evolution of the shelf regime through time?

## CHARACTERISTICS OF WOODBINE TABULAR SHELF SAND DEPOSITS

### Woodbine Formation in Damascus Field

Geology.--Damascus Field is located in northernmost Polk County slightly southwest of the Sabine uplift and approximately 24 km (15 mi) updip of the Lower Cretaceous Edwards reef trend (Fig. 1). The Woodbine--Eagle Ford interval in this area consists of 49 to 61 m (160 to 200 ft) of dominantly shale, but contains numerous 2 to 7 m (6 to 23 ft) single to multistory thin sandstone bodies (Siemers, 1981). Age data based on palynomorph zonations (R. Christopher, 1984, personal communication) confirms Siemers' stratigraphic correlations, indicating that only the lower Woodbine section is preserved between the Rapides Shale and Buda Limestone at Damascus Field. Microfaunal data (C. T. Siemers, personal communication) and regional seismic data indicate an inner to middle shelf depositional setting for these Woodbine sandstones.

A core from the Hinton Dorrance No. 7A well in Damascus Field (Fig. 23) displays four distinct facies units bounded above and below by bioturbated, silty, shelf shales. These facies units are; (1) graded, medium to very-fine grained, massive to laminated sandstone beds (2) contorted, soft-sediment deformed intervals, (3) swirled and sheared siltstone beds, and (4) thin diamict conglomerate (shale pebble conglomerate) beds. The sandstones occur as thin 1 to 2 m (3 to 6 ft) fining-upward bedsets, characterized

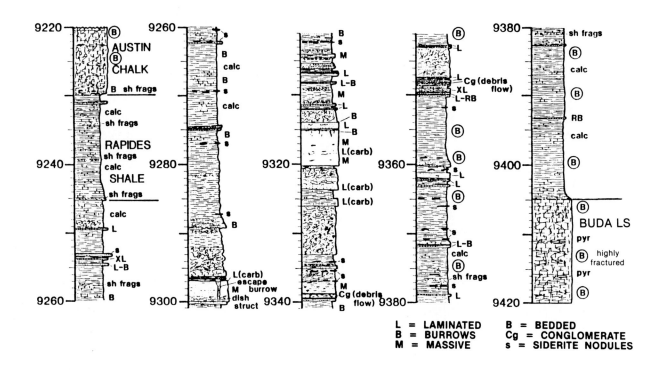

Fig. 23. Core lithology for Hinton Dorrance No. 7A showing vertical sequence of lithologies and sedimentary structures (C.T. Siemers, personal communication).

by sharp basal contacts, shale clasts, a Bouma-like (or Aigner-like) succession of primary sedimentary structures and abundant contorted bedding.

### Depositional Model, Damascus Field

Net sandstone isopachs, and subsurface cross-sections (Fig. 24) indicate that the Damascus sandstones trend northeast-southwest, across the strike of the Lower Cretaceous paleo-shelf edge, and have a tabular (terminology of Krynine, 1948) morphology. These characteristics have led Siemers (1981) to infer seaward transport of sand by debris flows and turbidity currents. Siemers model (Fig. 25) depicts the sandstones as prodelta, gravity flow, lobe deposits that are modified by storm and flood events.

We present an alternative interpretation, constrained by recent continental shelf oceanographic observations. In this alternative model, Damascus sands are suggested to have been deposited by alongshelf storm currents (Fig. 26). As discussed previously, studies of modern shelves typically show that sediment is transported alongshelf by geostrophic storm currents, ut density underflows have not been observed to occur on modern shelves. These e strophic currents deposit upward-fining beds (tempestites) with Bouma-like (Aigner-like) sequences and sharp basal contacts. In our alternative model, geostrophic storm deposition, which has been well documented on modern shelves is h sen as an actualistic deposotional mechanism which can explain the ser e Damascus sandstone characteristics.

The alongshelf storm flow model can also adequately account for the fact that the Damascus Field sandstones apparently trend across shelf contours, rather than parallel to the shelf contours, as do sand ridges. The sands may have been deposited as a series f localized, discontinuous tabular to sheet-like bodies. Sand eposition may have occurred in a cross-shelf zone where the shelf widened in a downcurrent direction around a deltaic projection in the shoreline. This widening of the ⁻helf effectively increased the cross-sectional area for flow, forcing storm currents to expand and decelerate (Fig. 26). On modern shelves, the shelf configuration formed by the relative positions of the shoreline and continental shelf e e a ts as a

quantities except under special circumstances. I approximation, the shelf flow is a closed system, analogous to a river. As a result, where the shelf configuration narrows in a downcurrent durection (as in front of a deltaic bulge in the shoreline), the flow must contract and accelerate, and where the shelf widens, the flow must expand and decelarate. Over repeated storm events, sheet sands would tend to be deposited in flow-transverse zones of deceleration (Fig. 26). Thus sands would be elongated across the strike f the shelf, and perpendicular to the current transport direction. Consequently, a seaward or dip-oriented sediment transport direction is not necessarily implied by a geometry that is elongate normal to the strike of the shelf.

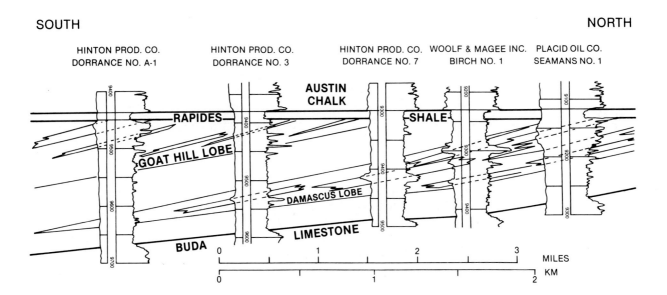

Fig. 24. Dip cross-section of Woodbine--Eagle Ford interval, Damascus Field, Polk County, Texas. Section illustrates dip correlation of sandstones and "lobe" designations interpreted by Siemers (C.T. Siemers, personal communication).

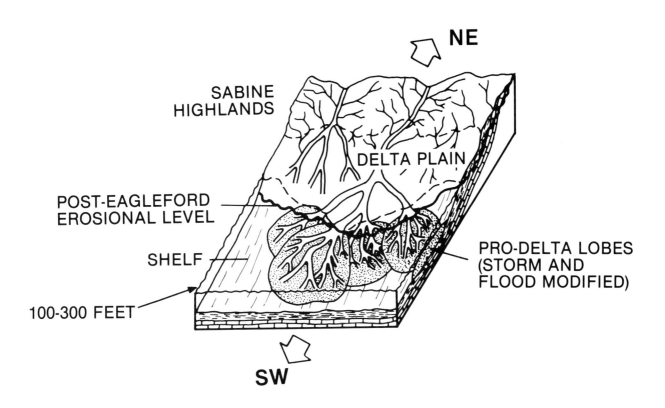

Fig. 25. Conceptual model for prodelta lobe interpretation of Damascus sandstones. Block diagram shows delta-front, flood-and storm-influenced shelf deposits of Woodbine--Eagle Ford sediments, northern Polk County, Texas (from Siemers, 1981).

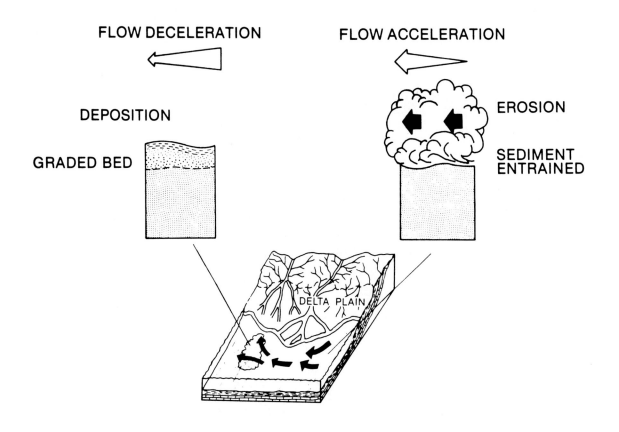

Fig. 26. Conceptual model for deposition of Damascus Field sandstones in cross-shelf zones due to deceleration of alongshelf storm flows in an area of shelf widening. Diagram shows inferred areas of flow contraction (acceleration and erosion) and flow expansion (deceleration and deposition of graded bed).

There is a close relationship between the flow deceleration model proposed above for the tabular sandstones at Damascus Field and the plume model proposed on previous pages as an alternative to the lobe switching model for Kurten Field sand ridge deposits. The plume model is a special case of the model for flow deceleration in a zone of shelf-widening. In this case, the shelf widens so abruptly downcurrent of the promontory, and the alongshore pressure gradient is so steep, that flow response goes beyond expansion and deceleration; the coastal boundary layer separates from the coast in the lee of the promontory and an eddy develops. There are differences between the models as well as similarities. In the flow deceleration model, the source of the sand is the upcurrent shelf, where fine sand is disseminated through the shelf floor mud. It is winnowed out and preferentially deposited in the deceleration zone. In the plume model the sand source is the shoreface, not the sandy mud of the shelf floor. Sand leaks out onto the plume from the shoreface, but the shelf floor is not necessarily sufficiently sandy for the deceleration zone to accumulate sand seaward of the plume.

Both Kurten Field and Damascus Field were demonstrably deposited in prodelta shelf settings. Data presented in the literature is not sufficiently detailed to distinguish with certainty among the plume, the lobe switching, and the flow deceleration models for the sandstones in these areas. However, in view of the evidence, the lobe switching model appears to be the best fit for the Kurten Field sand ridge deposits. Sand ridges in both modern and ancient settings are associated with transgressions in which winnowing of the substrate is intense and sustained. A thick mantle of lag sand forms and is swept into ridges characterized by distinctive facies. Of the three models evaluated, probably only the transgressive, lobe switching model produces sufficient winnowing to create the Kurten ridges.

The plume and flow deceleration models are more appropriate for the tabular sandstones of Damascus Field. Flow deceleration sands, like transgressed lobe sands, are also the product of winnowing, but the deceleration is a consequence of shelf geometry and can occur during a period of general regression. Plume formation is a response to coastal geometry and may also occur during a regression. The very fine-grained, upward-fining Damascus sands are responses to single-flow events, and represent at most intermittent winnowing during a period of general accumulation. The abundant evidence for slumping suggests that the storm sands were deposited on rapidly accumulating, unstable, prodelta clays. The coarse-grained, cross-bedded, upcurrent facies found in transgressive sand ridge deposits (Gaynor and Swift, in preparation) is totally missing. Damascus Field is thus compatible with the flow deceleration and plume depositional models. In order to distinguish between these, it would be necessary to determine whether the source for the Damascus beds was the upcurrent shoreface (plume model) or the upcurrent shelf as a whole (flow deceleration model).

# CHARACTERISTICS OF EAGLE FORD LENTICULAR SHELF SAND DEPOSITS

## Eagle Ford in Grimes County

Sandstones of the uppermost Eagle Ford Group in Grimes County (Fig. 1), are representative of the third distinct type of shelf sand deposit recognized in the Woodbine--Eagle Ford section of East Texas. These sandstones are referred to as "Sub-Clarksville" (Forgotson, 1958; Nichols, 1964; Barton 1982a). The term has been used to refer to a series of sands which develop at various stratigraphic levels in different parts of the Eagle Ford interval, but it has not been consistently applied. Generally, the Sub-Clarksville sands occur in the upper Eagle Ford just below the Eagle Ford-Austin boundary. Cores of the Sub-Clarksville sandstones in the Iola--Hill--Martin's Prairie area (Fig. 1) display vertical sequences of composition, texture and sedimentary structures (Barton, 1982a,b) which are significantly different from both those of the sand ridge deposits, and from the tabular or sheet-like deposits described previously.

In the Grimes County area, the Woodbine--Eagle Ford interval has a maximum thickness of approximately 210 m (700 ft), consisting of largely marine shelf deposits with some interbedded deltaic deposits. The Iola Field itself is located approximately 40 km (25 mi) updip from the Lower Cretaceous Edwards shelf-slope break. However, the configuration of the late Eaglefordian shelf may have been significantly different from that of the Lower Cretaceous shelf, due to the Cenomanian-Turonian clastic progradation of the shelf margin. At Iola, the Sub-Clarksville is a relatively thin, highly-quartzose sandstone directly below the Austin Chalk (Fig. 27). It reaches a maximum thickness of approximately 6 m (20 ft) at Iola Field and 3 m (10 ft) at adjacent Martin's Prairie (Barton, 1982b). The Sub-Clarksville here is reported to be gradationally overlain by the Austin Chalk and separated from the underlying Woodbine interval by a sharp basal contact which constitutes an erosional unconformity that is present throughout the area. Barton (1982b) described the general sequence of bedding units from bottom to top, as consisting of a clean, massive to laminated sandstone facies overlain by a bioturbated sandstone facies.

## Sedimentary Structures and Textures

Barton (1982b) subdivided the Sub-Clarksville into three units based on differences in sedimentary structures. (1) The lowermost unit consists of massive sandstone containing abundant shale clasts and rock fragments in a basal lag zone, immediately overlying black shale in sharp contact. This lower structureless sandstone grades upward into cross-bedded sandstone, overlain by

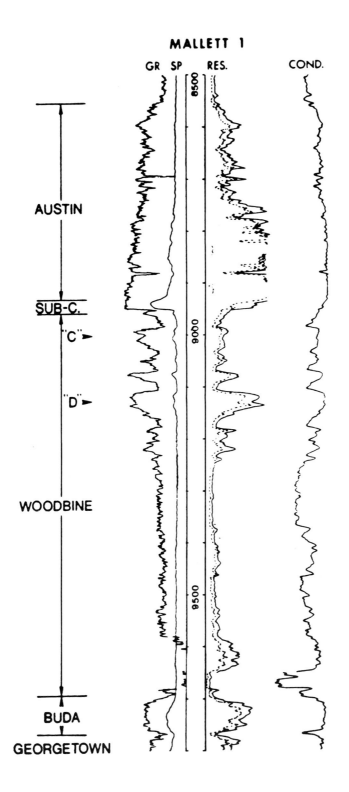

Fig. 27. Electric log of the Mallett 1 well showing typical response and stratigraphic relationships of the Sub-Clarksville (Sub-C) and Woodbine sandstones at Iola Field (from Barton, 1982a). Vertical scale is labelled in increments of 500 feet.

horizontally bioturbated sandstone with wavy (ripple?) laminae. (2) The next unit in the sequence is "characterized by clean, massive to laminated sandstone without bioturbation and is the thickest of the three units" (Barton, 1982b). These two lower units appear to represent an amalgamated sandstone bedding sequence. (3) The uppermost unit consists of intensely bioturbated (wavy-bedded?) shaley sandstone which grades over several feet into the Austin lime mudstones above.

Texturally, the Sub-Clarksville sandstones typically fine upwards, decreasing in mean grain size from the upper to lower limits of the medium sand range (Fig. 28). The overall mean grain size is 0.38 mm (medium grained) and the average maximum quartz grain size for the sandstones is 0.76 mm (coarse sand), the absolute maximum recorded is 1.21 mm (very coarse sand; Barton, 1982b). This textural gradation and associated vertical sequence of sedimentary structures indicate deposition from decreasing velocity currents. Barton (1982b) citing Hjulstrom (1935) notes that current velocities required to erode and transport sand grains are on the order of 10-20 cm sec$^{-1}$, for the moderately coarse grain size range, and 30 cm sec$^{-1}$ or greater for the very coarse sand. Barton (1982b) concluded current flow was intermittent and of varying velocity, based on the presence of alternating grain size trends observed in some of the Sub-Clarksville cores. The current velocities that Barton (1982b) cites are of similar magnitude to those typically recorded on modern shelves during winter storms (Lavelle et al., 1978). In any case the storm flow regime on the continental shelf contains a high frequency wave orbital component as well as a mean flow component, and as a result the mean threshold velocities for grain size classes are substantially reduced (Swift et al., in press a). This suggests that storm currents could easily have transported even the coarsest particles present in the Sub-Clarksville sandstones.

## Depositional Model, Grimes County

Primary rock properties, paleogeographic setting and foraminiferal paleoecology of the Sub-Clarksville sandstones suggest that they were deposited by storm-generated bottom-currents in an inner to middle neritic shelf environment (Barton, 1982a,b). The Sub-Clarksville sandstone sequences exhibit several characteristics indicative of rapid deposition by waning current flows. These include (1) sharp basal contacts (2) basal beds of massive sandstone containing shale clast lags, (3) upward decrease in grain size (4) upward transition in sedimentary structures from massive to laminated to rippled sandstones, and (5) general lack of bioturbation except in the uppermost part of the sandstone sequences. Barton (1982a,b) described the mechanism of deposition as storm-generated bottom currents. This interpretation is consistent with both the sandstone characteristics observed in the cores and the fluid dynamical observations of modern continental shelves. Although Barton did recognize the currents as storm generated, he called upon the storm surge mechanism described by Hayes (1967) to explain sandstone deposition. We have discussed the objections to

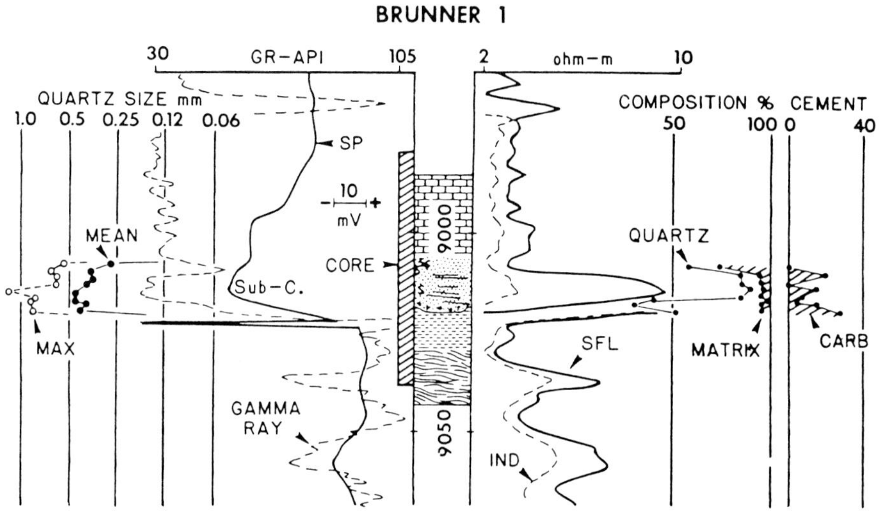

Fig. 28. Plot of texture, composition and log character of the Sub-Clarksville sandstone in the Brunner 1 well, Iola Field (from Barton, 1982a).

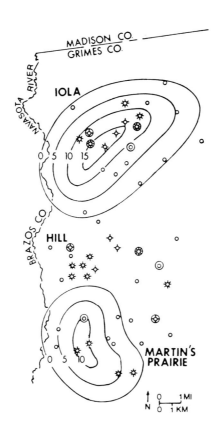

Fig. 29. Isopach of lenticular Sub-Clarksville sandstones in Grimes County, Texas. Location of Iola, Hill and Martin's prairie fields are shown. Contour interval is 5 feet (from Barton, 1982a).

this storm surge hypothesis on previous pages and suggest that wind-forced geostrophic flows provide a more reasonable mechanism.

Sub-Clarksville sand deposition apparently represented a distinct change in physical processes from those responsible for deposition of the underlying sediments. The sandstones occur immediately above a black, fissile shale unit containing pyrite and lacking bioturbation. This unit may represent a period of slow and possibly anoxic sedimentation prior to Sub-Clarksville deposition. Similar anoxic conditions during Eagle Ford deposition have been documented in the exposed section along the Balcones Fault Zone in Central Texas (Charvat and Grayson, Jr., 1981). Alternatively, the black shales may owe their color to postdepositional reducing conditions caused by low permeability related to fine grain size (Stehli et al., 1972). In either case, the period of low clastic influx and low energy suspension deposition evidenced by the shales was succeeded by high-energy current deposition of the much coarser sand-sized Sub-Clarksville sediments (Barton 1982a, b). We suggest that the vertical sequence of sedimentary structures and sandstone composition of the Sub-Clarksville sands indicate deposition during a regional transgression.

Modification of the shelf regime during transgression was probably characterized by (1) the transformation of deltas into sediment-trapping estuaries, (2) remobilization of the substrate by wind-forced storm currents forming broad submarine erosional surfaces, and (3) deposition of sand-dominated sequences swept into zones of flow deceleration. A transgressive setting for the Sub-Clarksville is further substantiated by the presence of phosphate ooids in the sandstones in amounts of up to 19 percent of the detrital composition (Barton 1982b). Phosphate formation was probably favored by both the of slow clastic sedimentation rate characteristic of transgressive settings and the presence of significant shelf topography related to contemporaneous movement on the adjacent Hill salt dome.

The relatively coarse mean grain size of the Sub-Clarkesville, and the presence in it of pebble size terrigenous clasts is problematic. Absence of coarse material lower in the Eagle Ford--Woodbine section indicates that until this time, the successive subaerial depositional zones of the coastal plain had filtered out coarse clastics from the load delivered to the sea. The change in textural type may indicate that the transgression overran the braid plain beyond the sediment-trapping coastal meander belt, and breached it by the mechanism of erosional shoreface retreat (Swift et al., in press b).

The morphology of the Sub-Clarksville sandstones is strongly lenticular (Fig. 29). Barton (1982a,b) inferred that sandstone deposition was restricted to bathymetric lows that flanked the Hill dome (Fig. 29). Near-bottom currents decelerated in topographic lows on the continental shelf, resulting in sediment deposition. However, flows interact in a similar manner with topographic highs: they accelerate on the up current side of the high then decelerate over the crest, resulting in deposition there (Swift and Rice, 1984).

The hypothesis of Sub-Clarksville sand deposition by flow deceleration in lows is a reasonable one, but it is difficult to unequivocably demonstrate that the present-day topographic lows flanking the Hill dome actually had negative relief at the time of Sub-Clarksville deposition. In view of the transgressive setting of the Sub-Clarksville, it would be reasonable to expect sand to also be deposited as shelf sand ridges, in which sand deposition and zones of flow deceleration generate each other through a feedback mechanism (Figueriredo et al., 1981). Consequently, it is possible that sand was deposited both as sand ridges and in pre-existing lows and highs on the Sub-Clarksville shelf (Fig. 30).

## CONCLUSIONS

At least three distinct types of shelf sandstone bodies can be recognized in the Woodbine--Eagle Ford based on external geometry and internal facies associations. They are:

1) <u>sand ridge deposits</u> comprised of four major facies; highly quartzose flaser- to cross-bedded sandstones, thin-bedded sandstones with intervening silty shale laminae, bioturbated sandstones, and lenticular bedded silty shales.

2) <u>tabular</u> or <u>sheet sandstones</u> comprised of four major facies; massive to laminated sandstones, contorted thin-bedded (?) sandstones, swirled and sheared siltstones, thin diamict conglomerates.

3) lenticular sandstones comprised of an amalgamated massive to laminated sandstone facies and a bioturbated (wavy-bedded?) sandstone facies.

The primary factors controlling sandstone deposition and distribution are: (1) episodic wind-forced shelf storm current deposition, (2) shelf configuration, (3) shelf topography and (4) transgressive or regressive setting.

Woodbine--Eagle Ford sand ridge deposits (e.g. Kurten Field) occur in inner to middle shelf settings in association with deposits of a wave-dominated delta. These sand bodies have asymmetric geometries and facies distributions similar to sand ridges described in the Western Interior Seaway (Gaynor and Swift, in preparation). However, it is considered likely that the sand ridges developed from local transgressive reworking of abandoned delta lobes during an overall regressive shelf progradational phase. Eagle Ford sand ridge deposits may have formed on the central and outer shelf during a regional transgression.

Deposition of tabular sands (e.g. Damascus field) in the Woodbine interval was probably controlled by shelf configuration. A narrow shelf with a shoreline perturbation due to a prograding

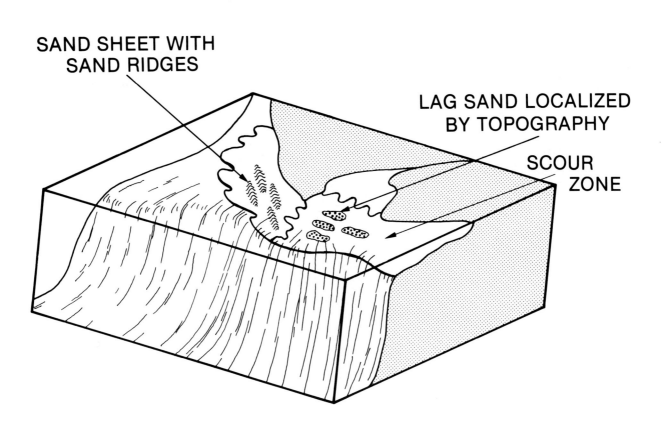

Fig. 30. Conceptual model for deposition of Sub-Clarksville sandstones as transgressive sands occurring as (1) lag sands localized by shelf topography and (2) thin sand sheets swept up into sand ridge deposits.

delta may have caused sand deposition across the shelf by local flow deceleration of alongshelf storm currents in the zone where the shelf widened downcurrent of the delta. Development of a coastal eddy and the resulting formation of sand plume deposits (as on the Nile delta shelf) may be involved, but this class of deposit has not yet been positively identified in the subsurface. Theoretically, this type of flow deceleration mechanism could result in deposition of reservoir-quality shelf sands in either a transgressive or regressive setting, but the fine-grained nature of the Woodbine deposits suggests that they were regressive in nature.

Lenticular sands in the upper Eagle Ford interval appear to have been deposited during a regional transgression. Evidence for a transgressive setting includes deposition of sands on broad erosional surfaces, a relatively coarser grain size and significant content of phosphate ooids. The distribution of the lenticular sandstones was strongly influenced by local shelf topography. These sandstones are probably accumulations of beds deposited by localized deceleration of intermittent storm flows. Eagle Ford transgressive sandstones may also occur as sand ridges or widespread "patchy" sheet sands.

Research currently in progress on the sedimentology and sequence stratigraphy of the different Upper Cretaceous shelf sandstone types is aimed at providing integrated depositional models for shelf sedimentation on clastic progradational passive continental margins.

## ACKNOWLEDGEMENTS

We would like to express our gratitude to Chuck Seimers for discussing with us the depositional origin of the Woodbine sandstones at Damascus Field. Chuck also made it possible for us to utilize the Hinton Dorrance 7A core and contributed his interpretation of the area for inclusion in this manuscript. We are also grateful for the cooperation of the ARCO Exploration district personnel in Dan Nedland and Bob Tabbert's groups. We would like to extend special thanks to Hilary Johns and Glenn Zinter for their contributions and many hours of discussion on the Woodbine over the last year which have benefited us significantly in developing a better understanding of East Texas Upper Cretaceous stratigraphy. We also thank Rod Tillman and Chuck Siemers for their comments on the manuscript. This research has been part of an ongoing sedimentologic and seismic stratigraphic study of the Woodbine--Eagle Ford funded by the Fannie and Johh Hertz Foundation and ARCO Exploration and Technology.

REFERENCES CITED

AGER, D. V., 1974, Storm deposits in the Jurassic of the Moroccoan high atlas: Paleogr., Paleoclimat. Paleontol., v. 15, p. 83-93.

AIGNER, T., 1982, Calcareous tempestites: Storm-dominated stratification in Upper Muschelkalk limestones (Middle Trias, SW-Germany), in G. Einsele and A. Seilacher, (eds) Cyclic and Event Stratification: Berlin, Springer-Verlag, p. 180-198.

AIGNER, T. and H. E. REINECK, 1982, Proximality trends in modern storm sands from the Helgoland Bight (North Sea) and their implications for basin analysis: Senckenbergiana Maritima, v. 14, p. 183-215.

BAGNOLD, R. A., 1962, Autosuspension of transported sediment, turbidity currents: Royal Soc. London Proc., v. A265, p. 315-319.

BAGNOLD, R.A., 1963, Mechanics of marine sedimentation, in N. M. Hill, (ed) The Sea: New York, Wiley Interscience, p. 507-582.

BALSLEY, J., 1982, Cretaceous wave-dominated delta systems: Book Cliffs, east central Utah: Mitchell Energy Co., Denver, 219 p.

BARTON, R.A., 1982a, Contrasting depositional processes of Sub-Clarksville and Woodbine reservoir sandstones: Gulf Coast Association Geological Societies Transactions, v. 32, p. 121-136.

BARTON, R.A., 1982b, Contrasting depositional processes of Sub-Clarksville and Woodbine reservoir sandstones, Grimes County, Texas: Master's thesis, Texas A & M University, College Station, Texas, 138 p.

BATES, R.L. and J.A. JACKSON, (eds) 1980, Glossary of Geology, second edition: Falls Church, Virginia, American Geological Institute, p. 575-576.

BELL, W.A., 1980, The Kurten "Woodbine" Field, Brazos County, Texas, in Facts and Principles of World Petroleum Occurrence: Canadian Society of Petroleum Geologists Memoir 6, p. 913-950.

BOUMA, A.H., 1962, Sedimentology of some flysh deposits: Amsterdam, Elsevier Publishing Co., 168 p.

BOYLES, J.M. and A.J. SCOTT, 1982, A model for migrating shelf-bar sandstones in Upper Mancos Shale (Campanian), northwestern Colorado: AAPG Bulletin, v. 66, p. 491-508.

CHAMBERLIN, C.K., 1978, Recognition of trace fossils in cores, in P.B. Basan, (ed) Trace Fossil Concepts: SEPM Short Course No. 5, p. 133-184.

CHARVAT, W.A., and R.C. GRAYSON, Jr., 1981, Anoxic sedimentation in the Eagle Ford Group (Upper Cretaceous) of Central Texas (abs): Gulf Coast Association Geological Societies Transactions, v. 31, p. 256.

CLARKE, T.L., B. LESHT, R.A. YOUNG, D.J.P. SWIFT, and G.L. FREELAND, 1982, Sediment resuspension by surface-wave action: an examination of possible mechanisms: Marine Geology, v. 49, p. 43-59.

COLEMAN, J.M., H.H. ROBERTS, S.P. MURRAY, and H. SALAMA, 1981, Morphology and dynamic sedimentology of eastern Nile delta shelf, in C. A. Nittrouer, (ed), Sedimentary Dynamics of Continental Shelves: New York, Elsevier, p. 301-326.

CUOMO, R. F., 1984, The geologic and morphologic evolution of Ship Shoal, Northern Gulf of Mexico: Master's thesis, Louisiana State University, 248 p.

DOTT, R.H., Jr. and J. OURGEOIS, 1982, Hummocky stratification: significance of its variable bedding sequences: Geol. Soc. America Bulletin, 93, p. 663-680.

DRAKE, D. E., D.A. CACCHIONE, R.D. MUENCH, and C.H. NELSON, 1980, Sediment transport in Norton Sound, Alaska: Marine Geology, v. 36, p. 97-126.

FIGUEIREDO, A.G., D.J.P. SWIFT, W.C. STUBBLEFIELD, and CLARKE, T., 1981, Sand ridges on the inner Atlantic Shelf of North America: Morphometric comparisons with Huthnance stability model: Geomarine Letters, v. 1, p. 157-191.

FORGOTSON, J.M., 1958, The basal sediments of the Austin Group and the stratigraphic position of the Tuscaloosa Formation of central Louisiana: Gulf Coast Association Geological Societies Transactions, v. 8, p. 117-125.

FOSS, D.C., 1978, Depositional environment of Woodbine sandstones, Polk, Tyler and San Jacinto counties, Texas: Master's thesis, Texas A&M University, College Station, Texas, 234 p.

FOSS, D.C., 1979, Depositional environment of Woodbine sandstones Polk County, Texas: Gulf Coast Association Geological Societies Transactions, v. 29, p. 83-94.

GALLOWAY, W.E., T.E. EWING, C.M. GARRETT, N. TYLER, and D.G. BEBOUT, 1983, Atlas of Major Texas Oil Reservoirs: Austin Bureau of Economic Geology, 139 p.

GAYNOR, G.C. and D.J.P. SWIFT, in preparation, Shannon sandstone depositional model: sand ridge formation on the Campanian western Interior shelf.

HALBOUTY, M.T. and J.J. HALBOUTY, 1982, Relationships between East Texas Field region and Sabine uplift in Texas: AAPG Bulletin, v. 66, p. 1042-1054.

HAMBLIN, A.P. and R.G. WALKER, 1979, Storm dominated shallow marine deposits: the Fernie-Kootenay (Jurassic) transition, Southern Rocky Mountains: Canadian Journal Earth Science, v. 16, p. 1673-1689.

HAYES, M.O., 1967, Hurricanes as geological agents: case studies of Hurricanes Carla, 1961 and Cindy 1963: Texas Bureau Economic Geology Report of Investigations 61, 54 p.

HJULSTROM, F., 1935, Studies of morphological activity of rivers as illustrated by the River Fyris: Upsala University, Mineralogisk-Geologiska Institute Bulletin, v. 25, p. 221-227.

HOLMES, C.W., 1982, Geochemical indicies of fine sediment transport, northwest Gulf of Mexico: Journal of Sedimentary Petrology, v. 52, p. 307-321.

INTERNATIONAL OIL SCOUTS ASSOCIATION, 1983, International Oil and Gas Development Yearbook (Review of 1977), v. 48, part II, p. 39, 55, 56.

KENYON, N. H., R. H. BELDERSON, A. H. STRIDE and M. A. JOHNSON, 1981, Offshore tidal sand banks as indicators of net sand transport and as potential deposits in Holocene Marine Sedimentation in the North Sea Basin, p. 257-268, in S. D. Nio, R. T. E. Schuttenhelm and T. C. E. van Weering, (eds): Int. Assoc. Sedimentologists, Special Publ. 5., 515 pp.

KRYNINE, P.D., 1948, The megascopic study and field classification of sedimentary rocks: Journal Geology, v. 56, p. 130-165.

LAVELLE, J.W., D.J.P. SWIFT, P.E. GADD, W.L. STUBBLEFIELD, F.N. CASE, H.R. BRASHEAR, and K.W. HUFF, 1978, Fair weather and storm transport on the Long Island, New York, inner shelf: Sedimentology, v. 25, p. 825-842.

MORTON, R.A., 1981, Formation of storm deposits by wind-forced currents in the Gulf of Mexico and the North Sea: Special Publication International Association of Sedimentologists No. 5, p. 303-396.

NELSON, C.H., 1982, Modern shallow-water graded sand layers from storm surges, Bering Shelf: mimic of Bouma sequences and turbidite systems: Journal of Sedimentary Petrology, v. 52, p. 537-545.

NICHOLS, P.H., 1964, The remaining frontiers for exploration in northeast Texas: Gulf Coast Association of Geological Societies Transactions, v. 14, p. 7-22.

OLIVER, W.B., 1971, Depositional systems in the Woodbine Formation (Upper Cretaceous) northeast Texas: Texas University Bureau Economic Geology Report of Investigations 73, 23 p.

PANTIN, H. M., 1979, Interaction between velocity and effective density in turbidity flow: Phase plume analysis, with criteria for autosuspension: Marine Geology, v. 31, p. 59-99.

PARKER, G., 1982, Conditions for the ignition of catastrophically erosive turbidity currents: Marine Geology, v. 46, p. 307-327.

PARKER, G., N. LANFREDI, and D.J.P. SWIFT, 1982, in press. Substrate response to flow in a southern hemisphere ridge field: Argentine Inner Shelf: Sedimentary Geology, v. 33, p. 195-216.

PATTERSON, J. E., Jr., 1983: Exploration potential and variations in shelf plume sandstones, Navarro Group (Maestrechtian, East Central Texas): Master's thesis, Univ. Texas Austin, 91 p.

PRIOR, D.B. and J.M. COLEMAN, 1978a, Disintegrating retrogressive landslides on very-low-angle subaqueous slopes, Mississippi Delta: Marine Geotechnology, v. 3, p. 37-60.

PRIOR, D.B. and J.M. COLEMAN, 1978b, Submarine landslides on the Mississippi River delta-front slope: Geoscience and Man, v. 19, p. 41-53.

REINECK, H.E. and F. WUNDERLICH, 1968, Classification and origin of flaser and lenticular bedding: Sedimentology, v. 11, p. 99-104.

RICE, D.D., 1980, Coastal and deltaic sedimentation of Upper Cretaceous Eagle Sandstone: relation to shallow gas accumulations, north central Montana: AAPG Bulletin, v. 64, p. 316-338.

RICE, D.D. and SHURR, G., 1980, Shallow, low permeability reservoirs of northern Great Plains - assessment of their natural gas resources: AAPG Bulletin, v. 64, p. 969-987.

SCRUTON, P. C., 1960, Delta building and the deltaic sequence, p. 82-102 in Shepard, F. P. Phleger, F. B., and Van Andel, T. H., (eds), Recent Sediments, Northwest Gulf of Mexico: Tulsa, Am. Assoc. Petroleum Geologists, 394 p.

SHARAF AL DIN, S. H., 1977, Effect of the Aswan High Dam on the Nile flood and on the estuarine and coastal circulation pattern along the Mediterranean Egyptian coast: Limnology and Oceanography, 1977, p. 194-207.

SIEMERS, C.T., 1978, Submarine fan deposition of the Woodbine-Eagle Ford interval (Upper Cretaceous), Tyler County, Texas: Gulf Coast Association Geological Societies Transactions, v. 28, p. 493-533.

SIEMERS, C.T., 1981, Recognition of "shallow-water" updip shelf and "deep-water" downdip slope deposits of the subsurface Woodbine-Eagle Ford interval (Upper Cretaceous) in East Texas: Second Annual Research Conference, Gulf Coast Section, SEPM, p. 57-61.

SIEMERS, C.T. and P.C. HUDSON, 1981, Delta-front shelf storm deposits of subsurface Woodbine-Eagle Ford interval (Upper Cretaceous), Damascus Field, northern Polk County, Texas: success from combined development geology and sedimentologic core analysis (abs): AAPG Bulletin, v. 65, p. 992.

SLATT, R.M., 1984, Continental shelf topography: key to understanding the distribution of shelf sand ridge deposits from the Cretaceous Western Interior Seaway: AAPG Bulletin, v. 68, p. 1107-1120.

SOUTHARD, J. P., AND M.E. MACKINTOSH, 1981, Experimented test of autosuspension: Earth Surface Processes and Landforms, v. 6, p. 103-111.

SPEARING, D.R., 1976, Upper Cretaceous Shannon Sandstone: an offshore shallow-marine sand body: Wyoming Geological Association 28th Annual Guidebook, p. 65-72.

STEHLI, F.G., W.B. CREATH, C.F. UPSHAW, and J.M. FORGOTSON, Jr., 1972, Depositional history of Gulfian Cretaceous of the East Texas embayment: AAPG Bulletin, v. 56, p. 38-67.

SWIFT, D.J.P., G. HAN, and C.E. VINCENT, in press, Fluid process and sea-floor response on a modern storm-dominated shelf: Middle Atlantic Shelf of North America. Part I: The storm current regime, in R. J. Knight, (ed), Shelf Sands and Sandstone Reservoirs: Can. Soc. Petr. Geol. Symposium Volume.

SWIFT, D.J.P., J.A. THORNE, and G. F. Ortel, in press b, Fluid process and sea-floor response on a modern storm-dominated shelf: Middle Atlantic Shelf of North America. Part II: Response of the shelf floor, in R. J. Knight, (ed), Shelf Sands and Sandstone Reservoirs: Can. Soc. Petr. Geol. Symposium Volume.

SWIFT, D.J.P. and D.D. RICE, 1984, Sand bodies on muddy shelves: A model for sedimentation in the Western Interior Seaway, North America, in R.W. Tillman and C.T. Siemers, (eds), Siliciclastic Shelf Sediments: SEPM Special Publication 34, p. 43-62.

THEISS, R.M., 1983, Environment of deposition of Woodbine and Eagle Ford sandstones, Leon, Houston, and Madison counties, Texas: Master's thesis, Texas A&M University, College Station, Texas, 119 p.

TILLMAN, R.W. and R.S. MARTINSEN, 1984, The Shannon shelf-ridge sandstone complex, Salt Creek Anticline Area, Powder River Basin, Wyoming, in R.W. Tillman and C.T. Siemers, (eds), Siliciclastic Shelf Sediments: SEPM Special Publication 34, p. 85-142.

TURNER, J.R. and S.J. CONGER, 1981, Environment of deposition and reservoir properties of the Woodbine sandstone at Kurten Field, Brazos County, Texas: Gulf Coast Association Geological Societies Transactions, v. 31, p. 213-232.

TURNER, J.R. and S.J. CONGER, 1984, Environment of deposition and reservoir properties of the Woodbine sandstone at Kurten Field, Brazos County, Texas, in R.W. Tillman and C.T. Siemers, (eds), Siliciclastic shelf sediments: SEPM Special Publication 34, p. 215-249.

VINCENT, C.E., R.A. YOUNG, and D.J.P. SWIFT, 1982, On the relationship between bedload and suspended sand transport on the inner shelf, Long Island, New York: Journal of Geophysical Research, v. 87, p. 4163-4170.

WALKER, R.G., 1985, Geological evidence for storm transportation and deposition on ancient shelves, in R.W. Tillman, D.J.P. Swift and R.G. Walker, (eds), Shelf Sands and Sandstone Reservoirs: SEPM Short Course Notes.

TOCITO SANDSTONE CORE, HORSESHOE FIELD,
SAN JUAN COUNTY, NEW MEXICO

R. W. Tillman
Cities Service Research
Tulsa, Oklahoma

## INTRODUCTION

The Solar Petroleum Navajo F-151 (NW Sec. 10 T31N R17W), which is the subject of this discussion, is located in Horseshoe Field in the San Juan Basin in San Juan County, New Mexico (Fig. 1). It produces from the Tocito Sandstone Lentil of the Mancos Shale which has been incorrectly included by many workers in the Gallup Sandstone of Turonian and Coniacian (Upper Cretaceous) age (Fig. 2 and 3). The Tocito Sandstone Lentil is typically interbedded with or lies at the base of the upper Mancos (Niobrara) Shale in the western and central parts of the San Juan Basin (Molenaar, 1973, 1983a).

Substantial amounts of the production attributed to the Gallup Sandstone should probably now be attributed to the Tocito Sandstone Lentil (125 million barrels, to date, Fassett, 1981). The most conclusive evidence for differentiating the Tocito Sandstone from the Gallup Sandstone is obtained from outcrop studies and from cores such as the Solar Petroleum Navajo F-151.

## GALLUP SANDSTONE

The more seaward portions of the true Gallup Sandstone are generally interpreted to be beach or shoreface deposits in the subsurface and in some outcrops along the west flank of the San Juan Basin (Fig. 1). Shoreline sandstones occur in the lower part of the Gallup in even the most landward (southwesterly) outcrops (Molenaar, personal communication, 1984). The upper part of the Gallup consists of continental and fluvial facies southwest of the upper Gallup shoreline facies (Molenaar, 1973, 1983b).

The Gallup prograded from southwest to northeast prior to deposition of the upper part of the upper Mancos (Niobrara) Shale and Tocito Sandstone Lentil. A major unconformity separates many of the lower Tocito sandstones from underlying formations. The Tocito overlies units ranging from the Juana Lopez Member of the Mancos Shale on the northeast to the true Gallup Sandstone on the southwest (Figs. 2 and 3).

## TOCITO SANDSTONE LENTIL OF THE MANCOS SHALE

The Tocito contains many facies which are similar to those observed in the Upper Cretaceous Shannon Sandstone of Wyoming. In describing the Tocito in the Solar Navajo F-151 core the same facies designation are used as for the Shannon Sandstone in the Powder River basin of Wyoming, (Tillman and Martinsen, 1984). This facies terminalogy is used somewhat hesitantly since it is doubtful that thick sand-ridge accumulations which are typical of some Shannon sandstones are common in the Tocito.

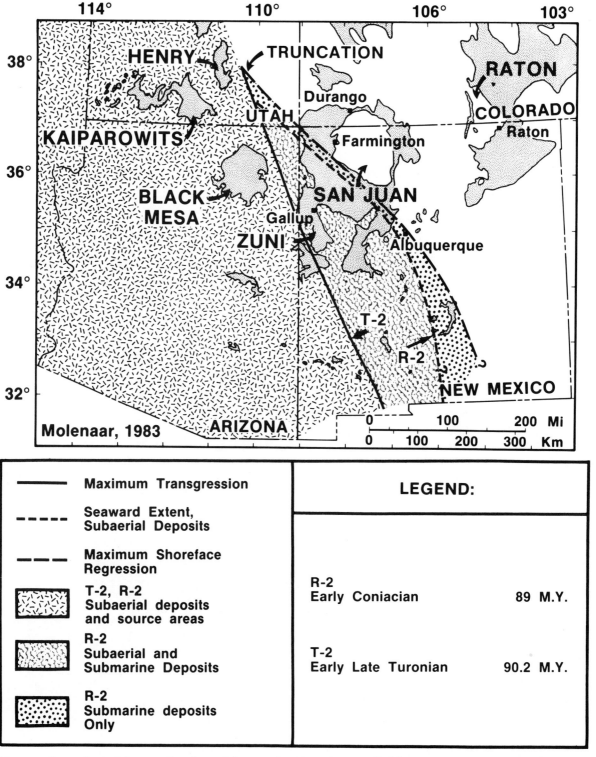

Figure 1. Location of shoreline of the true Gallup Sandstone during late Turonian to early Coniacian in northwestern New Mexico. Regression R-2, which includes the Gallup, and Transgression T-2 are as designated by Molenaar (1983). R-2 and T-2 are also shown in Figure 3. Horseshoe field lies just northeast of the position of maximum Gallup shoreface regression.

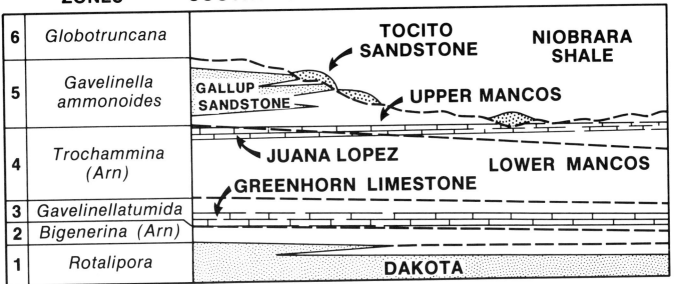

Figure 2. Cretaceous foraminiferal zones. Note that foraminiferal zones are indicated to be oblique to the lithologic units. The Gallup Sandstone, Juana Lopez and Greenhorn Members are interpreted by Lamb (1968) to climb stratigraphically to the northeast. Molennar (personal communication, 1985) believes that the Greenhorn and Juana Lopez do not climb stratigraphically. A major unconformity is present below the Tocito and upper Mancos (Niobrara) Shale.

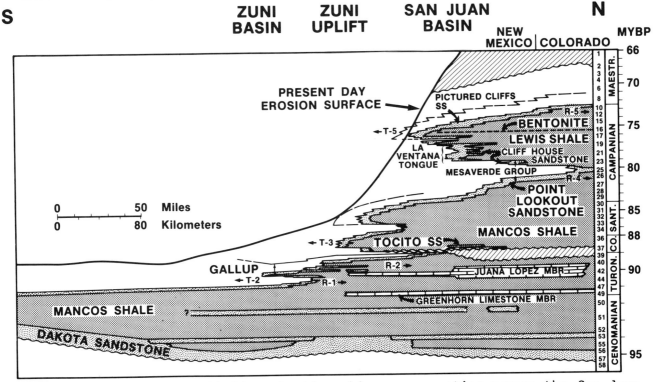

Figure 3. South to north time-stratigraphic cross section across the San Juan Basin and extending south into the Zuni Basin. The Gallup Sandstone and Dilco Coal Member of the Crevasse Canyon Formation are the youngest pre-unconformity units. The pre-Tocito and post-Gallup unconformity is shown at about 89 mybp in the early Coniacian. Modified from Molenaar (1983a).

Production from the several benches of sandstone has been established at Horseshoe field (McCubbin, 1969). Horseshoe field is very long and narrow and trends NW-SE (Fig. 4). Production at Horseshoe field is from a medium- to coarse-grained glauconitic sandstone which lies at or near the base of the upper part of the Mancos (Niobrara) Shale and from similar sandstones up to 100 feet higher in the section (Fig. 5). These sandstones, which occur well above the base of the Upper Mancos, are commonly completely surrounded by shale. The relationships of the sandstones to the basal unconformity and to the portion of the Mancos Shale which lies between the lenses of Tocito sandstone are shown in Figure 5, a west to east cross section through the field. A single core, the Solar Petroleum F-151 Navajo, was selected for this study to show the typical facies sequence at Horseshoe field.

As can be seen in the sketch of Tocito facies (Fig. 6) in the Solar Petroleum F-151, the Gallup-Juana Lopez contact at 944.6 feet is very sharp. The contact is marked by strongly contrasting lithologies and facies types (Fig. 7) and this is interpreted as an unconformity. The Mancos Shale, which is below the unconformity, consists of rippled to horizontally laminated interbedded shaly siltstones and silty fine-grained sandstones which are interpreted to be inner(?) shelf deposits.

## TOCITO FACIES

The Tocito, which overlies the unconformity, may be described in the core as an interbedded trough-bedded and ripple-bedded highly-glauconitic sandstone containing interbeds of silty bioturbated sandstone. The ratio of ripples to troughs increases upward. The sandstone contains scattered pebbles and Inoceramus fragments. Thin sets, 0.1-0.3' thick, of burrowed to bioturbated sandstone are interbedded within the main producing interval which is designated as Unit 2. The perforated interval is from 910-34 feet (log depth), which includes most of Unit 2. Most of the Tocito in this core may be designated as a Bar Margin Facies (Type 2) following Tillman and Martinsen (1984) and as modified from Porter (1976). Overlying this unit (Figs. 6 and 7) are several moderate to low energy sandy facies also assigned to the Tocito. In the top several feet of the core the amount of burrowing increases dramatically from a burrowed rippled shaly sandstone to a thoroughly burrowed (bioturbated) silty sandstone.

The Tocito is interpreted to have been deposited during a transgression or stillstand several tens of miles offshore by shelf currents which moved fine- to medium-grained sediments from northwest to southeast (McCubbin, 1969). The fine- to medium-grained Tocito sandstones were not sourced by Gallup beach and shoreface deposits which are almost all fine grained.

A cross section (Fig. 8) constructed by Penttilla (1964) across the northwest end of Horseshoe Field (Fig. 9) shows the topography on which the lower Tocito was deposited. Although not directly in the line of cross section, the Solar Petroleum core would project into the section at about the location as well number 4. A true Central Bar Facies of the type recognized in the Shannon was not deposited in the cored well.

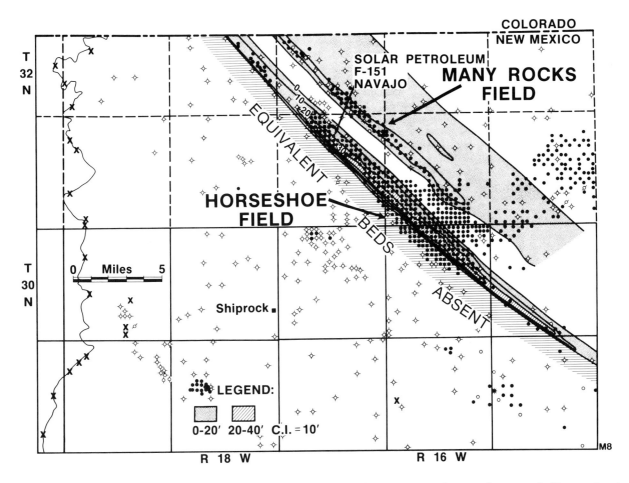

Figure 4. Isopach map of lower Tocito Sandstone at Horseshoe and Many Rocks fields, San Juan Basin, northwestern New Mexico (McCubbin, 1969). Sandstones trend NW-SE as indicated by closely spaced contours. The sandstones in Horseshoe Field thicken abruptly perpendicular to the trend of the contours. The contour interval is 10' and areas of thickest Tocito Sandstone deposition are indicated by the diagonally ruled pattern. Location of cored well, the Solar Petroleum F-151 Navajo, is indicated.

---

The cross section constructed by McCubbin (1969) shows that the lower sand at Horseshoe field is concentrated in an area which is topographically low, trends NW-SE, and was deposited offshore on a shallow shelf (Fig. 10). The location of the Solar Petroleum F-151 Navajo, which produces from the lower Tocito Sandstone, is shown in Figures 4 and 9.

Logs from the cored well (Fig. 11) show that the cored interval includes almost all of the lower Tocito Sandstone. The lower unconformable contact is at the very abrupt change to siltstone and shale at the base (944 feet, core depth) of the sandstone. The gamma ray log run on the core is included and was utilized to determine the differences in depth between the core and subsurface gamma-ray log (-6.6'). Additional core to log corrections are shown at the top of Figure 11.

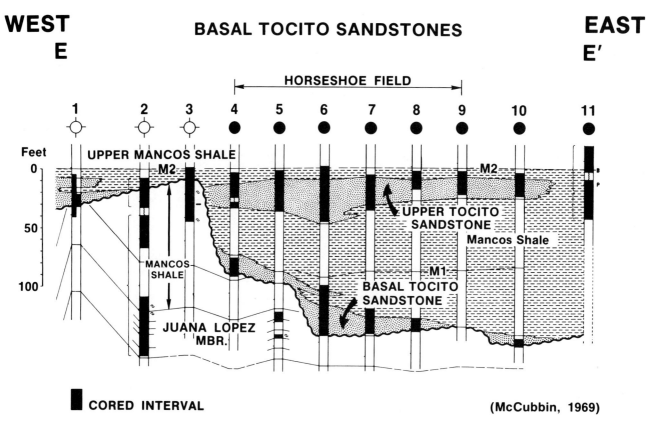

Figure 5. Upper and lower Tocito sandstones at Horseshoe field. Section is hung on M2 bentonite marker and shows the relationship of the Tocito sandstones to the underlying topographic surface which is cut into (1) Juana Lopez Member of the Mancos Shale and (2) the shaley to silty portion of the Mancos Shale. Cored intervals, on which lithologic and environmental interpretations are based, are indicated.

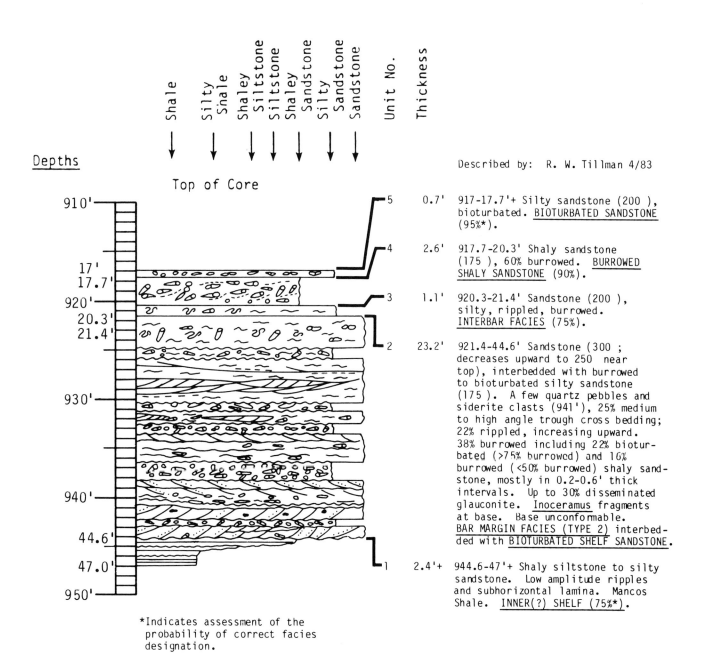

Figure 6. Core description Solar Petroleum Navajo F-151, 917.7-947.0'.

Figure 7 (A-D). Photographs of slabbed core, Solar Petroleum F-151 Navajo, which includes the Tocito Sandstone and the Mancos Shale. The contact at 944.6 feet (core depth) is a significant unconformity.

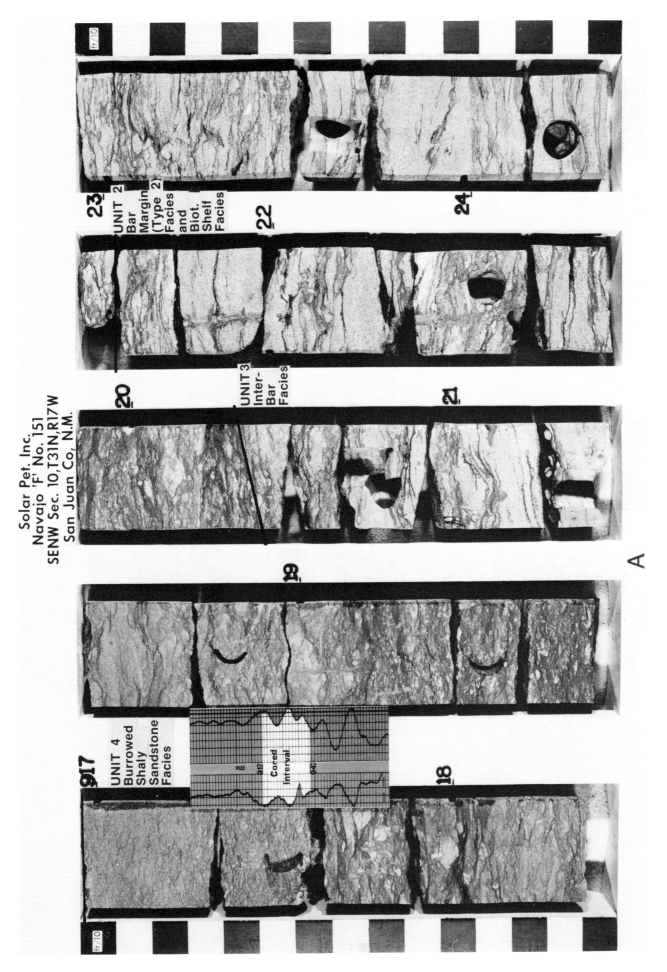

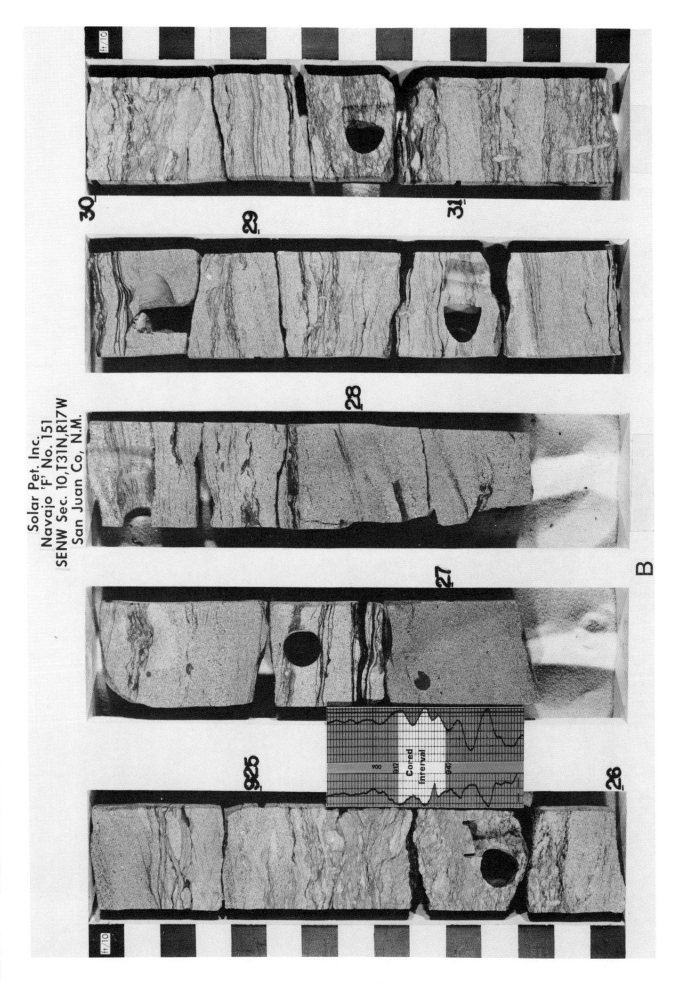

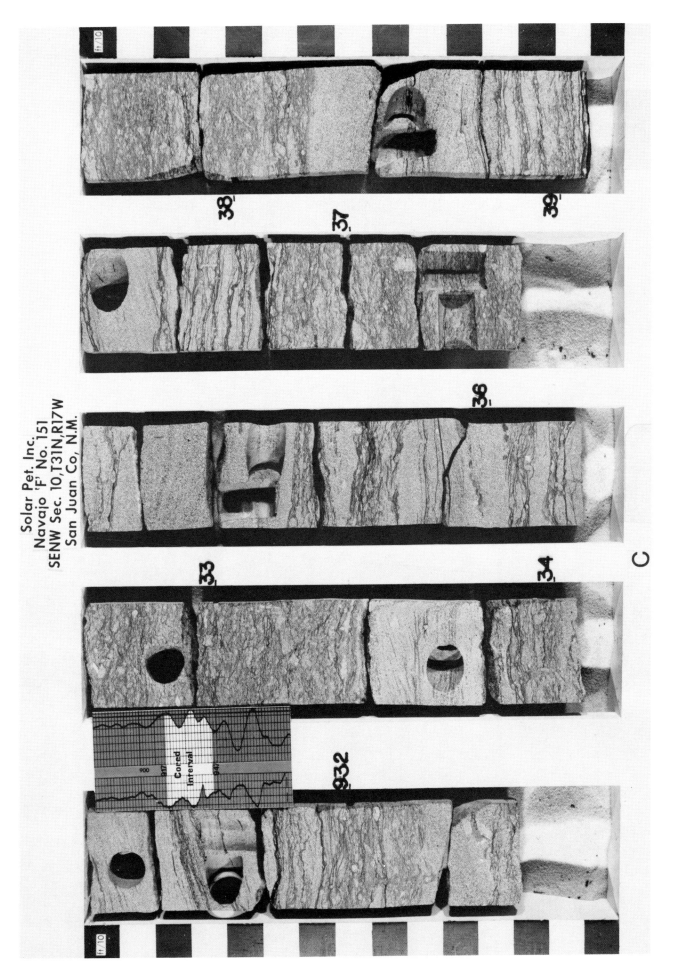

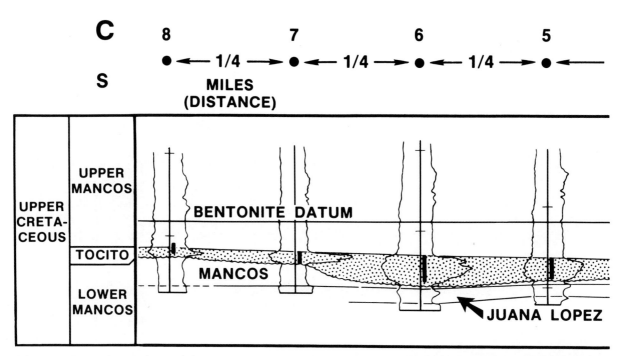

Figure 8. Stratigraphic cross section through northwest end of Horseshoe field. Location of section shown in Figure 9. Note variation in thickness of Tocito, which is in part dependent on topography of underlying surface. The Solar Petroleum F-151 Navajo core lies very close to this section and could be projected into the section at about the location of well number 4. For details on individual wells see Penttila (1964).

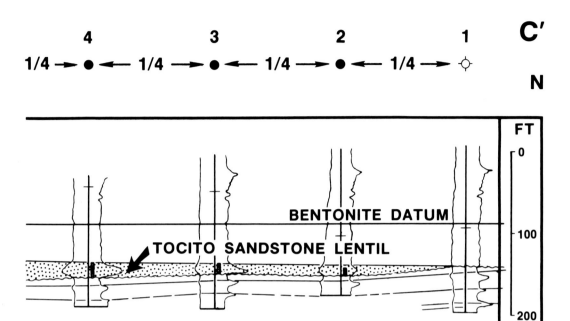

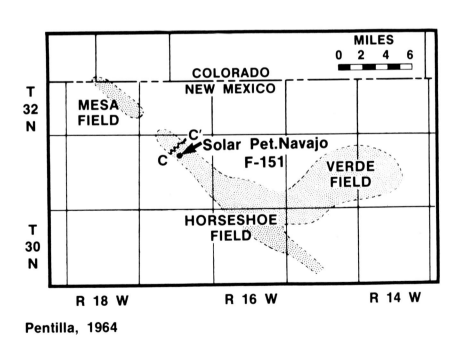

Figure 9. Location map of Horseshoe, Verde and Mesa fields, San Juan Basin, New Mexico (Penttila, 1964). Section C-C' is Figure 8 of this paper.

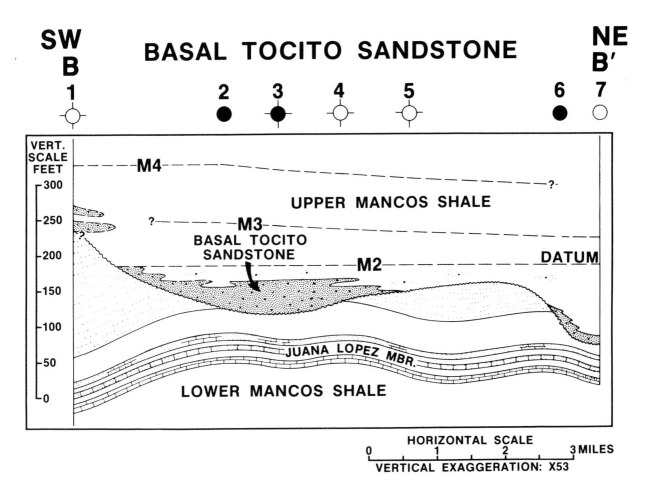

Figure 10. Lower Tocito Sandstone deposited in topographic lows on or near unconformity (from McCubbin, 1969). Sandstone "thicks" are elongate NW-SE as shown in Figure 4.

---

ACKNOWLEDGEMENTS

The core was prepared by Donna Woodruff and photographed by Fred Mason. The manuscript was reviewed by C. M. Molenaar, R. L. Doty, D. J. P. Swift, R. G. Walker and D. Nummedal. Jayme Kline and Janice Brewer typed the manuscript. The slabbed core was donated to Cities Service by the U.S.G.S. core repository in Denver, Colorado. One slab of this core is in permanent storage at that facility.

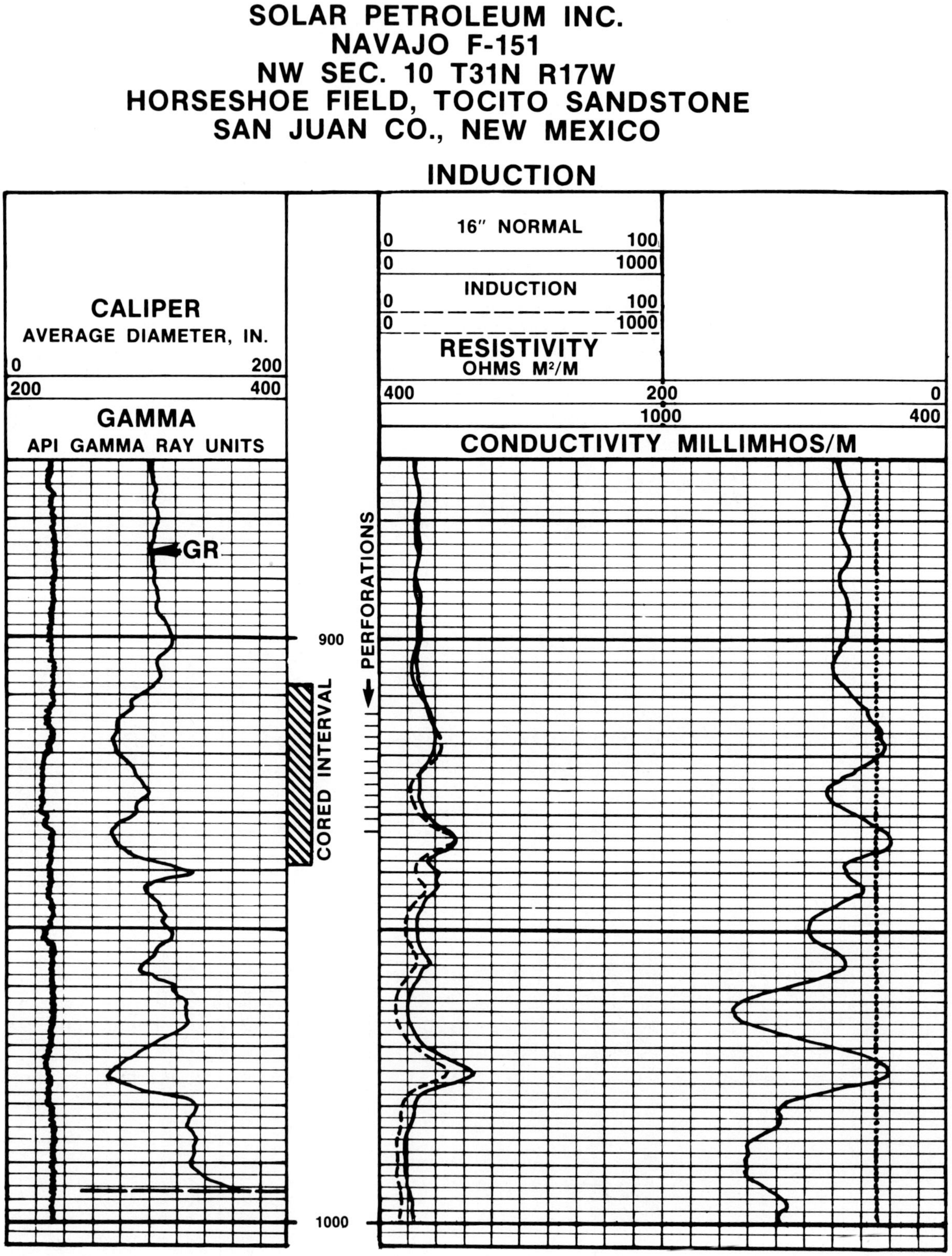

**SOLAR PETROLEUM INC.
NAVAJO F-151
NW SEC. 10 T31N R17W
HORSESHOE FIELD, TOCITO SANDSTONE
SAN JUAN CO., NEW MEXICO**

Figure 11. Subsurface logs and core-gamma logs. Cored interval (hachured) is 917-947 feet (log depth) and includes most of the lower Tocito Sandstone Lentil. The Tocito is unconformable on the Mancos Shale. Core-gamma was the primary tool used to correlate core with subsurface logs.

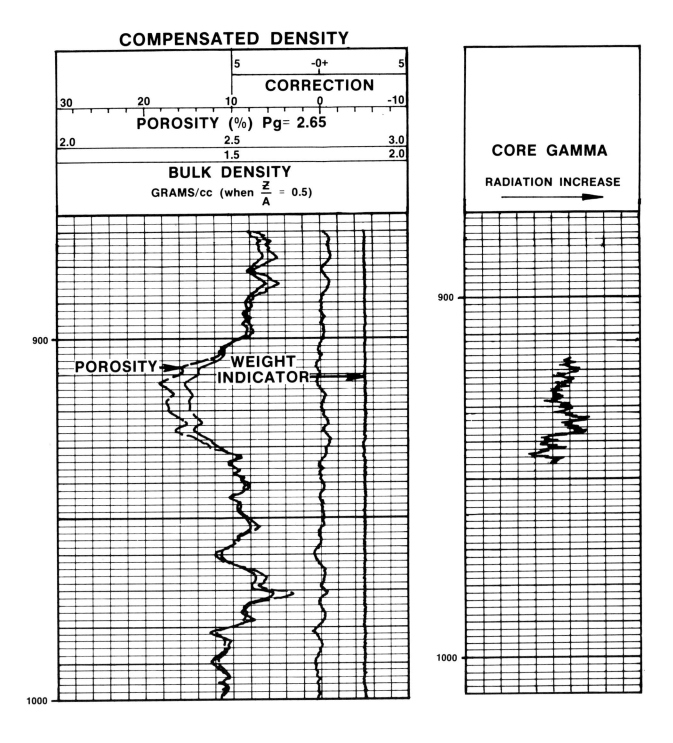

# REFERENCES

Campbell, C. V., 1973, Offshore equivalents of Upper Cretaceous Gallup beach sandstone, northwestern New Mexico, in Cretaceous and Tertiary rocks of the southern Colorado Plateau, J. E. Fassett (ed.): Four Corners Geological Society Memoir, p. 78-84.

Fassett, J. E., 1981, Upper Cretaceous Tocito Sandstone Lentil of Mancos Shale, San Juan Basin, New Mexico--Is there any oil left? (abs): American Association of Petroleum Geologists Bulletin, v. 65, p. 559.

McCubbin, D. G., 1969, Cretaceous strike valley sandstones reservoirs, northwestern New Mexico: American Association of Petroleum Geologists Bulletin, v. 53, p. 2114-40.

Molenaar, C. M., 1973, Sedimentary facies and correlation of the Gallup Sandstone and associated formations, northwestern New Mexico, in Cretaceous and Tertiary rocks of the southern Colorado Plateau, J. E. Fassett (ed.): Four Corners Geological Society Mem., p. 85-110.

_____ 1983a, Major depositional cycles and regional correlations of Upper Cretaceous rocks, southern Colorado Plateau and adjacent areas, in Mesozoic paleogeography of west-central United States: M. W. Reynolds and E. D. Dolly (eds.), Rocky Mountain Section, Society of Economic Paleontologists and Mineralogists, p. 201-224.

_____ 1983b, Principal reference section and correlation of Gallup Sandstone, northwestern New Mexico, in Contributions to mid-Cretaceous paleontology and stratigraphy of New Mexico, part II: New Mexico Bureau of Mines and Mineral Resources Circular 185, p. 29-40.

Penttila, W. C., 1964, Evidence for the Pre-Niobrara unconformity in the northwestern part of the San Juan Basin: Mountain Geologist, v. 1, p. 3-14.

Porter, K. W., 1976, Marine shelf model, Hygiene Member of the Pierre Shale, Upper Cretaceous Denver Basin, Colorado, in Studies in Colorado Field Geology, R. C. Epis and R. J. Weimer (eds.): Professional Contributions of Colorado School of Mines, p. 251-263.

Tillman, R. W. and R. S. Martinsen, 1984, The Shannon shelf-ridge sandstone complex, Salt Creek anticline area, Powder River Basin, Wyoming, in R. W. Tillman and C. T. Siemers (eds.), Siliciclastic Shelf Sedimentation: Society of Economic Paleontologists and Mineralogists Special Publication 34, p. 85-142.

Tillman, R. W. and R. S. Martinsen (1985), Shannon Sandstone, Hartzog Draw field core study, in R. W. Tillman, D. J. P. Swift and R. G. Walker (eds.), Shelf Sands and Sandstone Reservoirs, Society of Economic Paleontologists and Mineralogists Short Course Notes, 57 p.

# SHANNON SANDSTONE
# HARTZOG DRAW FIELD CORE STUDY

R. W. Tillman[1] and R. S. Martinsen[2,3]

## SUMMARY

Hartzog Draw field is a stratigraphically controlled oil reservoir which produces from the Upper Cretaceous Shannon Sandstone at depths from 9000 to 9600 feet. The producing interval consists of a large mid-shelf sand-ridge (bar) complex deposited below effective normal wave base more than 100 miles from shore. The productive (net pay) interval in the bar complex has a maximum thickness of 60 feet, is 22 miles long, and is one to four miles wide. The field was discovered in 1975 and 177 producing wells were completed on 160 acre spacing during the primary production phase of development. Initial oil in place was calculated to be 350,000,000 barrels.

The shelf sand-ridge complex is competely enveloped in shale, has a solution gas drive, no water table and no produced formation water. Net pay is primarily a product of porosity, permeability and thickness of the sandstone, and is related primarily to sedimentary facies and the degree of diagenesis.

Of the six to eight facies observed in cores, only one, the Central Ridge Facies, a high-angle trough cross-bedded glauconitic quartz sandstone, is consistently high-quality reservoir. Two others, High-Energy Ridge-Margin Facies, a predominantly trough cross-bedded highly glauconitic sandstone with abundant shale and siderite clasts, and Low-Energy Ridge-Margin Facies, interbedded trough and rippled sandstone, also may be good quality reservoirs. Inter-Ridge Facies which consist of rippled interbedded sandstone and shale, generally are poor quality to non-reservoirs.

Values from the Central Ridge Facies from three of the cores taken early in the development of the northern part of the field have significantly better average porosity and permeability (12.7%, 6.4 md) than the Ridge Margin Facies (8.1%, 2.7 md) or Inter-Ridge Facies (6.2%, 2.1 md) (Martinsen and Tillman, 1979). Wells with a very thick (35') Central Ridge Facies such as the Cities Service Fed. AK-1 have higher average porosities and permeabilities. Average values from this well are: Central Ridge 17%, 30 md; High-Energy Ridge-Margin 14%, 17 md; Low-Energy Ridge-Margin 9%, 13 md. Recognition of the facies, as well as understanding their distribution and interrelationships, are prerequisites to developing a program which will maximize oil recovery from shelf-ridge (bar) fields such as Hartzog Draw field.

## INTRODUCTION

Hartzog Draw is the largest of the Shannon Sandstone reservoirs in the Powder River Basin (Fig. 1). The field is one of several elongate fields of similar age which trend northwest-southeast. Heldt Draw which lies immediately

---

1. Consulting Sedimentologist, 4555 S. Harvard, Tulsa, Oklahoma 74102
2. Consultant, 3901 Gray Gables Rd., Laramie, Wyoming 82070
3. Both authors were formerly with Cities Service Oil Company.

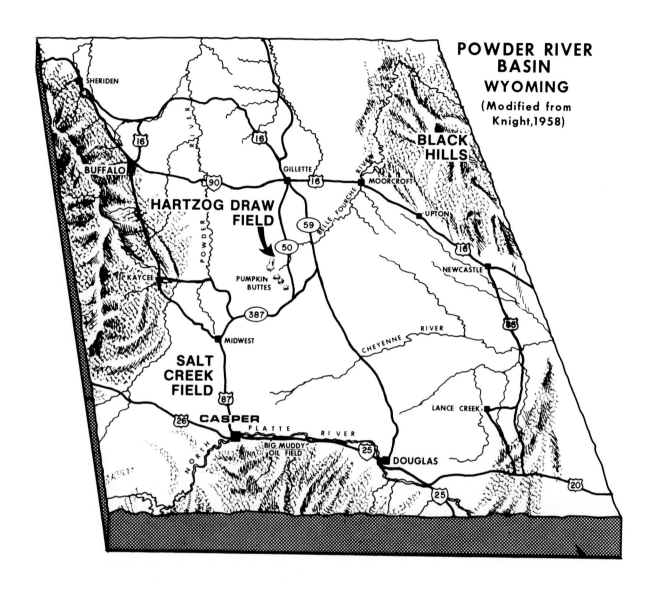

Figure 1. Hartzog Draw Field, Powder River Basin, Wyoming.

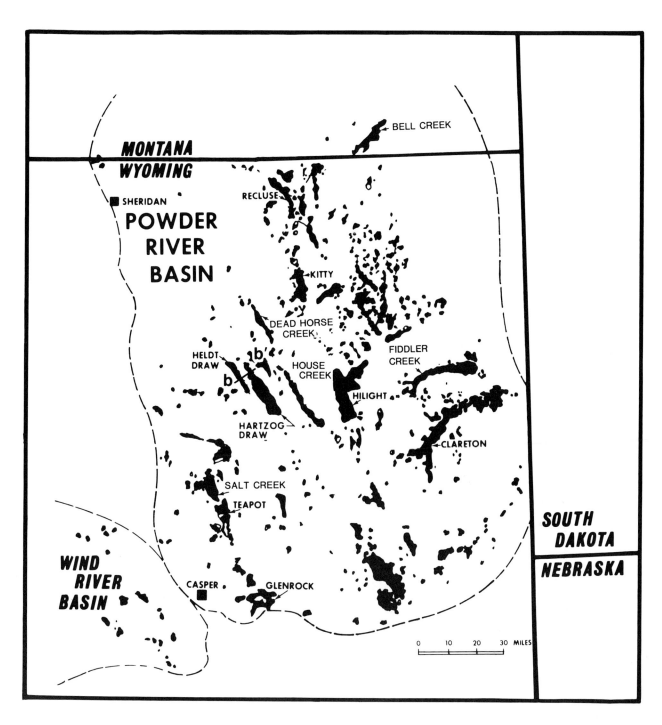

Figure 2. Oil fields of Powder River Basin. Hartzog Draw and Heldt Draw fields produce from the Shannon Sandstone. Cross section B-B' is Figure 7.

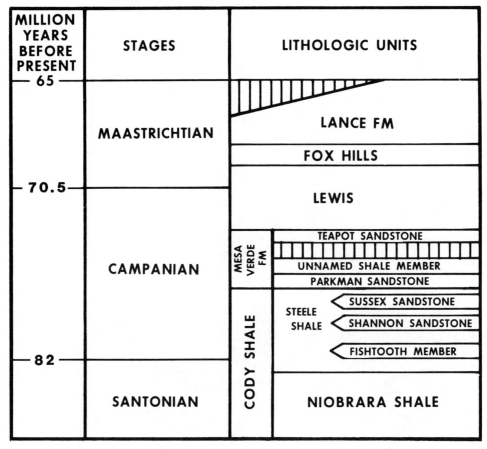

Figure 3. Stratigraphic section. Age of Shannon is 80.5 mybp. Dates from Gill and Cobban, 1973.

west of Hartzog Draw field (Fig. 2) also trends northwest-southeast and produces from the Shannon Sandstone (Fig. 3), but like most other Shannon fields, is thinner and covers less area than Hartzog Draw field. In addition to the Shannon, the slightly younger Sussex Sandstone also produces from northwest-southeast trending fields such as House Creek (Fig. 2).

The model developed for Hartzog Draw (Martinsen and Tillman, 1978, 1979 and Tillman and Martinsen, 1979) involves a very wide shelf (Fig. 4) which was probably developed by slope progradation as described by Asquith (1970) and Spearing (1975). Hartzog Draw was deposited approximately 100 miles offshore on the middle shelf in water depths approximating 200 feet. The field lies slightly oblique to the shoreline direction projected for Shannon time (80.5 mybp). The net sand distribution within the field is shown in Figure 5. The map shows relatively uniform changes from 0-50 feet in thickness.

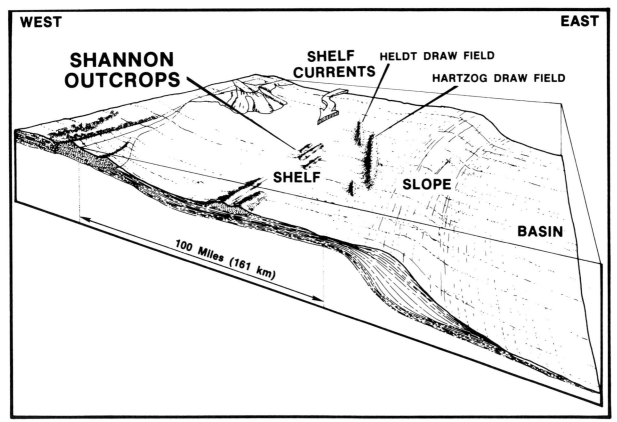

Figure 4. Shelf-slope to basin paleogeography during deposition of the Shannon Sandstone. Note wide shelf and prograding slope as originally described by Asquith (1970). Hartzog Draw field lies oblique to and far offshore of the interpreted time equivalent Shannon shoreline. Oriented core data at Hartzog Draw field and in outcrops southwest of the field indicate a southerly transport direction lightly oblique to the axis of the oil fields.

---

Two cored wells were selected to illustrate the facies and producing characteristics in the field. The Cities Service AB-1A, the confirmation well for the field, is located in the northern part of the field, contains a varied and thick sequence of facies, and has a minimal reservoir thickness.

The Cities Service Federal AK-1 is located just east of the axis of the field, had high initial producing capacity (1750 BOPD) and contains a very thick producing interval (45'). The Federal AK-1 is typical of the high initial production capacity wells, most of which lie in a northwest-southeast trend near the axis or slightly northeast of the axis of the field (Hobson et al, 1984).

The major producing facies in the field are the Central Ridge Facies and High-Energy Ridge-Margin Facies. These as well as other Shannon facies are summarized in Figure 6 and 6B and in Tillman and Martinsen (1984).

The areal and vertical distribution of the various facies is complex and is summarized in Figure 7, which is a stratigraphic section for the north part of Hartzog Draw field and the south part of Heldt Draw field.

Trough and planar-tabular bedding is the most significant sedimentary structure of the Central Ridge Facies (Fig. 8). This facies normally contains only a few shale rip-up clasts and/or siderite clasts, is moderately glauconitic and was deposited by southward flowing high energy traction currents. Lenses

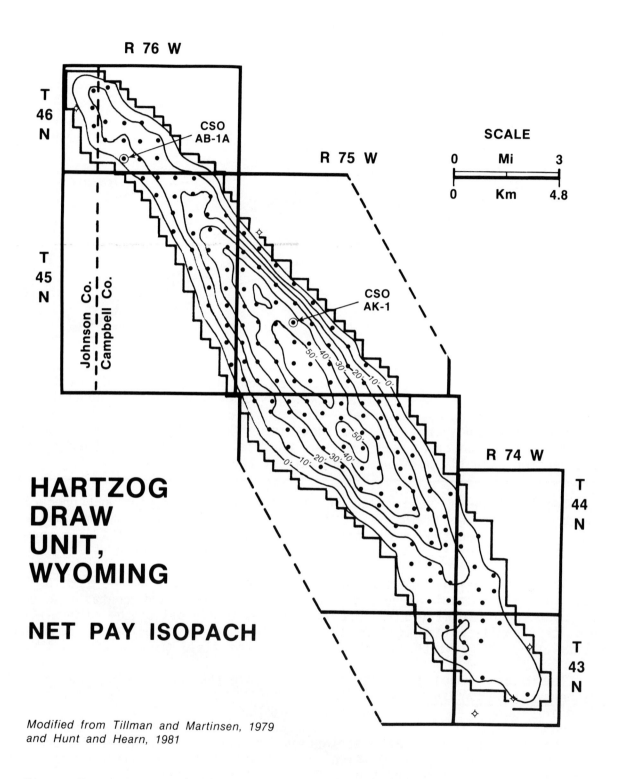

Figure 5. Hartzog Draw field net pay isopach map. Hartzog Draw unit is outlined by offsetting rectangles. Locations of cored wells, Cities Service Fed. AB-1A and Federal AK-1, are indicated. CI = 10'.

## SHANNON FACIES SUMMARY

| | CENTRAL-RIDGE FACIES | CENTRAL-RIDGE (PLANAR LAMINATED) FACIES | HIGH-ENERGY RIDGE-MARGIN FACIES | LOW-ENERGY RIDGE-MARGIN FACIES |
|---|---|---|---|---|
| LITHOLOGY | Fine to medium grained quartzose sandstone, moderately glauconitic; rare siderite clasts and shale rip-up clasts. | Fine to medium grained quartzose sandstone. | Predominately medium grained sandstone, abundant shale and limonite rip-up clasts and lenses, commonly very glauconitic. | Fine-grained sandstone with only rare shale interbeds. Fewer clasts and lenses and less glauconitic than High-Energy Ridge-Margin Facies. |
| SEDIMENTARY STRUCTURES | Predominantly moderate angle trough and planar-tangential cross bedding. Trough sets commonly horizontally truncated. | Mostly sub-horizontal plane-parallel laminated sandstone, 0.5'-thick laminasets. Minor shale and sandstone ripples. | Mostly moderate angle troughs, some current ripples, shale clasts rarely show preferred orientations. | Sequences of several beds of troughs interbedded with sequences of several rippled beds. |
| BURROWING | Sparse | Sparse | Sparse | Sparse |
| RESERVOIR POTENTIAL | Excellent | Limited? | Good | Moderate to Good |
| SUBSURFACE OCCURENCES HARTZOG DRAW FIELD | Common | Very uncommon | Common | Common |

| | INTER-RIDGE FACIES (SHALEY) | INTER-RIDGE SANDSTONE FACIES | BIOTURBATED SHELF-SANDSTONE FACIES | BIOTURBATED SHELF-SILTSTONE FACIES | SHELF SILTY-SHALE FACIES |
|---|---|---|---|---|---|
| LITHOLOGY | Thinly interbedded fine to very fine-grained silty sandstone and silty shale, slightly glauconitic. | Fine-grained sandstone. Virtual absence of silty shale. Slightly glauconitic | Silty, fine-grained sandstone. Up to 15% shale, primarily associated with burrows. Slightly glauconitic. | Shaly, slightly sandy dark gray siltstone, traces to moderate amounts of glauconite. | Silty dark gray shale; rare thin (1/8" thick) silty sandstone lenses. |
| SEDIMENTARY STRUCTURES | Predominantly horizontal ripple-form bedding surfaces marked by interbedded shales. Trace of wave ripples; current ripples predominate | Predominantly horizontal ripple-form bedding surfaces. Bedding commonly indistinct. Trace of wave ripples; current ripples predominate. | Few physical structures preserved. Mottled appearance. Some ripple-form horizontal beds up to 8 inches thick. Trace of distinct ripples and small troughs. | Few physical structures preserved. Scattered thin rippled sand and horizontal laminasets. Bedding commonly destroyed. | Common sub-horizontal laminae. Bedding surfaces indistinct, horizontal. Rare current ripples. |
| BURROWING | Moderate to locally high | Low to Moderate | Mottled to distinctly burrowed. More than 75% burrowed. | More than 75% Burrowed | Low to moderate |
| RESERVOIR POTENTIAL | Limited | Limited | Limited | None | None |
| SUBSURFACE OCCURENCES HARTZOG DRAW FIELD | Common | Uncommon | Common | Moderately Common | Common |

Figure 6A. Shannon facies summary. All facies shown are recognized at Hartzog Draw field except the <u>Inter-Ridge Sandstone</u> and <u>Bioturbated Shelf Sandstone Facies</u>. These later two facies are prominent in outcrops of the Shannon and, these as well as the other facies, are discussed in Tillman and Martinsen (1984). This figure is modified from Tillman and Martinsen (1984).

| Present Terminology | Tillman and Martinsen, 1984 | Spearing, 1976 |
|---|---|---|
| Central-Ridge Facies | Central Bar Facies | Cross bedded Sandstone Facies |
| Central-Ridge (Planar-Laminated) Facies | Central Bar (Planar Laminated) Facies | |
| High-Energy Ridge-Margin Facies | Bar Margin Facies (Type 1) | |
| Low-Energy Ridge-Margin Facies | Bar Margin Facies (Type 2) | |

Figure 6B. Comparison of facies names used by Tillman and Martinsen (1984), Spearing (1975) and usage in this paper.

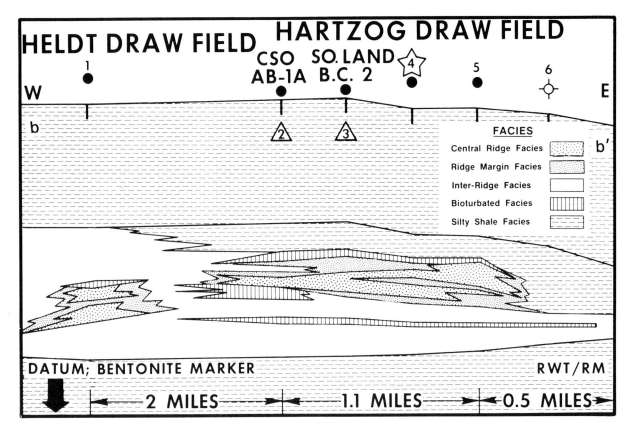

Figure 7. Stratigraphic cross section through Hartzog Draw and Heldt Draw fields. Location of section shown in Figure 2. Data points in line of cross section are indicated. Note location of Cities Service Fed. AB-1A. Datum is bentonite marker at base; three other subhorizontal bentonite markers are also shown. Note the mirror image asymmetry of the two fields and that both dip away from a vertical line separating the high-energy facies in the two fields. Rippled sandstones and shales of the Inter-Ridge Facies separate the fields and form most of the sandy section below the fields. The major reservoir facies are the Central Ridge Facies and the Ridge Margin Facies.

---

of siderite (Figs. 9A, B) are the source of siderite clasts (Fig. 9C, D) commonly found in cross bedded facies of the Shannon. The degree of cementation in the facies is variable and is commonly either calcite or siderite cemented (Fig. 9D).

High-Energy Ridge-Margin Facies are similar to Central Ridge Facies except they are generally a little coarser grained and contain more shale rip-up clasts and siderite pebbles. Typical facies of this type are illustrated in Figure 10. Low-Energy Ridge-Margin Facies consist of interbedded rippled beds and trough beds. The ripples are generally sandstone, but may be shaley as shown in Figure 11. Inter-Ridge Sandstone Facies physical structures are almost all ripples; the degree of burrowing within the facies is variable but less than 50% by volume.

A facies that would have been initially designated as an Inter-Ridge Facies may be changed to a Burrowed or, if 75% burrowed, to a Bioturbated Shelf Facies. In Figure 13 units designated 2A-2B and 3A-3B reflect the change in the upper of the paired units from what was initially Inter-Ridge Facies to Bioturbated Shelf Facies. The facies designations followed in this study (Fig. 11A) are modified from Porter (1976) and Tillman and Martinsen, 1984.

Figure 8. Typical cross bedding of Central Ridge Facies, Cities Service Fed. AK-1. Width of slabbed core 4 inches.
A., B. Laminae converging from left to right; typical of curved-tangential (trough) cross bedding which makes up most of the Central Ridge and High-Energy Ridge-Margin Facies at Hartzog Draw. Note angular elongate clay rip-up clasts and concentrations of glauconite (dark grains) on some lamina. Unit 4A, 9368.1', 9370.7'.
C. Planar-tangential to planar-tabular cross bedding. Within area of photo, laminae do not converge but are parallel as is typical of planar tabular bedding. Unit 4B, 9347.7'.
D. Oversteepened trough cross bedding. Note upper contact of oversteepened set is base of nearly horizontal set of laminae containing small locally derived shale rip-up clasts. Unit 4A, 9369.7'.

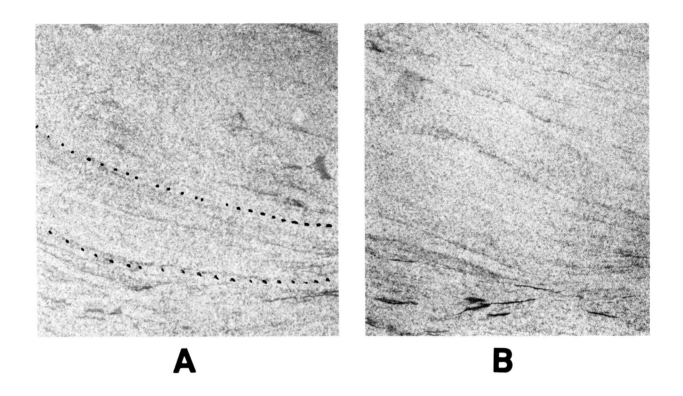
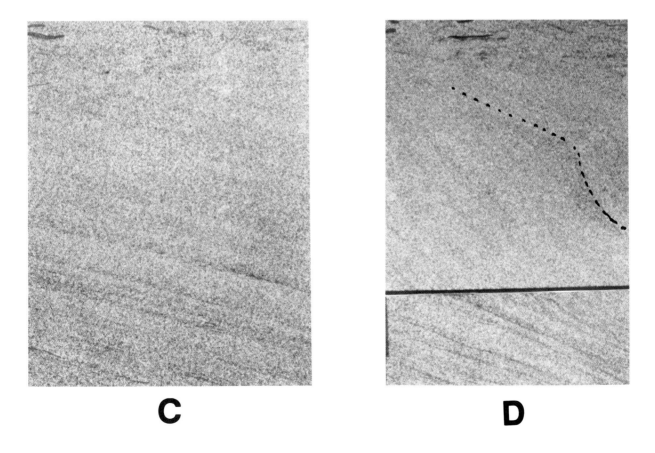

Figure 9. Siderite lenses, rip-up clasts, and siderite cemented sandstone in Shannon Sandstone in Cities Service Fed. AK-1. Width of the core photos is 4 inches.
A., B. Lens (bed) of siderite (between arrows) draping over reactivation surface. Note that lower contact of siderite may be sharp (A) or slightly gradational (B). Upper 1/4" of lense in A may be "reworked" since it contains added sand. Note early "fracturing" of lens in B. These lenses are the source for rounded siderite clasts. Unit 3, High-Energy Ridge-Margin Facies. A. 9378.5'; B. 9385.8'.
C. Typical rounded oblate siderite clasts forming "edgewise conglomerate" where long axis of clasts are oriented parallel to laminations. Unit 4B, Central Ridge Facies, 9353.6'.
D. Rounded siderite clasts in highly siderite cemented sandstone. Here siderite clasts are parallel to oversteepened laminations of trough. Unit 3, High-Energy Ridge-Margin Facies, 9371.7'.

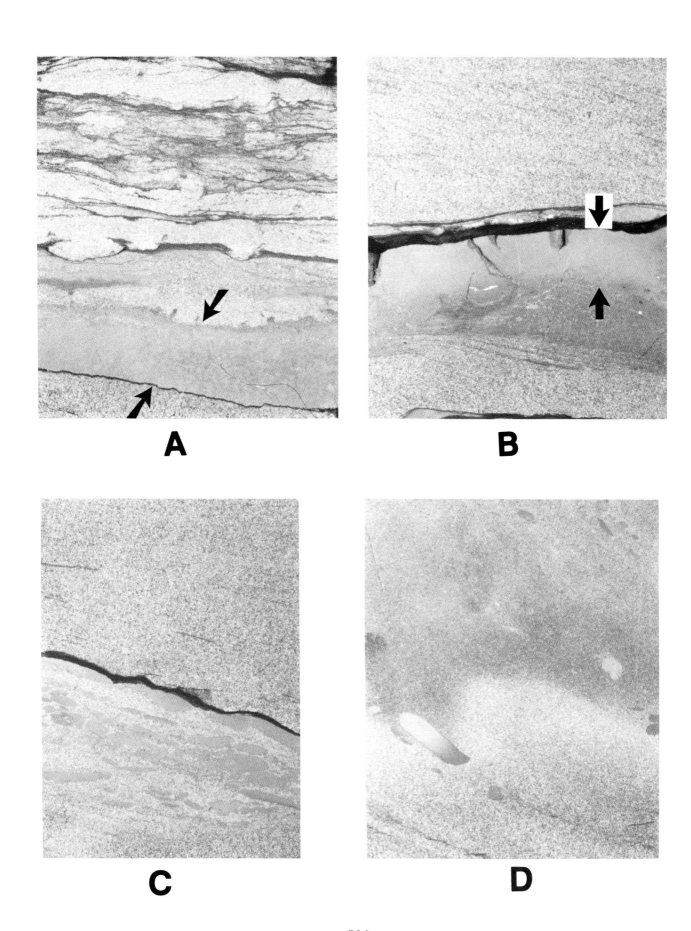

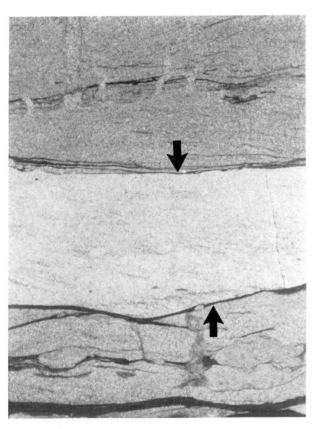 

A  B

Figure 10. High-Energy Ridge-Margin Facies in sandstones, Cities Service Fed. AK-1. Width of cores is 4 inches.
A. Small troughs and ripples; shale drapes separate ripples. Note minor soft sediment deformation of ripples below lower arrow. Vertical burrows are observed in upper portion and a trace of burrows can be seen in the rippled area between arrows. Whiter color between arrows is highly calcite cemented interval. (Contrast with siderite cement in Figure 9). Unit 3, High-Energy Ridge Margin Facies, 9385.0'.
B. Troughs containing shale rip-up clasts typical of High-Energy Ridge-Margin Facies Note that shale rip-ups are angular, elongate, partially deformed and subparallel to laminations. Lower trough described as planar-tangential. Unit 3, High-Energy Ridge-Margin Facies, 9373.1'.

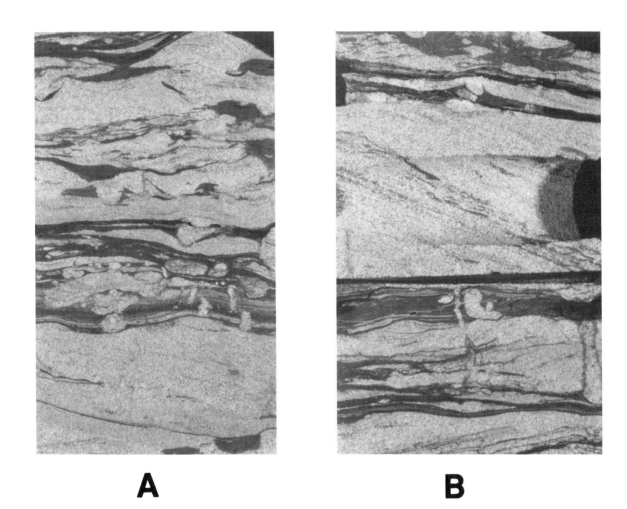

Figure 11. <u>Low-Energy Ridge-Margin Facies</u> in Cities Service Fed. AK-1. Width of cores is 4 inches.
A. Rippled sandstone with interbedded shale drapes; small trough at base. Note rare angular to deformed shale rip-up clasts. Unit 2, 9387.8'.
B. Thin trough (upper middle) interbedded with sandstone ripples draped by shale. Interbedded ripples and troughs is typical of this facies. Unit 2, 9388.2'.

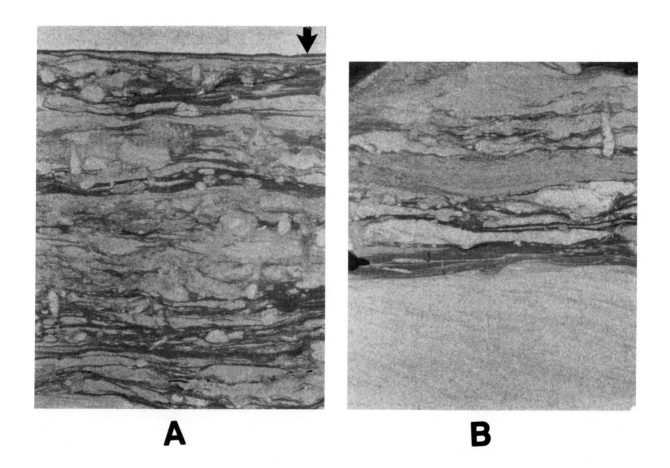

Figure 12. <u>Inter-Ridge Facies</u>, Fed. AK-1, Unit 1, 9392'. Width of cores is 4 inches.
A. Rippled interbedded sandstone and shale. In area of photograph 50% burrowed. If burrowed interval were thicker, this interval would be assigned to a <u>Burrowed Shelf Facies</u>, but because this burrowed interval is less than one foot thick (0.4') and is interbedded with dominantly physically bedded units, it is included in the <u>Inter-Ridge Facies</u>. Note sharp contact with rippled sandstone near top of photo (arrow). The upper part of this thin low relief trough forms the lower half of photo B. B. Rippled interbedded shale and sandstone in sharp contact with thin low relief trough sandstone similar to thin sandstones described in the North Sea as storm deposits by Reineck et al (1971) and Tillman and Reineck (1975).

# CITIES SERVICE FEDERAL AB-1A (6334)

The Cities Service Federal AB-1A well is located in the northwestern part of the field (Fig. 5). This well was the confirmation well for the field and was the first well to be cored. The well is located near the western boundary of the field outside the area of good production.

## Geological Facies Analysis

Geological core analysis identified 5.2' of Central Ridge Facies (Table 1) and 3.2' of High-Energy Ridge-Margin Facies within the core (Fig. 13). The remainder of the facies encountered are essentially nonreservoir facies. As is indicated in Figure 13, there is a thick (41') succession of rippled to burrowed facies composed of Inter-Ridge Facies and Burrowed to Bioturbated Shelf Siltstone Facies below the reservoir at this location. The variable vertical succession of facies in this well is typical for this part of the field. The base of the shelf-ridge complex is drawn at the base of the Inter-Ridge Facies (9299.3').

The complete sequence of shelf-ridge (bar) complex facies, as well as overlying and underlying shales, were cored in the 82' core from this well. Photographs of the slabbed core are included as Figures 14-A through 14-O. Well laminated Cody Shale was cored above the shelf-ridge complex and more diffusely laminated, more silty shale was cored below the lowest Shannon Inter-Ridge Facies.

## Log Signatures

Log signatures are very apparent for several of the facies recognized in cores within the field (Fig. 15). The Gamma Ray log is usually the most lithologically diagnostic single survey. The SP, induction and density logs are also useful in varying degrees for identifying facies boundaries. The importance of these logs are approximately in the order listed. The contact between the Inter-Ridge and Shelf Silty Shale Facies which lies below the Shannon is easily recognized in this well (9289.3' log depth; 9299.3' core depth) and most others in the field. The strong deflection of the gamma-ray curve to the left opposite the Central Ridge Facies (9243.2-9248.4') is typical of most wells in the field.

## Production Characteristics

This well is typical of many of the wells in the northern part of the field, especially those along the northwestern margin of the field. Production characteristics are shown graphically in Figure 16 and are tabulated in Table 2. As can be seen in Figure 16, permeability varies directly with the facies type. Average permeabilities in decreasing order are: Central Ridge 3 md, High-Energy Ridge-Margin 0.8 md, Low-Energy Ridge-Margin 0.2 md; all other facies have even lower permeabilities. The decrease of porosity values among these facies is in the same order of decrease as the permeability. Mean facies porosity values are: Central Ridge 12%, High-Energy Ridge-Margin 10%, Low-Energy Ridge-Margin 8%. As will be seen in the discussion of the Federal AK-1 well, this sequence of porosity and permeability decrease also holds true in the more productive parts of the field, however, the absolute values for all the facies is higher in the areas of the thicker sandstones deposited under high energy conditions.

# TABLE 1
## STATISTICAL SUMMARY OF FACIES COMPONENTS

CITIES SERVICE FEDERAL AB-1A (6334)
SE SEC. 33 T46N R76W, HARTZOG DRAW FIELD
CAMPBELL CO., WYOMING (SHANNON SANDSTONE)

Columnized Data Format: Shannon Sandstone
%/Max. set thickness/Additional notes

Described by: R. W. Tillman
R. Martinsen 1977

CENTRAL RIDGE FACIES
UNIT 4 (5.2')
Core depth: 9253.2-58.4'
(Log depth: 9243.2-48.4')

Lithology
|  |  |
|---|---|
| 1. Sandstone (%) | 91%/0.5' (250) |
| 2. Siltstone (%) | tr/0.05' |
| 3. Shale (%); a) Laminated; b) Clasts | 2%/a,b |
| 4. Siderite (%); a) Laminated; b) Clasts | 6%/b, 1/4-2" dia. |
| 5. Glauconite (%); a) Disseminated; b) Laminated | 5-10%/3% a |

Physical Structures
|  |  |
|---|---|
| | 99%/0.1' |
| 1. High angle cross-bedding (20°+); a) Troughs; b) Planar-tabular; c) Planar-tangential; d) Curved-tangential | 7%/0.4'/a |
| 2. Moderate angle cross-bedding (10-20°); a) Troughs; b) Planar | 45%/0.3'/a |
| 3. Subhorizontal to low angle bedding (<10°); a) Trough; b) Planar-tabular; c) Planar-tangential; d) Curved-tangential; e) Planar | 10%/0.1'/a |
| 4. Horizontal laminations; a) Bedding | // tr |
| 5. Rippled; a) Sandstone; b) Shale | 5%/0.05'/a,b |
| 6. Rippled interbedded sandstone and shale | tr |
| 7. Ripples superimposed on troughs | 30%/0.2' |
| 8. Reworked: a) By waves & currents; b) Bedding destroyed, massive; c) Soft sediment deformed; | tr/a/at base |

Biogenic Sedimentary Structures
|  |  |
|---|---|
| | tr/0.1'/at base |
| 1. Identified burrows; a) Asterosoma; c) Chondrites; d) "Donut burrows"; g) Gastropod tracks; p) Plural curving tubes; s) Skolithos; t) Teichichnus; th) Thallasinoides | - |
| 2. Distinct burrows: a) <1/8"; b) 1/8"-1/4"; c) 1/4"-1/2"; d) >1/2"; e) Silt filled; f) Clay filled; g) Sand filled; h) Silt lined; i) Clay lined; j) Spreiten; k) Vertical; l) Oblique; m) Horizontal | tr-b, g, m |
| 3. Bioturbated (75% + burrowed) (%) a) Distinct; b) Mottled | - |
| 4. Total interval burrowed (%, footage) | 3.2%/0.2' |
| 5. Diversity (Number of burrow types); 1-4 low, 5-8 moderate; >8 high | 1-low |

Contact Relations
|  |  |
|---|---|
| 1. Upper; a) Very sharp (<0.05' transition); b) Sharp 0.05-0.1' transition); c) Transitional (0.1-0.3' transition); d) Gradational (>0.3' transition); e) Contact, erosional (truncated) angular; f) Contact erosional parallel; g) Covered | d |
| Lower; a), b), c), d), e), f), g) | c |

Figure 13. Facies Analysis of Cities Service Federal AB-1A (6334), Hartzog Draw field, Wyoming.

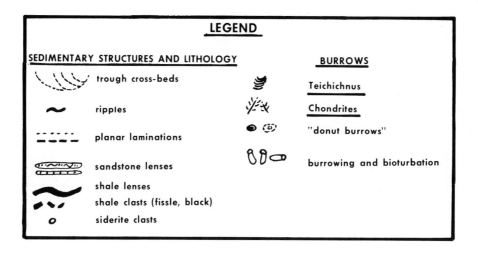

CITIES SERVICE FED. AB-1A
NW SE Sec. 33 T46N R76W,
HARTZOG DRAW FIELD, CAMPBELL CO., WYOMING

SHANNON SANDSTONE;
CORED INTERVAL:
86.2'; 9220-9306.2'

Lithology columns (left to right): Shale, Silty Shale, Shaley Siltstone, Siltstone, Shaley Sandstone, Silty Sandstone, Sandstone

Described by:
R. Tillman, R. Martinsen
1977, 1983

| Unit No. | Thickness | Description |
|---|---|---|
| 9 | 20.1'+ | 9220-40.1' Shale and siltstone interlaminated, 10% very fine grained sandstone (100), 2% glauconite. 85% physical structures including 60% ripples (15% flaser bedded) and 25% horizontal laminations. 15% burrowed, moderate diversity (6) increasing toward base, trace of "donut" burrows. Very low energy facies. SHELF SILTY SHALE FACIES (95%). |
| 8 | 3.8' | 9240.1-43.9' Shaly to sandy siltstone, 5% glauconite. 75% burrowed, moderate diversity (7), mostly horizontal to oblique 1/4" diameter silt filled. Trace of 2" diameter burrows. Rippled to flaser bedded, less burrowed near base. Low energy facies. BIOTURBATED SHELF SILTSTONE FACIES (90%). |
| 7 | 2.7' | 9243.9-46.6' Sandstone (250) with interlaminated shale (40%), 45% ripples, 20% small troughs, 10% physically reworked, 5% subhorizontally laminated. 5% shale rip-up clasts, 15% glauconite mostly disseminated, also concentrated on lamina. 20% burrowed, low diversity (4). Moderate energy facies. LOW-ENERGY RIDGE-MARGIN FACIES (90%). |
| 6 | 3.4' | 9246.6-50.0' Siltstone, shaley to sandy (125u) interlaminations increasingly sandier and less burrowed towards the base. 60% burrowed mostly oblique to horizontal, 1/16-3/4" diameter, 3% Chondrites. 40% physical structures including 15% flaser bedding and 10% additional ripples; 15% horizontal laminations. Transitional into underlying unit. Low energy facies. BURROWED SHELF SILTSTONE FACIES (85%). |
| 5 | 3.2' | 9250.0-53.2' Sandstone, fine grained (200), up to 75% glauconite on laminations. 25% dark grey shale rip-up-clasts (1/2-2 1/2"); 8% siderite, 97% physical structures including 40% trough cross-beds, 40% ripples (some superimposed on troughs). 3% burrowed. Very high energy facies. HIGH-ENERGY RIDGE-MARGIN FACIES (90%). |
| 4 | 5.2' | 9253.3-58.4' Sandstone, medium grained (300), 83% trough cross beds; 15% ripples, some superimposed on troughs. Trace of burrowing. 5% disseminated glauconite. 8% rounded siderite pebbles (1/2"-2" diameter). Sharp lower and upper contacts. High energy facies. CENTRAL RIDGE FACIES (95%) |
| 3B | 9.6' | 9258.4-68.0' Siltstone, sandy to shaley, 25% sandstone lenses (rippled). 75% burrowed, mostly oblique, very diverse suite of burrows including 3% Teichichnus, trace of Chrondrites, trace of silt lined "donut" burrows. 23% rippled, 2% subhorizontal laminations. Low energy facies. BIOTURBATED SHELF SILTSTONE FACIES (90%) |
| 3A | 11.1' | 9268.0-79.1' Sandstone (very fine grained, 100) (35%) interbedded with shale (35%) and siltstone (30%). 60% physical structures, mostly current ripples, some symetrical (wave) ripples, 10% horizontal to subhorizontal laminated sandstones up to 2" thick. Upper contact drawn where physical structures no longer dominent. Basal contact is sharp. Moderately low energy facies. INTER-RIDGE FACIES (95%) |
| 2B | 3.5' | 9279.1-82.6' Siltstone (40%), sandy (20%) to shaley (40%). 80% burrowed, diverse suite of burrows (10) including traces of Chondrites and "donut" burrows. Most burrows oblique. 15% ripples, 5% horizontal to subhorizontal laminations. Upper and lower contacts picked at unburrowed sandy ripples. Low energy facies. BIOTURBATED SHELF SILTSTONE FACIES (95%) |
| 2A | 15.7' | 9182.6-99.3' Sandstone (45%), very fine grained (100), interbedded with shale (40%) and siltstone (15%). Some coarser sandstone lenses. 63% physical structures including 57% ripples, and 5% small troughs. 37% burrowed, mostly horizontal to oblique, high diversity (8) including shale lined and silt lined ("donut") burrows. Low energy facies. INTER-RIDGE FACIES (90%). |
| 1 | 6.9'+ | 9299.3-9306.2 Shale with disseminated silt in lenses and as burrow filling, mostly subhorizontally laminated, some ripples. 35% burrowed, mostly silt filled horizontal to oblique, trace of "donut" burrows, moderate diversity (6). Very low energy facies. SHELF SILTY SHALE FACIES (90%) |

Figure 14 (A-O). Photographs of Cities Service Federal AB-1A (6334) slabbed core. Unit boundries and facies names are designaged. Cored interval is 82 feet and extends from 9220 to 9306 feet. Core to subsurface log correction requires subtraction of 10.0 feet from core depth as is tabulated in Figure 15. A summary description of facies types is given in Figure 6 and a detailed description of this core is given in Figure 13.

## CITIES SERVICE FED. AB-1A (6334)
## SE SEC.33 T46N R76W, HARTZOG DRAW FIELD
## CAMPBELL CO., WYO.

9220
UNIT 9
SHELF
SILTY
SHALE

A

# CITIES SERVICE FED. AB-1A (6334)
## SE SEC.33 T46N R76W, HARTZOG DRAW FIELD
## CAMPBELL CO., WYO.

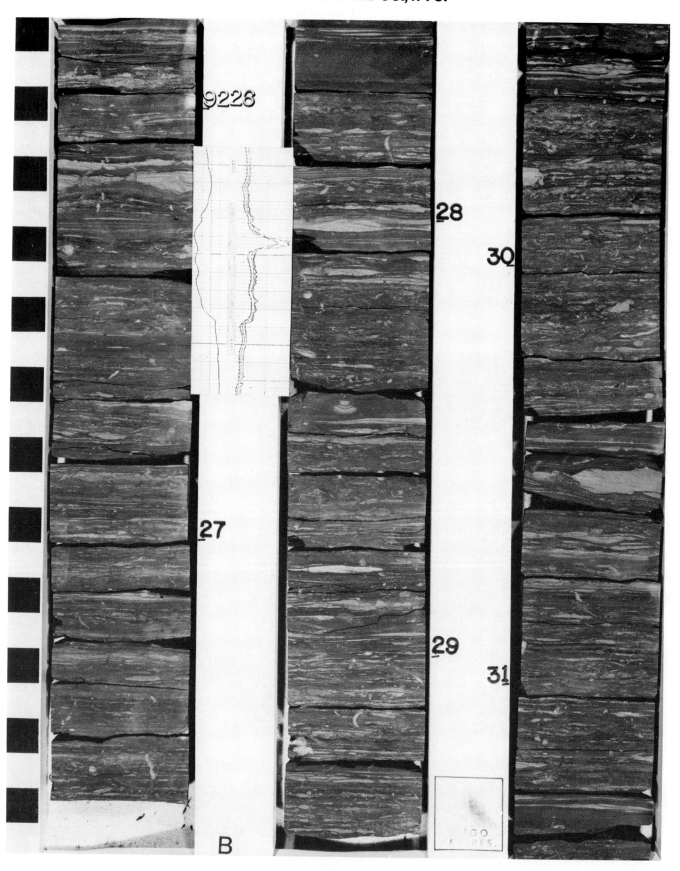

# CITIES SERVICE FED. AB-1A (6334)
## SE SEC.33 T46N R76W, HARTZOG DRAW FIELD
## CAMPBELL CO., WYO.

## CITIES SERVICE FED. AB-1A (6334)
## SE SEC.33 T46N R76W, HARTZOG DRAW FIELD
## CAMPBELL CO., WYO.

UNIT 8
BIOTURBATED
SHELF
SILTSTONE

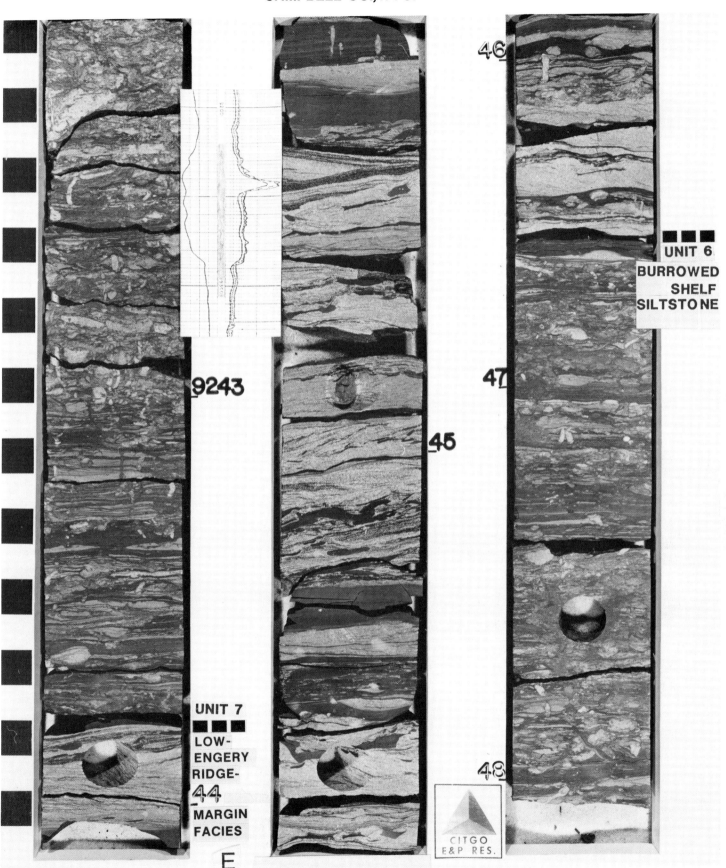

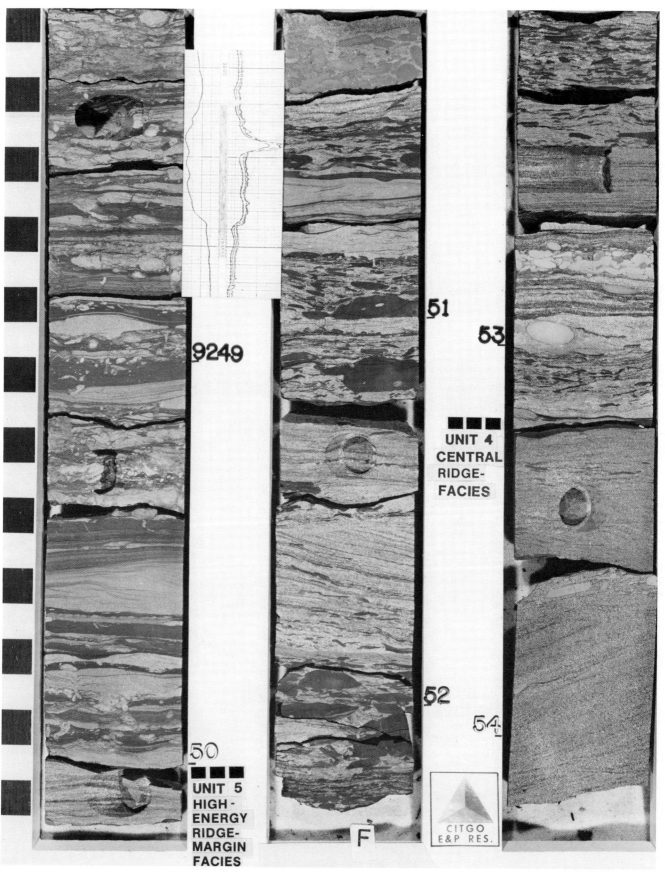

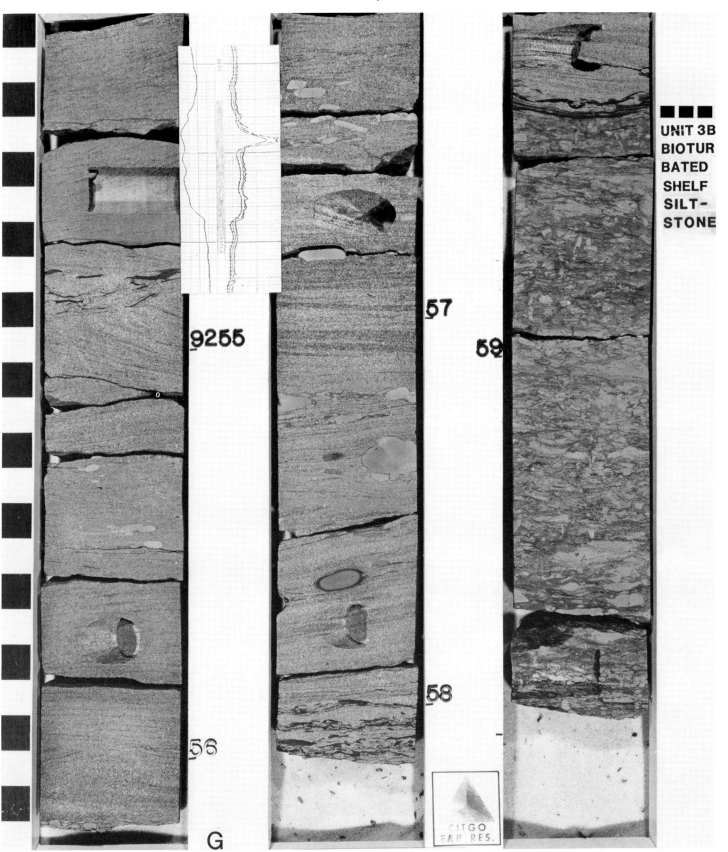

# CITIES SERVICE FED. AB-1A (6334)
## SE SEC.33 T46N R76W, HARTZOG DRAW FIELD
## CAMPBELL CO., WYO.

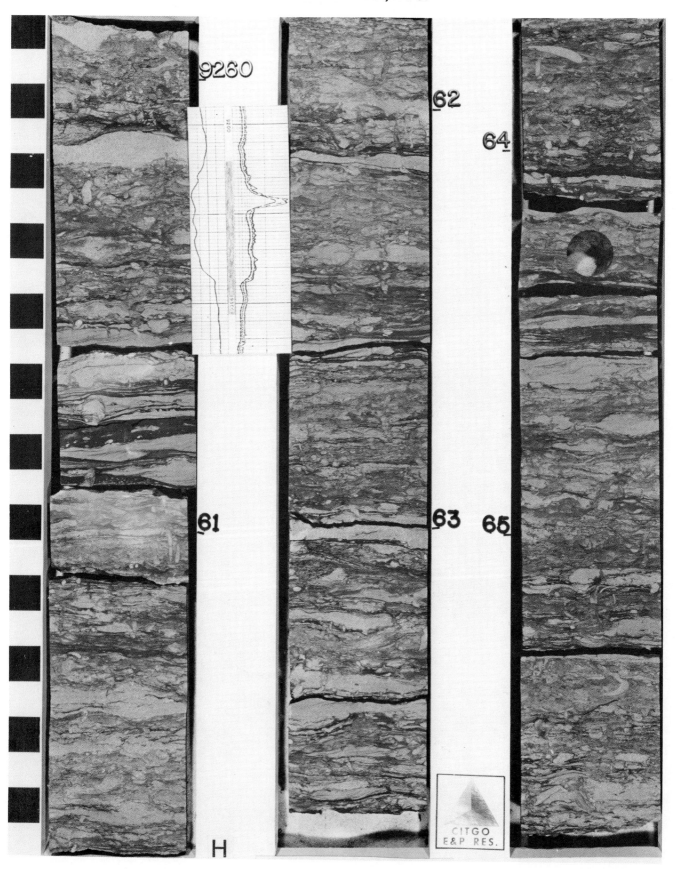

# CITIES SERVICE FED. AB-1A (6334)
## SE SEC.33 T46N R76W, HARTZOG DRAW FIELD
## CAMPBELL CO., WYO.

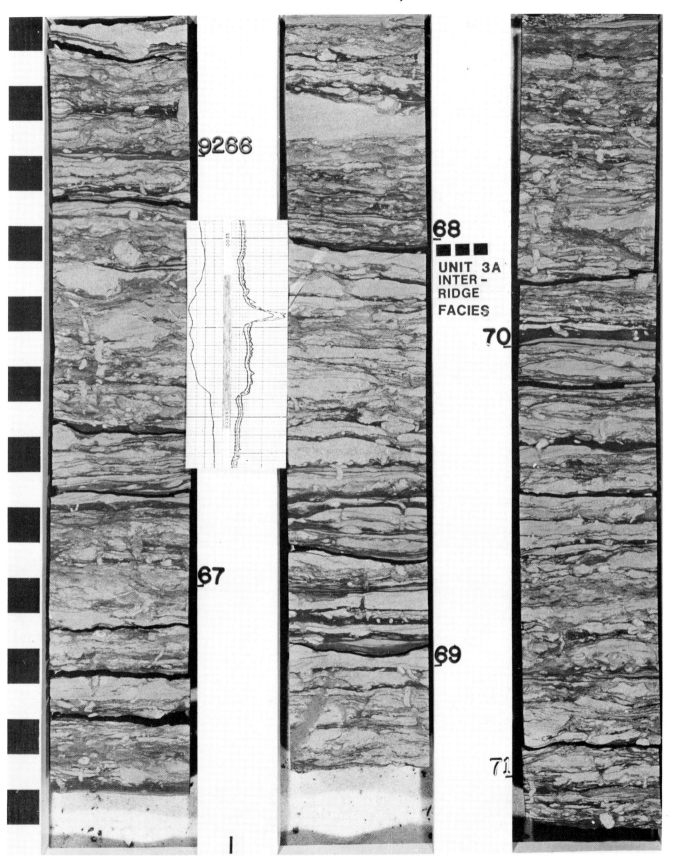

UNIT 3A INTER-RIDGE FACIES

## CITIES SERVICE FED. AB-1A (6334)
## SE SEC.33 T46N R76W, HARTZOG DRAW FIELD
## CAMPBELL CO., WYO.

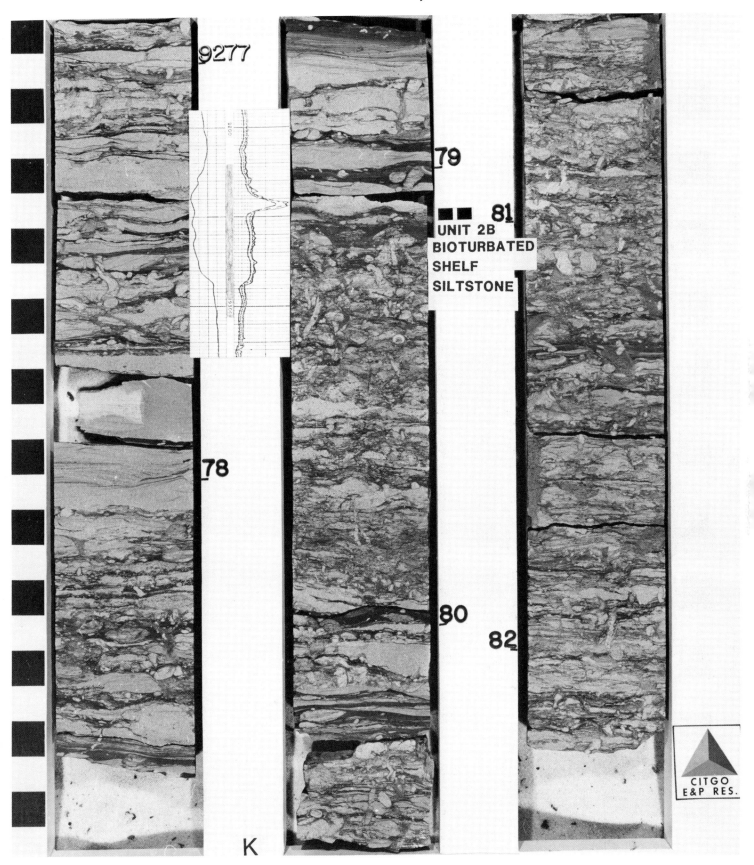

CITIES SERVICE FED. AB-1A (6334)
SE SEC.33 T46N R76W, HARTZOG DRAW FIELD
CAMPBELL CO., WYO.

UNIT 2B
BIOTURBATED
SHELF
SILTSTONE

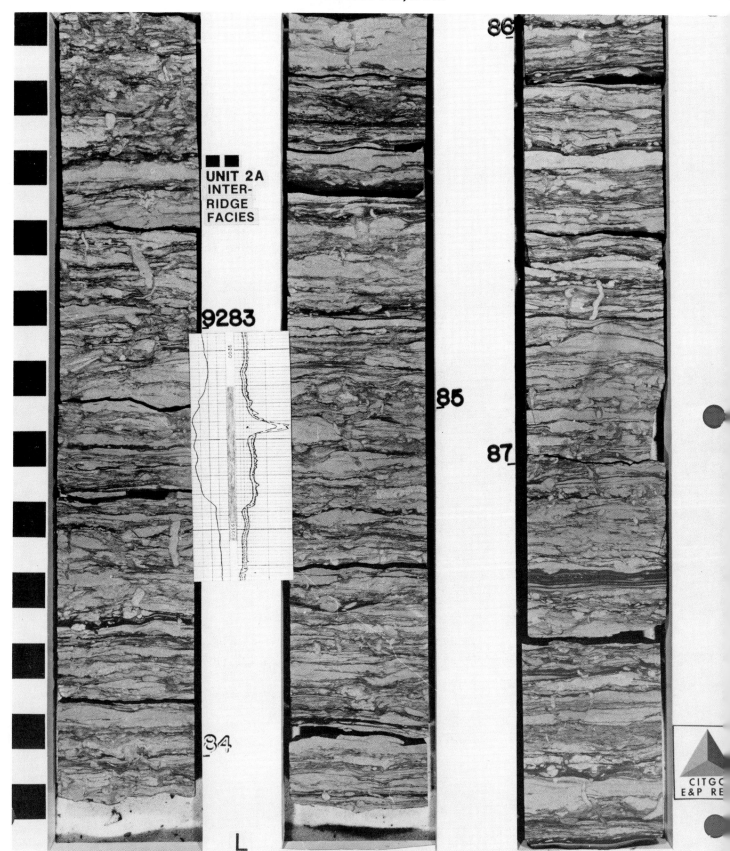

CITIES SERVICE FED. AB-1A (6334)
SE SEC.33 T46N R76W, HARTZOG DRAW FIELD
CAMPBELL CO., WYO.

UNIT 2A INTER-RIDGE FACIES

# CITIES SERVICE FED. AB-1A (6334)
## SE SEC.33 T46N R76W, HARTZOG DRAW FIELD
## CAMPBELL CO., WYO.

# CITIES SERVICE FED. AB-1A (6334)
## SE SEC.33 T46N R76W, HARTZOG DRAW FIELD
## CAMPBELL CO., WYO.

# CITIES SERVICE FED. AB-1A (6334)
## SE SEC.33 T46N R76W, HARTZOG DRAW FIELD
## CAMPBELL CO., WYO.

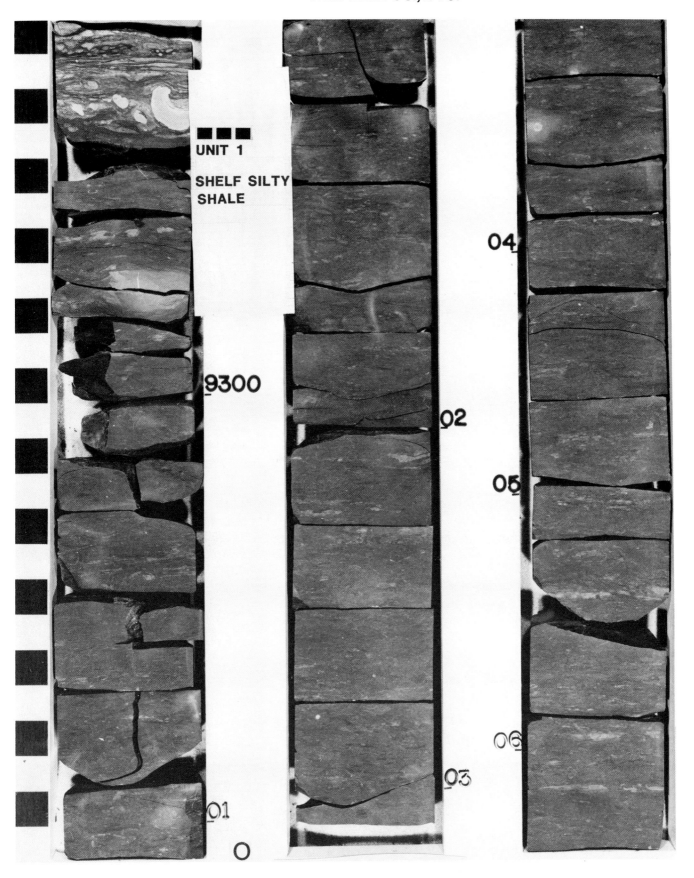

| FACIES (ENVIRONMENT OF DEPOSITION) | UNIT NO. | THICK- NESS | GR LOG DEPTH | CORE DEPTH |
|---|---|---|---|---|
| SHELF SILTY SHALE | 9 | 20.1' | 9210' | 9220' |
| BIOTURBATED SHELF SILTSTONE | 8 | 3.8' | 9230.1' | 9240.1' |
| LOW-ENERGY RIDGE MARGIN | 7 | 2.7' | 9233.9' | 9243.9' |
| INTER-RIDGE FACIES | 6 | 3.4' | 9236.6' | 9246.6' |
| HIGH-ENERGY RIDGE-MARGIN | 5 | 3.2' | 9240.0' | 9250.0' |
| CENTRAL RIDGE FACIES | 4 | 5.2' | 9243.2' | 9253.2' |
| SHELF SILTSTONE, BURROWED | 3B | 9.6' | 9248.4' | 9258.4' |
| INTER-RIDGE FACIES | 3A | 11.1' | 9258.0' | 9268.0' |
| BIOTURBATED SHELF SILTSTONE | 2B | 3.5' | 9269.1' | 9279.1' |
| INTER-RIDGE FACIES | 2A | 15.7' | 9272.6' | 9282.6' |
| SHELF SILTY SHALE | 1 | 6.9' | 9289.3' | 9299.3' |

Figure 15. Subsurface logs of Cities Service Federal AB-1A (6334). Facies interpreted from slabbed core are designated. Note that log to core correction required is +10 feet. Units are the same as those in Figure 13 and Table 1.

# CITIES SERVICE FED. AB 1A

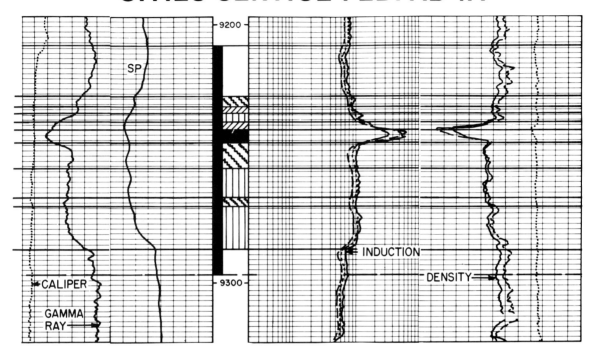

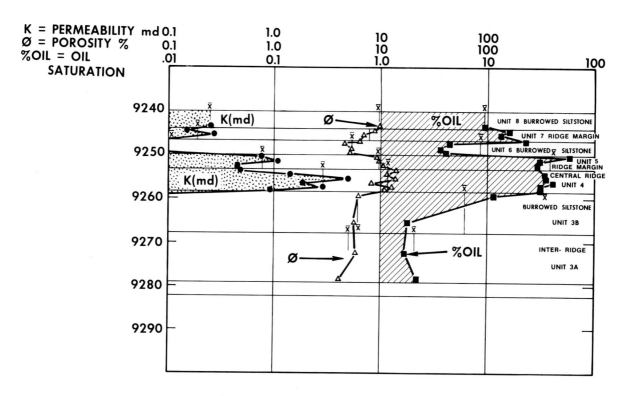

Figure 16. Production characteristics of individual facies recognized in slabbed core. Values for permeability are circles; porosity, triangles; oil saturation, squares. Mean values for each unit are indicated by X. Facies are the same as described in Figure 13.

TABLE 2
PRODUCTION CHARACTERISTICS
CITIES SERVICE OIL COMPANY FEDERAL AB-1A NW SE, SEC. 3 45N 76W
HARTZOG DRAW FIELD, SHANNON SANDSTONE CAMPBELL CO., WYOMING

CORE ANALYSIS (PLUGS) (CORE LABORATORIES, INC.)

Unit 9    SHELF SILTY SHALE FACIES (9220-40.1')

Unit 8    BIOTURBATED SHELF SILTSTONE FACIES (9240.1-43.9')

|       | 9243-44 | 0.25 | 9.8 | 9.3 | 76.2 |
|---|---|---|---|---|---|
| MEANS |         | 0.25 | 9.8 | 9.3 | 76.2 |

Unit 7    LOW-ENERGY RIDGE-MARGIN FACIES (9243.9-46.6')

|       | 9244-45 | 0.10 | 8.8 | 16.0 | 64.1 |
|---|---|---|---|---|---|
|       | 9245-46 | 0.27 | 7.0 | 13.1 | 72.9 |
| MEANS |         | 0.19 | 7.9 | 14.5 | 68.5 |

Unit 6    BURROWED SHELF SILTSTONE FACIES (9246.6-50.0')

|       | 9246-47 | 0.03 | 6.4 | 22.5 | 67.4 |
|---|---|---|---|---|---|
|       | 9247-48 | 0.01 | 4.7 | 4.5  | 89.5 |
|       | 9248-49 | 0.04 | 5.4 | 3.8  | 89.5 |
|       | 9249-50 | 0.01 | 5.2 | 4.1  | 90.3 |
| MEANS |         | 0.02 | 5.4 | 8.7  | 84.2 |

Unit 5    HIGH-ENERGY RIDGE-MARGIN FACIES (9250.0-53.2')

|       | 9250-51 | 0.77 | 9.3  | 58.0 | 19.3 |
|---|---|---|---|---|---|
|       | 9251-52 | 1.1  | 9.7  | 30.7 | 40.9 |
|       | 9252-53 | 0.46 | 10.8 | 29.3 | 47.5 |
| MEANS |         | 0.78 | 9.9  | 39.3 | 35.9 |

Unit 4    CENTRAL RIDGE FACIES (9253.2-58.4')

|       | 9253-54 | 4.9  | 14.0 | 29.6 | 25.6 |
|---|---|---|---|---|---|
|       | 9254-55 | 1.4  | 11.7 | 34.6 | 29.7 |
|       | 9255-56 | 5.5  | 13.9 | 35.0 | 21.5 |
|       | 9256-57 | 1.8  | 8.3  | 40.8 | 28.8 |
|       | 9257-58 | 2.8  | 12.8 | 31.2 | 9.7  |
|       | 9258-59 | 0.92 | 11.0 | 30.3 | 32.0 |
| MEANS |         | 2.8  | 11.9 | 33.5 | 31.1 |

|  | Depth (Feet) | Permeability (MD) Hor'z. | Porosity | Fluid Saturation Oil | Water |
|---|---|---|---|---|---|
| Unit 3B | BIOTURBATED SHELF SILTSTONE FACIES (9528.4-68.0') | | | | |
|  | 9259-60 | 0.03 | 6.3 | 11.5 | 82.5 |
|  | 9264-65 | 0.01 | 5.7 | 1.8 | 93.7 |
| MEANS |  | 0.02 | 6.0 | 6.7 | 88.1 |
| Unit 3A | INTER-RIDGE FACIES (9268.0-79.1') | | | | |
|  | 9271-72 | 0.01 | 5.9 | 1.7 | 93.4 |
|  | 9277-78 | 0.01 | 4.3 | 2.4 | 95.3 |
| MEANS |  | 0.01 | 5.1 | 2.1 | 94.3 |

## CITIES SERVICE FEDERAL AK-1 (5292)

The Cities Service Federal AK-1 is located near the center of Hartzog Draw field. It lies on the eatern side of the maximum thickness (55') trend of the reservoir (Fig. 5) and contains a vertical facies sequence and thickness of individual facies typical of that observed in many wells in the central eastern part of the field. In the area where the well is located, the typical vertical sequence from bottom to top is (1) Shelf Silty Shale, (2) Inter-Ridge Facies, (3) Ridge-Margin Facies, (4) Central Ridge Facies, (5) Shelf Shale Facies (Fig. 17). The Shelf Shale Facies are designated as Cody Shales (Fig. 3). The facies which contain predominantly sandstone are included in the Shannon. Detailed percentages of the components which make up and are used to define each facies are tabulated in Table 3.

### Geological Facies Analysis

The High- and Low-Enerrgy Ridge-Margin Facies are subdivided on the basis of variations in percentages of their components (Table 3).

The Central Ridge Facies, which is very thick and highly productive, in this well is subdivided into two subunits primarily on the basis of reservoir quality. Unit 4A (Fig. 17) contains significantly more highly to totally cemented intervals (37% vs. 3%) and as such has a significantly inferior reservoir quality; both siderite and calcite cement occur. The top of Unit 4B was not cored, but on the basis of log characteristics, appears to be a noncemented high quality reservoir. The physical and biological components which make up the two subfacies are very similar (Table 3) except for the degree of cementation. The complete sequence of slabbed core photographs from this core are included as Figure 18A through 18H. Details of the Central Ridge Facies are also indicated in Figure 8.

The Inter-Ridge Facies (Fig. 17, Table 3) observed in this well is typical of this facies throughout the field, although it is somewhat thinner at this location than farther to the west. Details of this facies may be observed in Figures 12 and 18G and H.

The Ridge-Margin Facies are subdivided into High- and Low-Energy Ridge-Margin Facies on the basis of the percentage of interbedded rippled beds. For purposes of this study ripples are defined to include only cross laminated units less than 0.2 feet thick; all thicker cross laminated sets are termed small scale troughs or small scale planar-tabular sets. Where the percentage of interbedded ripples (exclusive of "ripples on troughs") exceed 33 percent a High-Energy designation is used. Other factors which commonly change from High- to Low-Energy Ridge-Margin Facies are

1) Bedding thickness diminished

2) Average grain size decreases

3) Galuconite percentage diminished

4) Percentage of shale lamina increases

5) Percentage of shale rip-up clasts decreases

Figure 17. Facies analysis of Cities Service Federal AK-1 (5292) based on geological facies recognized in slabbed core (Fig. 18). Core to log correction requires subraction of 2.5 feet from core depth.

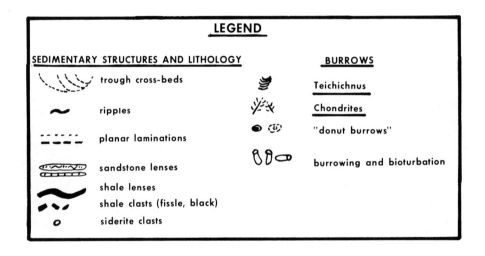

CITIES SERVICE FEDERAL AK-1
NE 1/4 SEC. 29 T45N R75W
HARTZOG DRAW FIELD, WYOMING
CAMPBELL CO., WYOMING
SHANNON SANDSTONE (CRETACEOUS)

Described by:
R. Tillman, R. Snyder,
R. Scott (2-78)

Cored Interval:
9344-9402 (Core Depths)
(58 feet)
(9341.5-9400.5') Log Depth

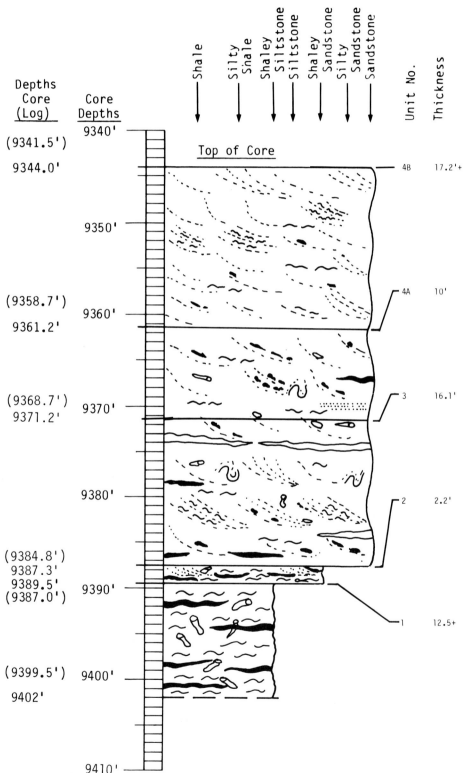

Interval and Description

9344-9361.2' Sandstone (200), 10% glauconite. 76% moderate to low angle (5-20°) troughs 12% high angle troughs (>20°). 12% planar-tangential cross bedding. 5% ripples on troughs and 2% ripples. 3% shale rip up clasts and a trace of rounded siderite clasts. A trace of shale laminations. Non burrowed. Lower contact very sharp. 2% siderite cemented, trace of calcite cement. CENTRAL RIDGE FACIES (95%).

9361.2-71.2' Sandstone (250), 10% glauconite, 77% low to moderate angle troughs, 11% high angle troughs, 4% shale clasts scattered uniformly through unit. 2% laminated shale and 2% siderite clasts. 7% rippled including 2% wave ripples. 5% oversteepened beds. No burrowing. 37% cemented, 25% calcite, 12% siderite. CENTRAL RIDGE FACIES (95%)

9371.2-87.3' Sandstone (225). 15% glauconite, up to 50% on some lamina. 60% moderate to low angle (5-20°) troughs, 28% ripples, 6% high angle troughs. 2% siderite (clasts and layered), 4% shale clasts, 5% laminated shale. 3% 1/8-1/4" dia. burrows. 7% siderite cemented. Lower contact very sharp, erosional. HIGH-ENERGY RIDGE-MARGIN FACIES (90%)

9387.3-89.5' Sandstone (125), 65%; 35% shale (30% laminated, 5% clasts), 18% siderite (10% lenses, 8% clasts). Highly glauconitic (20%). 55% ripples (30% superimposed on troughs), 35% low to high angle troughs. 60% interbedded shale and sandstone ripples. 10% burrowed (5% <1/8", clay filled, vertical; remainder oblique to horizontal, low diversity. Lower contact very sharp, erosional. LOW-ENERGY RIDGE-MARGIN FACIES (95%)

9389.5-9402.0' Sandstone, 78% (140) and 21% shale interbedded. 2% glauconite. 63% ripples, mostly sandstone. Trace of trough cross bedding. 34% burrowed, 13 types of burrows, mostly <1/4" diameter; high diversity. 94% of vertical interval contains one or more burrows. Lower contact not cored; on logs sharp.
INTER-RIDGE FACIES (95%).

TABLE 3
STATISTICAL SUMMARY OF FACIES COMPONENTS

FACIES SUMMARY DEVELOPED FROM GEOLOGIC
CORE INTERPRETATIONS; CSO AK-1

The following data summary is compiled to allow a systematic detailed description of the lithology, physical structures and biologic structures of the various facies observed in Hartzog Draw Field cores. The amount of detail included here is not of general interest to most geologists but the detail is necessary for geologists involved in doing similar descriptions of cores.

Under "Lithology" the percentage values should total to 100%. Percentages are based on the percentage of rock volume occupied by a feature. Line entries which have a "slashed box" on the far left are descriptors and are not included in the tabulation to total 100%. Designation of boxes to be treated as "descriptors" and having "slashed boxes" may vary from study to study.

"Physical structures" and "Biogenic Sedimentary Structures" together total to 100%. Individual entries under each sub-section total to the percentage indicated opposite the major heading.

Under all categories, the percent (by volume) for each component is followed by a value indicating the maximum set thickness for that feature. The area following the second slash is for notes.

Burrow types which cannot be identified by name are included in section 2. In sections 1 and 2 the volume of burrowing is indicated followed by descriptors concerning size, type of filling, and orientation of individual burrows. The percentage and footage of vertical interval burrowed is indicated and the diversity of burrow types is given.

The contact relations for the units are described on the basis of (1) the thickness of the transition from one unit to another, (2) the degree of erosion, and (3) the degree of parallelism of underlying and overlying units. The thickness numbers assigned to each of the descriptors for indicating the thickness of transition may be varied from one study to another.

GEOLOGIC CORE DESCRIPTIONS
CITIES SERVICE OIL COMPANY
FEDERAL AK-1 (5292) NE SEC. 29 T45N R75W
HARTZOG DRAW FIELD, SHANNON SANDSTONE,
CAMPBELL CO., WYOMING

Described by: R. W. Tillman (1979, 1983)
R. Scott (1979)

Columnized Data Format: Shannon
%/Max. set thickness/Additional notes

| | INTER-RIDGE FACIES (SHALEY) |
|---|---|
| | Unit 1 (12.5') |
| | Core: 9389.5-9402.0' |
| | (Log: 9387.0-9499.5') |

Lithology

| | |
|---|---|
| 1. Sandstone (%) | 78%/0.15' (140) |
| 2. Siltstone (%) | tr/0.01 |
| 3. Shale (%); a) Laminated; b) Clasts | 21%/0.06'/a |
| 4. Siderite (%); a) Laminated; b) Clasts | - |
| 5. Glauconite (%); a) Disseminated; b) Laminated | 2%a |

Physical Structures

| | |
|---|---|
| | 66%/0.15' |
| 1. High angle cross-bedding (20°+); a) Troughs; b) Planar-tabular; c) Planar-tangential; d) Curved-tangential(trough) | - |
| 2. Moderate angle cross-bedding (10-20°); a) Troughs; b) Planar-tabular; c) Planar-tangential; d) Curved-tangential (trough) | tr/0.1'/d,a |
| 3. Subhorizontal to low angle bedding (<10°); a) Trough; b) Planar-tabular; c) Planar-tangential; d) Curved-tangential; e) Planar | tr/0.05'/e/subhorizontal |
| 4. Horizontal laminations; a) Sandstone; b) Shale | tr/0.1'/9384.6' |
| 5. Rippled; a) Sandstone; b) Shale; c) Current; d) Wave | 63%/0.15'/mostly a |
| 6. Rippled interbedded sandstone and shale | // 40%/0.05' |
| 7. Ripples superimposed on troughs | - |
| 8. Reworked: a) By waves & currents; b) Bedding destroyed, massive; c) soft sediment deformed | // |
| | // 20% a |

Biogenic Sedimentary Structures

| | |
|---|---|
| | 34%/0.3' |
| 1. Identified burrows; a) Asterosoma; c) Chondrites; d) "Donut burrows"; g) Gastropod tracks; p) Plural curving tubes; s) Skolithos; t) Teichichnus; th) Thallasinoides | - |
| 2. Distinct burrows; a) <1/8"; b) 1/8"-1/4"; c) 1/4"-1/2"; d) >1/2"; e) Silt filled; f) Clay filled; g) Sand filled; h) Silt lined; i) Clay lined; j) Spreiten; k) Vertical; l) Oblique; m) Horizontal | 3%-adgl  tr-afgl  tr-aadl<br>2%-bdl   tr-bdj   tr-bdgi<br>3%-adgj  tr-bdk   tr-adgli<br>2%-afgi  tr-adgk<br>2%-afgk  tr-aadj |
| 3. Bioturbated (75% + burrowed) (%); a) Distinct; b) Mottled | 12% |
| 4. Total interval burrowed (%, footage) | 94%, 11.8' |
| 5. Diversity (Number of burrow types); 1-4 low, 4-8 moderate; >8 high | 13 - high |

Cemented Intervals; a) Siderite; b) Calcite | tr/0.1'/b |

Contact Relations

| | |
|---|---|
| 1. Upper; a) Very sharp (<0.5' transition); b) Sharp (0.5-1' transition); c) Transitional (1-3' transition); d) Gradational (>3' transition); e) Contact, erosional (truncated) angular; f) Contact erosional parallel; g) Covered | a, f |
| 2. Lower; a), b), c), d), e), f), g) | NA |

RWT/jk 1/84

TABLE 3 (cont'd)

CITIES SERVICE OIL COMPANY
FEDERAL AK-1 (5292) NE SEC. 29 T45N R75W
HARTZOG DRAW FIELD, SHANNON SANDSTONE,
CAMPBELL CO., WYOMING

Described by: R. W. Tillman (1979, 1983)
R. Scott (1979)

Columnized Data Format: Shannon
 %/Max. set thickness/Additional notes

| | LOW-ENERGY RIDGE-MARGIN<br>Unit 2 (2.2')<br>Core: 9387.3-89.5'<br>(Log: 9384.8-8387.0') |
|---|---|
| **Lithology** | |
| 1. Sandstone (%) | 65%/0.2' (125 ) |
| 2. Siltstone (%) | - |
| 3. Shale (%); a) Laminated; b) Clasts | 35%/0.1'/30%a, 5%b |
| 4. Siderite (%); a) Laminated; b) Clasts | 18%/0.01'/9%a, 9%b |
| 5. Glauconite (%); a) Disseminated; b) Laminated | 20%/0.1'/15% b, 5% a |
| **Physical Structures** | 90%/0.5' |
| 1. High angle cross-bedding (20°+); a) Troughs; b) Planar-tabular; c) planar-tangential; d) Curved-tangential (trough) | 10%/0.2'/d |
| 2. Moderate angle cross-bedding (10-20°); a) Troughs; b) Planar-tabular; c) Planar-tangential; d) Curved-tangential (trough) | 15%/0.2'/d |
| 3. Subhorizontal to low angle bedding (<10°); a) Trough; b) Planar-tabular; c) Planar-tangential; d) Curved-tangential; e) Planar | 10%/0.2'/d |
| 4. Horizontal laminations; a) Sandstone; b) Shale | - |
| 5. Rippled; a) Sandstone; b) Shale; c) Current; d) Wave | 41%/0.5' |
| 6. Rippled interbedded sandstone and shale | // 60%/0.2' |
| 7. Ripples superimposed on troughs | 30%/0.2' |
| 8. Reworked: a) By waves & currents; b) Bedding destroyed, massive; c) Soft sediment deformed | 14%//d |
| **Biogenic Sedimentary Structures** | 10%/0.1' |
| 1. Identified burrows; a) Asterosoma; c) Chondrites; d) "Donut burrows"; g) Gastropod tracks; p) Plural curving tubes; s) Skolithos; t) Teichichnus; th) Thallasinoides | - |
| 2. Distinct burrows; a) <1/8"; b) 1/8"-1/4"; c) 1/4"-1/2"; d) >1/2"; e) Silt filled; f) Clay filled; g) Sand filled; h) Silt lined; i) Clay lined; j) Spreiten; k) Vertical; l) Oblique; m) Horizonal | 5%-afk<br>2%-afj<br>tr-afl<br>tr-cfl |
| 3. Bioturbated (75% + burrowed) (%); a) Distinct; b) Mottled | - |
| 4. Total interval burrowed (%, footage) | 12%/0.1' |
| 5. Diversity (Number of burrow types); 1-4 low, 5-8 moderate; >8 high | 4 - low |
| Cemented Intervals; a) Siderite; b) Calcite | - |
| **Contact Relations** | |
| 1. Upper; a) Very sharp (<0.5' transition); b) Sharp (0.5-1' transition); c) Transitional (1-3' transition); d) Gradational (>3' transition); e) Contact, erosional (truncated) angular; f) Contact erosional parallel; g) Covered | a, f |
| 2. Lower; a), b), c), d), e), f), g) | a, f |

RWT/jk  1/84

## TABLE 3 (cont'd)

| HIGH-ENERGY RIDGE-MARGIN FACIES | CENTRAL RIDGE FACIES | CENTRAL RIDGE FACIES |
|---|---|---|
| Unit 3 (16.1') | Unit 4A (10') | Unit 4B (17.2'+) |
| Core: 9371.2-87.3' | Core: 9361.2-71.2' | Core: 9344-9361.2' |
| (Log: 9368.7-9384.8') | (Log: 9358.7-9368.7') | (Log: 9341.5-9358.7') |
| 88%/0.6' (200-250 ) | 92%/0.6' (250 ) | 95%/1.9'/ (200 ) |
| tr/0.01'/ | - | tr/0.1' |
| 9%/0.02'/5%a, 4%b | 6%/0.01'/2%a, 4% b | 4%/0.05'/tr-a, 3% b |
| 2%/0.1'/tr a, tr b | 2%/0.2'/b, 1 x 1/2" | tr/0.02'/b |
| 15%/3% a, 12% b (50%) | 10%/a, b | 10%/3% a, 7% b |
| 97%/0.5' | 100%/0.6' | 100%/1.0' |
| 6%/0.3'/d | 11%/0.6'/d | 12%/1.0'/a |
| 25%/0.5'/d | 26%/0.3'/d, tr-c(9373.5') | 26%/0.9'/a |
| 34%/0.5'/d, tr-e | 45%/0.2'/d | 38%/0.2'/39% d, tr-c, 2% e |
| - | - | - |
| 14%/0.5'/ 8% c, 6% d | 7%/0.2'/5% c, 2% d | 2%/0.1'/tr-a, tr-b |
| // 17%/0.5' | // 4%/0.2' | // tr |
| 11%/0.2' | 5%/0.2'/current | 5%/0.2' |
| over- 7%/0.5'/tr-d; a, @71.3'/steep | over- 11%/0.5'/5% a, 6% d/steep | 5%/4% d, tr-a |
| 3%/0.2' | 0 | 0 |
| - | - | - |
| tr-bfk<br>tr-bfl<br>tr-bfj | - | - |
| - | - | - |
| 2%/0.01' | - | - |
| - | - | - |
| 7%/0.5'/a | 37%/0.3'/12%a, 25%b | 3%/0.5 c/2% a, tr-b |
| d, f | d | Not cored |
| a, f | d | d |

Figure 18 (A-H). Photographs of Cities Service Federal AK-1 (5292) slabbed core. Unit boundaries and facies names are desigated. Interval cored is 58 feet and extends from 9344 to 9402 feet. Core to subsurface log correction requires subtraction of 2.5 feet from core depths to obtain log depths. A detailed statistical description of each facies is given in Table 3.

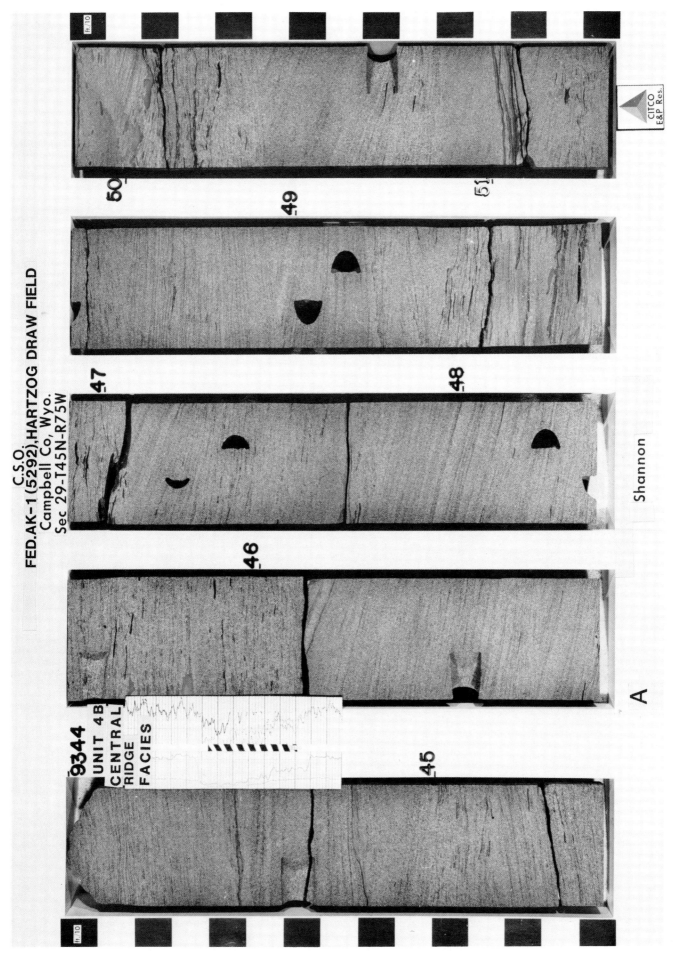

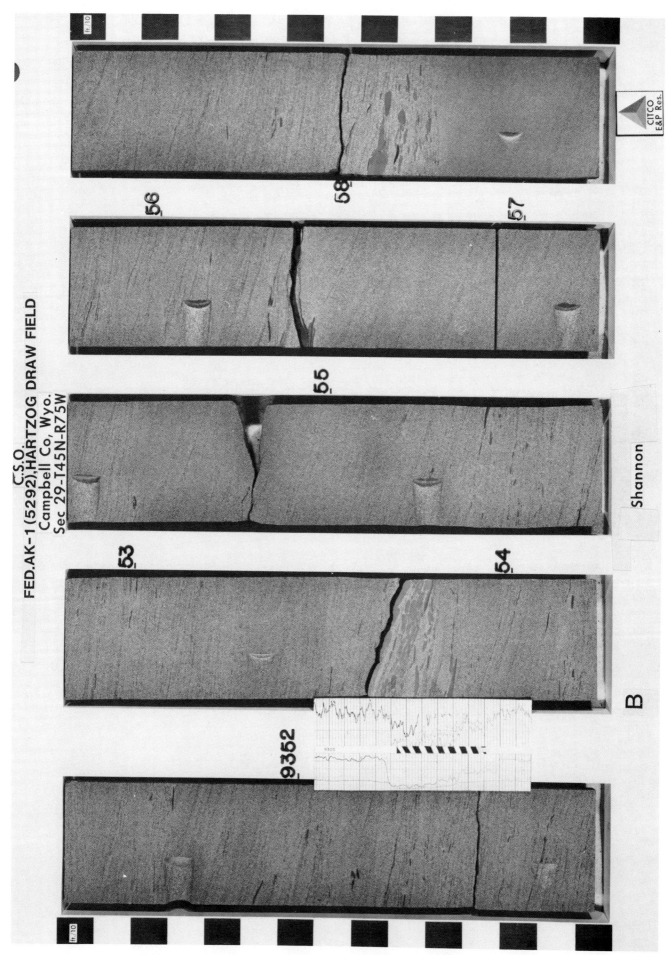

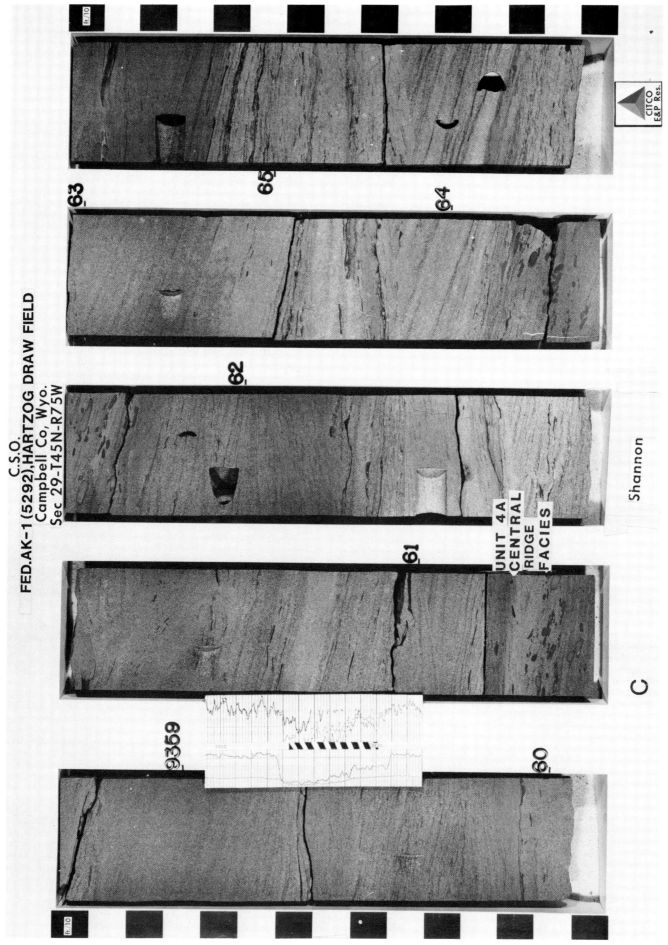

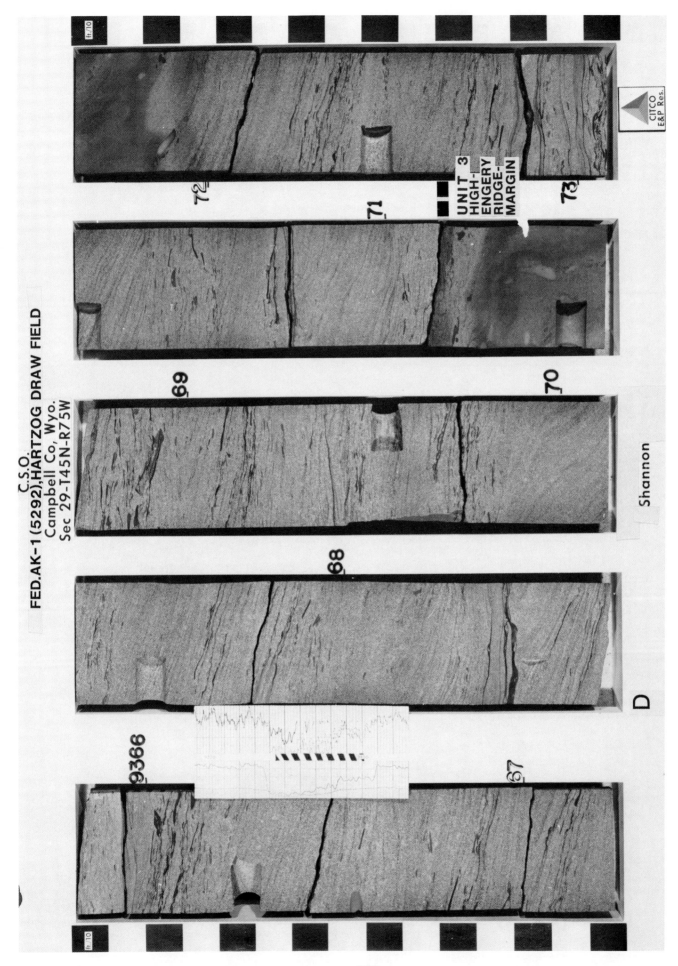

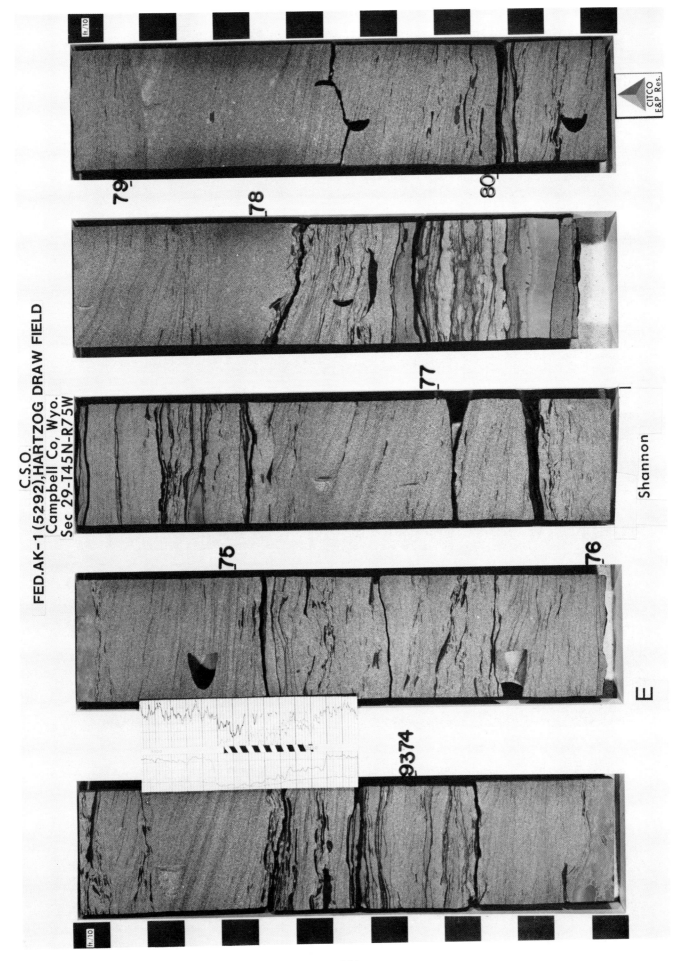

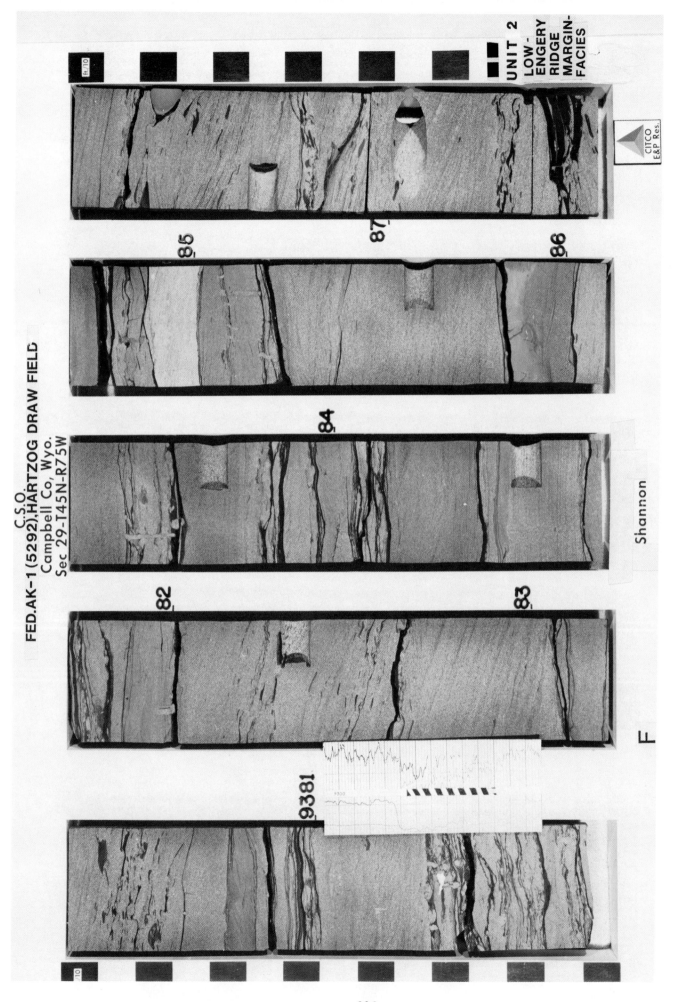

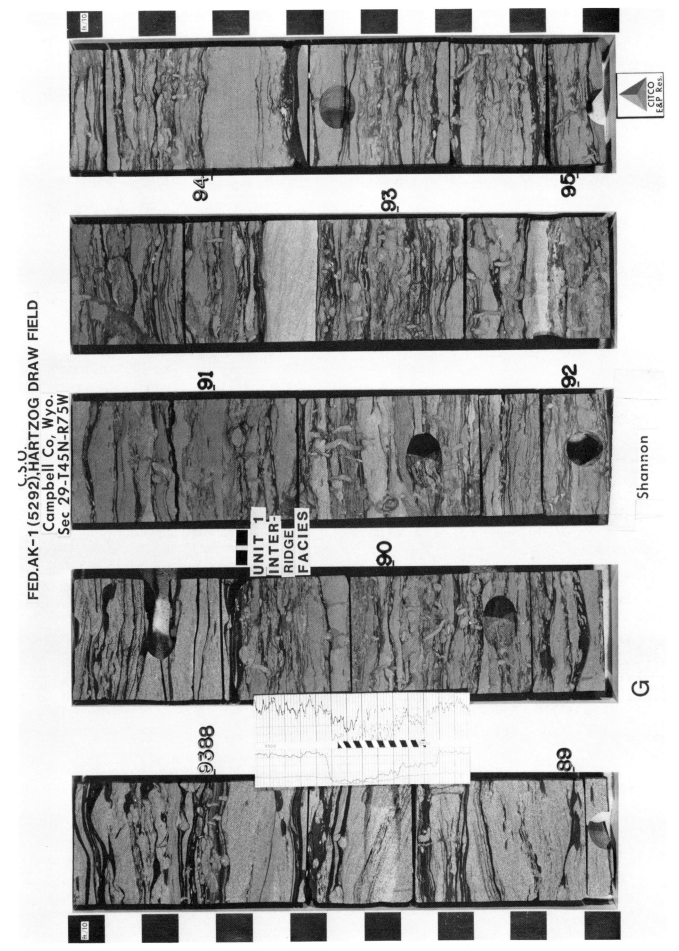

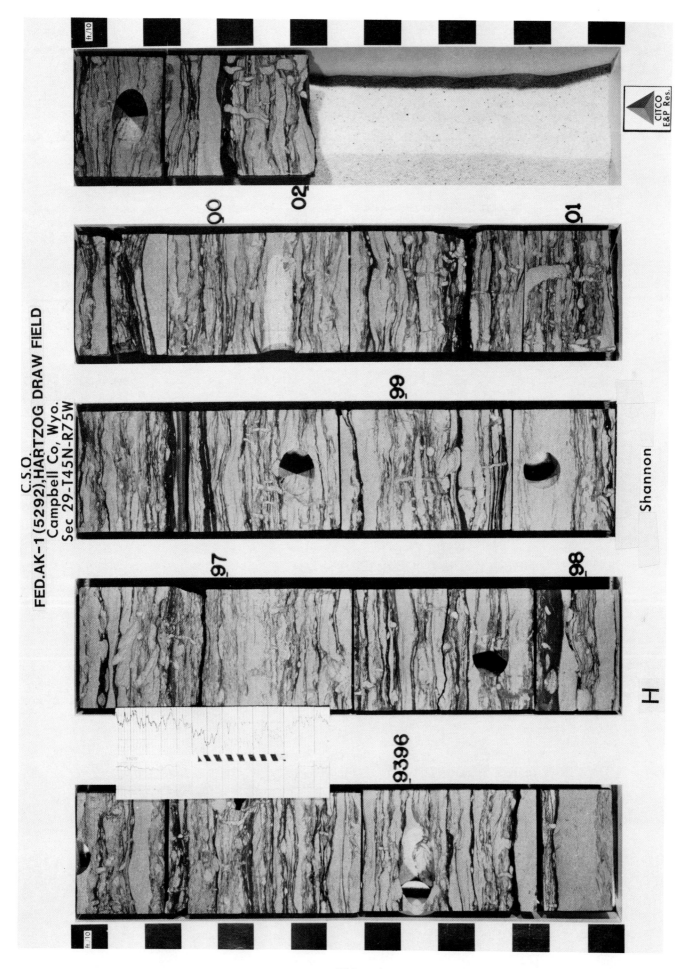

6) Percentage of siderite lenses and rip-up clasts decreases

7) Percentage of ripples on troughs diminishes.

Fairly typical of these facies types, where they occur in the thicker parts of the field, are Units 2 and 3. The percentages of the components for each facies are tabulated in Table 3. Details of these facies are shown in Figures 9, 10 and 17.

## Log Signatures

Log signatures in this well (Fig. 19) are similar to most of the wells in the central part of the field. The sharp blocky lower contact between the underlying Cody Shale and the Inter-Ridge Facies marks the change from silty shale to interbedded sandstone and shale. The transition from Low-Energy Ridge-Margin Facies to High-Energy Ridge-Margin Facies is gradual on both the GR and SP logs. The upper subunit of the Central Ridge Facies has a relatively constant left deflection, while the lower unit, probably as a result of the increasing cement, decreases in value downward on both SP and GR logs. The upper contact is sharp on all log surveys, and is typical for conditions where the Central Ridge Facies is overlain directly by the Cody Shale. Core-gamma ray responses are shown in Figure 19A. These core-gamma logs may be used to calculate the log to core correction.

The major differences between this well and the Cities Service Federal AB-1A are that, in the AK-1, there is:

1) Greater overall thickness of shelf-ridge (bar) complex (72' vs. 60')

2) Greater Central Ridge Facies thickness

3) Thicker reservoir facies

4) More organized vertical coarsening upward facies sequence

5) Fewer individual units; thicker units

6) Better reservoir quality for all producing facies

7) More variability in porosity due to cementation (Unit 4A).

## Production Characteristics

The Cities Service Federal AK-1 is typical of many of the wells in the central part of the field, especially those east of the area of maximum thickness. The variations in reservoir properties are shown graphically in Figure 20 and are tabulated in Table 4. Table 4 contains data for each foot from three different analyses and is organized to take into account the high degree of cementation in certain intervals of the Central Ridge and Ridge-Margin Facies. Analyses were run by three different laboratories to determine if a systematic variation in calculated results could be observed. In general the values are systematically highest in the Central Ridge Facies for the Cities

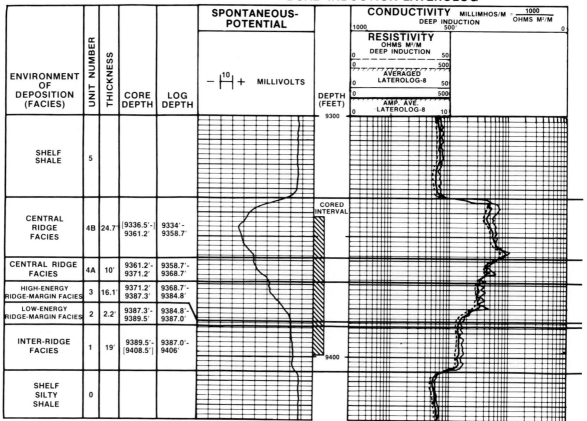

Figure 19. Subsurface logs of Cities Service Federal AK-1 (5292). 5292 is the well designation applied for use in the Hartzog Draw production unit. Individual facies are delineated. See text for discussion of key features of logs as related to facies recognition. Depth corrections required (upper right) have been made.

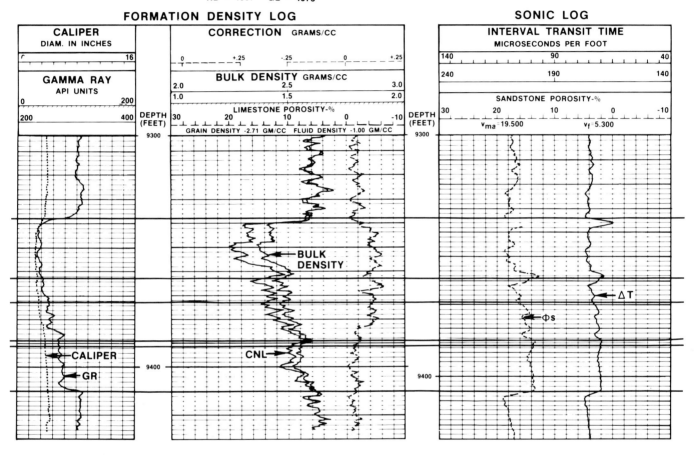

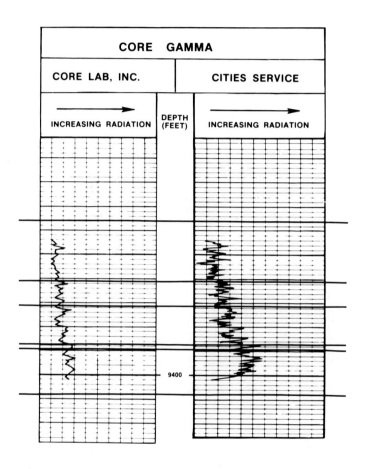

Figure 19A. Core-gamma logs done at Core Lab Inc. and Cities Service. These core-gamma logs were used to determine log to core correlation; subsurface gamma-ray log was correlated with core-gamma (Fig. 19).

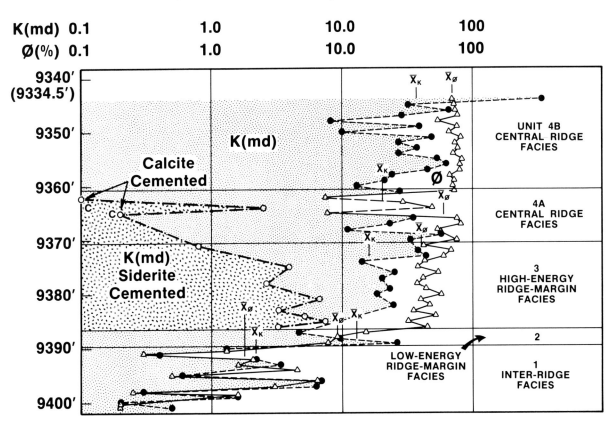

Figure 20. Production characteristics of individual facies recognized in slabbed core. Depths are core depths Values for "normal" permeability are black circles.

Interbedded in the "normal" facies are 3 inch or thicker intervals which are highly cemented. These values are plotted separately. Open circles are values for highly cemented intervals. Open circles with C are highly calcite cemented; remainder of the open circles are highly siderite cemented intervals.

Porosity is designated by triangles. Mean values for each facies are indicated by $X_k$ (permeability) and $X\phi$ (porosity). Facies correspond to those described in Figure 17 and Table 4.

## TABLE 4

### CITIES SERVICE OIL AND GAS CORPORATION
### FEDERAL AK-1, NE SEC. 29 T45N R75W
### HARTZOG DRAW FIELD, SHANNON SANDSTONE
### CAMPBELL COUNTY, WYOMING

Date: 10-9-78
Elevation: 4975'

| Feet | Permeability to Air (md) | | | Porosity (%) | | | Grain Density | Bulk Density |
|------|------|------|------|------|------|------|------|------|
|      | *    | **   | ***  | *    | **   | ***  |      |      |

Unit 4B  CENTRAL RIDGE FACIES (9344-9361.2'; 17.2') ("Net Pay" - 14.7')

| Feet | * | ** | *** | * | ** | *** | Grain Density | Bulk Density |
|------|------|------|------|------|------|------|------|------|
| 9344 | 135 | 134 | 120 | 17.0 | 17.2 | 17.4 | 2.66 | 2.21 |
| 9345 | 32  | 32  | 28  | 17.3 | 16.6 | 15.9 | 2.68 | 2.22 |
| 9346 | 66  | 56  | 56  | 17.5 | 18.1 | 17.4 | 2.66 | 2.20 |
| 9347 | 29  | 28  | 26  | 17.8 | 17.3 | 17.2 | 2.68 | 2.21 |
| 9348 | 8   | 8   | 8   | 15.4 | 14.6 | 14.5 | 2.70 | 2.29 |
| 9349 | 39  | 35  | 32  | 17.7 | 17.1 | 17.2 | 2.68 | 2.21 |
| 9350 | 10  | 10  | 8   | 16.7 | 15.5 | 16.0 | 2.70 | 2.25 |
| 9351 | 49  | 42  | 43  | 18.1 | 17.7 | 17.5 | 2.68 | 2.19 |
| 9352 | 27  | 25  | 22  | 17.4 | 16.9 | 17.0 | 2.69 | 2.22 |
| 9353 | 37  | 34  | 32  | 17.8 | 17.2 | 17.4 | 2.67 | 2.20 |
| 9354 | 27  | 25  | 22  | 17.2 | 16.7 | 16.7 | 2.68 | 2.22 |
| 9355 | 53  | 47  | 45  | 18.4 | 17.8 | 17.7 | 2.67 | 2.18 |
| 9356 | 63  | 53  | 50  | 18.1 | 17.5 | 16.9 | 2.68 | 2.20 |
| 9357 | 45  | 40  | 37  | 18.2 | 17.6 | 17.5 | 2.68 | 2.19 |
| 9358 | 24  | 23  | 21  | 16.8 | 16.3 | 16.2 | 2.70 | 2.25 |
| 9359 | 21  | 21  | 19  | 17.3 | 16.2 | 16.4 | 2.70 | 2.26 |
| 9360 | 13  | 12  | 11  | 17.1 | 16.4 | 15.9 | 2.69 | 2.23 |
| MEANS | 40 | 37 | 34 | 17.4 | 16.9 | 16.8 | 2.68 | 2.21 |
| RANGE | 8-135 | 8-134 | 8-120 | 15.4-18.2 | 14.6-18.1 | 15.9-17.7 | 2.66-2.70 | 2.19-2.26 |
| DIFFERENCE | 127 | 126 | 112 | 2.8 | 3.5 | 1.8 | 0.40 | 0.07 |

% of unit cemented: 5% siderite, 1% calcite.  "Net pay" - 14.7'
No  plugs taken in cemented intervals in this facies.

*Cities Service Lab     **Core Lab Inc. (Aurora)     ***Core Lab Inc. (Casper)

| Feet | Permeability to Air (md) | | | Porosity (%) | | | Grain Density | Bulk Density |
|---|---|---|---|---|---|---|---|---|

Unit 4A   CENTRAL RIDGE FACIES (9361.2-71.2';10')  ("Net pay" = 6.7')

| Feet | * | ** | *** | * | ** | *** | Grain Density | Bulk Density |
|---|---|---|---|---|---|---|---|---|
| 9361 | 27.5 | 25.0 | 22.0 | 17.3 | 16.6 | 16.8 | 2.70 | 2.24 |
| 9362(c) | 0.1 | 0.8 | 0.2 | 7.5 | 5.9 | 6.4 | 2.72 | 2.51 |
| 9363 | 16.9 | 7.9 | 15.0 | 12.9 | 12.0 | 12.4 | 2.68 | 2.33 |
| 9364(s) | 2.5 | 1.4 | 2.5 | 14.9 | 12.3 | 12.8 | 2.76 | 2.35 |
| 9365(c) | 0.2 | 0.5 | 3.1 | 7.8 | 7.3 | 6.6 | 2.69 | 2.48 |
| 9366 | 34.5 | 32.0 | 30.0 | 17.7 | 16.6 | 17.0 | 2.69 | 2.21 |
| 9367 | 23.1 | 21.0 | 19.0 | 18.1 | 16.9 | 16.9 | 2.71 | 2.22 |
| 9368 | 11.2 | 16.0 | 9.3 | 15.4 | 13.3 | 13.8 | 2.73 | 2.31 |
| 9369 | 59.0 | 27.0 | 22.0 | 15.8 | 15.2 | 15.2 | 2.68 | 2.25 |
| 9370 | 33.0 | 30.0 | 28.0 | 16.8 | 16.5 | 16.5 | 2.68 | 2.23 |
| MEAN | 20.8 | 16.2 | 15.1 | 14.4 | 13.3 | 13.4 | 2.70 | 2.31 |
| NON-CEMENTED MEAN | 26.0 | 20.1 | 18.5 | 17.8 | 16.1 | 15.1 | 2.69 | 2.26 |
| NON-CEMENTED RANGE | 2.5-34.5 | 1.4-30.0 | 2.5-30.0 | 12.9-17.7 | 12.0-16.9 | 12.4-17.0 | 2.68-2.76 | 2.21-2.35 |
| NON-CEMENTED DIFFERENCE | 32.0 | 29.6 | 27.5 | 4.8 | 4.9 | 4.6 | 0.08 | 0.14 |

% of unit cemented:   10% Siderite (s), 23% Calcite (c), "Net pay" = 6.7'

Unit 3   LOW-ENERGY RIDGE-MARGIN FACIES (9371.3-9387.3'; 16.0')  ("Net pay" = 14.7')

| Feet | * | ** | *** | * | ** | *** | Grain Density | Bulk Density |
|---|---|---|---|---|---|---|---|---|
| 9371(s) | 0.8 | 0.6 | 0.7 | 14.4 | 13.9 | 14.0 | 3.03 | 2.59 |
| 9372 | 38.4 | 35.0 | 34.0 | 16.9 | 15.5 | 15.7 | 2.70 | 2.24 |
| 9373 | 44.0 | 40.0 | 39.0 | 16.0 | 15.4 | 15.4 | 2.68 | 2.25 |
| 9374 | 14.0 | 14.0 | 11.0 | 14.4 | 13.8 | 13.6 | 2.67 | 2.29 |
| 9375"s" | 3.8 | 3.9 | 3.3 | 13.7 | 12.2 | 12.8 | 2.72 | 2.34 |
| 9376 | 24.9 | 25.0 | 21.0 | 15.5 | 15.2 | 15.1 | 2.68 | 2.26 |
| 9377 | 20.2 | 27.0 | 22.0 | 14.2 | 14.2 | 14.5 | 2.66 | 2.28 |
| 9378"s" | 2.6 | 18.0 | 22.0 | 13.8 | 11.5 | 12.2 | 2.74 | 2.36 |
| 9379 | 23.3 | 24.0 | 21.0 | 14.9 | 14.9 | 14.6 | 2.67 | 2.27 |
| 9380 | 18.2 | 18.0 | 15.0 | 15.6 | 14.9 | 15.3 | 2.68 | 2.26 |
| 9381"s" | 6.7 | 13.6 | 13.0 | 14.0 | 13.6 | 13.0 | 2.68 | 2.30 |
| 9382 | 24.5 | 14.9 | 14.7 | 15.2 | 14.9 | 14.7 | 2.66 | 2.26 |
| 9383"s" | 3.2 | 12.2 | 12.5 | 13.5 | 12.2 | 12.5 | 2.70 | 2.34 |
| 9384"s" | 5.1 | 14.4 | 14.2 | 15.2 | 14.4 | 14.2 | 2.69 | 2.28 |
| 9385"s" | 7.2 | 13.0 | 12.9 | 13.4 | 13.0 | 12.9 | 2.67 | 2.31 |
| 9386"s" | 3.2 | 13.2 | 13.1 | 14.5 | 13.2 | 13.1 | 2.70 | 2.31 |
| MEAN (TOTAL UNIT) | 15.0 | 17.9 | 16.8 | 14.7 | 13.9 | 14.0 | 2.71 | 2.31 |
| NON-CEMENTED MEAN | 25.9 | 19.0 | 17.8 | 15.6 | 13.9 | 14.0 | 2.68 | 2.29 |
| | 2.6-44.0 | 3.9-40.0 | 3.3-39.0 | 13.5-16.9 | 11.5-15.5 | 12.2-15.7 | 2.66-2.74 | 2.24-2.36 |
| NON-CEMENTED DIFFERENCE | 41.4 | 36.1 | 35.7 | 3.4 | 4.0 | 3.5 | 0.08 | 0.12 |

% of unit cemented:   8% Siderite (s), 0% Calcite (c), "Net pay" = 14.7'

| Feet | Permeability (md) | Porosity (%) | Saturation | | Grain Density |
|---|---|---|---|---|---|
| | | | Oil | Water | |
| Unit 2 LOW-ENERGY RIDGE-MARGIN FACIES (9387.3-89.5'; 2.2') | | | | | |
| 9387 | 4.6 | 11.0 | 13.3 | 32.7 | 2.65 |
| 9388 | 9.6 | 9.3 | 10.0 | 46.7 | 2.67 |
| 9389 | 26.0 | 7.8 | 11.4 | 48.0 | 2.68 |
| MEAN | 13.4 | 9.4 | 11.6 | 42.4 | 2.66 |
| RANGE | 21.4 | 3.2 | 1.6 | | 0.30 |
| Unit 1 INTER-RIDGE FACIES (9389.5-9402'; 12.5'+) | | | | | |
| 9390 | 1.6 | 1.3 | 15.3 | 60.7 | 2.66 |
| 9391 | 0.3 | 0.3 | 15.4 | 60.2 | 2.66 |
| 9392 | 2.2 | 2.1 | 14.2 | 58.1 | 2.66 |
| 9393 | 3.4 | 1.6 | 14.8 | 53.4 | 2.66 |
| 9394 | - | 4.5 | 13.3 | 62.4 | 2.67 |
| 9395 | 0.6 | 0.4 | 14.3 | 61.5 | 2.64 |
| 9396 | 6.9 | 6.5 | 15.4 | 41.3 | 2.65 |
| 9397 | 6.3 | 3.0 | 14.3 | 56.3 | 2.65 |
| 9398 | 0.3 | 0.3 | 13.4 | 60.5 | 2.65 |
| 9399 | 1.6 | 1.6 | 14.0 | 47.5 | 2.64 |
| 9400 | 0.2 | 0.2 | 14.0 | 55.8 | 2.65 |
| 9401 | 0.5 | 0.2 | 12.1 | 59.2 | 2.65 |
| MEAN | 2.2 | 1.8 | 14.3 | 56.4 | 2.65 |
| RANGE | 6.7 | 4.2 | 3.3 | | 0.02 |

Service data, and lowest for Core Laboratory Inc. (Casper). Variations within the other facies are not as systematic.

Net pay was determined for each facies by identification of the cemented intervals in the core and subtraction of these intervals from the total core interval for each facies. The percentage of "net pay" is significantly higher in Central Ridge Facies, Unit 4B than in Unit 4A (14.7'/17.2' = 85% for Unit 4B; 6.7'/10' = 67% for Unit 4A). The permeability of the two Central Ridge Facies subunits are significantly different ($\bar{X}$ = 40 md., Unit 4B vs. 21 md, Unit 4A). However, even when highly cemented zones are eliminated from the calculations (Table 4), the average permeability of Unit 4A is significantly less (26 md) than Unit 4B suggesting that cementation and diagenesis have affected Unit 4A to a greater degree.

As was the case in the Federal AB-1A well, a significant decrease in reservoir quality can be observed when comparing Central Ridge Facies and Low- to High-Energy Ridge-Margin Facies. Porosities and permeabilities in three of the productive units in this well are:

| Mean Values | Permeability | Porosity |
|---|---|---|
| Central Ridge Facies (Unit 4B) | 40 md | 17% |
| Central Ridge Facies (Unit 4B; non-cemented) | 26 md | 18% |
| High-Energy Ridge-Margin Facies (Unit 3) | 18 md | 14% |
| Low-Energy Ridge-Margin Facies (Unit 2) | 13 md | 9% |

High energy deposits of the shelf-ridge sandstones at Hartzog Draw field have distinctive physical, biological and production characteristics as seen in the two cored wells discussed in this study.

## ACKNOWLEDGEMENTS

Thanks are extended to the Wyoming Geological Association for permission to publish substantial portions from Tillman and Martinsen, 1979. Fred Mason did all core photography and dark room work. Dick Scott and Rich Snyder aided in describing the AK-1 core. Jayme Kline, Nancy Arnold and Janice Brewer typed the manuscript. John Hobson, Charles Hearn, James Ebanks and Robert Tye aided by discussion and/or editing. Don Swift and Roger Walker reviewed portions of the paper.

## REFERENCES CITED

Asquith, D. O., 1970, Depositional topography and major marine environments, Late Cretaceous, Wyoming: AAPG Bulletin, v. 54, p. 1184-1224.

Gill, J. R. and W. A. Cobban, 1973, Stratigraphy and geologic history of the Montana Group and equivalent rocks, Montana, Wyoming and North Dakota: U. S. Geological Survey Professional Paper 776, 37 p.

Hobson, J., M. Fowler and R. W. Tillman, 1984, Asymmetry of form and lithology, Late Cretaceous shelf sandstone complexes, Horse Creek and Hartzog Draw fields, Powder River Basin, Wyoming (abs.): Shelf Sands and Sandstones Symposium, Calgary, Alberta.

Hunt, R. D. and C. L. Hearn, 1982, Reservoir management of the Hartzog Draw field: Journal of Petroleum Technology, July, 1982, p. 1575-1582.

Martinsen, R. S. and R. W. Tillman, 1978, Hartzog Draw, new giant oil field (abs.), AAPG Bulletin, v. 62, p. 540.

_____ and R. W. Tillman, 1979, Facies and reservoir characteristics of shelf sandstones, Hartzog Draw field, Powder River Basin Wyoming (abs.), AAPG Bulletin, v. 63, p. 491.

Porter, K. W., 1976, Marine shelf model, Hygiene Member of the Pierre Shale, Upper Cretaceous Denver Basin, Colorado, in Studies in Colorado Field Geology, R. Epis and R. Weimer (eds.): Professional Contributions of Colorado School of Mines (Annual Meeting GSA, Denver), p. 251-263.

Reineck, H. E. and I. B. Singh, 1971, Genesis of laminated sand and graded rhythemites in storm-sand layers of shelf mud: Sedimentology, v. 18, p. 123-128.

Spearing, D. R., 1975, Shannon Sandstone, Wyoming, in SEPM Short Course No. 2, Depositional environments as interpreted from primary sedimentary structures and stratification sequences, Dallas, p. 104-114.

Tillman, R. W. and H. E. Reineck, 1975, Discrimination of North Sea sand environments with population-derived grain size parameters, in Resumes des Publications, IX ME Congress International de Sedimentologie, Nice, Theme 6, p. 217-220.

_____ and R. S. Martinsen, 1979, Hartzog Draw field, Powder River Basin Wyoming, in R. F. Flory (ed.), Rocky Mountain High, Wyoming Geological Association 28th Annual Meeting: Core Seminar Core Book, p. 1-38.

_____ and R. S. Martinsen, 1984, The Shannon shelf-ridge sandstone complex, Salt Creek anticline area, Powder River Basin, Wyoming, in R. W. Tillman and C. Siemers (eds.), Siliciclastic shelf sedimentation: SEPM Special Publication No. 34, p. 85-142.

# CARDIUM AND VIKING SANDSTONE CORES

R. G. Walker
Department of Geology
McMaster University, Hamilton, Ontario

Three shelf sandstone cores from western Alberta are discussed in terms of lithology, bedding types, log character, and processes of deposition. Two of the cores are Cardium cores, one each from Ricinus and Caroline Fields. A single Viking Sandstone core from Stettler Field is discussed.

## GULF RICINUS 13-26-36-9W5

The general geology of the Cardium field at Ricinus is given in Walker (this volume: "Upper Cretaceous (Turonian) Cardium Formation, Southern Foothills and Plains, Alberta"). The Cardium Formation is a 50-100 m thick sandstone within the dominantly shaley Alberta (= Colorado) Group, and is Turonian (U. Cretaceous) in age. The geologic setting at Ricinus is summarized in Walker (this volume, Figs. 1, 12, 14, 15, 18); the field is basically a channel fill, the channel being at least 45 km long, 4-5 km wide (before palinspastic reconstruction), and 20-40 m deep. Ricinus sits on top of the junction between the Plains (no deformation) and Foothills (multiple imbricate thrusts); in some wells the sandstone is repeated up to about 6 times.

The Gulf Ricinus 13-26 well is in the northern part of Ricinus Field, which is structurally simpler than the southern part. The gamma-sonic log (Fig. 1) and SP-resistivity log show the sand to be abrupt and sharp-based; the top is also abrupt. Serrations in the gamma ray indicate either thin mudstone partings, or zones of ripped-up mud clasts (e.g., 9011.5, 9017, 9019, 9020 feet).

As can be seen in the photographs of the slabbed cores (Fig. 2, A-D) the base of the sandstone (9023.6 feet) rests sharply on somewhat bioturbated dark mudstones. There is a scattering of mud clasts and gravel near the base, grading up into rather massive sandstone (9023-9020.3 feet). There are several units of ripped-up mud clasts and gravel, and suggestions of grading (9017-9017.75, 9016-9017, 9014-9016 feet) but there are no indications of cross bedding or ripple cross lamination. The interval between 9011 and 9014 feet also contains some gravel, and scattered ripped-up clasts. In the core it is difficult to isolate individual depositional events. There is another well-graded bed (9011-9009.5 feet) in which the delicacy and definition of lamination improves upward. Yet another graded bed extends from about 9007 to 9009.5 feet. Near the base are gravel and ripped-up mud clasts, and cross lamination is observed at the top. The uppermost part of the core, 9000-9007 feet is gravelly, with ripped-up mud clasts, and some low angle stratification (maximum dip at 9005 feet is about 20°). Some of this stratification may be medium scale cross bedding, but mostly the low dips and regional setting (see Walker "Upper Cretaceous (Turonian) Cardium Formation, this volume) suggest that it might be hummocky cross stratification.

Although the evidence is not shown in this core, the regional setting, along with grading and Bouma sequences in other cores (Walker, this volume, Figs. 36-39) suggest possible deposition from turbidity currents. The grading

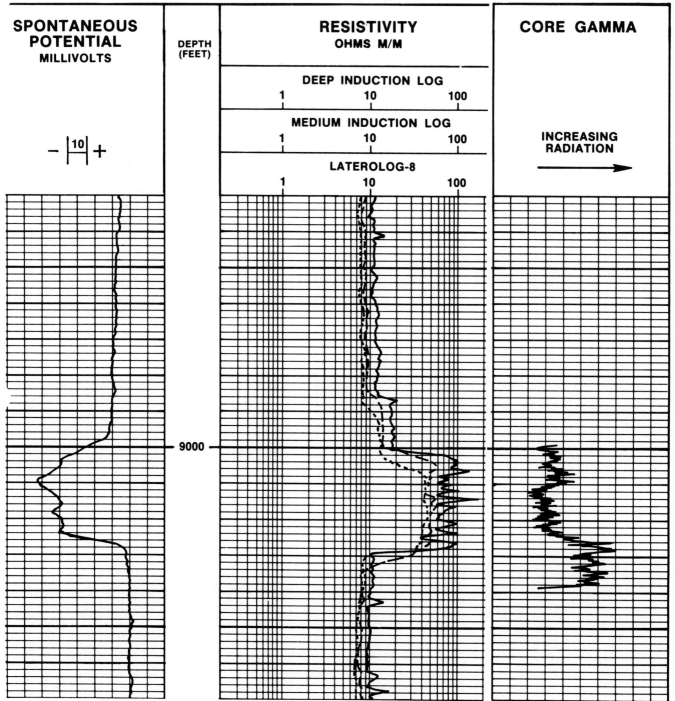

Figure 1A. SP-Resistivity logs. Cored interval 9000-9034'.

Figure 1B. Core-Gamma and subsurface Gamma-Ray Sonic logs through Cardium Sandstone interval. Core-Gamma used to correlate log and core depths; no correction required. Cored interval 9000-9034'.

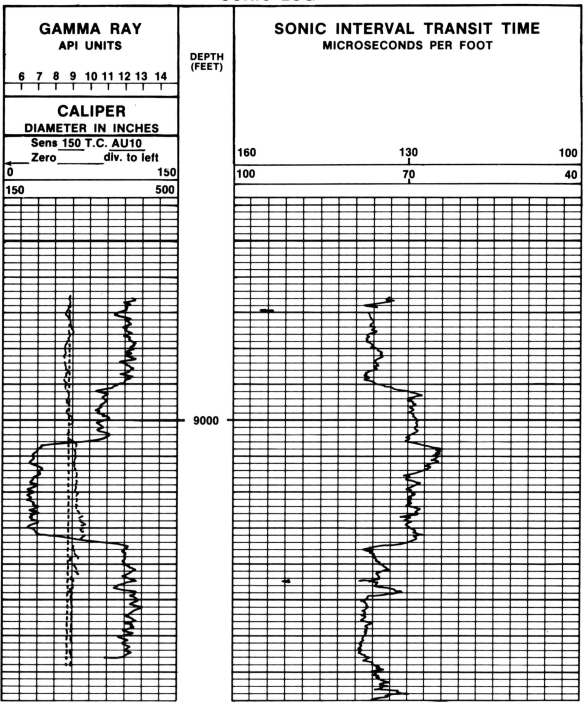

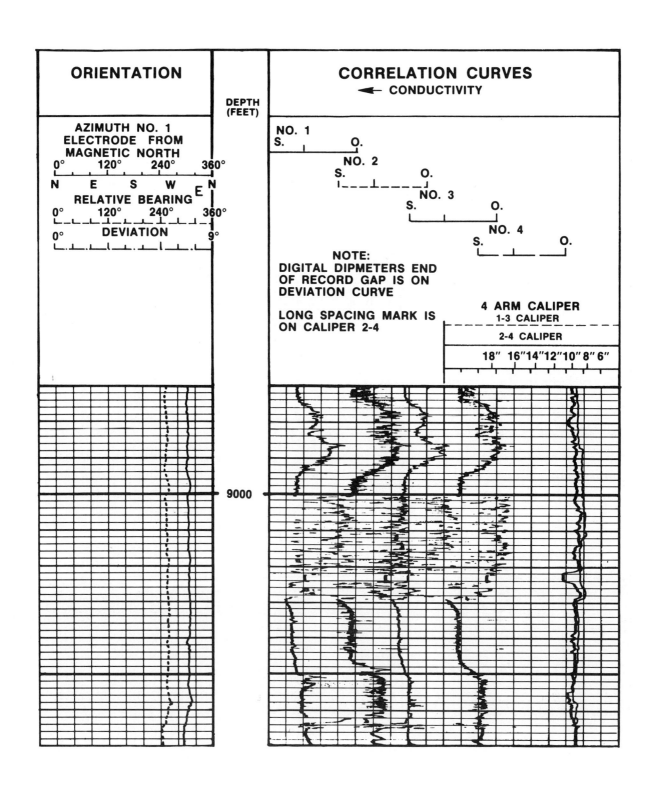

Figure 1C. Dipmeter correlation curves. Note "shotgun" appearance through Cardium interval. Data from these curves used to calculate "tadpole plot" (Figure 1D). Note correction required between core and dipmeter.

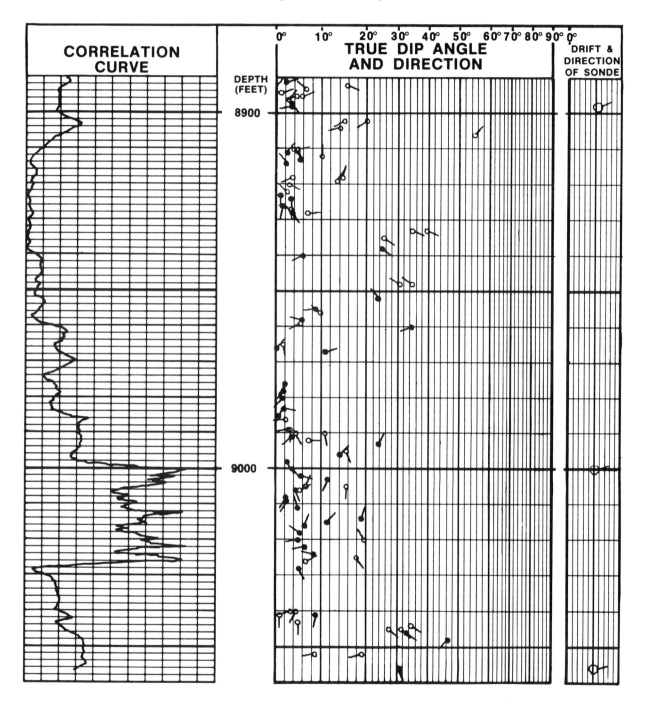

Figure 1D. Dipmeter survey. "Higher quality" data indicated by solid "tadpoles".

Figure 2 (A-E). Slabbed core photographs of Gulf Ricinus 13-26-39-9 Cardium Sandstone, Alberta, Canada. Depths are in feet. Cored interval 9000-9060'; interval photographed 9000-9034'.

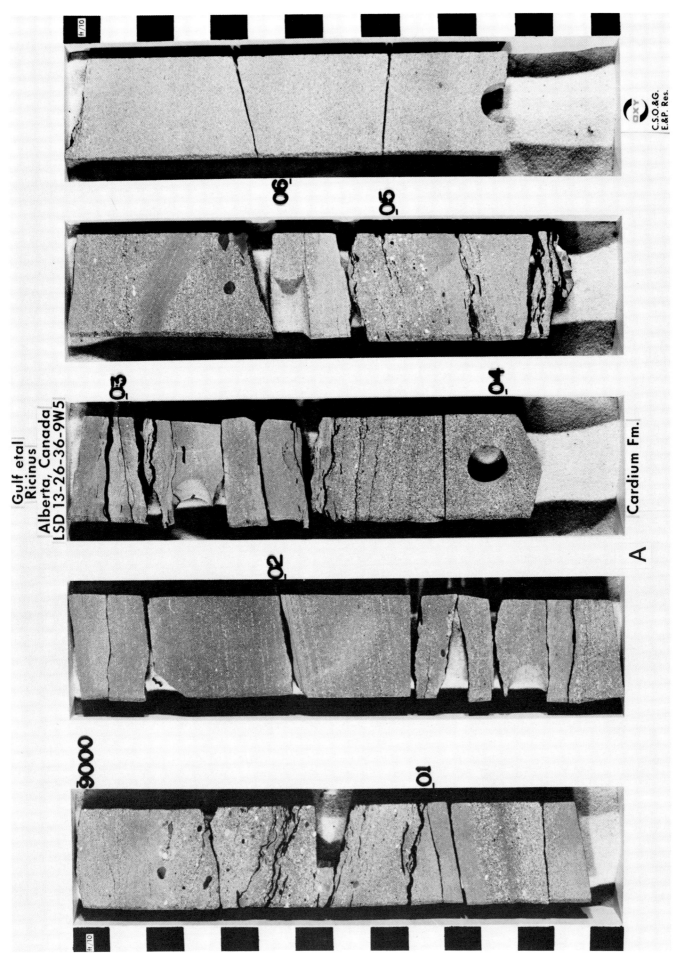

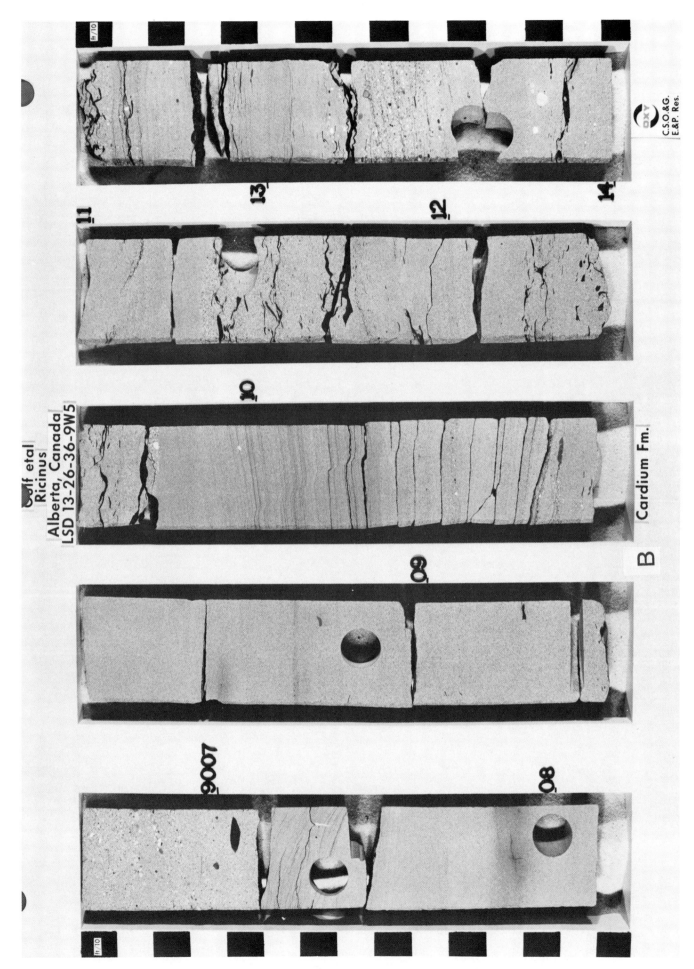

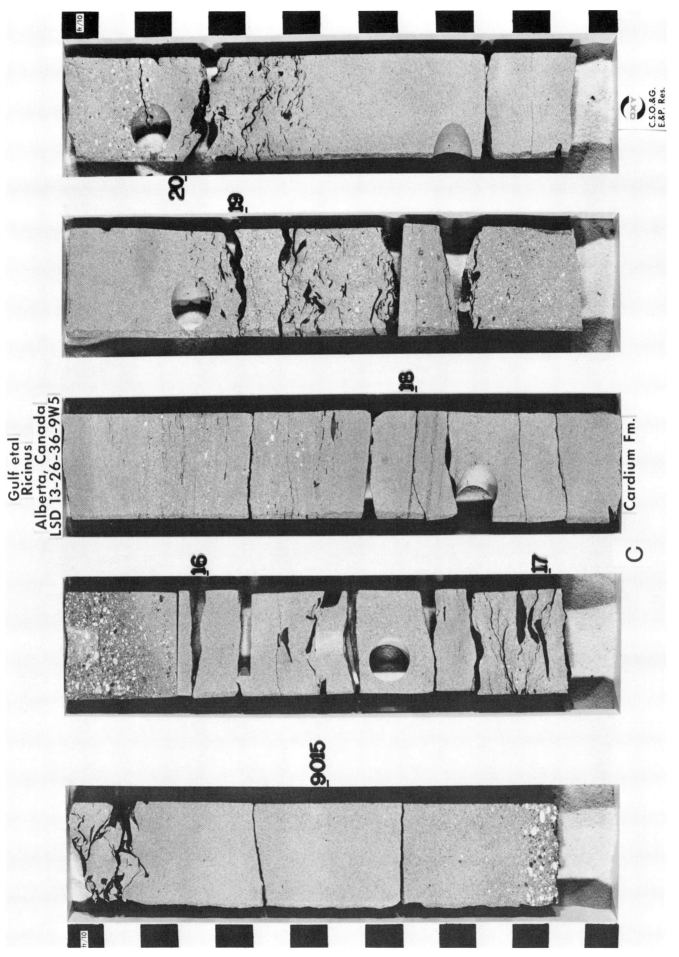

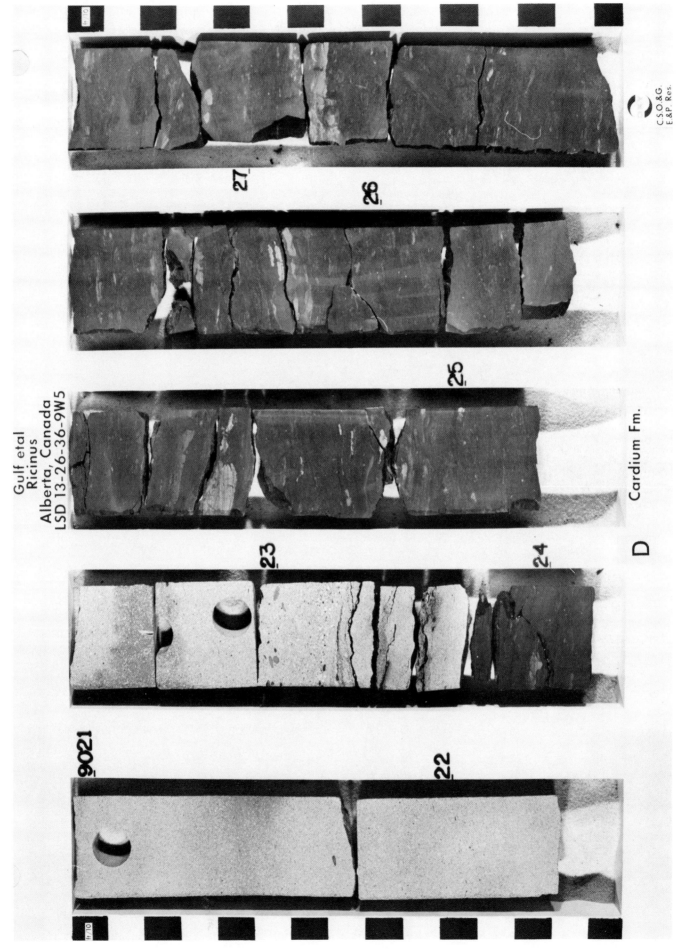

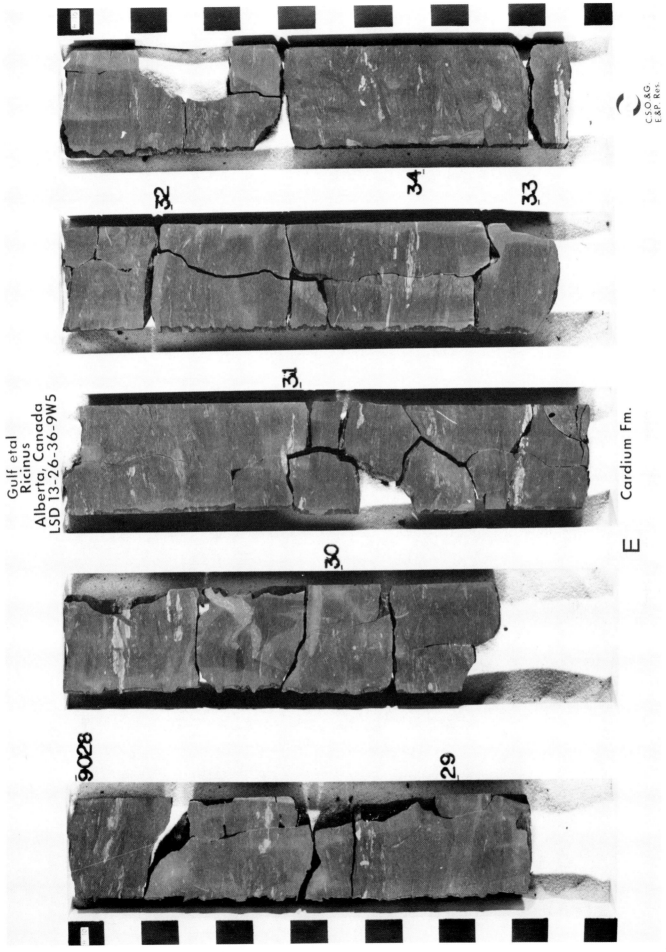

(pebbles into massive sandstones) and ripped-up mud clasts here indicate episodic deposition, sudden erosive flows which gradually died away, and deposition from suspension (massive sands) rather than by bed load rolling (which would be expected to form more cross lamination and medium scale cross bedding).

I am indebted to Gulf Canada for permission to slab this core, and to Greg Nadon for arranging the slabbing.

## PACIFIC CAROLINE 5-22-35-7W5

The Cardium field at Caroline is parallel to, and adjacent to that at Ricinus (Fig. 12 in Walker, "Upper Cretaceous (Turonian) Cardium Formation, this volume). However, the geology is very different. At Caroline, the Cardium consists of two coarsening-upward sequences (Walker, this volume, Fig. 17), with no suggestions of channelling. The sandstone at Ricinus channels into the uppermost sandstone at Caroline (the A sandstone, or Raven River Member). The two coarsening-upward sequences can best be seen on the gamma-sonic log (Fig. 3) and to a lesser extent on the SP-resistivity -- the top of the A sandstone is at about 8145 feet and the top of the better-developed B sandstone is at 8251 feet.

The slabbed core photographs (Fig. 4, A-E) illustrate the upward coarsening sequence into the B sandstone, 8274-8247 feet. The lower mudstones (8274 to about 8264 feet) belong stratigraphically to the underlying Blackstone Formation. They are bioturbated, but not so silty as the overlying mudstones (8264-8258 feet). The sandstones (8258-8254) alternate with thin mudstone partings, and are bioturbated. There is a possible low angle stratification at 8258 feet (about 10°), but otherwise no physical sedimentary structures.

The main B sandstones (8254-8248) have no physical sedimentary structures except for possible grading (8249-8250.5). The gravel cap (8247.5-8248) is typical of Cardium coarsening-upward sequences. Note how abruptly the Cardium B sand is overlain by a completely massive black mudstone -- facies 1 of Walker (1983; see Walker, this volume, Figs. 17, 18). This mudstone is a regional marker lithology that can be traced between several Cardium fields, notably Garrington, Caroline, underwater Ricinus, Willesden Green, and possibly as far north as Carrot Creek (see Walker, this volume, Fig. 1).

I am indebted to Petro Canada for permission to slab this core, and to Dale Leckie for arranging the slabbing.

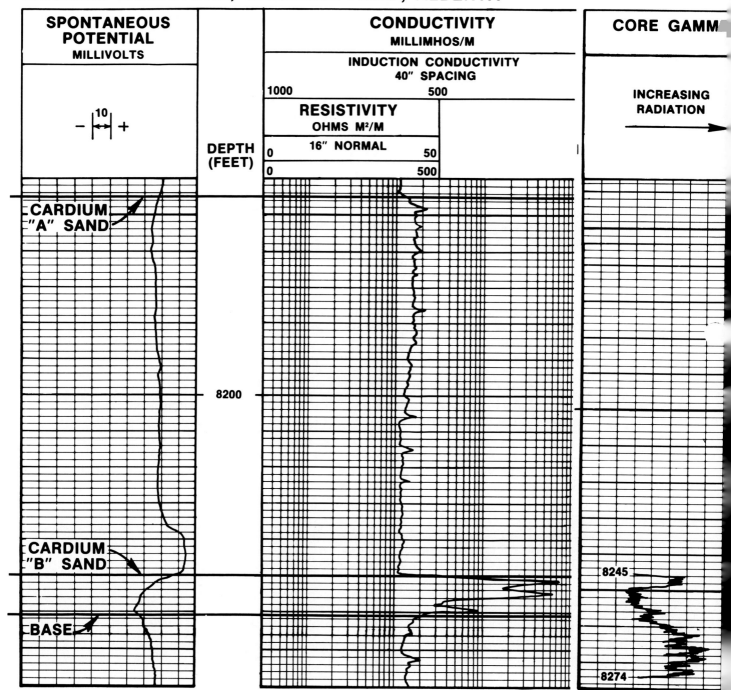

Figure 3. A. SP-Resistivity logs through Cardium Sandstone interval. Tops of Cardium-A and Cardium-B sandstones are indicated. Total interval of Cardium B sandstone was cored. Cored interval (log depths) is 8255-77'. B. Core-gamma and subsurface Gamma-Ray acoustic log. Core-gamma used to establish core to log correction (+3').

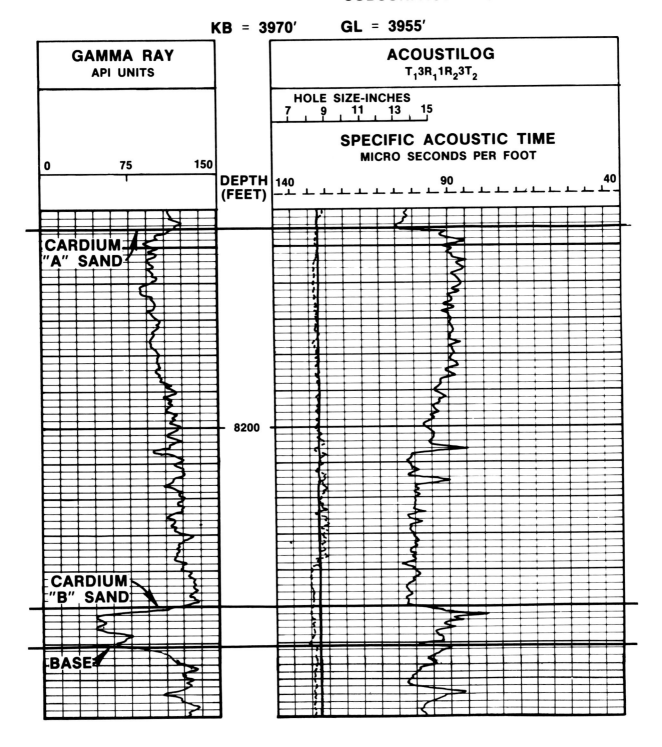

Figure 4. Photographs of slabbed core of Pacific Caroline 5-22-35-7 Cardium Sandstone, Alberta, Canada. Cored interval 8235-86'. Interval photographed 8244-74'. No core to log correction required.

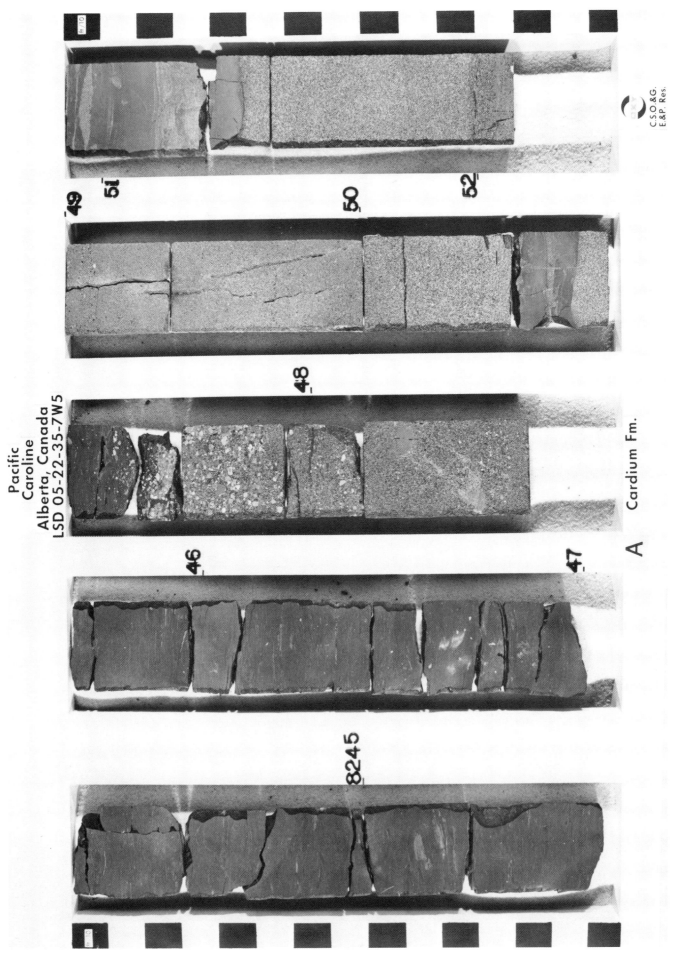

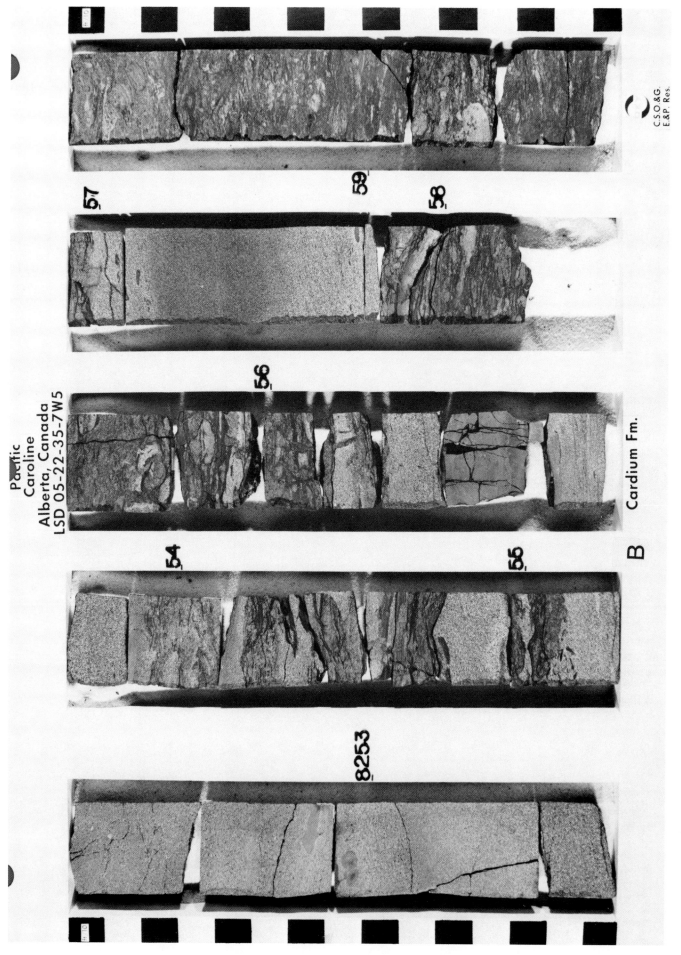

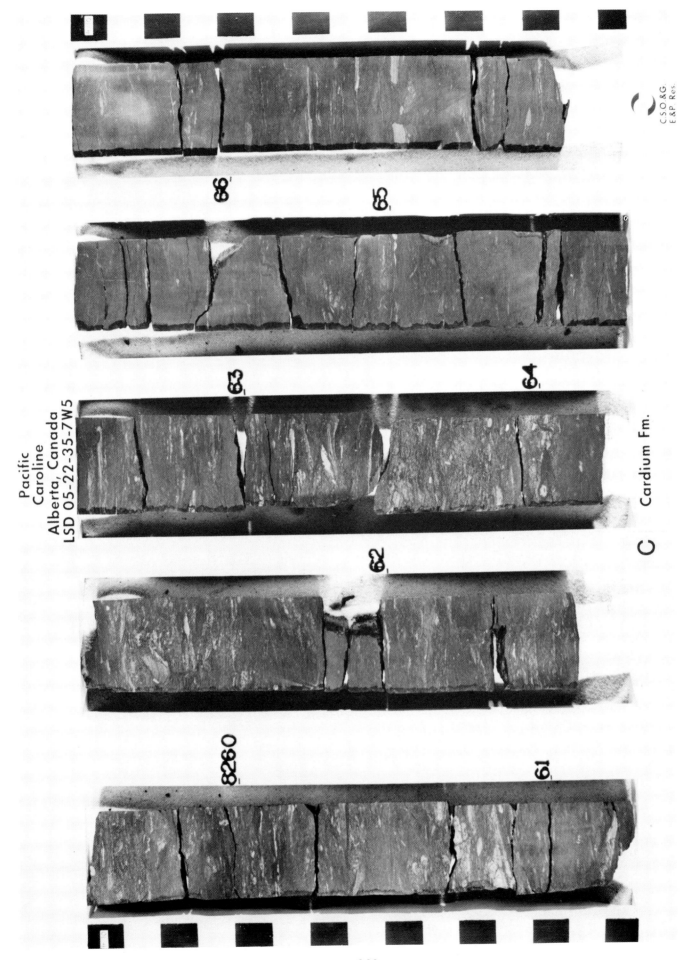

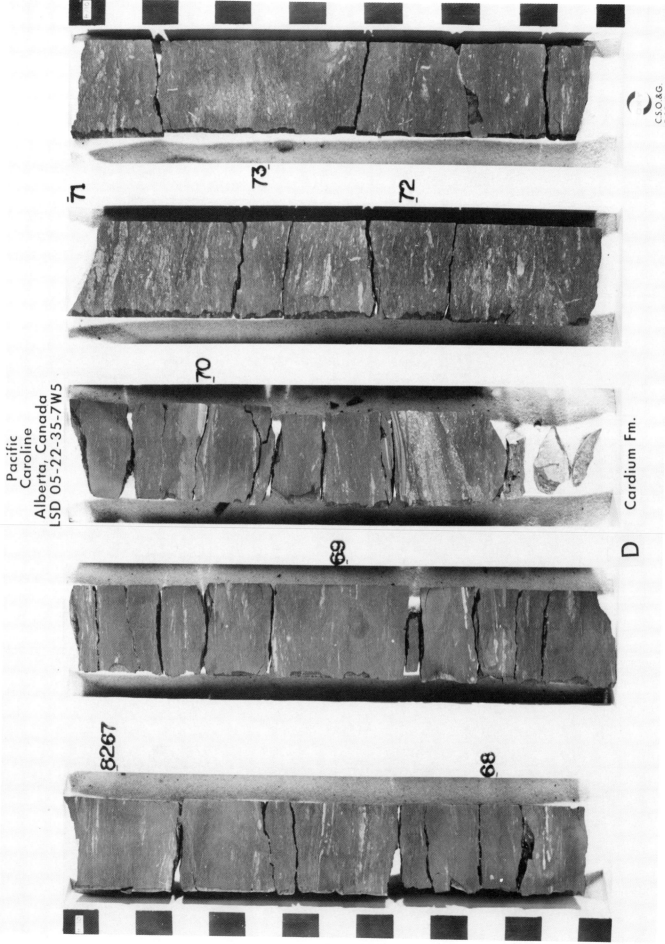

## GULF STETTLER 7-22V-38-20W4

The Gulf Stettler 7-22V-38-20 is from the Lower Cretaceous (late Albian) Viking Formation. The Viking does not outcrop but in the subsurface it occurs over a wide region in the southeastern quarter of the Province of Alberta, and the southern half of Saskatchewan. It is mostly a shallow marine sandstone or group of sandstones, typically occurring in coarsening-upward sequences capped by gravels. Some Viking fields are very similar to those that produce from the Cardium and are dominated by hummocky cross stratification. However, in the Viking there is abundant angle-of-repose medium scale cross bedding -- quite unlike the Cardium. Viking fields tend to be elongate, both NW-SE and E-W, but with rather variable orientations. Little is known about dispersal directions and shoreline positions.

Stettler Field lies a little north of a group of elongate Viking fields, from northwest to southeast these are Gilby (Koldijk, 1976), Joffre (Reinson et al., 1983), Mikwan and Fern. The logs for this well are in Figure 5. Northeast of this trend, there is one main Viking sandstone. In the slabbed core photographs (Fig. 6, A-G) this main Viking sandstone is illustrated (top at 1155 m), caps a progressively coarsening-upward sequence (1169-1155 m) which is 14 m (46 feet) thick. A lower sandstone (1171-1169 m) was not cored.

The bioturbated mudstones become progressively sandier upward through the interval 1168 m to about 1161 m. Above this point, sandstones and mudstones are interstratified -- the mudstones and the tops of the sandstones are bioturbated, but the sandstone layers mostly have abrupt, sharp bases and a flat to low-angle inclined stratification that is probably hummocky cross stratification (HCS). There is no medium scale angle-of-repose cross bedding, but there may be some small scale (1-2 cm) ripple cross lamination (see 1156.9 and 1157.5 m, for example). These thin sandstones may well be incrementally-emplaced storm deposits: there is no strong evidence here for turbidity currents (see Walker, this volume, "Geological Evidence for Storm Transportation and Deposition on Ancient Shelves").

Note the very glauconitic nature of the sandstone at about 1157.90 m, and the (unusual) absence of a distinct conglomeratic cap at the top of the sandstone at about 1154.70 m. The uppermost shales contain a few silty laminations, but mostly suggest slow, quiet deposition with surprisingly little bioturbation.

## REFERENCES

Koldijk, W. S., 1976, Gilby Viking B: a storm deposit, in M. M. Lerand (ed.), The sedimentology of selected clastic oil and gas reservoirs in Alberta: Canadian Society of Petroleum Geologists, p. 62-77.

Reinson, G. E., A. E. Foscolos and T. G. Powell, 1983, Comparison of Viking Sandstone sequences, Joffre and Caroline fields, in J. R. McLean and G. E. Reinson (eds.), Sedimentology of selected Mesozoic clastic sequences: Corexpo '83, Canadian Society of Petroleum Geologists, p. 101-117.

# GULF OIL CANADA LTD., GULF STETTLER
# STETTLER FIELD
# VIKING SANDSTONE, ALBERTA, CANADA

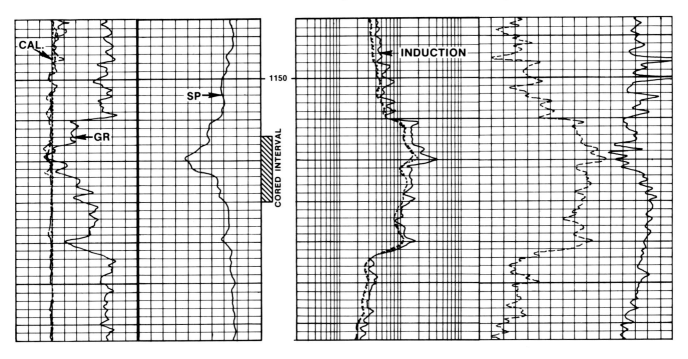

Figure 5. Gamma-Ray SP Induction log through Viking Sandstone interval in Gulf Stettler 7-22-38-20-4. Cored interval is indicated.

Figure 6 (A-E). Photographs of slabbed core of Gulf Stettler 7-22-38-20-4 Viking Sandstone, Alberta, Canada. Cored interval 1152-66 meters.

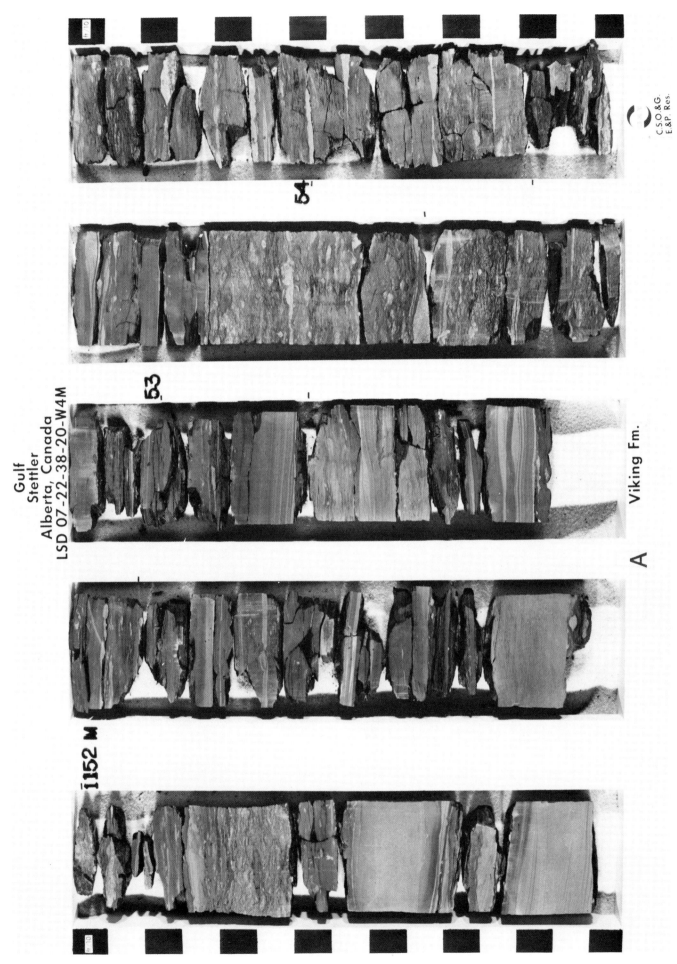

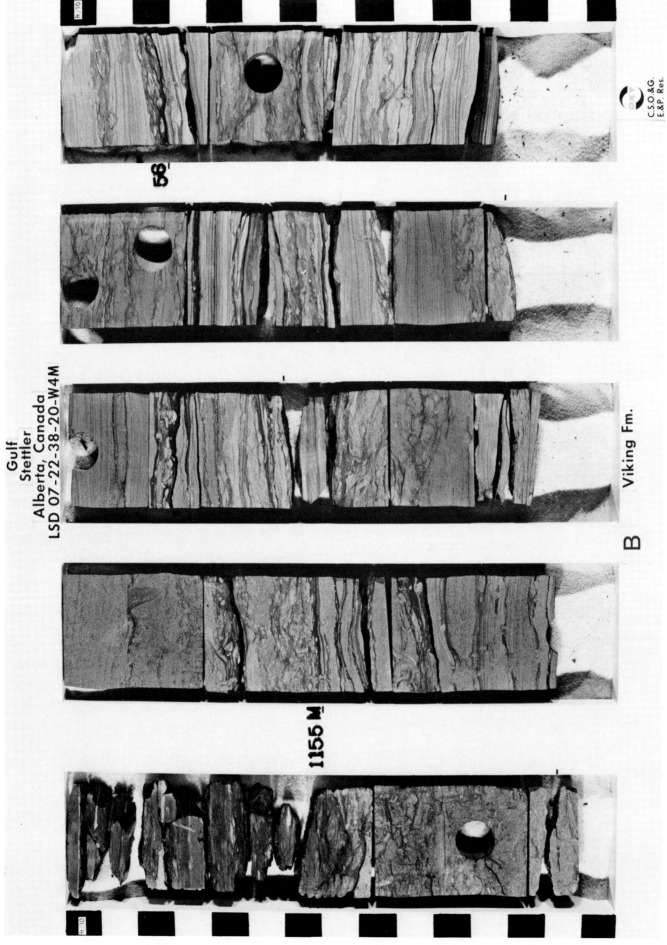

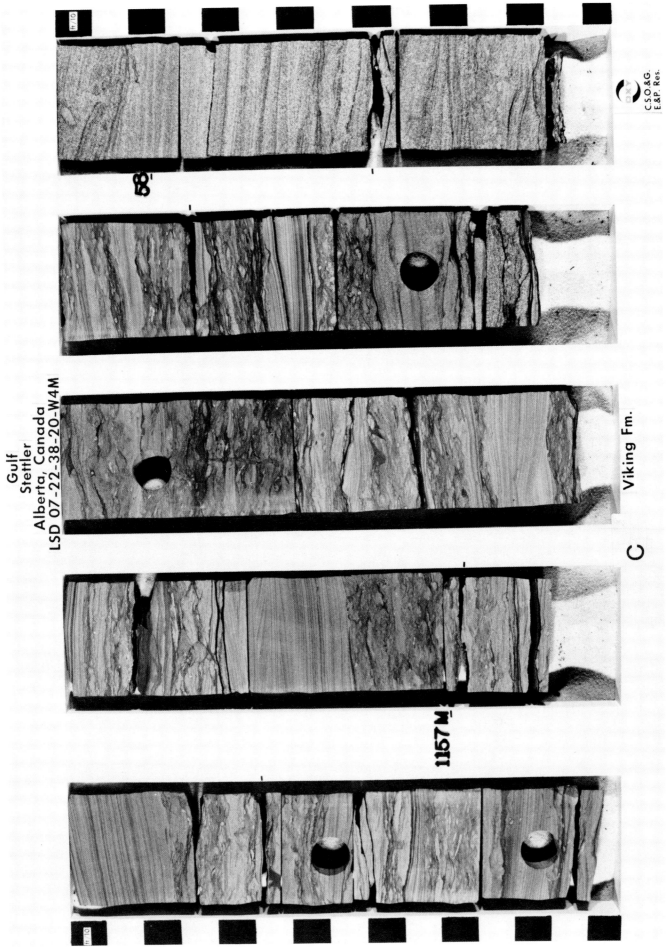

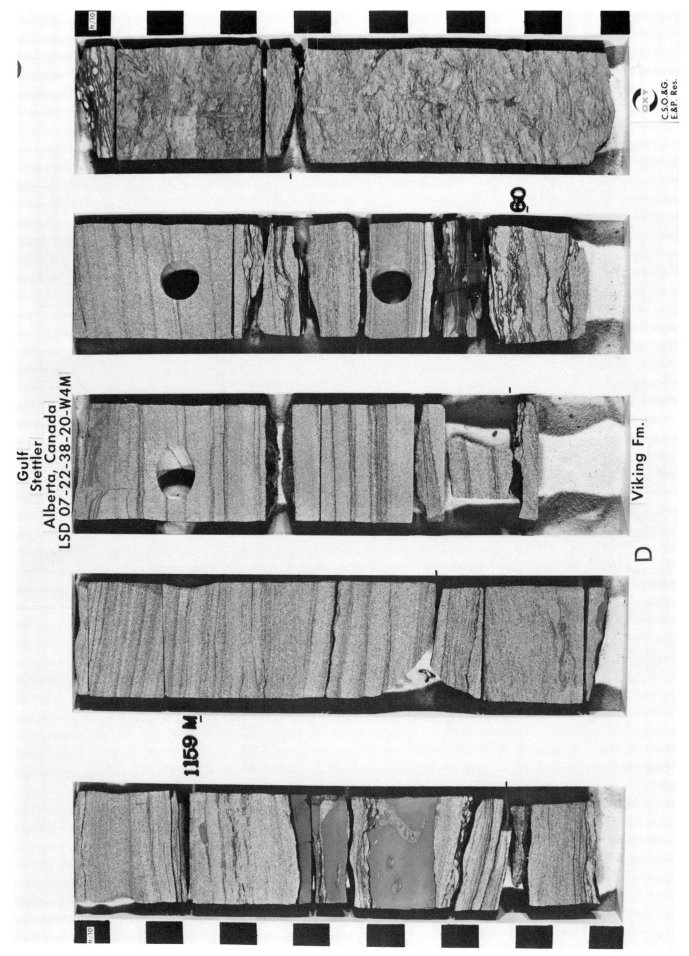

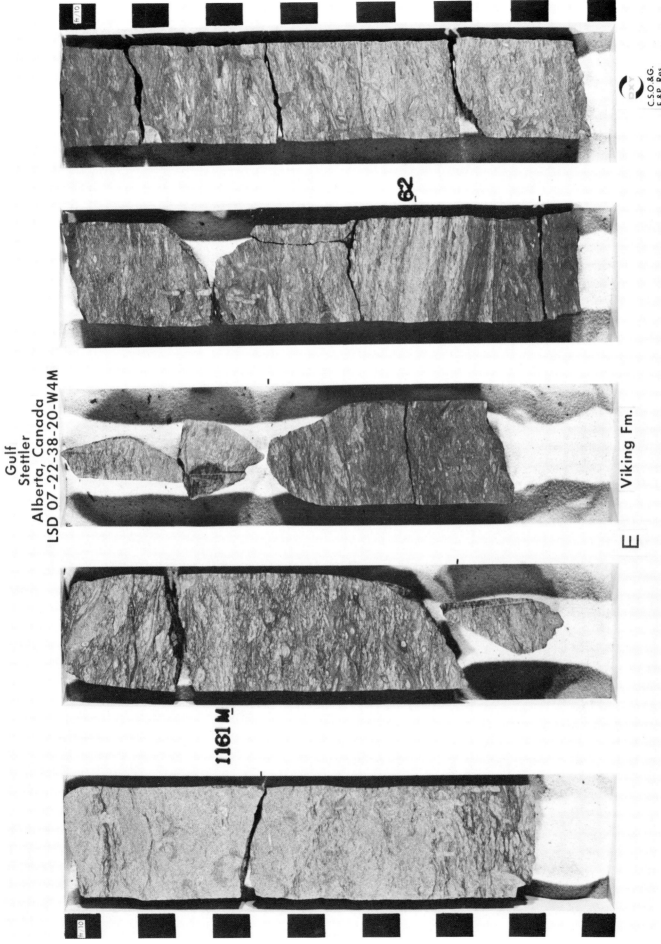

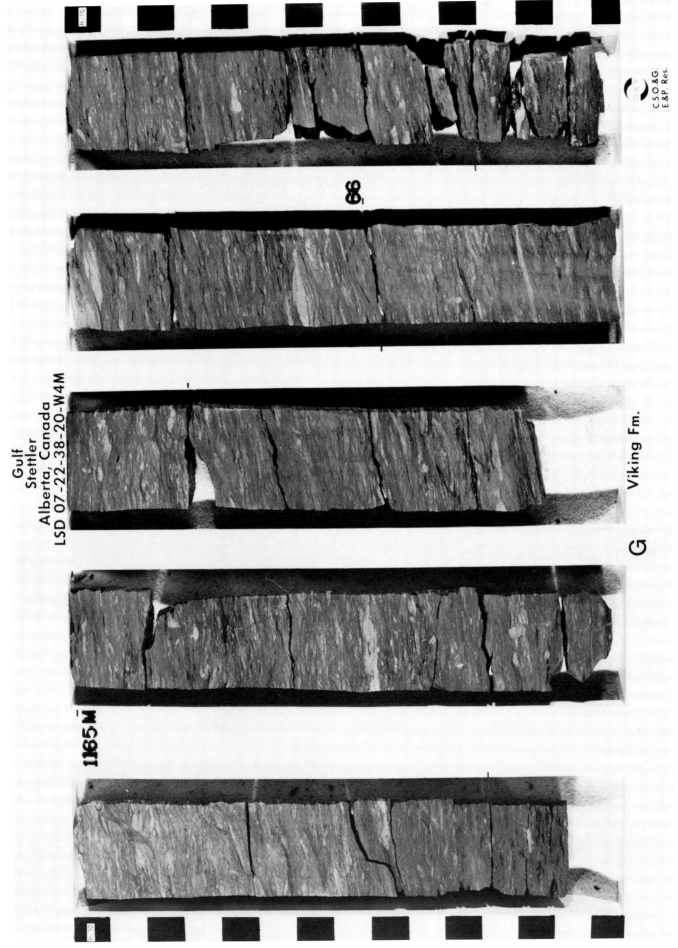

WOODBINE SANDSTONE CORE (Hinton Dorrance 7A)

## DELTA-FRONT SHELF STORM DEPOSITS OF SUBSURFACE WOODBINE-EAGLE FORD INTERVAL (UPPER CRETACEOUS), DAMASCUS FIELD, NORTHERN POLK COUNTY, TEXAS: SUCCESS FROM COMBINED DEVELOPMENT GEOLOGY AND SEDIMENTOLOGIC CORE ANALYSIS

Siemers, Charles T., Sedimentology, Boulder, Colorado
and
P. Connie Hudson, Hinton Production Co., Mt. Pleasant, Texas

Gas production from several, 6 to 23 ft. (2 to 7 m), single to multi-story sandstone bodies of the Woodbine-Eagle Ford interval, 160 to 200 ft. (49 to 61 m) thick at 9,000 to 9,600 ft. (2,743 to 2,926 m) in the Damascas field has been developed since discovery in 1976. Subsequent offset drilling resulted in a few gas wells and several dry holes. In February 1979 the entire Woodbine-Eagle Ford interval was cored in the No. 7A Dorrance well. Sedimentologic core study generated a predictive depositional model which has guided field development of the subtle stratigraphic traps at a 5 to 1 well success ratio. Present gas reserves are 40 Bcf with 440,000 bbl of condensate.

The productive area is located slightly southwest of the Sabine uplift and just updip from the Lower Cretaceous continental shelf edge. Seismic sections and foraminiferal paleoecology establish a middle-shelf depositional setting. Bioturbated, silty, shelf shales comprise the upper and lower Woodbine-Eagle Ford interval. The middle is a complex of (1) graded, medium to very fine-grained, massive to laminated sandstone beds; (2) contorted, soft-sediment-deformed intervals; (3) swirled and sheared siltstone beds; and (4) thin diamict conglomerate beds. Genetic units indicate periodic rapid deposition by debris flows and low to high-concentration density currents. The several distinct productive sandstone bodies (porosities 9 to 14%; permeabilities 2 to 10 md) are northward-thickening, dip-oriented lobes.

The localized deposition in the shelf setting was controlled by delta development slightly to the north. Periodic major storms generated delta flooding which contributed high-energy reservoir-quality deposits to the shelf. Similar shelf sand buildups should occur throughout the area; however, recognition must rely on detailed sedimentologic study of core sequences.

<u>Reference</u>: AAPG Bulletin, 1981, v. 65, p. 992.

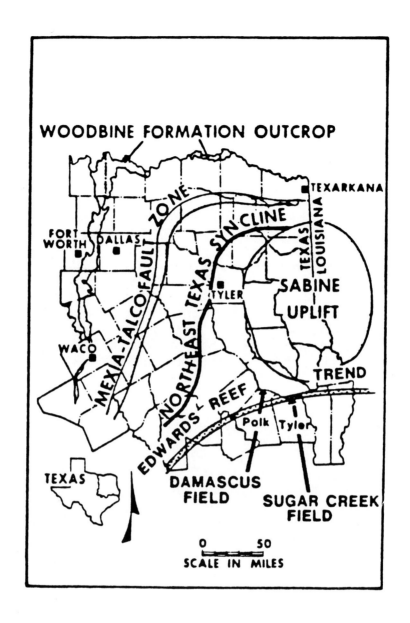

Figure 1. Location of Damascus field.

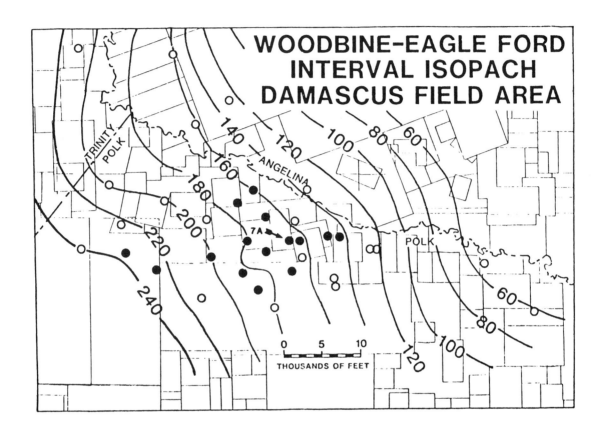

Figure 2. Interval isopach of Woodbine-Eagleford interval.

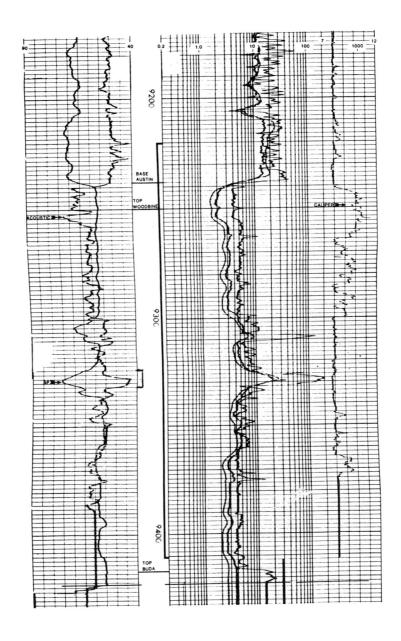

Figure 3. Well log from the cored interval of Dorrance Hinton 7A well.

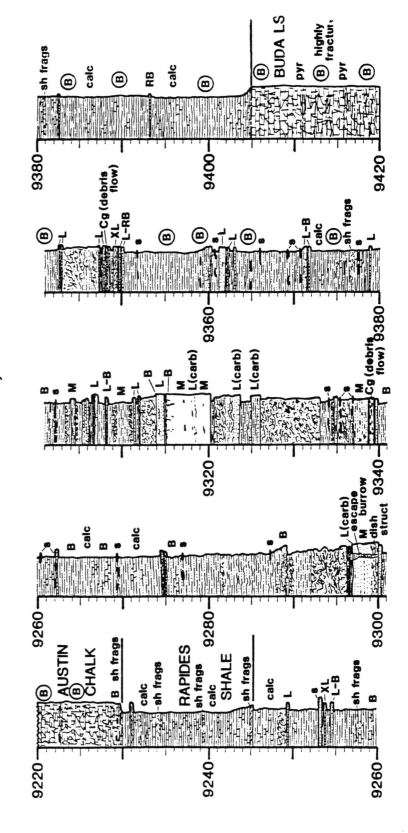

SHANNON SHELF-RIDGE SANDSTONES FIELD TRIP

R. W. Tillman, R. M. Martinsen,
N. Gaynor, D. J. P. Swift

## INTRODUCTION

This day in the field is designed to allow you to observe the various facies which occur in one of the many varieties of shelf sandstones. The Shannon Sandstone outcrops of Campanian age (Fig. 1) in the Salt Creek area contain two stacked shelf sand ridges.

The four or five stops to be included on the trip have a somewhat similar coarsening-up sequence at each location. The locations of the stops are shown in Figure 2. Several of the stops have extensive Central Ridge Facies while the others contain, predominently High- and Low-Energy Ridge- Margin Facies as their highest energy facies.

## GEOLOGIC SUMMARY

The Shannon sandstones in the Powder River Basin (Fig. 3) of Wyoming were deposited as shelf-ridge complexes on a wide shelf (Fig. 4A) during early Campanian time in the interior seaway stretching from Alaska to Mexico. The shelf-ridge complexes are considered by different workers to have been deposited in a variety of settings ranging from regression to sea level still-stand to transgression. The wide shelf model proposed by Asquith (1970), in which the shelf-slope boundary can be recognized, is believed to be applicable for the time of deposition of the Shannon (Fig. 4B).

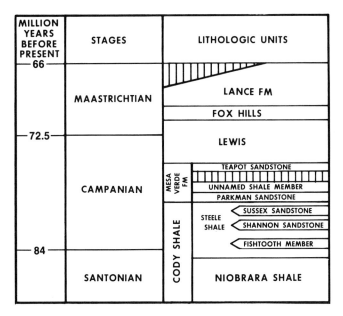

Figure 1. Upper Cretaceous time-stratigraphic section, western Powder River Basin, Wyoming.

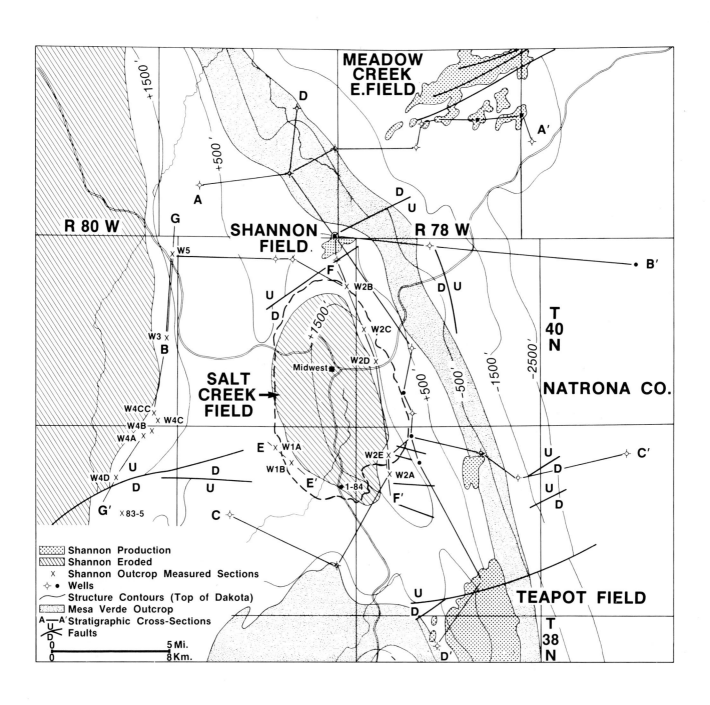

Figure 2. Map showing the details of the Salt Creek anticline area. Measured section locations are shown as are oil fields which produce from the Shannon Sandstone. Sections to be visited include W2A (Stop 1), W2B (Stop 2), W3 (Stop 3), 83-5 (Stop 4) 1-84, (Stop 5).

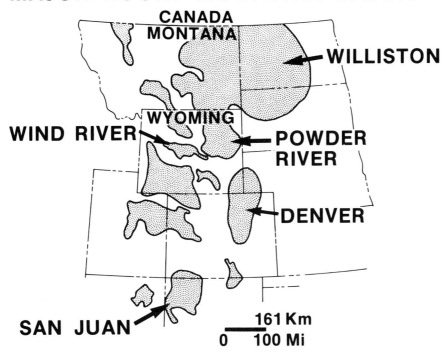

Figure 3. Major Basins of the Rocky Mountains. The Powder River Basin is located in northeastern Wyoming and can be differented along the northern boundary from the Williston Basin.

---

Two vertically stacked shelf-ridge (bar) complexes in the Shannon Sandstone member of the Cody Shale (designated upper and lower sandstones) crop out in the Salt Creek anticline of the Powder River Basin, Wyoming. The shelf-ridge complexes are composed primarily of moderately to highly glauconitic, fine to medium-grained lithic sandstones that attain thicknesses of over 70 feet. The shelf-ridge complexes were deposited at least 70 miles from shore at middle shelf depths by south to southwest-flowing shore-parallel currents intensified periodically and frequently by storms (Fig. 4B). The ridges in both the upper and lower Shannnon at Salt Creek trend north-south, slightly oblique to current flow. A possible source of sediments for the shelf ridges was the Eagle Sandstone shoreline and deltaic deposits of southern Montana 200 miles to the northwest.

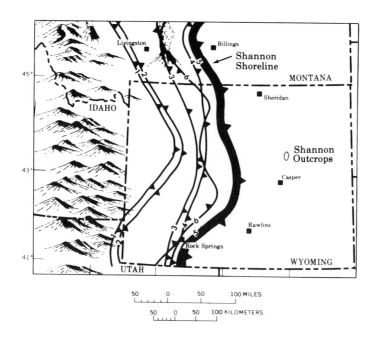

Figure 4A. Interpreted shoreline locations for Lower Campanian-Upper Santonian time. This period is commonly referred to as the Telegraph Creek-Eagle regression (and minor transgression). Barbs indicate direction of shoreline movement. Shoreline 5, Baculites sp., (smooth) existed during Shannon Sandstone deposition. Shoreline 6, Baculites sp. (weak flank ribs) existed during Sussex deposition. To the south near Rawlins, Wyoming several sandstone members of the Haystack Mountains Formation, which are primarily offshore deposits, are time equivalent to these same shorelines. (Modified from Gill and Cobban, 1973).

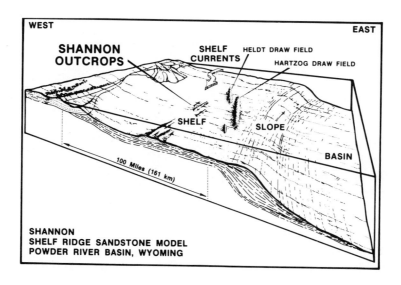

Figure 4B. Shannon Sandstone shelf model (Tillman and Martinsen, 1984).

## SHANNON SHELF FACIES

Methods for designation of shelf sandstone facies types include at least two points of view. The senior author follows the method of genetic facies designation first suggested for Cretaceous shelf sandstones by Porter (1976). Genetic facies are discussed by Siemers and Tillman (1981, p. 38) and are utilized in an extended description of the Shannon outcrops at Salt Creek by Tillman and Martinsen (1984). A more descriptive format (Spearing, 1976, Seeling, 1978, and Gaynor and Swift, this paper) emphasizes the sedimentary features in the names given to different types of shelf sandstone facies. The outcrops to be visited on this trip are described in a format which emphasizes the position of the facies on or relative to the sand ridge (bar).

The Shannon Sandstone shelf-ridge (bar) complexes in the Salt Creek area consist of a series of interfingering facies which have abrupt vertical transitions and are laterally variable. Eleven facies were defined in outcrop on the basis of physical and biologic sedimentary structures and lithology (Fig. 5; and Tillman and Martinsen, 1984). Facies name synonyms are summarized in Figures 6A and 6B. Where the Shannon crops out in the Salt Creek area, the facies are stacked in coarsening-upward sequences in which the Central-Ridge Facies commonly immediately overlie Inter-Ridge Sandstone Facies. High quality porous and permeable potential reservoir facies include: Central-Ridge Facies, a clean, cross-bedded, moderately glauconitic sandstone (Figs. 7 and 8) and High-Energy Ridge-Margin Facies, a highly glauconitic, cross-bedded sandstone containing abundant shale and limonite (after siderite) rip-up clasts and lenses (Fig. 9). The Low-Energy Ridge-Margin Facies, consisting of interbedded cross-bedded to rippled sandstone, is a lower quality reservoir. These three cross-bedded facies are differentiated by Tillman and Martinsen (1984 and this paper) on the basis of (1) percentage of clay clasts, (2) percentage of limonite clasts and lenses, and (3) the percentage of interbedded rippled beds within the cross-bedded units. The high-angle cross bedding in these facies was formed by medium- to large-scale sand waves and troughs which prograded across the tops of the shelf ridges. Storm-flow deposited Central-Ridge (Planar Laminated) Facies (Figs. 10 to 11) are rare but occur at several locations including Measured Section 85-3 (Stop 4) and Location 1-84 (Stop 5).

Average grain sizes for the more common sandstone facies are: 150 microns for Bioturbated Shelf-Sandstone and Shelf-Sandstone Facies; 175 microns for Inter-Ridge Sandstones; 200 microns for Low-Energy Ridge-Margin Facies; 225 microns for High-Energy Ridge-Margin Facies; and 250 microns for Central-Ridge Facies, (Tillman and Martinsen, 1984).

## SHANNON FACIES SUMMARY

| | CENTRAL-RIDGE FACIES | CENTRAL-RIDGE (PLANAR LAMINATED) FACIES | HIGH-ENERGY RIDGE-MARGIN FACIES | LOW-ENERGY RIDGE-MARGIN FACIES |
|---|---|---|---|---|
| LITHOLOGY | Fine to medium grained quartzose sandstone, moderately glauconitic; rare siderite clasts and shale rip-up clasts. | Fine to medium grained quartzose sandstone. | Predominately medium grained sandstone, abundant shale and limonite rip-up clasts and lenses, commonly very glauconitic. | Fine-grained sandstone with only rare shale interbeds. Fewer clasts and lenses and less glauconitic than High-Energy Ridge-Margin Facies. |
| SEDIMENTARY STRUCTURES | Predominantly moderate angle trough and planar-tangential cross bedding. Trough sets commonly horizontally truncated. | Mostly sub-horizontal plane-parallel laminated sandstone, 0.5'- thick laminasets. Minor shale and sandstone ripples. | Mostly moderate angle troughs, some current ripples, shale clasts rarely show preferred orientations. | Sequences of several beds of troughs interbedded with sequences of several rippled beds. |
| BURROWING | Sparse | Sparse | Sparse | Sparse |
| RESERVOIR POTENTIAL | Excellent | Limited? | Good | Moderate to Good |
| SUBSURFACE OCCURENCES HARTZOG DRAW FIELD | Common | Very uncommon | Common | Common |

| | INTER-RIDGE FACIES (SHALEY) | INTER-RIDGE SANDSTONE FACIES | BIOTURBATED SHELF-SANDSTONE FACIES | BIOTURBATED SHELF-SILTSTONE FACIES | SHELF SILTY-SHALE FACIES |
|---|---|---|---|---|---|
| LITHOLOGY | Thinly interbedded fine to very fine-grained silty sandstone and silty shale, slightly glauconitic. | Fine-grained sandstone. Virtual absence of silty shale. Slightly glauconitic | Silty, fine-grained sandstone. Up to 15% shale, primarily associated with burrows. Slightly glauconitic. | Shaly, slightly sandy dark gray siltstone, traces to moderate amounts of glauconite. | Silty dark gray shale; rare thin (1/8" thick) silty sandstone lenses. |
| SEDIMENTARY STRUCTURES | Predominantly horizontal ripple-form bedding surfaces marked by interbedded shales. Trace of wave ripples; current ripples predominate. | Predominantly horizontal ripple-form bedding surfaces. Bedding commonly indistinct. Trace of wave ripples; current ripples predominate. | Few physical structures preserved. Mottled appearance. Some ripple-form horizontal beds up to 8 inches thick. Trace of distinct ripples and small troughs. | Few physical structures preserved. Scattered thin rippled sand and horizontal laminasets. Bedding commonly destroyed. | Common sub-horizontal laminae. Bedding surfaces indistinct, horizontal. Rare current ripples. |
| BURROWING | Moderate to locally high | Low to Moderate | Mottled to distinctly burrowed. More than 75% burrowed. | More than 75% Burrowed | Low to moderate |
| RESERVOIR POTENTIAL | Limited | Limited | Limited | None | None |
| SUBSURFACE OCCURENCES HARTZOG DRAW FIELD | Common | Uncommon | Common | Moderately Common | Common |

Figure 5. Shannon Facies Characteristics (Modified from Tillman and Martinsen, 1984).

| % SAND | | REINECK AND WUNDERLICH, 1968 | TILLMAN AND MARTINSEN, 1984 | THIS REPORT |
|---|---|---|---|---|
| 100 | | SANDSTONE | CENTRAL BAR FACIES<br>BAR MARGIN FACIES (TYPE 1)<br>BAR MARGIN FACIES (TYPE 2) | CROSS-BEDDED SANDSTONE FACIES<br>TABULAR CROSS-BEDDED AND PLANAR LAMINATED SUBFACIES<br>TROUGH CROSS-BEDDED GLAUCONITIC SUBFACIES<br>TROUGH CROSS-BEDDED SUBFACIES |
| 95 | | | | |
| 90 | | | | |
| 80 | | FLASER-BEDDED SANDSTONE | INTERBAR FACIES<br>INTERBAR SANDSTONE FACIES | THIN-BEDDED SANDSTONE FACIES |
| 70 | | WAVY-BEDDED SANDSTONE WITH INTERBEDDED SHALE | SHELF SANDSTONE | |
| 60 | | | | |
| –50– | | | | |
| 40 | | LENTICULAR-BEDDED SANDSTONE WITH INTERBEDDED SHALE | BIOTURBATED SHELF SANDSTONE FACIES | BIOTURBATED SANDSTONE FACIES |
| 30 | | | | |
| 20 | | SHALE WITH LENTICULAR-BEDDED SANDSTONE STREAKS | SHELF SILTSTONE FACIES<br>BIOTURBATED SHELF SILTSTONE FACIES | LENTICULAR-BEDDED SILTSTONE FACIES |
| 10 | | | | |
| 5 | | SHALE | SHELF SILTY SHALE FACIES | |
| 0 | | | | |

Figure 6A. Comparison of Gaynor and Swift facies (right column) with those of Reineck and Wunderlich, 1969 and Tillman and Martinsen, 1984.

| Present Terminology | Tillman and Martinsen, 1984 | Spearing, 1976 |
|---|---|---|
| Central-Ridge Facies | Central Bar Facies | Cross bedded Sandstone Facies |
| Central-Ridge (Planar-Laminated) Facies | Central Bar (Planar Laminated) Facies | |
| High-Energy Ridge-Margin Facies | Bar Margin Facies (Type 1) | |
| Low-Energy Ridge-Margin Facies | Bar Margin Facies (Type 2) | |

Figure 6B. Comparison of facies nomenclature. See also Figure 6A.

Figure 7. Cross laminated to horizontally laminated 42-foot-thick <u>Central Ridge</u> Facies. Measured Section 83-5 (Gaynor), Stop 4.

Figure 8. Highly-planar cross-bedded sandstone (above arrow) overlying horizontally bedded sandstones. Note bedding boundries are mostly horizontal and individual beds are mostly less than 1 meter (3 feet) thick. <u>Central Ridge Facies</u>, Measured Section 83-5 (Gaynor), Stop 4

Figure 9. Planar-tangential sand waves or troughs in High-Energy Ridge-Margin Facies. Abundant siderite rip-up clasts especially at "finger level." Note minor erosion (arrow) at top of bed. Measured Section W3 (STOP 3) Unit 7B.

Figure 10. Horizontally bedded and horizontally laminated Central-Ridge (Planar Laminated) Facies. Horizontal bedding is planar and tracable laterally over twenty meters. Shannon Sandstone, Location 1-85 (Stop 5) east side of Highway 87 South of Midwest, Wyoming.

Figure 11. Horizontal laminations in horizontally bedded Central-Ridge (Planar Laminated) Facies. Beds are traceable laterally for 20 meters. Location 1-85 (Stop 5).

## PALEOCURRENT FLOW DIRECTIONS

In outcrops of the Shannon, in the subsurface at Salt Creek and in Hartzog Draw Field, detailed mapping indicates that the shelf-ridge complexes trend nearly north-south, slightly oblique to current flow (Tillman and Martinsen, 1984 and Fig. 12).

Transport directions, determined from abundant high-angle cross beds (mostly sand waves or troughs), indicate a south-southwest transport direction (188°) for current deposition of the high energy facies. The overall range of variation in transport direction at individual outcrops is relatively small (60°) (for details see Tillman and Martinsen, 1984). Many current ripples in the Inter-Ridge Sandstone Facies also indicate a southerly transport direction although ripples are much less reliable for determining paleocurrent directions. In the top foot or two of some Central-Ridge and Ridge-Margin Facies at Stop 3, trough orientations locally indicate transport directions strongly oblique (northeast) to the general south-southwest flow direction. Flow directions which have other than a southerly component are rare in the Shannon.

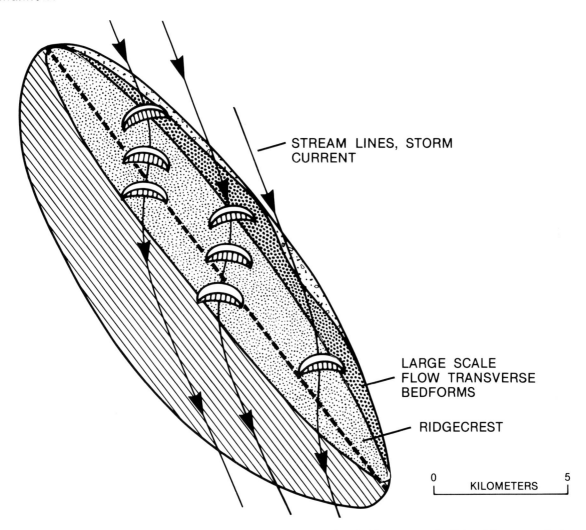

Figure 12. Gaynor and Swift plan view model of Shannon shelf-ridge sandstone. Flow lines and resultant sedimentary features are indicated. See Figure 20 for facies legend.

SHANNON OUTCROP

DESCRIPTIONS

## STOP 1. MEASURED SECTION W2A

### Location

SW NW Sec. 8 T39N R78W "Castle Rocks", Wyoming (Fig. 2). This outcrop is east of the highway which runs north-south through Midwest and is accessable on an unpaved road by turning east near the Shannon outcrop where it crosses the highway on the south side of the anticline. The road runs east about 1/4 mile before it turns north. The stop is about 1/2 mile north. Access to the outcrop is obtained by proceeding west on foot on a trail (road?) which runs to the base of the section. An easy climb of about two hours duration through the section should allow recognition of most of the major facies. The facies and units are designated in Figure 13. Units 2 through 4 are viewed by climbing in a northerly direction. Units 5 through 7 are above units 1 through 4 and may be observed by walking south parallel to the cliff face. Unit 7 is best observed on the southeast corner of the upper "plateau" surface. Note that the facies are numbered in sequence beginning at the base of the section. The location of the section (W2A) is shown in Figure 5.

### Geologic Highlights

<u>Unit 2, Bioturbated Shelf-Siltstone Facies.</u> This is the first unit encountered. It is designated as a    This outcrop is very typical of this facies. Features to note are:

(1) Mean grain size 90 microns.

(2) Uniformly high degree of bedding disturbance due to burrowing.

(3) Indistinct physical structures and horizontal bedding.

(4) Association of clay with individual burrows.

(5) Slightly recessive nature of the outcrop.

<u>Units 3A and 3B, Inter-Ridge Facies and Inter-Ridge Sandstone Facies.</u> Both facies contain high precentages of current ripples and differ mainly in their percentage of interbedded shale (see Table 1). Features which should be noted are:

(1) Horizontal "ripple-form" bedding surfaces.

(2) 1-3" thick beds composed predominently of current ripples.

(3) Very low percentages of glauconite.

(4) "Lower fine" mean grain size (175 microns).

(5) Interbedded silty shale in Unit 3B (<u>Inter-ridge Facies</u>).

MEASURED SECTION W2A-77, SHANNON SANDSTONE
SWNW SECTION 8 T39N R78W
CASTLE ROCKS, NATRONA CO., WYOMING

Jackstaff Section:
Measured by: R. Martinsen and R. Scott
Described by: R. Tillman and R. Wolff
October, 1977 (8-80)
Salt Creek and Edgerton Quadrangle Maps
Structural Strike: N30°E Dip 6°SE

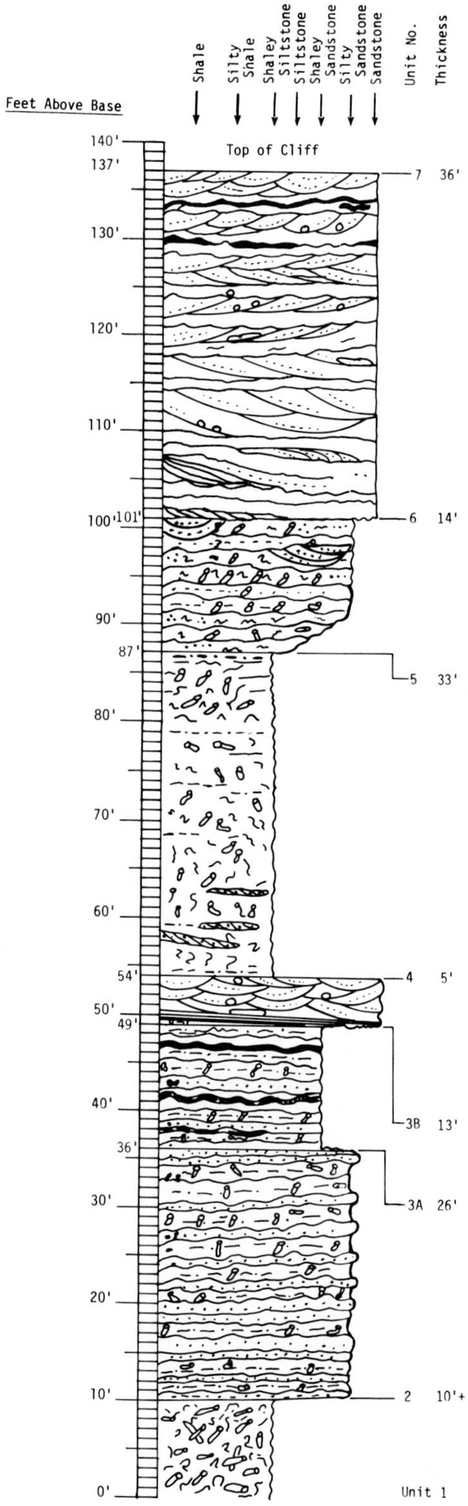

| Unit No. | Thickness | Description |
|---|---|---|
| 7 | 36' | 101-137' Sandstone (200μ). 65% troughs, maximum 1' thick, 40% high angle. Rippled beds interbedded with troughs. 30% rippled beds including 15% ripples on troughs. Ripple bedded shales up to 1/2" thick, especially near the top. 5% subhorizontal to slightly rippled sandstone (10% shale). Shale on cross bed lamina and as lenses between troughs. Lenses of limonite (less than Unit 4); scattered limonite clasts (trace amount). Some burrowing (5%), trace of Thalassinoides. Trace of sand-filled white horizontal burrows. Some 1/4" diameter upward flaring vertical burrows; also some with concentric clay lined cylinders similar to Asterosoma. Locally top 3" calcite cemented. Glauconite concentrated on laminations. LOW-ENERGY RIDGE-MARGIN FACIES (90%). |
| 6 | 14' | 87-101' Sandstone (175μ). Ripple bedded; up to 0.2' thick sets. Increase in number of 0.3' beds of sandstone (125) upward. Above base no shale. 5% glauconite. Siltier and shalier part uniformly burrowed locally up to 60%. Mottled in shaley to silty laminasets. Some clay filled 1/8-1/4" diameter burrows. Total burrowing 30%. Lower 5' transitional with unit below. Near top interbedded highly glauconite sandstone containing vertical burrows. Upper contact taken as lowest trough cross bedding. INTER-RIDGE SANDSTONE FACIES (95%). |
| 5 | 33' | 54-87' Siltstone (88μ). Shaley. Medium gray cliff former to recessive. Almost totally bioturbated except for a few thin 0.3' sand lenses (175μ) near base. Sand lenses pinch out laterally in 10-20'. Glauconite (5-10%). Shale associated with burrows and intermixed. Trace of coaly fragments. BIOTURBATED SHELF-SILTSTONE (95%). |
| 4 | 5' | 49-54' Sandstone (200μ). Some planar laminations at base with angle of 12° with horizontal. All medium to large scale troughs and rippled troughs above, (85%). Resistive, quartz cemented. 5% lenses and clasts of limonite. No shale. 5% burrowed. HIGH-ENERGY RIDGE-MARGIN FACIES (90%). |
| 3B | 13' | 36-49' Sandstone (40%, 175μ). Silty sandstone and silty shale interbedded. 60% of sandstone beds burrowed. Rippled, uniform interbedding. Nonresistant. Weathers back, recessive. INTER-RIDGE FACIES (95%). |
| 3A | 26' | 10-36' Sandstone, fine grained (175μ). Interbedded sandstone and slightly more recessive burrowed shaley silty sandstone. Ripple bedded 1-3" thick. 10% glauconite in sandier beds. Shale lined burrows in recessive beds. INTER-RIDGE SANDSTONE FACIES (95%). |
| 2 | 10'+ | 0-10' Siltstone (90μ); shaley (20%) to sandy (10%). Medium gray, traces of faint subhorizontal bedding (ripples?). Totally bioturbated, uniformly mottled. Trace of "donut burrows". Trace of glauconite. Upper contact gradational through 6". Uniformly recessive unit. BIOTURBATED SHELF-SILTSTONE (90%). |
| Unit 1 | | Covered |

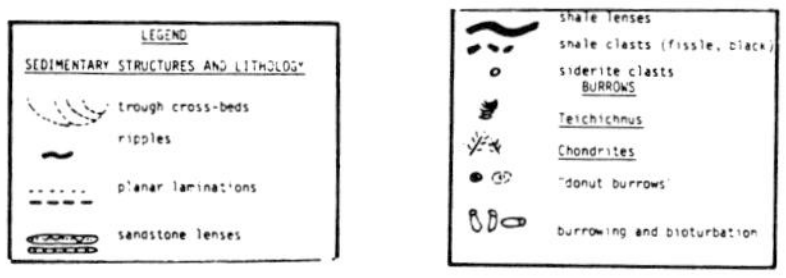

Figure 13. Measured Section W2A; Shannon Sandstone, Stop 1.

Unit 4, High-Energy Ridge-Margin Facies. This unit forms the top of the lower shelf-ridge complex at this location. Notice the following:

(1) Preponderance of planar-tangential cross bedding.

(2) The presence of ripples on some planar-tangential "foresets".

(3) Relatively high percentages of glauconite and siderite clasts (and lenses).

(4) Coarser grain size than unit below; "upper fine" (200 microns) mean grain size.

(5) Rare low angle planar laminations at base of unit.

Unit 5, Bioturbated Shelf-Siltstone. This unit is very similar to unit 2 and forms the lower part of the upper shelf ridge coarsening-up sequence. Things to notice:

(1) Items listed for unit 2.

(2) Lenses of unburrowed sandstone near the base.

(3) Mean size (90 microns), very fine-grained sand to silt.

Unit 6, Inter-ridge Sandstone Facies. This unit is very similar to unit 3A. This facies has a coarser mean grain size than the underlying unit but is finer grained (175 microns) than the facies which forms the top of this sand ridge (200 microns). Although these sizes (175 and 200 microns) seem to be very similar, by using a grain size comparator a consistent mean grain size within facies types can usually be recognized.

Unit 7, Low-Energy Ridge-Margin Facies.

This is the thickest facies at this location (36 feet). It is characterized by alternating subhorizontal beds of rippled sandstones and planar-tangential sandstones. The mean grain size for the facies as a whole is 200 microns, although the rippled beds may be slightly finer grained. Ripples on relatively steeply dipping planar-tangential laminations are also common. Burrowing and siderite are minor at this location and in this facies in general. This facies was designated by Tillman and Martinsen (1984) and in Table 2 as a Low-Energy Ridge-Margin Facies which contrasts with that of Unit 4 a High-Energy Ridge-Margin Facies at the top of the lower sandstone. Things to note in this facies:

(1) Horizontal master bedding surfaces.

(2) Alternating beds of rippled sandstones and beds of planar-tangential sandstone.

(3) Relatively less glauconite and siderite than in High-Energy Ridge-Margin Facies (i.e., Unit 4).

(4) Relatively uniform grain size of planar tangential cross-laminated beds and rippled beds.

(5) Uniform southerly transport direction of moderately high angle planar-tangential cross laminations.

(6) Variable orientations of ripples, including some long crested wave(?) ripples.

## STOP 2. MEASURED SECTION W2B

### Location

NW SE Sec. 12 T40N R79W, Salt Creek Field Area.

The location of the section is shown in Figure 2. Access is by paved and gravel road Natrona County 116. Turn north from highway running north through Midwest at about location north of Midwest where the highway begins to turn west; drive north until road climbs through Shannon outcrops, then turn west on oil well access road and park on river flood plain. The outcrop occurs on two sides of an east-west trending gulch.

### Geologic Highlights

The facies (Fig. 14) and units for this measured section are designated in Figure 15. Vertical facies sequences here are simlar to those at measured section W2A (Stop 1); however, there are thick sections of Central-Ridge Facies in addition to a relatively thick High-Energy Ridge-Margin Facies (Fig. 14). Typical very highly glauconitic (30-50%) and siderite clast-rich facies occur in the lower High-Energy Ridge-Margin Facies sequence (Unit 3). Good examples of planar-tabular beds with reverse flow toeset ripples may be observed in Unit 7 (see Figure 16D, Tillman and Martinsen, 1984).

Figure 14. Lower Shannon sandstone. Bioturbated Shelf Siltstone Facies, Unit 1 (below arrow); 13 feet thick. Above arrow is Inter-Ridge Sandstone Facies, Unit 2; 24 feet thick. High-Energy Ridge-Margin Facies, Unit 3, is at top of outcrop. Surface above arrow is interpreted by Swift and Gaynor (see Fig. 20) as an unconformity. Measured Section W2B-77, Stop 2.

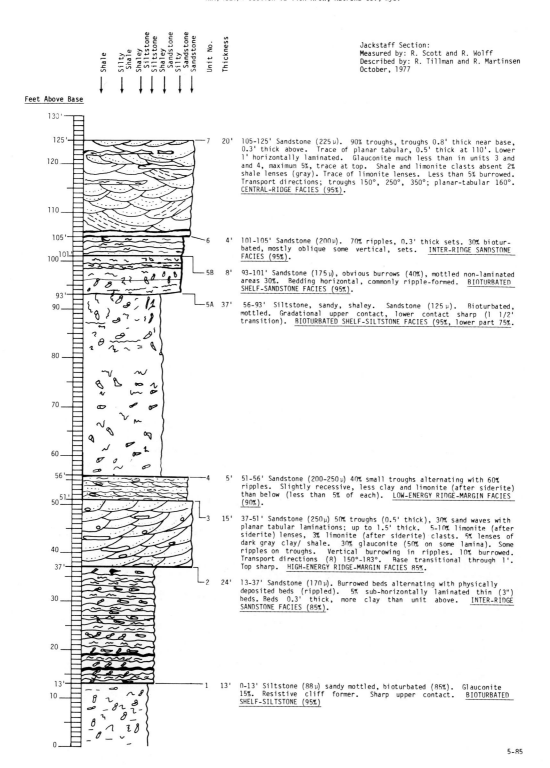

Figure 15. Measured Section W2B. Shannon Sandstone Stop 2.

## STOP 3. MEASURED SECTION W3

### Location

NE Sec 24 T40N R80W. West facing monocline west of Salt Creek anticline Access is from gravel road extending west from access road perpendicular to Interstate highway (Fig. 2). Outcrop is 1/8 mile north of gravel road.

### Discussion

This stop is designed to allow you to test your ability to recognize the shelf-ridge facies observed in outcrops visited earlier in the day. Diagrams are provided for sketching an outcrop columnar stratigraphic section (Figs. 16 and 17). The exercise will be done in groups of four or five people. Each group should complete a columnar stratigraphic section and should nominate a secretary and a spokesperson to discuss the description and interpretation of the section.

Footages above the base of the section are indicated by red paint on the outcrop. It is recommended that you sketch the facies by first picking facies contacts and indicating their boundaries on the stratigraphic sketch. Then determine the lithology and extend facies boundary lines to the right on the diagram to correspond to the lithology observed. Thirdly, list and indicate by symbols the salient features of each facies including: lithology(s), sedimentary structures, degree and orientation of burrowing, mean grain size (microns) and percentages of glauconite. A detailed form which lists most salient shelf sandstone features is Figure 18. After identifying the salient features, assign a facies name (Table 1) to each unit. It may be useful to also list your second choice of facies name. Where planar-tangential cross-laminations are present determine their approximate transport directions and note the directions in your descriptions (the face of the outcrop is approximately north-south). Record transport directions at at least three locations.

Approximately 1 1/2 hours will be allowed for the description and interpretation of the columnar section and 1/2 hour for discussion. In describing the section begin at the base (or near the middle if assigned to do so) and describe the section from bottom to top. Be sure to progress through the whole section; don't get hung up in discussion of too much detail. The field trip leaders will be happy to answer questions concerning observable features, but will not varify interpretations of facies until the scheduled discussion involving all the groups.

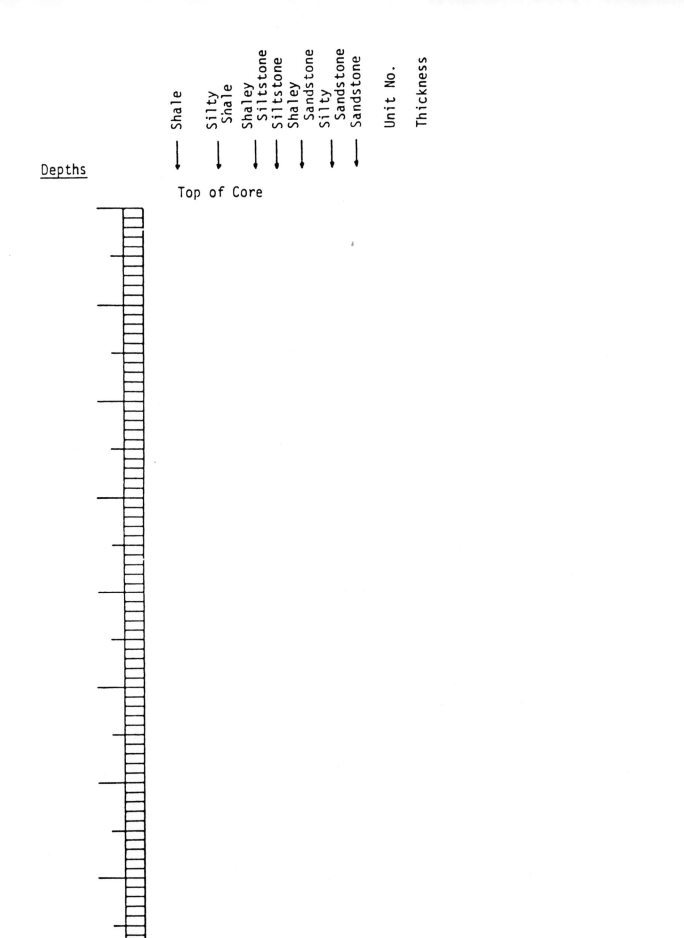

Figure 16. Stratigraphic section form designed by R. W. Tillman.

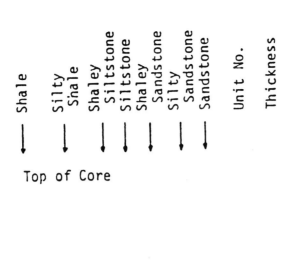

Figure 17. Stratigraphic section form designed by R. W. Tillman.

Columnized Data Format:
  %/Max. set thickness/Additional notes

                                                                    Unit
                                                              Core Depth:
                                                              (Log Depth:         )

Lithology
1. Sandstone    (%)
2. Siltstone    (%)
3. Shale        (%); a) Laminated; b) Clasts
4. Siderite     (%); a) Laminated; b) Clasts
5. Glauconite   (%); a) Disseminated; b) Laminated

Physical Structures
1. High angle cross-bedding (20°+); a) Troughs; b) Planar-tabular; c) Planar-tangential; d) Curved-tangential (trough)
2. Moderate angle cross-bedding (10-20°); a) Troughs; b) Planar-tabular; c) Planar-tangential; d) Curved-tangential (trough)
3. Subhorizontal to low angle bedding (<10°); a) Trough; b) Planar-tabular; c) Planar-tangential; d) Curved-tangential; e) Planar
4. Horizontal laminations; a) Sandstone; b) Shale
5. Rippled; a) Sandstone; b) Shale; c) Current; d) Wave
6. Rippled interbedded sandstone and shale
7. Ripples superimposed on troughs
8. Reworked: a) By waves & currents; b) Bedding destroyed massive; c) soft sediment deformed; d) clasts

Biogenic Sedimentary Structures
1. Identified burrows; a) Asterosoma; c) Chondrites; d) "Donut burrows" (Terabellina); g) Gastropod tracks; p) Plural curving tubes; s) Skolithos; t) Teichichnus; th) Thallasinoides
2. Distinct burrows; a) <1/8"; b) 1/8"-1/4"; c) 1/4"-1/2"; d) >1/2"; e) Silt filled; f) Clay filled; g) Sand filled; h) Silt lined; i) Clay lined; j) Spreiten; k) Vertical; l) Oblique; m) Horizontal

3. Burrowing (non-bioturbated) (%)
4. Bioturbated, 75% (+) burrowed (%); % burrowed; a) Distinct, b) Mottled
5. Total interval burrowed (%, footage)
6. Diversity (Number of burrow types); 1-4 low, 4-8 moderate; >8 high

Cemented Intervals; a) Siderite; b) Calcite

Contact Relations (core contacts)
1. Upper; a) Very sharp (<0.05' transition); b) Sharp (0.05-0.1' transition); c) Transitional (0.1-0.3' transition); d) Gradational (>0.3' transition); e) Contact,erosional (truncated) angular; f) Contact erosional parallel; g) Covered or not covered
2. Lower; a), b), c), d), e), f), g)

Figure 18. Detailed statistical data form designed by R. W. Tillman for shelf and other types of sandstones. This form may be referred to as a check list of features on stratigraphic column form.

## STOP 4. MEASURED SECTION 83-5

### Location

SE SW Sec 14 39N, R80W, Big Gulch, Natrona County, Wyoming (Fig. 2).

This outcrop lies on the westward-facing monocline west of Salt Creek anticline proper (Fig. 19). Access is on foot from the south bound lane of Interstate 25, up the valley of Big Gulch. This section (Fig. 19) was measured by Gaynor as a part of his PhD dissertation study.

In this outcrop, the planar cross-bedded facies, <u>Central Ridge Facies</u> of Tillman and Martinsen (1984 and this volume), is especially well developed. The planar crossbedded facies is the principal reservoir facies at Hartzog Draw field. See Figures 5 and 6 for comparisons of Reineck and Wunderlich (1968) Tillman and Martinsen (1984, 1985), Spearing, 1976 and Gaynor and Swift. At this location the planar cross-bedded facies is 12.9 m (42.3 feet) thick. This outcrop compares closely with the Cities Service Federal AK-1 which is located in the thick producing interval at Hartzog Draw field and is described by Tillman and Martinsen in this volume.

An important feature of this outcrop is what is interpreted to be a disconformable surface at 2.4 meters on the measured section. The surface is observed in the big meander bend of Big Gulch and can be traced from there for almost a kilometer, west towards the measured section. It is considered by Gaynor and Swift to be the same disconformity as one near the base of the Castle Rock outcrop (Stop 1). Gaynor and Swift consider this to be a regional marine erosion surface on which the Shannon sandstone was deposited (see model, Fig. 20).

# SHANNON SANDSTONE

## MEASURED SECTION 83-5
### SE SW 14 T39N R80W

Measured by Gaynor and Swift 1983

| INTERVAL (METERS) | THICKNESS | DESCRIPTION |
|---|---|---|

### UPPER SHANNON SANDSTONE

| | | |
|---|---|---|
| 34.7-38.1 | 3.4 | SANDSTONE, TROUGH CROSS-BEDDED, BED SETS 10-30 CM, POORLY EXPOSED, FINE TO MEDIUM (200-250 µM), BETWEEN 36.3-37.8 M SCATTERED OUTCROPS OF TROUGH CROSS-BEDDED SANDSTONE. |
| 31.4-34.7 | 3.3 | COVERED INTERVAL WITH RIPPLE CROSS-LAMINATED BED OUTCROPPING AT 32.3-32.6 M. |
| 30.2-31.4 | 1.2 | SANDSTONE, THIN-BEDDED, BED SETS 3-10 CM, RIPPLE CROSS LAMINATED, FINE GRAINED (150-200 µM) |
| 29.0-30.2 | 1.2 | SANDSTONE, BIOTURBATED, TRACE WAVY BEDDING, BED SETS ≈ 5 CM, VERY FINE TO FINE (100-125 µM) |
| 22.9-29.0 | 6.1 | COVERED INTERVAL |

### LOWER SHANNON SANDSTONE

| | | |
|---|---|---|
| 16.1-22.9 | 6.8 | SANDSTONE, MEDIUM TO THICK BEDDED, BED SETS 30-100 CM, TROUGH CROSS-BEDDED WITH TABULAR-PLANAR CROSS-BEDDING DOMINANT IN THE INTERVAL 16.1-18.9 M FINE TO MEDIUM (180-300 µM, DOM. 250 µM), GLAUCONITIC. **CENTRAL RIDGE FACIES** |
| 10.0-16.1 | 6.1 | SANDSTONE, MEDIUM BEDDED, BED SETS 10-20 CM, LOW ANGLE CROSS-BEDDED TO PLANAR LAMINATED, LOCALLY LIMONITIC (AT 11.6 M), GLAUCONITIC, LOWER CONTACT GRADATIONAL, ABOVE 13.7 M — BEDS DOMINANTLY PLANAR LAMINATED, WITH BED THICKNESS INCREASING TO 20-30 CM., FINE GRAINED (150-250 µM). **CENTRAL RIDGE FACIES** |
| 6.2-10.0 | 3.8 | SANDSTONE, THIN-BEDDED, PARTLY BIOTURBATED, RIPPLE CROSS-LAMINATED, BED SETS (2-8 CM), TRACE LOW ANGLE TROUGH CROSS-BEDDING NEAR TOP OF INTERVAL, MINOR GLAUCONITE, FINE GRAINED (125-180 µM). **SHELF SANDSTONE FACIES** |
| 2.4-6.2 | 3.8 | SANDSTONE, SILTY, PARTLY BIOTURBATED, THIN-BEDDED, WAVY TO FLASER BEDDED (BEDSETS ≈ 5 CM), FINE GRAINED (125 µM). **BURROWED TO BIOTURBATED SANDSTONE FACIES** |
| 0-2.4 | 2.4 | SANDSTONE, V. SILTY, BIOTURBATED, TRACES WAY BEDDING, MOTTLED, VERY FINE TO FINE GRAINED, (100-125 µM). **BIOTURBATED SILTSTONE FACIES** |

Figure 19. Shannon Sandstone, Measured Section 83-5. Description by Gaynor and Swift. Facies names as interpreted by Tillman.

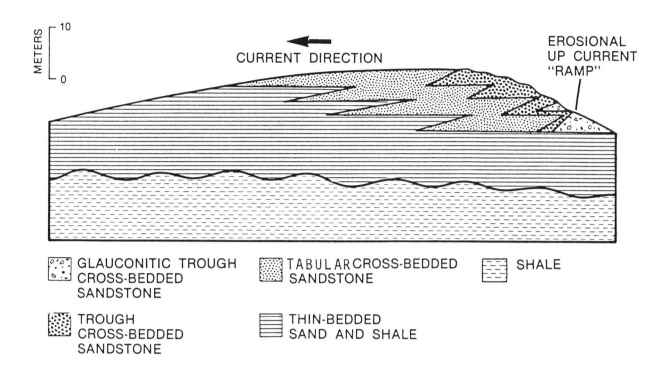

Figure 20. Gaynor and Swift model for Shannon Sandstone facies distribution. Section oriented southwest(left)-northwest. Note unconformity below thin bedded sand and shale (Inter-Ridge Sandstone Facies). A similar surface is marked by an arrow in Figure 14. See Figure 12 for plan view of model.

## STOP 5. LOCATION 1-84

### Location

C Sec. 18 T39N R79W, Wyoming (Fig. 2).

This outcrop is on the right side of Highway 87 where the southern most Shannon Sandstone outcrops cross the highway south of Midwest, Wyoming.

### GEOLOGIC HIGHLIGHTS

This stop is intended to be a brief stop to observe one type of sedimentary structure, horizontal laminations. Relatively thick horizontally laminated intervals are rare in the Shannon Sandstone, however these features are well exposed at this location.

Laminasets containing horizontal planar laminations are traceable over twenty meters laterally (Fig. 10). Commonly one to two feet of horizontally laminated section is observed, however ripples may be interbbedded with the horizontal laminae.

## References Cited

Asquith, D. O., 1970, Depositional topography and major marine environments, Late Cretaceous, Wyoming: American Association Petroleum Geologists Bulletin, v. 54, p. 1184-1224.

Gill, J. R., F. A. Merewether and W. A. Cobban, 1970, Stratigraphy and Nomenclature of Some Upper Cretaceous and Lower Tertiary Rocks in South Central Wyoming: USGS Prof. Paper 667, 53 p.

Gill, J. R., and W. A. Cobban, 1973, Stratigraphy and geologic history of the Montana Group and equivalent rocks, Montana, Wyoming and North Dakota: U.S. Geological Survey Professional Paper 776, 37 p.

Porter, K. W., 1976, Marine shelf model, Hygiene Member of the Pierre Shale, Upper Cretaceous Denver Basin, Colorado, in Studies of Colorado Field Geology, R. Epis and R. Weimer (eds.): Professional Contributions of Colorado School of Mines (Annual Meeting GSA, Denver), p. 251-263.

Reineck, H. E. and F. Wunderlich, 1969, Classification and origin of flaser and lenticular bedding: Sedimentology, v. 11, p. 99-104.

Seeling, A., 1978, The Shannon Sandstone, a further look at the environment of deposition at Heldt Draw Field, Wyoming: The Mountain Geologist, v. 15, no. 4, p. 133-144.

Siemers, C. T. and R. W. Tillman, 1981, Recommendations for the proper handling of cores and sedimentological analysis of core sequences: in Deep Water Clastic Sediments, a core workshop: SEPM Core Workshop No. 2, p. 20-44.

Spearing, D. R., 1976, Upper Cretaceous Shannon Sandstone: an offshore, shallow marine sand body: Wyoming Geological Association, 28th Annual Field Conference, p. 65-72.

Tillman, R. W. and R. S. Martinsen, 1984, The Shannon shelf-ridge sandstone complex, Salt Creek anticline area, Powder River Basin, Wyoming, in R. W. Tillman and C. T. Siemers (eds.): Siliciclastic Shelf Sedimentation, SEPM Special Publication 34, p. 85-142.

Tillman, R. W. and R. S. Martinsen, 1983, Upper Cretaceous Shannon and Haystack Mountains Formation Field Trip No. 11, in 1985 SEPM Midyear Meeting Field Guides, Golden, Colorado: Rocky Mountain Section of Society of Economic Paleontologists and Mineralogists, p. 11-1 to 11-97.